# GEOLOGY TODAY
# UNDERSTANDING OUR PLANET

# GEOLOGY TODAY

## UNDERSTANDING OUR PLANET

BARBARA W. MURCK

University of Toronto

BRIAN J. SKINNER

Yale University

JOHN WILEY & SONS, INC.

NEW YORK • CHICHESTER • WEINHEIM • BRISBANE • TORONTO • SINGAPORE

*We dedicate this book to Stella Kupferberg, an inspiration to all who knew her.*

| | |
|---|---|
| Acquisitions Editor | Cliff Mills |
| Senior Developmental Editor | Nancy Perry |
| Production Editor | Sandra Russell |
| Marketing Manager | Catherine Beckham |
| Cover and Text Designer | Karin Gerdes Kincheloe |
| Photo Editor | Kim Khatchatourian |
| Illustration Editor | Anna Melhorn |

Background: © Greg Pease Insets from left to right: (Rainbow over field)
© John Perret/Tony Stone Images, NY. (Bald Cypress trees, Merchants
Millpond Swamp, North Carolina) © Jeff Gnass. (Eroded sandstone
formations, Vermillion Cliffs Wilderness) © Willard Clay. (Iceberg, Glacier
Bay National Park, Alaska) © Jeff Gnass.

This book was typeset in 10/12 Sabon by Ruttle, Shaw & Wetherill, Inc.
and printed and bound by Von Hoffmann Press, Inc. The color
separations were prepared by Color Associates, Inc. The cover was
printed by The Lehigh Press.

The paper in this book was manufactured by a mill whose forest
management programs include sustained yield harvesting of its timberlands.
Sustained yield harvesting principles ensure that the number of trees
cut each year does not exceed the amount of new growth.

This book is printed on acid-free paper.

*Library of Congress Cataloging-in-Publication Data:*
Murck, Barbara Winifred, 1954-
 Geology today : understanding our planet / Barbara W. Murck, Brian
J. Skinner.
  p.    cm.
 Includes bibliographical references and index.
 ISBN 0–471–16733–9 (pbk. : acid-free paper)
 1. Historical geology.   I. Skinner, Brian J., 1928–
II. Title.
QE28.2.M87   1999                      98–23728
550—dc21                               CIP

Printed in the United States of America

10  9  8  7  6  5  4  3  2  1

# • ABOUT THE AUTHORS •

The authors, Barbara W. Murck and Brian J. Skinner, have been privileged to have been involved in the development, study, and teaching of the emerging view of our home planet, called Earth system science. They bring to this project a love of geology and of teaching, a wealth of professional experience on all the continents of the world, and a relaxed but skillful writing style that makes even difficult concepts seem accessible to students.

As an undergraduate, *Barbara Murck* was a confirmed nonscientist, until an introductory geology course changed her plans. Since then her professional focus has ranged from igneous geochemistry and ore-deposit petrography to alternative energy sources and state-of-the-environment reporting. Her current work focuses primarily on environmental management training for decision makers in developing countries.

*Brian Skinner* was born and raised in Australia on the edge of what is now called degraded land—some call it desert. That is where he gained his earliest interest in the complex interactions that make up our environment. He has been privileged to work on every continent and to be involved in the beginnings of what we now call the Space Age. His interests are most centrally involved with mineral and energy resources, but because these materials are found in a great many geological environments, his professional interests are diverse.

# • PREFACE •

## What *Geology Today* Is About

This introductory geology book is about the fascinating interrelated processes that make the Earth the special place it is and about the wonderfully balanced ways by which it works and has worked over the vast geologic ages. *Geology Today* is also a book about continuing discoveries, because there is much about the Earth that we still know imperfectly or don't know at all. We continue to make discoveries about plate tectonics and how plate tectonic processes rearrange and renew the Earth's surface; about rainfall, winds, ice caps, and all the other processes that continually erode away the rocks that have been brought to the surface by volcanism and plate tectonics; about climate and the global interactions that drive climatic change; about the influence of life on geologic processes, and vice versa; and about how humans can survive and flourish on a geologically active planet. It is, in short, a book about the interactions among the many parts of the Earth system and about the new approaches now being used to study and assess these interactions.

It is our enthusiasm for the insights of the Earth system science approach that has led us to write an introductory geology text that builds on this approach and to do so at a level that is accessible to all. This is not a book about the tougher and more quantitative parts of Earth science. It is a book about understanding how planet Earth works. The most confident homeowner is the one who knows how the home and its various support systems operate. Similarly, we become more confident and effective proprietors of the Earth when we understand how its various systems function and interact with one another and how geologic processes affect our lives on a daily basis. It is our hope that this book will bring the workings of our larger home, the Earth, into clearer and more useful focus for our readers.

## Organization

*Geology Today* is organized in five main parts.

In *Part One: The Third Planet*, students learn about the main characteristics of the Earth as a planet and its place in the solar system. They read about the origin of modern geology and find out how geologists use scientific observations to draw conclusions about the history of the Earth and geologic processes. They learn about atoms, elements, minerals, and rocks, the fundamental building blocks of the Earth, and about the age of the Earth system and its parts.

In *Part Two: The Dynamic Earth*, students learn about geophysical processes that originate deep within the Earth, including earthquakes, volcanism, and plate tectonics. Students are invited to evaluate the plate tectonic model for themselves, on the basis of observational evidence. The chapters of *Part Two* are concerned primarily with the parts of the geologic cycle that involve internal Earth processes.

*Part Three: The Changing Earth* focuses on the Earth's surface, a dynamic interface between the geophysical activity of plate motion, seismicity, rock deformation, and volcanism within the Earth and the constant activity of the atmosphere and hydrosphere outside the Earth. The surface of the Earth is constantly modified by the processes of weathering, erosion, sedimentation, and rock deformation. The chapters of *Part Three* focus on the external or crustal part of the geologic cycle. In the last chapter of *Part Three*, we summarize the common rocks and rock-forming processes in the context of the tectonic environments in which they occur.

Water in all its forms dominates the Earth system. It shapes the surface of the land, controls weather and climate, and sustains life. In *Part Four: Water World*, students learn about all aspects of the hydrologic cycle and the various roles of water in the Earth system.

*Part Five: Living on Planet Earth* offers a brief synopsis of the history of life on the Earth and the impacts of life on the chemical evolution of the atmosphere and hydrosphere. We also examine our current reliance on the material resources of this planet. As the human population grows, it becomes even more important that we understand how such resources form, how and where they occur, and how they can be managed to ensure their continued viability. We close the book with a look at the continuing and changing role of geoscientists in the twenty-first century.

## Special Features

Many special features have been built into the book to make it a more useful and enjoyable educational companion for our readers.

### An Accessible, Engaging Style

Many introductory geology textbooks have claimed to be "accessible" and "readable"; this is the book that finally achieves it. We believe that students will genuinely enjoy reading and learning from it. *Geology Today* introduces students to the basic concepts of physical geology, within a broader context of Earth system science. Through the book's narrative style and historic perspective, students discover how scientists have used observation and deductive reasoning to arrive at some of the basic concepts that characterize our present-day understanding of the Earth system. See, for example, chapter 4 ("Plate Tectonics: A Unifying Theory"), where students follow the story of the theory of plate tectonics through conceptualization, controversy, observation, testing, refinement, and finally, general acceptance by the scientific community. Students learn that not all the answers have been found on the theory of plate tectonics, and so the final verdict remains out. Instead, they will find themselves in the middle of an ongoing process of scientific questioning, examination, and discovery.

### Chapter Openers and Part Openers

At the beginning of each of the five main parts of the book are a brief synopsis of the contents of the part and a broad overview of the main concepts covered by those chapters. Each chapter opener presents a short vignette showing the relevance of the chapter material to life on planet Earth. Each chapter opener ends with a brief list of the most important things students will be learning in that chapter.

### Strategic Questions

The inquisitive, problem-solving perspective of the text is reflected in the opening question and marginal puzzlers (denoted by question marks). These puzzlers are intended to pique students' interest by posing questions about geologic processes that they may previously have wondered about—questions that are addressed in the adjacent text.

### Artwork and Photographs

Wiley's Earth science textbooks are known for the quality of their artwork. In this book we have carefully designed the art program to complement the text and, in particular, to facilitate the learning process for introductory students. Line drawings are simple and clearly labeled. Caption "headlines" advertise the main content of each figure. Long, text-only tables are not used. Instead, the pedagogical effectiveness of both text and artwork has been enhanced by pairing what would formerly have been tabular material with line drawings and photos to illustrate concepts. And, of course, the Wiley tradition of spectacular instructional photographs is carried on in this book.

### Key Terms

Students will find key terms boldfaced in the text and defined in a handy marginal glossary, adjacent to the first appearance of the term in the text. At the end of each chapter is a list of these terms, The Language of Geology, with page references for easy review. Full definitions and page references are provided for all key terms in the Glossary at the back of the book. A list of italicized terms is also provided in the Instructor's Manual.

### Geology Around Us

Each chapter contains a focus box, entitled Geology Around Us. The boxes highlight examples and case studies of geology at work in our daily lives (e.g., "Minerals in Everyday Life," chapter 2). Some of the boxes provide an in-depth look at a geologic concept (e.g., "The Hawaiian Islands: A Record of Plate Motion," chapter 4). Some of the boxes simply tell interesting stories related to the chapter topic (e.g., "Darwin's Atlantis," chapter 15).

### Summary and Review: Text and Diagrams

Each chapter presents a summary of the chapter's most important concepts, the Chapter Highlights. Questions for Review provide an opportunity for students to test what they have learned from their reading. Questions for Thought and Discussion are intended to be more open-ended and are appropriate for research projects, homework or laboratory assignments, or group discussions. In both types of questions, you will find references to figures that appear in the text. Many beginning science students have trouble using and interpreting graphs and diagrams; we hope that by integrating the use and interpretation of diagrams into the end-of-chapter questions, we may help instructors to address this problem.

### New Media

Our goal has been to bring the powers of interactive media to bear on the exploration of physical geology. At the end of appropriate chapters, references are made to two new-media components of the program: GEOSCIENCES IN ACTION CD-ROM, where students find themselves as "virtual interns" exploring certain problems as a geologist would, and GEOSYSTEMS TODAY: An Interactive Casebook on the WWW, where students can take virtual tours exploring cases in geoscience.

### Geology in Art and Literature

Sandwiched between each of the main parts of the book is a two-page spread with artistic and literary references to geology. These "coffee breaks," as we have come to refer to them, are intended to provide a visual and conceptual break for readers. To most geologists it will not come as a surprise that the Earth and its ways have served as inspiration for artists, poets, and writers throughout the ages and in all cultures. We hope the "coffee breaks" will hold special appeal for students with a background in nonscientific fields, many of whom now populate introductory-level university and college courses in geology and Earth science.

### Appendixes

The Appendixes provided at the end of the book present useful information on elements and minerals, units and conversions, and geologic maps. In addition, we append a more detailed discussion of Bowen's reaction series than the one that appears in the main text. Some teachers of geoscience consider this material to be too advanced for an introductory course; it is included here for use at the instructor's discretion.

## Supplements

**Geosciences in Action.** This CD-ROM accompanies each text and allows students to become "virtual interns" in geology, whether exploring the source of a certain pollutant or determining the volcanic hazard at an island resort. These "Virtual Internships" were authored by David DiBiase, Thomas Bell, and Hobart King, and developed by the Deasy Geographic Labs at Pennsylvania State University.

**Geosystems Today.** This casebook and interactive WWW site provides students with eight cases from around the world in which to see and explore the interaction of people and their environment. Authored and developed by Robert Ford, Westminster University, and James Hipple, University of Missouri.

**Geology Today Media Resource Manager.** This CD-ROM, free to adopters of the text, contain all of the line illustrations and many of the photos from the text for lecture projection, as well as several animations showing key geologic processes.

**The Student Companion.** This student study guide is authored by Barbara Murck and contains study hints, further explanations of key concepts, and self-tests.

**Take Note!** This free supplement for students contains all of the line illustrations from the text in a black-and-white format for students to use to take notes.

**Instructors Manual and Test Bank.** This resource is also authored by Barbara Murck and contains course material suggestions and test questions (also available in a computerized format).

**Transparencies.** The text figures are presented on transparency acetates for lecture enhancement.

## Acknowledgments

As authors we could not ask for a more talented, patient, and caring team of professionals than those who came together to apply their expertise to this project. Our grateful thanks include our colleagues at John Wiley & Sons as well as those in the extended Wiley family who contributed to the book in many valuable ways.

The idea for a truly modern, readable, and accessible introductory physical geology text originally came from Wiley Geology Editor Cliff Mills, who has continued to be the creative force behind the project. The stamp of approval for the project came from Nedah Rose and Kaye Pace; we thank them warmly for their continuing support. From the beginning, the project has been nurtured with the greatest of care and skill by Senior Development Editor Nancy Perry. Nancy's guidance, enthusiasm, and friendship led to an enjoyable project and a highly professional finished product. Sandra Russell, Senior Production Editor, saw the book through its production schedule with great competence and boundless good humor.

Special mention is due to Fred Schroyer, who put much effort into developing a particularly accessible and visually engaging art program. Fred's creative input is responsible for much of the pedagogical effectiveness of the illustrations used in the book. Thanks also to Dan

Botkin, who helped the authors correct and clarify the difficult concepts presented in chapter 15.

As we strive to expand beyond the traditional boundaries of the printed page, new media are becoming more and more important as pedagogical tools. The "Virtual Internships" of the CD-ROM, *Geosciences in Action,* were developed by David DiBiase, Hobart King, and Thomas Bell. Robert Ford and James Hipple contributed the interactive case studies that are available through Wiley's web page. Many thanks to these colleagues for developing these exciting new media explorations and to Fadia Stermasi who coordinated the development of new media for this project.

Others who contributed their considerable talents to the project include (in no particular order): Cathy Donovan, who provided ever-present helpfulness on all fronts; Kim Khatchatourian and Alexandra Truitt, who (as usual) managed to find the most remarkable photos; Jennifer Yee, who coordinated the supplements; Bridget O'Lavin, editorial assistant; Carl Spector, permissions editor; Carolyn Smith, whose familiar green pen once again tightened and improved the text; Ishaya Monokoff and Anna Melhorn, who coordinated the art program; Karin Kincheloe, who contributed an inspired design; and Catherine Beckham, who coordinated marketing for the book.

We also extend warm thanks to J. Marion Wampler. Many teachers of geoscience are familiar with Marion's regular feature article in the *Journal of Geoscience Education,* entitled "Geomythology: A Column about Errors in Geoscience Textbooks." Instead of waiting for *Geology Today* to show up in Marion's column, we decided to take proactive measures by asking him to review each chapter for accuracy and clarity. Marion's dedication to the task went beyond the call of duty; he combed each chapter in search of errors, omissions, inaccuracies, and other sources of confusion. The arrival of one of his reviews invariably elicited cries of anguish, irritation, and sometimes embarrassment ("How could I possibly have let *that* slip by . . . ?"). The book is both pedagogically and scientifically more sound—and certainly more readable—for Marion's painstaking efforts. Of course, the authors take responsibility for any errors that still managed to slip through.

Finally, we recognize with gratitude the contributions of Steve Porter, co-author of *Environmental Geology, The Dynamic Earth,* and *The Blue Planet,* among others. Steve's research and administrative responsibilities prevented him from participating in this project. However, his influence can be detected throughout the book, from his wonderful photographs to the echo of his thoughts and words on the printed page. Thanks for your input, Steve.

Geology is an interdisciplinary science, encompassing many areas of expertise. For this reason, we sought the input of colleagues who collectively represent a wide range of experience in all aspects of geology research and teaching. The careful reading and extensive commentary by these colleagues improved the book immeasurably. Their thoughtful suggestions touched on every aspect of the book, from the overall organization to the tiniest details. Through their comments, our reviewers made available to us their many years of collective experience in conveying both the knowledge and the love of geology to beginning students. Thank you to those who assisted us by reviewing all or part of the manuscript. They are:

Robert W. Baker
*University of Wisconsin-River Falls*

Theodore J. Bornhorst
*Michigan Technical University*

John Callahan
*Appalachian State University*

John A. Campbell
*Fort Lewis College*

Murlene W. Clark
*University of South Alabama*

Robert Douglas
*University of Southern California-Los Angeles*

Ernest H. Gilmour
*Eastern Washington University*

Thomas H. Giordano
*New Mexico State University*

Peter W. Goodwin
*Temple University*

Mickey Gunter
*University of Idaho*

Craig Bond Hatfield
*University of Toledo*

Robert W. Hinds
*Slippery Rock University*

Paul F. Hudak
*University of North Texas*

Ronald Konig
*University of Arkansas-Fayetteville*

David R. Marchant
*Boston University*

Francine McCarthy
*Brock University*

Sally McGill
*California State University-San Bernardino*

Charles Merguerian
*Hofstra University*

Ken Morgan
*Texas Christian University*

Donald W. Neal
*East Carolina University*

Gerald Osborn
*University of Calgary*

Donald Pair
*University of Dayton*

Bruce C. Panuska
*Mississippi State University*

Robert W. Reynolds
*Central Oregon Community College*

Peter Saccocia
*Bridgewater State College*

John D. Stanesco
*Red Rocks Community College*

Our thanks also go to the following geology professors and their students, who provided very constructive feedback on the Geosciences in Action CD-ROM:

Alan I. Benimoff
*The College of Staten Island*

J. V. Chemosky
*University of Maine*

Brooks Ellwood
*University of Texas-Arlington*

Joseph Nadeau
*Rider University*

Catherine Rigsby
*East Carolina University*

Kevin Stewart
*University of North Carolina-Chapel Hill*

Bryan Tapp
*University of Tulsa*

Karl Wirth
*Macalester College*

Barbara W. Murck
Brian J. Skinner

# • BRIEF CONTENTS •

# • CONTENTS •

Geology Around Us: Soil for Food   204

## Chapter 8 • From Sediment to Rock: Rocks That Form Near the Earth's Surface   217

Geology Around Us: The Gravel Page: Forensic Geology   228

## Chapter 9 • Folds, Faults, and Geologic Maps   241

Geology Around Us: The World's Oldest Geologic Map   260

# PART ONE
# THE THIRD PLANET

**Geology is the scientific study of the Earth,** the third planet from the Sun and our home planet. People have been studying the Earth for thousands of years, but the formal discipline of geology—the scientific study of the Earth—is relatively new. Modern geology began in the 1800s, when natural scientists began to make systematic, detailed observations about rocks and fossils. They used this information to draw conclusions about the history of the Earth and geologic processes.

Today geologists have a new way of studying the Earth. Instead of studying various aspects of the planet in isolation, we try to look at the Earth system as an integrated whole. In Part One of *Geology Today: Understanding Our Planet,* we look at the Earth in the context of the solar system. In so doing, we find a planet that is similar to others in many ways, yet unique in its ability to support life. We also look at the fundamental building blocks of the Earth—atoms, elements, minerals, and rocks. The Earth system is very old—4.56 billion years old. Geologists use the rock record to determine the ages of different parts of the Earth and the rates at which geologic processes occur.

> ▶ All of these topics and more are addressed in the chapters of Part One:
>
> Chapter 1: Earth in Space: The Science of the Earth System
>
> Chapter 2: Earth's Materials: Atoms, Elements, Minerals, and Rocks
>
> Chapter 3: Earth in Time: The Rock Record and Geologic Time

*Scar from Ancient Impact*

Gosses Bluff, a 5-km-wide ring of twisted and broken rocks west of Alice Springs, Australia was the site of a great meteorite impact about 130 million years ago. Impact craters remind us that the Earth always faces the possibility of an unexpected visitor from space.

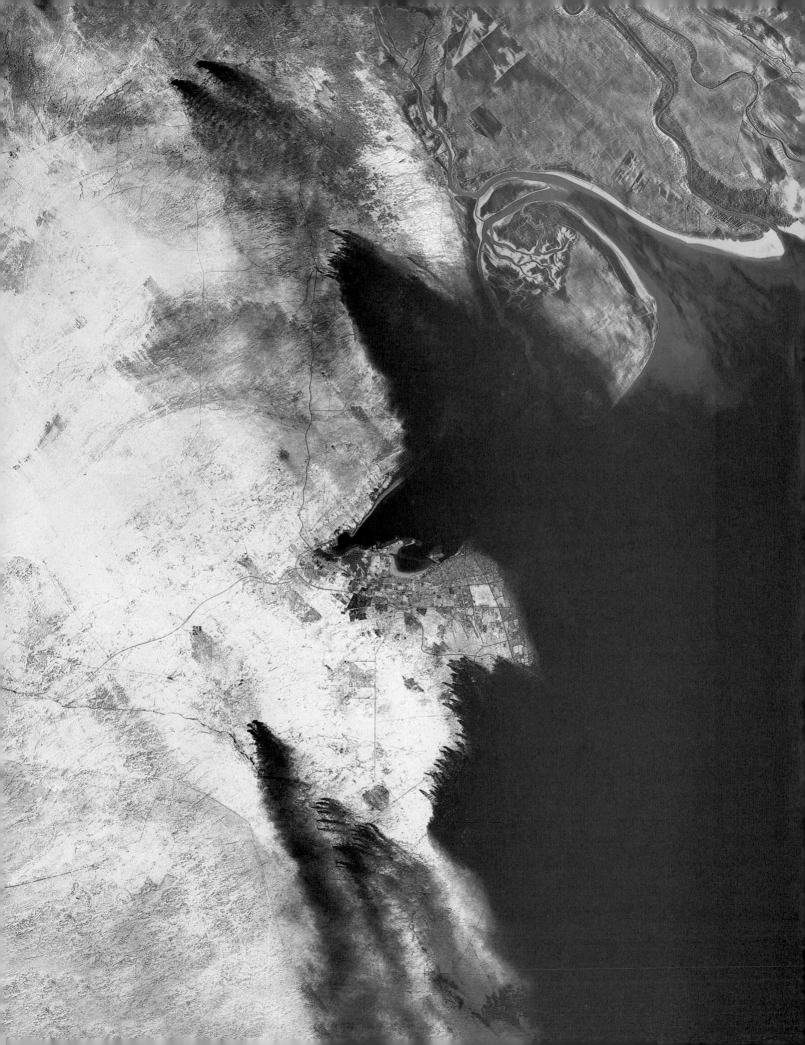

# · 1 ·

# EARTH IN SPACE: THE SCIENCE OF THE EARTH SYSTEM

**Even a science as young as geology** can have revolutions, and that is what happened in the 1960s. At that time, plate tectonics emerged as a unifying theory that completely changed our understanding of geologic processes, as you will learn in this book. The tools, the methods, and even the language of geology changed as a result of that scientific revolution. Geology is currently undergoing another, more subtle revolution, which is driven by the ability of scientists to observe and collect information about the Earth as a whole, using instruments mounted on satellites. This ability is quite new; remember that no one had *ever seen* a picture of the whole Earth until the 1960s, when the first photograph was taken of the Earth from a spaceship.

Today, satellite images and data collected from outer space provide the scientific foundation for our study of the Earth as an integrated system. Earth system science, as this approach is called, is not new in philosophy, but its tools and techniques are very new. These new tools are used in a wide range of applications, from weather forecasting to the monitoring of changes in the muddiness of rivers, measuring the flow of polar ice, locating mineral resources, planning land use, documenting the extent of oil spills, and many others. Through Earth system science, geologists are contributing more than ever to our understanding of the Earth as a whole, how the Earth changes over time, and the impacts of human actions on the Earth system.

◆

### In this chapter you will learn about

- Geology, the scientific study of the Earth
- Earth system science, an integrated approach that makes use of global data and imagery obtained by satellites
- The origin of the Earth and its place in our solar system
- The basic structure of the Earth, inside and out
- The forces of plate tectonics, which make the outer crust of the Earth a dynamic, ever-changing place

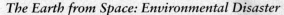

*The Earth from Space: Environmental Disaster*

This satellite image, taken in late February, 1991, shows oil well fires in Kuwait during the Persian Gulf War. Thick plumes of black smoke from the fires spread over hundreds of kilometers. Each tiny red dot represents an individual oil well fire.

**geology** The scientific study of the Earth.

**geologist** A scientist who studies the Earth.

**physical geology** A branch of geology that is concerned with geologic processes and materials.

**historical geology** A branch of geology that is concerned with geologic events that occurred in the past.

# WHAT IS GEOLOGY?

**Geology** is the scientific study of the Earth. The word comes from two Greek roots: *geo-,* meaning "of the Earth," and *-logis,* meaning "study" or "science." The science of geology encompasses the study of our planet: how it formed; the nature of its interior; the materials of which it is composed; its water, glaciers, mountains, and deserts; its earthquakes and volcanoes; its resources; and its history—physical, chemical, and biological. Scientists who study the Earth are called **geologists** (Fig. 1.1).

The teaching of geology is traditionally divided into two broad subject areas: *physical geology* and *historical geology.* **Physical geology** is concerned with understanding the *processes* that operate at or beneath the surface of the Earth and the *materials* on which those processes operate. Some examples of geologic processes are mountain-building, volcanic eruptions, earthquakes, river flooding, and the formation of ore deposits. Some examples of materials are minerals, soils, rocks, and water.

**Historical geology,** on the other hand, is concerned with geologic *events* that have occurred in the past (Fig. 1.2). These events can be read from the rock record. Through the applications of historical geology, scientists seek to resolve questions such as when the oceans formed, why the dinosaurs died out, when the Rocky Mountains rose, and when and where the first trees appeared. Historical geology gives us a perspective on the past. It also establishes a context for thinking about present-day changes in our natural environment. *Geology Today* is concerned mainly with physical geology, but it also deals with some aspects of historical geology.

**Figure 1.1**
**GEOLOGISTS AT WORK**
These geologists are descending into the crater of an active volcano in Java, Indonesia, to learn more about how volcanoes work.

There are many other areas of study within the traditional domains of physical and historical geology. Some of them reflect our dependence on the Earth's material resources. *Economic geology,* for example, is concerned with the formation and occurrence of valuable mineral deposits. Other aspects of geology reflect our changing attitudes toward the Earth and the natural environment. *Environmental geology* focuses on how materials and processes in the natural geologic environment affect—and are affected by—human activities. Today many geologists work in highly specialized fields. For example, *volcanologists* study volcanoes and eruptions, past and present. *Seismologists* study earthquakes. *Mineralogists* undertake the microscopic study of minerals and crystals. *Paleontologists* study fossils and the history of life on Earth. *Structural geologists* study how rocks break and bend. This specialization is needed because geology encompasses such a broad range of topics.

Geologists are scientists who make a career out of the scientific study of the Earth. Yet to a certain extent we are all geologists. Everyone living on this planet relies on geologic resources: water, soil, building stones, metals, fossil fuels, gemstones, plastics (from petroleum), ceramics (from clay minerals), salt (the mineral *halite*), and many others. We are affected by geologic processes every day. We all interact with and influence the geologic environment through our daily activities. *Geology Today* will help you find the information and the tools to become more aware of the environment that surrounds you. As a result, you will be better equipped to make decisions about the materials and processes that affect your life.

Accompanying this textbook is a CD-ROM entitled *Geosciences in Action.* The CD-ROM will immerse you in a guided discovery of applied and basic research problems confronted by professional geologists on a daily basis. As you work through these real-world geology challenges on your personal computer, you will use the fundamental concepts and basic terminology of geology to fulfill concrete, practical objectives. Through these *Virtual Internships,* you will become a geologist in action. The Virtual Internships are denoted by an icon that looks like this: Interactive case studies located on our web page will help you explore geology even further. They are denoted by this icon:

# THE HOME PLANET

As geologists we mainly study our home planet, the Earth. But before we begin our study of the Earth itself, we can broaden our perspective and ask a basic question: What makes the Earth a special and unique place?

We can begin to answer this question by contemplating a photograph of the Earth from space—perhaps the most powerful image of all time (Fig. 1.3). This image, first seen in the 1960s, made it possible as never before for ordinary people to recognize the astronomer's view of the Earth—a small planetary body in orbit around an ordinary, medium-sized star. To fully appreciate "Spaceship Earth," to comprehend its complexities, its limits, and its vulnerabilities, it is helpful to look at the Earth's place among its neighbors in space.

## The Solar System

The Earth is one of nine planets in our **solar system,** the Sun and the group of objects in orbit around it. In addition to the Sun and the planets, the solar system also includes 64 known moons, a vast number of asteroids, millions of comets,

**Why are there so many kinds of geologists?**

Figure 1.2
**CLUES IN THE ROCK RECORD**
These geologists are helping to reconstruct the history of the Earth by examining fossil remains preserved in the rock record. These particular fossils are the bones of dinosaurs discovered in China's Gobi Desert.

**solar system** The Sun and the group of objects in orbit around it.

Figure 1.3
THE HOME PLANET
The Earth from space. The blue
ocean, the fluffy, water-rich
clouds, and the green, plant-cov-
ered land areas make it clear that
the Earth is a unique planet.

**Are all planets alike?**

**terrestrial planets** The inner planets of the solar system: Mercury,
Venus, Earth, and Mars.

**jovian planets** The "gas giant"
outer planets: Jupiter, Saturn,
Uranus, and Neptune.

and innumerable fragments of rock and dust called *meteoroids*.[1] All the objects
in our solar system move through space in smooth, regular orbits, held in place
by gravitational attraction. The planets, asteroids, comets, and meteoroids orbit
the Sun, and the moons orbit the planets.[2]

The planets can be separated into two groups on the basis of their physical
characteristics and distances from the Sun (Fig. 1.4). The innermost planets—
Mercury, Venus, Earth, and Mars—are small, rocky, and relatively dense. They
are similar in size and chemical composition, and are called **terrestrial planets** because they resemble *Terra* ("Earth" in Latin). With the exception of Pluto, the
outer planets are much larger than the terrestrial planets, yet much less dense.
These **jovian planets**—Jupiter, Saturn, Uranus, and Neptune—take their name
from *Jove,* an alternate name for Jupiter. The jovian planets probably have small
solid centers that may resemble terrestrial planets, but much of their planetary
mass is contained in thick atmospheres of hydrogen, helium, and other gases.
The atmospheres are what we actually see when we observe these planets (Fig.
1.5). Pluto—the smallest of the nine principal planets—doesn't fit into either of
these planetary groups: it is much smaller than the jovian planets but much less
dense than the terrestrial planets. In many respects, Pluto is more like a large
comet than a planet.

## The Origin of the Solar System

How did the solar system form? We may never know the precise answer to this
question, but we can discern the outlines of the process from evidence obtained
by astronomers and Earth scientists, and from the laws of physics and chemistry.
The process began in a part of space that was not entirely empty because earlier

---

[1] While still in orbit, these objects are called meteoroids. If they pass through the Earth's atmosphere and strike the surface, they are called *meteorites.*

[2] As immense as the planets may seem to us, most of the solar system actually consists of empty
space. The vast distances from one planet to the next are almost beyond comprehension, and any de-
piction of the planets (such as Fig. 1.4) must greatly minimize the distances and exaggerate the sizes
of the planets.

Figure 1.4
FAMILY PORTRAIT OF OUR SOLAR SYSTEM
Our solar system's nine planets, shown in proper size scale against the Sun. (Note, however, that the distances between the planets are much greater than shown, and the planets never line up neatly like this.) The terrestrial planets are the four small, rocky ones closest to the Sun. The jovian planets are the large gaseous bodies distant from the Sun. Images from space satellites reveal ring systems around all four jovian planets, though only Saturn's rings are bright enough and large enough to depict on a diagram of this scale. Pluto is the "odd planet out"; it is much smaller and lacks the huge gaseous envelope characteristic of the other jovian planets.

stars had exploded, scattering matter across vast distances of space. Most of the matter in this cloud of interstellar gas and dust was composed of the element hydrogen, but small percentages of all the other chemical elements were present too. Everything in the solar system was eventually constructed from such matter. In a sense, the Earth and everything on it is made of star dust.

Figure 1.5
SATURN, A JOVIAN PLANET
The jovian planets are all extremely massive but have very low densities. Saturn, for example, is 95 times more massive than the Earth but has a density less than that of water. In other words, an object with the same density as Saturn would actually float in water.

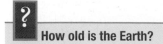

**How old is the Earth?**

The swirling cloud of dust and gas (Fig. 1.6A) began to thicken over a very long period, as the atoms in it gravitated to each other. As the atoms moved closer together, the rotating gas cloud flattened into a disk (Fig. 1.6B). Near the center of the disk, pressure and temperature were extremely high. There, in the newly forming Sun, hydrogen atoms were subjected to such high pressures and temperatures that they began to undergo *nuclear fusion*. The fusion of hydrogen atoms to form heavier helium atoms is still going on inside the Sun; this is the source of the Sun's radiant energy. However, nuclear fusion within the Sun does not produce many elements heavier than helium. Almost all of the heavier elements in our solar system were inherited from the explosion of earlier generations of stars that were much more massive than our Sun.

Eventually, the outer portions of the cosmic gas and dust cloud cooled enough to allow solid particles to condense, in the same way that snowflakes condense from water vapor (Fig. 1.6C). The materials that condensed from the cloud eventually formed the planets, moons, and other solid objects of the solar system, including the Earth (Fig. 1.6D). The gas and dust cloud is called the *solar nebula*, and this story of the birth of the solar system and its component parts is called the *nebular theory*.

Condensation of the gas cloud is only the first part of the planetary birth story. Condensation formed innumerable small, dust-sized particles, but the particles still had to be joined together to form a planet. This happened as a result of

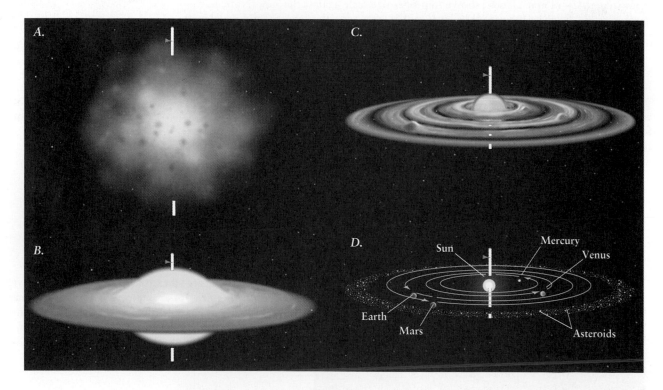

Figure 1.6
**BIRTH OF A SOLAR SYSTEM**
*A.* The gathering of matter in space created a cloud of dust and gas, which began to rotate. *B.* As the cloud of gas and dust became smaller and denser, it flattened, and its center eventually collapsed inward to become the Sun. The planets were formed by condensation from the gas cloud (*C, D*) and accretion of the condensed particles. Today the planets all orbit the Sun in the same direction (blue arrows).

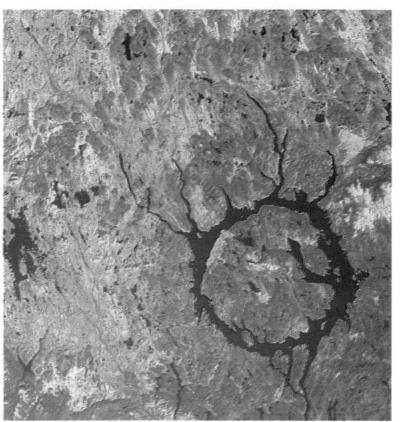

**Figure 1.7**
**MESSENGER FROM SPACE: A METEORITE**
This boy is examining a large iron meteorite in the American Museum of Natural History. Information from meteorites like this help scientists interpret the chemical history of the Earth and other planets in our solar system.

**Figure 1.8**
**SCARS FROM AN ANCIENT IMPACT**
Manicouagan Crater in Quebec was created 210 million years ago when a large meteorite struck the Earth. The original crater, now marked by a ring lake, was 75 to 100 km (50 to 67 mi) in diameter.

impacts between fragments drawn together by random collisions and gravitational attraction. The largest masses slowly swept up more and more of the condensed dust particles, growing into even larger *planetesimals* and eventually becoming the planets. The largest of these bodies held huge gaseous envelopes—primordial atmospheres—in their gravitational grasp.

This growth process, the gradual gathering of more and more bits of solid matter from surrounding space, is called *planetary accretion.* Ancient rocky fragments that are left over from that long-ago process still exist in space and still fall to Earth from time to time; we call them **meteorites** (Fig. 1.7). Meteorites and the scars of ancient impacts (Fig. 1.8) provide evidence of the way the terrestrial planets grew to their present sizes. The formation of the Earth and other planetary bodies through the processes of condensation and planetary accretion was essentially complete 4.56 billion years ago. (How do we know this date? Chapter 3 will explain.)

**meteorite** A fragment of extraterrestrial material that falls to Earth.

### The Family of Planets

The nine planets and other planetary bodies (meteoroids, asteroids, moons, and comets) are like siblings; all were born of the same processes that gave rise to the

> **?** How are the terrestrial planets alike? How does the Earth differ from the others?

**crust** The outermost compositional layer of the solid Earth.

**mantle** The middle compositional layer of the Earth, between the core and the crust.

**core** The innermost compositional layer of the Earth.

whole solar system family. Some of the planets are more alike than others. The Earth and its close neighbor Venus, for example, are so much alike in size, density, and chemical composition that they are almost "twins."

As a group, the terrestrial planets have many things in common beyond their small sizes and rocky compositions. They have all been hot and, indeed, were partially molten at some time early in their histories. During the period of partial melting, all of the terrestrial planets separated into three layers of differing chemical composition: a relatively thin, low-density, rocky **crust,** the outermost layer; a rocky, intermediate-density **mantle;** and a metallic, high-density **core** (Fig. 1.9). This separation process is called *planetary differentiation.*

There are other similarities as well, three of which are very important. First, all of the terrestrial planets have experienced volcanic activity. The volcanism is dominated by the formation of a special kind of volcanic rock called *basalt* (Fig. 1.10). The second similarity is that all of the terrestrial planets passed through a period of intense meteorite impacts and surface modification by cratering processes that continue today, though fortunately at a much slower pace. The third similarity arises from the size and closeness of the terrestrial planets to the Sun. Unlike the jovian planets, the terrestrial planets were too small to hold on to their original envelopes of gas, which were swept away by intense eruptions of the Sun early in the history of the solar system. The three terrestrial planets that ended up with atmospheres (Earth, Mars, and Venus) evolved new gaseous envelopes from material that slowly leaked out from their interiors via volcanoes.

To summarize, when we look at the solar system we see a group of planets and other objects that are related by the way they were formed and by their asso-

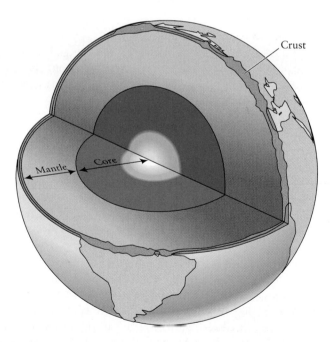

**Figure 1.9**
**A DIFFERENTIATED PLANET**
Early in its history, the Earth—like the other terrestrial planets—separated into three layers of differing physical properties and chemical composition. The outermost layer is the low-density, rocky crust. Next is the mantle, which is still rocky but of intermediate density. The innermost layer is the high-density, metallic core. This process of separating into layers is called planetary differentiation.

**Figure 1.10**
BASALT: THE MOST COMMON ROCK IN THE SOLAR SYSTEM
Basaltic lava erupts from Mauna Loa volcano in Hawaii. Basalt is the most common
rock type on the Earth and in the solar system as a whole.

ciation with the Sun. Within this system is a smaller group, the terrestrial planets,
which are related even more closely by their similar origins and planetary characteristics. The Earth is the way it is because of all the things that have happened
during its long history. This history is different enough from those of the other
terrestrial planets to make the Earth habitable, while the others are not.

## What Makes the Earth Unique?

The image of our blue and white planet taken from space (Fig. 1.3) reveals much
of what makes the Earth unique. Although the terrestrial planets are alike in
many ways, they differ greatly in the composition of their atmospheres, the characteristics of their surfaces, and the presence or absence of water and life.

   The Earth has an overall blue-and-white hue because it is surrounded by an
**atmosphere** of gases, predominantly nitrogen, oxygen, argon, and water vapor,
with small amounts of some other gases, such as carbon dioxide. No other planet
in the solar system has such an atmosphere. The Earth's atmosphere contains
clouds of condensed water vapor. The clouds form because water evaporates
from the hydrosphere, another unique feature of the Earth. The **hydrosphere**
("watery sphere") consists of the oceans, lakes, and streams; underground water;
and snow and ice. Planets farther from the Sun are too cold for liquid water to
exist on their surfaces. Planets closer to the Sun are so hot that any surface water
evaporated long ago. Other planetary bodies have ice on the surface and water
vapor in the atmosphere, but only the Earth has the right surface temperature to
maintain a hydrosphere consisting of liquid water, ice, and water vapor.

   Another reason the Earth is special is the **biosphere** ("life sphere"), which includes the totality of the Earth's living matter. When the Earth is viewed from
space, the biosphere is revealed by the blankets of green plants on some of the

**?** **Why does the Earth look
blue-and-white from space?**

**atmosphere** The envelope of gases
that surrounds the Earth.

**hydrosphere** The oceans, lakes,
and streams; underground water;
and snow and ice.

**biosphere** The totality of living
matter on the Earth.

**Figure 1.11**
**THE EARTH'S SURFACE**
This is a composite of numerous satellite images, each selected to be cloud-free. It is unrealistic because, at any moment, half of the Earth is in nighttime darkness and much of it is cloud-covered. But this beautiful image lets us view the entire surface at once. It shows densely vegetated regions in green, dry deserts in yellow or brown, and ice-covered regions in white.

land masses (Fig. 1.11). The biosphere embraces innumerable living things, large and small, which belong to millions of different species. It also includes dead plants and animals that have not yet been completely decomposed.

The nature of the Earth's solid surface is another special characteristic. The Earth is covered by an irregular blanket of loose debris formed as a result of *weathering*—the chemical alteration and mechanical breakdown of rock caused by exposure to the atmosphere, hydrosphere, and biosphere. This layer is called *regolith* (from the Greek *rhegos,* blanket, and *lithos,* stone). Soils, muds in river valleys, sands in the desert, rock fragments, and all other unconsolidated debris are part of the regolith. Other planets and planetary bodies with rocky surfaces have regolith too, but in those cases the regolith has formed primarily from endless pounding by meteorite impacts. The Earth's regolith is unique because it is formed by complex interactions among physical, chemical, and biological processes, usually involving water. The Earth's regolith is also unique because it teems with life; most plants and animals live on or in the regolith or in the hydrosphere.

The regolith is like a loose blanket that lies on top of the rocks that make up the outermost part of the Earth. This tough, rocky outermost part of the Earth is called the **lithosphere.** The lithosphere is defined by strength and rigidity, not by composition. It is about 100 km (67 miles)[3] thick, and consists of the crust and the outermost part of the mantle. The Earth differs from all other known planets because of the unique relationship between its lithosphere and the hotter, weaker, putty-like rocks that lie immediately below, in the **asthenosphere.** The lithos-

**lithosphere** The tough, rocky, outermost part of the Earth, comprising the crust and the uppermost part of the mantle.

**asthenosphere** A weak layer within the mantle, just below the lithosphere.

[3]A kilometer is 1000 meters, and 100 km = 67 miles. See Appendix A for further information about scales and conversions.

phere–asthenosphere boundary (that is, the strong rock–weak rock boundary) lies in the upper mantle and does not coincide with the compositional boundary between the crust and the mantle.

The Earth's lithosphere is very thin relative to the Earth as a whole. The solid rock that makes up the lithosphere is strong, but not strong enough to withstand the constant movement of hotter, weaker material in the underlying mantle. Consequently, the lithosphere has broken into a set of enormous rocky fragments we call *plates*. These plates move around on top of the asthenosphere, driven by the movement of hot material in the mantle. They collide, split apart, and slide past one another. This generates a lot of geologic activity, such as earthquakes and mountain-building, much of which occurs along the boundaries where plates interact with one another. Collectively, these processes are referred to as **plate tectonics**.

**plate tectonics** The horizontal movement and interactions of large fragments of lithosphere at the surface of the Earth.

Plate tectonic activity has been an important process throughout much of the Earth's history. It is responsible for the creation of mountains (Fig. 1.12), the locations of volcanoes, the intensities of earthquakes, and the shapes of continents and deep ocean basins. It has influenced the formation of the atmosphere, the development of climatic zones, and the evolution of life. Because of plate tectonics, the Earth has developed two fundamentally different kinds of crust: The **oceanic crust** that underlies the deep ocean basins is relatively thin (average thickness 8 km, or 5.4 mi) and dense; it is composed mostly of *basaltic* rocks. The **continental crust** is much thicker (average thickness 45 km, or 30 mi) and is made of generally less dense rocks. As we will see in later chapters, the concept of plate tectonics enables us to understand many years' worth of observations of natural processes in a unified context.

**oceanic crust** The crustal rocks that underlie the world's deep ocean basins.

**continental crust** The crustal rocks that form the continents and continental shelves.

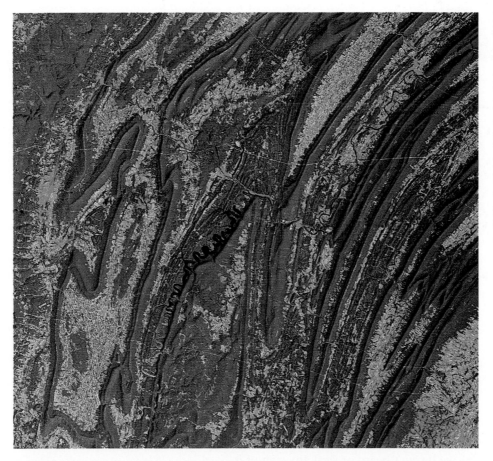

Figure 1.12
**MOUNTAINS SEEN FROM SPACE**
This view from space of the Appalachian Mountains in Pennsylvania shows layers of rock, once horizontal, that have been twisted and contorted as a result of collisions between lithospheric plates. The collisions occurred several hundred million years ago. The hills that now remain are the eroded remnants of a once much grander mountain range. The image is more than 100 km across.

**Nowhere is the fragile beauty of the Earth** more clearly visible than in photographs taken from space. Throughout more than three decades of manned space flight, astronauts have marveled at the awe-inspiring experience of viewing the planet from this unique perspective. Their words reveal the profound depth of the experience: the Earth is described as brilliant, intense, magical, incredibly fragile, shimmering, touchingly alone.

The very first image of the Earth from space was seen in the 1960s, at the dawn of the "space age." Before that, no one had ever actually seen the Earth in its entirety. In many respects, seeing this image changed our relationship to our planet forever. There it was—the whole planet in one sweeping view. We could see everything at a glance—the clouds, the oceans, polar ice caps, and continents—all at the same time and in their proper scale. For the first time in human history, it was abundantly clear just how small and isolated the Earth really is, and how unique.

American astronaut Rusty Schweickart went into space with the *Apollo 9* mission in 1969. Here is how he described his reaction to seeing the Earth from space:

*For me, having spent ten days in weightlessness, orbiting our beautiful home planet, fascinated by the 17,000 miles of spectacle passing below each hour, the overwhelming experience was that of a new relationship. The experience was not intellectual. The knowledge I had when I returned to Earth's surface was virtually the same knowledge I had taken with me when I went into space. Yes, I conducted scientific experiments that added new knowledge to our understanding of the Earth and the near-space in which it spins. But those specific extensions of technical details I did not come to know about until the data I helped to collect was analyzed and reported. What took no analysis, however, no microscopic examination, no laborious processing, was the overwhelming beauty . . . the stark contrast between bright colorful home and stark black infinity . . . the unavoidable and awesome personal relationship, suddenly realized, with all life on this amazing planet. . . Earth, our home.*

(Russell L. Schweickart, from Preface to *The Home Planet*, Kevin Kelley, ed., Addison-Wesley, 1988)

**Figure B1.1**
**RIVERS OF ICE**
This satellite image of Antarctica shows huge streams of ice flowing between rugged mountain ranges in Antarctica. This and the following satellite images are approximately 180 km (almost 300 mi) in their longest dimension.

What makes the Earth unique? We know of no other planet where plate tectonics has played, and continues to play, such an important role in forming the environment. We know of no other planet where the temperature permits water to exist near the surface in solid, liquid, and gaseous forms. We know of no other planet that would have been hospitable to the origin and evolution of life as we know it. There are billions upon billions of stars in the universe, so it is almost inevitable that there are billions of planets too; surely a few of those planets must be Earthlike and therefore might be capable of supporting life. However, if a relatively advanced civilization does exist on a planet somewhere out in space, so far we haven't heard or seen any sign of it.

Figure B1.2
PEOPLE LIVE HERE
In this satellite image, the black is the water of California's Salton Sea. The red areas are crops (healthy vegetation appears red in this type of satellite image). If you look closely at the red areas, you can see the circular and rectangular patches that are characteristic of irrigated regions.

Figure B1.3
DESERT DUNES
The subtle colors of the desert are seen in this satellite image, showing vast tracts of linear sand dunes in Namibia.

# EARTH SYSTEM SCIENCE

? Why should the Earth be studied as an integrated system?

The traditional way of studying the Earth has been to focus on separate units—the atmosphere, the oceans, or even a single mountain range—in isolation from the others. Now a new approach is taking hold in which the Earth is studied as a whole and viewed as a unified system. In particular, scientists today are focusing on the interactions and interrelationships among the various parts of the Earth system. Let's consider the concept of systems in greater detail.

15

## The System Concept

The system concept is a helpful way to break down a large, complex problem into smaller pieces that are easier to study, without losing sight of the interconnections between those pieces. A **system** can be defined as any portion of the universe that can be separated from the rest of the universe for the purpose of observing changes. By saying that *a system is any portion of the universe,* we mean that the system can be whatever the observer defines it to be. That's why a system is only a concept; you choose its limits for the convenience of your study. It can be large or small, simple or complex (Fig. 1.13). You could choose to observe the contents of a beaker in a laboratory experiment. Or you might study a lake, a fist-sized sample of rock, an ocean, a volcano, a mountain range, a continent, a planet, or the entire solar system. A leaf is a system, but it is also part of a larger system (a tree), which in turn is part of an even larger system (a forest).

The fact that a system has been *separated from the rest of the universe for the purpose of observing changes* means that it has conceptual boundaries that set it apart from its surroundings. The nature of those boundaries is one of the most important defining characteristics of a system, leading to three basic kinds of systems as shown in Figure 1.14. The simplest type of system to understand is an *isolated system*. In an isolated system, the boundaries prevent the system from exchanging either matter or energy with its surroundings. The concept of an isolated system is easy to understand, but such a system could only be imaginary. In the real world, it is possible to have boundaries that prevent the passage of matter, but it is impossible for any boundary to be so perfectly insulating that energy can neither enter nor escape from the system.

A second type of system, and the nearest thing to an isolated system in the real world, is a *closed system*. Such a system has boundaries that permit the exchange of energy, but not matter, with its surroundings. An example of a closed system would be a perfectly sealed oven, which would allow the material inside to be heated but would not allow any of that material to escape. (Note that in

**system** Any portion of the universe that can be separated from the rest of the universe for the purpose of observing changes.

**Figure 1.13**
**THE SYSTEM CONCEPT**
The river is a system, as is the lake it flows into. Together they form a larger system—the watershed. The small volumes of water and sediment indicated by boxes are examples of smaller systems.

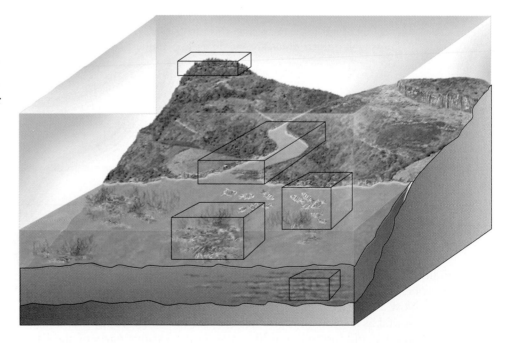

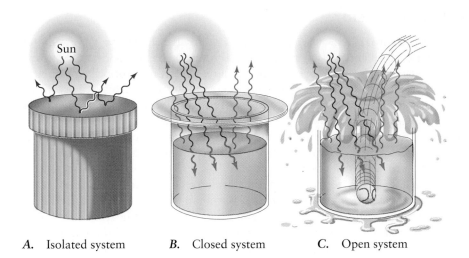

*A.*  Isolated system          *B.*  Closed system          *C.*  Open system

Figure 1.14
**THREE TYPES OF SYSTEMS**
*A.* An isolated system allows neither energy nor matter to cross its boundaries. Isolated systems don't really exist in nature. *B.* A closed system allows energy but not matter to cross its boundaries. Closed systems are rare; the Earth is a natural example of a closed system. *C.* An open system allows both energy and matter to cross its boundaries. Open systems are common in nature; most of the Earth's subsystems are open systems.

real life, ovens do allow some vapor to escape, so they are not perfect examples of closed systems.) The third kind of system, an *open system,* can exchange *both* matter and energy across its boundaries. An island offers a simple example of an open system. Matter (in the form of water) enters the system as precipitation and leaves by flowing into the sea or by evaporating back into the atmosphere. Energy enters the system as sunlight and leaves as heat (Fig. 1.15).

The system concept can be applied to both natural and artificial systems. For example, urban geographers and land use planners sometimes use a systems approach in the study of cities. Enormous flows of energy and materials occur in cities, and they are similar to natural systems in many respects. Geologists are increasingly using a systems approach, called **Earth system science,** to study the Earth and the relationships among its component "spheres."

**Earth system science** The application of a systems approach to the study of the Earth.

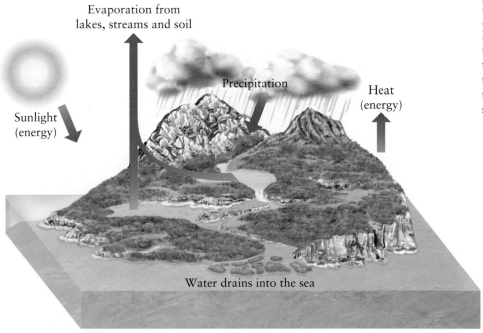

Evaporation from lakes, streams and soil

Precipitation

Heat (energy)

Sunlight (energy)

Water drains into the sea

Figure 1.15
**AN OPEN SYSTEM**
Energy (sunlight) and matter (water) reach an island from external sources. The energy leaves the island as heat. The water either evaporates or drains into the sea.

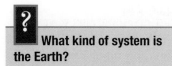

**What kind of system is the Earth?**

## The Earth System

The systems approach allows geologists to study the Earth as an integrated whole composed of interacting parts or *subsystems*. The four principal subsystems within the Earth system are the atmosphere, the hydrosphere, the biosphere, and the lithosphere (Fig. 1.16). You can think of these subsystems as huge *reservoirs* in which materials and energy are stored for a while before moving on. Each of the four main reservoirs can be further subdivided into smaller, more manageable units. For example, we can divide the hydrosphere into a number of subsystems (the oceans, glacial ice, streams, lakes, groundwater, and so on), each of which is a potential reservoir for water.

The Earth itself is a very close approximation to a closed system. Energy enters the Earth system as solar radiation. This energy is used in various biologic and geologic processes, then leaves the system in the form of heat. It is not quite correct to say that no matter crosses the boundaries of the Earth system, because we lose a small but steady stream of hydrogen and helium atoms from the outer-

**Figure 1.16**
**THE EARTH'S FOUR "SPHERES"**
The four principal subsystems of the Earth system: lithosphere, biosphere, atmosphere, hydrosphere. Materials and energy cycle among these subsystems, as shown by the arrows, making them open systems.

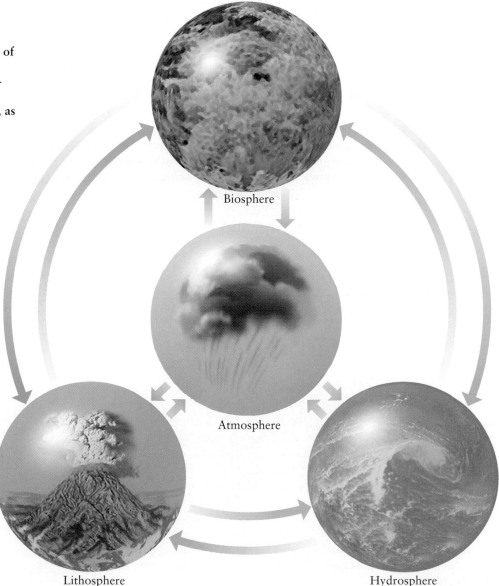

Biosphere

Atmosphere

Lithosphere

Hydrosphere

Figure 1.17
**THE CYCLING OF WATER AMONG THE EARTH'S RESERVOIRS**
The sun heats the ocean and causes water to evaporate into the atmosphere. Water vapor in the atmosphere condenses to form clouds, rain, or snow. Compacted snow forms the ice in a glacier. The glacier flows slowly down to the sea where huge bergs break off and slowly melt, rejoining the ocean. In this photo, icebergs are calving from the Hubbard Glacier in Alaska.

most part of the atmosphere, and we gain some material in the form of meteorites. However, the amount of matter that enters or leaves the Earth system on a daily basis is so minuscule compared with its overall mass that the Earth is essentially a closed system.

When changes are made in one part of a closed system, the results of those changes will eventually affect other parts of the system. The various subsystems of the Earth system are in a dynamic state of balance. When something disturbs one of the subsystems, the rest also change as the balance or *equilibrium* is reestablished. Sometimes an entire chain of events may ensue. For example, a volcanic eruption in Indonesia could throw so much dust into the atmosphere that it could initiate climatic changes leading to floods in South America and droughts in California, eventually affecting the price of grain in West Africa.

The fact that the Earth is a closed system has many implications. By definition, the amount of matter in a closed system is fixed and finite. Therefore, the resources on this planet are *all we have* and, for the foreseeable future, *all we will ever have*. This means that we must treat Earth resources with respect and use them wisely and cautiously. Another consequence of living in a closed system is that waste materials remain within the boundaries of the Earth system. As environmentalists sometimes say, "There is no *away* to throw things to."

## Cycles and Interactions

It is useful to envision interactions within the Earth system as a series of interrelated *cycles*, groups of processes that facilitate the movement of materials and energy among the Earth's reservoirs. Earth system science helps us study how materials and energy are stored and how they are cycled among the four principal reservoirs, as shown by the arrows in Figure 1.16. How long do materials and energy reside in each reservoir? The storage times can differ greatly. For example, an individual molecule of water may spend 100,000 years in a glacier, 1000 years in an underground reservoir, 7 years in a lake, 10 days in the atmosphere, or a few hours in an animal's body. Water is transferred from one of these reservoirs to another via any of innumerable pathways and processes that happen at different rates (Fig. 1.17). A single cycle includes processes that operate on several different time scales.

**?** **If materials and energy are constantly cycling, why do Earth systems appear to be stable and constant?**

Since this is a physical geology textbook, it focuses primarily on the lithosphere—that is, the outermost rocky portion of the solid Earth. However, the systems approach tells us that it is unreasonable—maybe even impossible—to consider one part of the Earth system in isolation from the rest. Physical geology also deals with interactions between the lithosphere and the interior of the Earth, as well as interactions with the hydrosphere, atmosphere, and biosphere. In this course, you not only study geology, but some oceanography, hydrology, meteorology, physics, chemistry, biology, and astronomy as well!

Using the concept of cycles, we can trace the movement of energy and materials from one subsystem to another. Some of the cycles that are particularly important in physical geology are:

**hydrologic cycle** A model that describes the movement of water through the Earth system.

- The **hydrologic cycle,** which describes the movement of water through the Earth system.

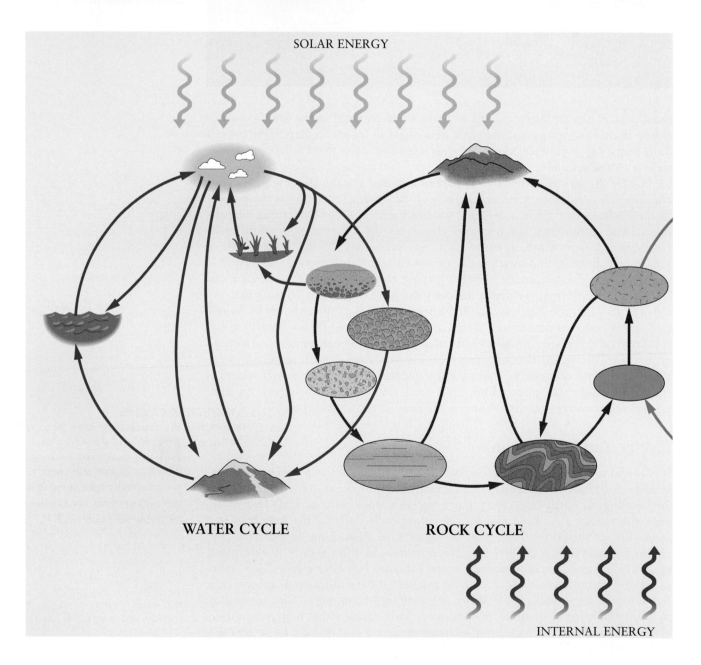

SOLAR ENERGY

WATER CYCLE  ROCK CYCLE

INTERNAL ENERGY

- The **rock cycle,** which describes all of the various processes by which rock is formed, modified, decomposed, and reformed by internal and external Earth processes.

- The **tectonic cycle,** which deals with the movements and interactions of lithospheric plates, and the internal processes that drive plate motion.

The rock cycle, tectonic cycle, and hydrologic cycle are closely linked together through physical, chemical, and biological processes (Fig. 1.18). Each will be discussed in greater detail in later chapters. Another useful group of cycles is the **biogeochemical cycles.** They trace the movements of chemicals that are essential to life, including carbon, oxygen, nitrogen, sulfur, and phosphorus. The processes that characterize all of these cycles are driven by energy that comes

**rock cycle** The set of crustal processes by which rock is formed, modified, transported, decomposed, and reformed.

**tectonic cycle** The movements and interactions of lithospheric plates, and the internal processes that drive plate motion.

**biogeochemical cycle** The movement through the Earth system of chemicals that are essential to life.

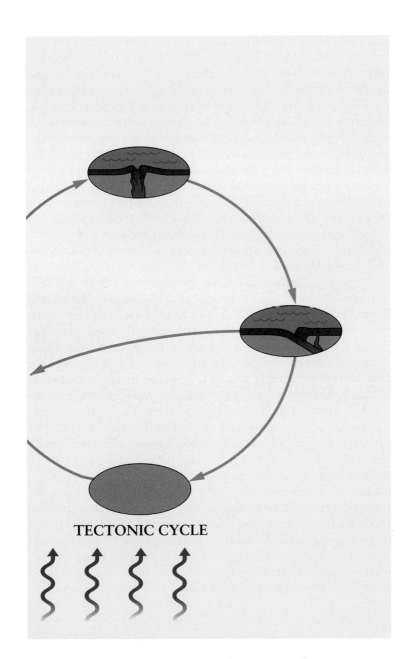

**TECTONIC CYCLE**

Figure 1.18
INTERCONNECTED CYCLES
The Earth system is characterized by interconnected cycles. The cycle on the righthand side of the diagram, which illustrates the internal Earth processes of the tectonic cycle, is driven primarily by heat from within the Earth. The middle cycle illustrates crustal processes through which rocks are formed, modified, broken down, and transformed; this cycle is driven partly by energy from within the Earth, and partly by energy from the Sun. The cycle on the lefthand side, the hydrologic cycle, is driven primarily by energy from the Sun. On this diagram we have left off most of the labels. As you proceed through the book you will learn about the processes that characterize each cycle, and we will fill in the labels little by little.

both from external sources (the Sun's energy and, to a much lesser extent, gravitational interaction with the Sun and Moon) and from the Earth's own internal (geothermal) energy. For geologists, the rock cycle is particularly important because it is the cycle that most directly connects internal and external Earth processes. The internal processes—igneous activity and plate tectonics—are driven primarily by heat from within the Earth. The external or crustal processes, including weathering, erosion, rock deformation, and metamorphism, act to modify and break down the products of igneous activity and tectonic uplift. In later chapters, we will revisit different parts of these cycles. Little by little, we will add labels to the cycle diagrams, as we build up a coherent picture of how all the different processes of the Earth system are related to one another. They work together to create the landscapes around us and the geologic environment in which we live.

## Uniformitarianism

Since material is constantly being transferred from one reservoir to another, you may wonder why these systems seem so stable. Why doesn't the sea become saltier or fresher? Why doesn't all the water in the world flow into the sea and stay there? Why should the chemical composition of the atmosphere be constant (as it has been for millions of years)? How can rock that is 2 billion years old have the same composition as rock that is being formed today? If mountains are constantly being worn down by erosion, why are there still high mountains? The answers to these questions are the same: Materials cycle from one system to another, but the systems themselves don't change noticeably because the different parts of the cycle balance each other—the amounts added approximately equal the amounts removed. While a mountain is worn down in one part of the cycle, a new mountain is being built up in another part. This cycling of materials has been going on since the Earth was formed, and it continues today.

A fundamental principle of geology, attributed to an eighteenth-century Scottish geologist named James Hutton, is based on this idea. It states that the processes operating in Earth systems today have operated in a similar manner throughout much of geologic time. This principle is called **uniformitarianism,** and it is a way of saying that "the present is the key to the past." We can examine any rock, however old, and compare its characteristics with those of similar rocks that are forming today. We can then infer that the ancient rock likely formed in a similar environment through similar processes and on a similar time scale. For example, in many deserts today we can see gigantic dunes formed from sand grains transported by the wind. Because of the way they form, the dunes have a distinctive internal structure (Fig. 1.19A). Using the principle of uniformitarianism, we can infer that a rock composed of cemented grains of sand and having the same distinctive internal structure as modern dunes is the remains of an ancient dune (Fig. 1.19B).

Hutton's principle of uniformitarianism provides a first step toward understanding the Earth's history. Geologists have used this principle to explain the Earth's features in a logical manner. In so doing, they have discovered that the Earth is incredibly old. An enormously long time is needed to erode a mountain range, or for huge quantities of sand and mud to be transported by streams, deposited in the ocean, and cemented into rocks, and for the rocks to be uplifted to form a mountain. Yet the cycle of erosion, formation of new rock, uplift, and more erosion has been repeated many times during the Earth's long history.

**uniformitarianism** The concept that the processes operating in Earth systems today have operated in a similar manner throughout much of geologic time.

Figure 1.19
SAND DUNES, ANCIENT AND MODERN
*A.* A distinctive pattern of wind-deposited sand grains can be seen in a hole dug in this modern sand dune near Yuma, Arizona. *B.* A similar pattern can be seen in sandstone rocks millions of years old, in Zion National Park, Utah. We can infer that these ancient rocks were once sand dunes.

*A.*

*B.*

**?** **Why is it important to study geology, even for those who don't plan to become geologists?**

# WHY STUDY GEOLOGY?

From this brief introduction to geology and the Earth system, you have probably deduced some of the reasons why it is important to study geology. We need to understand Earth materials because we are fundamentally dependent on them for the conduct of modern society. We depend on this small planet for all of our material resources—the minerals, rocks, and metals with which we construct our built environment; the energy with which we run it; the soil that sustains agriculture and other plant life; and the air and water that sustain life itself. Many Earth resources are finite and limited; they therefore require knowledgeable and thoughtful management. The materials of the Earth also have physical and chemical properties that affect us, such as their tendency to flow or fail during a landslide; their capacity to hold or transmit fluids such as water or oil; or their ability to absorb waste or prevent it from migrating.

We know that the Earth is essentially a closed system, which means that all materials remain within the system. Therefore, it is important to understand the processes whereby materials move from one reservoir to another. It is also important to understand the time scales that govern these processes in order to gain some perspective on the changes that we see occurring in the natural environment. Some Earth processes are hazardous—that is, damaging to human interests. These processes include earthquakes, volcanic eruptions, landslides, floods, and even meteorite impacts. The more we know about these hazardous processes, the more successful we will be in protecting ourselves from natural disasters (Fig. 1.20).

Finally, the Earth is our home planet. The features that make the Earth unique and the powerful geologic processes that characterize the Earth system are a constant source of awe and fascination to those of us who are privileged to study them. It makes good sense to work toward deepening and refining our understanding of the planet we live on.

From its beginnings a couple of centuries ago, geology has been an interdisciplinary science, because the Earth operates through the interactions of biologic, physical, and chemical processes. Yet we are discovering that the interactions are more complex and dynamic than we would have believed only a few decades ago. We are still learning about the complexities and interrelationships of subsystems such as climate, oceans, and shifting continents. We now appreciate more profoundly our own role in geologic change, and the need to study the Earth system as a whole rather than in separate fragments.

*Geology Today* starts your study of the Earth. If you are planning to become a geologist, this book will be an introduction to the many fascinating possibilities that await you in your career. If you are taking this course out of personal interest or to fulfill a degree requirement, you will emerge more aware of the geologic nature of our planet and better prepared to make informed decisions about the natural processes that affect your life on a daily basis. These decisions might range from selecting a home site with a good water supply to deciding whether to vote for a politician who supports nuclear energy.

# WHAT'S AHEAD

In the next two chapters of Part One, we look first at the materials of which the Earth is made and then at the Earth's history—how we can learn about past events by examining evidence preserved in the rock record. In Part Two, we take

Figure 1.20
LETHAL ERUPTION
Mount Pinatubo in the Philippines erupted in 1991. Volcanologists predicted this eruption, making possible the successful evacuation of thousands of residents from the area. The volcano sent this lethal cloud of searing, dust-laden gases rolling down its flanks, to spread rapidly across the surrounding plains. This particular car and driver escaped harm, but houses, trees, and fields were smothered with a blanket of hot volcanic dust.

a closer look at plate tectonics and at how earthquakes and volcanism reveal the inner workings of our dynamic planet. The processes discussed in Part Two concern the internal part of the tectonic cycle. Then, in Part Three, we revisit the crustal rock cycle, examining the processes of weathering and erosion, the formation of sedimentary rocks, and the processes through which rocks are altered and metamorphosed by exposure to elevated heat and pressure. We examine how the movement of lithospheric plates in the tectonic cycle affects the rocks of the Earth's crust, causing them to fold and fracture. The final chapter of Part Three puts the various pieces of the tectonic cycle and the rock cycle together by examining where different rocks form and why. In Part Four, our focus shifts to the hydrosphere and atmosphere, how they work, and how they interact to influence the Earth's climate system. Finally, in Part Five, we look at life on the Earth from the varied perspectives of Earth history, present-day use of Earth resources, and the future role of geoscientists.

## Chapter Highlights

**1. Geology** is the scientific study of the Earth. It encompasses the study of the planet, its formation, its internal structure, its materials and their properties, its chemical and physical processes, and its history.

**2. Physical geology** focuses on the materials and processes of the Earth system. **Historical geology** seeks to establish the chronology of events in the Earth's history. Today geology is a diverse science, with dozens of specialties.

**3.** The Earth is one of nine planets in the **solar system.** It is similar to the other planets—especially the rocky inner or **terrestrial planets**—in many ways. The terrestrial planets are all small, rocky, and relatively dense, and they all have experienced volcanic activity and extensive meteorite cratering. The outer or **jovian planets,** in contrast, consist of huge gaseous atmospheres with small solid cores, giving them very low densities overall.

**4.** Early in its history, the Earth differentiated into a dense, metallic **core,** a rocky **mantle,** and a brittle, rocky outer **crust.** There are two fundamentally different types of crust: the relatively thin, dense **oceanic crust** of basaltic composition, and the much thicker, less dense **continental crust** of granitic composition.

**5.** The Earth is the only planet we know of where **plate tectonics** plays such an important role in forming the environment. It is the only planet in our solar system where water is known to exist near the surface in solid, liquid, and gaseous forms, where regolith is formed by interactions among physical, chemical, and biological processes, and where life as we know it could exist.

**6.** Geologists think of the Earth as an integrated system. A **system** is a portion of the universe that can be separated from the rest of the universe for the purpose of studying changes in it. The four principal subsystems within the Earth system are the **atmosphere,** the **hydrosphere,** the **biosphere,** and the **lithosphere.** Materials and energy are stored for varying lengths of time in these great interconnected reservoirs and are cycled among them via innumerable pathways and processes.

**7.** The **hydrologic cycle, rock cycle, tectonic cycle,** and **biogeochemical cycles** help us describe the movement of materials and energy among the reservoirs of the Earth system. All of these cycles are linked through physical, chemical, and biological processes that operate at different rates. They are driven by energy that comes from both external and internal sources.

**8.** The principle of **uniformitarianism** is one of the fundamental concepts of geology. It is based on the idea that the processes we see operating in Earth systems today have operated in a similar manner throughout much of geologic time. Uniformitarianism says that "the present is the key to the past."

**9.** Earth materials and processes affect our lives through our dependence on Earth resources, through hazardous geologic processes, and through the physical and chemical properties of the natural environment. The Earth is our home planet, and it makes sense to refine and deepen our understanding of how it functions.

**10.** Geology is an interdisciplinary science because the Earth is characterized by many interacting biologic, physical, chemical, and geologic processes. Scientists have discovered that we need to study the Earth system as a whole rather than in separate fragments. **Earth system science** is an integrated approach to the study of the Earth and the relationships among its component parts.

## ▶ The Language of Geology

- asthenosphere 12
- atmosphere 11
- biogeochemical cycle 21
- biosphere 11
- continental crust 13
- core 10
- crust 10
- Earth system science 17
- geologist 4
- geology 4
- historical geology 4
- hydrologic cycle 20
- hydrosphere 11
- jovian planets 6
- lithosphere 12
- mantle 10
- meteorite 9
- oceanic crust 13
- physical geology 4
- plate tectonics 13
- rock cycle 21
- solar system 5
- system 16
- tectonic cycle 21
- terrestrial planets 6
- uniformitarianism 22

## ▶ *Questions for Review*

1. What are the two traditional areas of study within geology? What are their main concerns?
2. What are the nine principal planets in our solar system?
3. Describe the four basic processes that scientists believe were involved in the origin of the solar system.
4. What characteristics make the Earth's regolith, hydrosphere, and lithosphere unique in our solar system?
5. What is planetary differentiation?
6. What is a system? Give an example of a natural open system.
7. Summarize the steps in the formation of the solar system, starting with the initiation of collapse of the solar nebula and ending with the planets and other planetary bodies in their modern orbits.
8. What are the fundamental differences between oceanic crust and continental crust?
9. What are the main features that distinguish the inner, or terrestrial planets from the outer, or jovian planets?
10. What is the lithosphere? Where is the asthenosphere relative to the lithosphere? Is the asthenosphere part of the mantle? Is the lithosphere part of the mantle? (This is a slightly tricky question.)
11. What is Earth system science?

## ▶ *Questions for Thought and Discussion*

1. Do you think there may be life on a planet outside our solar system? What are the chances? What would the atmosphere be like? Would there be a hydrosphere?
2. Why is the systems approach so useful in studying both natural and artificial processes? Can you think of examples of artificial (that is, human-built) systems other than those given in the text? Are they open systems or closed systems? (Think about the materials and energy in them.)
3. How do you think the principle of uniformitarianism accounts for occasional, catastrophic events such as meteorite impacts, huge volcanic eruptions, or great earthquakes?
4. In this chapter we have suggested that the Earth is a close approximation of a natural closed system, and we have hinted at some of the ways that living in a closed system affects each of us. Can you think of some specific examples?

 For an interactive case study on the new tools of Earth system science, visit our Web site.

# · 2 ·

# EARTH'S MATERIALS: ATOMS, ELEMENTS, MINERALS, AND ROCKS

**Ours is a mineral world.** We mine minerals for metals, fertilizers, and materials for building and manufacture. We add salt (a mineral) to our food, and the food itself is grown in mineral-rich soil. Minerals are present in our bodies, and some minerals are essential for the maintenance of good nutrition.

Minerals, however, can have a bad side. Miners learned a long time ago that inhaling quartz-rich dust induces a lung disease called *silicosis.* Fortunately, silicosis is now rare as a result of dust control in mines. Everyone inhales quartz, because it is common among ordinary microscopic dust particles. However, we don't all get silicosis; our bodies seem to be able to handle the quartz particles at low levels of concentration.

*Chrysotile,* one of a group of minerals commonly referred to as *asbestos,* is currently at the center of a debate between scientists, lawmakers, and the public. Scientific evidence has proven that exposure to large amounts of chrysotile can cause lung diseases, but the law says, in effect, that even one tiny fiber of chrysotile is dangerous. Since chrysotile, like quartz, can be found in ordinary dust, all of us must inhale a little bit of it during our lifetimes, but we don't all suffer from asbestos-related diseases. Scientific evidence doesn't seem to support the law in this case.

Without minerals, it would be impossible to live as we do. If we keep studying minerals and the substances that compose them, we can learn to balance their positive and negative effects in our lives.

◆

## In this chapter you will learn about

- The simple substances of which all matter is composed
- How simple substances combine to form minerals, the fundamental building blocks of the Earth
- The properties and characteristics of common minerals
- The most important rock-forming minerals
- The most important families of rocks

▲

*Ancient Salt Mine*

Salt has been mined at Shahwah, Republic of Yemen, for thousands of years. The layer of salt, several meters thick, was originally deposited by the evaporation of seawater many millions of years ago. Miners dig out the salt using a pickaxe with a flat blade.

The Moon is very different from the Earth. One important difference is the virtual absence of an atmosphere. (There is a very thin atmosphere, but it is so tenuous that it is almost a perfect vacuum.) Another difference is that there are no liquids on the lunar surface—no water, no oil, no flowing lava. (There were lunar lava flows in the remote past, but today everything is solid.) By contrast, there are many liquids and gases on the Earth; ocean water and atmospheric gases are two important examples. On the Moon virtually all matter is solid, but on the Earth this is not the case. What has caused the difference?

**Can water, ice, and water vapor all exist together?**

# THE THREE STATES OF MATTER

We refer to liquids, solids, and gases as the three *states* of matter. On the Earth, unlike the Moon, matter occurs in all three states. As we discussed in chapter 1, one of the unique things about the Earth is that water, a vital component of all life-support substances, occurs in three states on the Earth's surface—as ice (a solid), as liquid water (a liquid), and as water vapor (a gas)(Fig. 2.1). Each of these states of $H_2O$ has an important influence on the environment in which we live: water in the ocean evaporates and forms water vapor; as the water vapor rises in the atmosphere, it condenses to form either rain or snow; the rainwater and snow meltwater then react with rock to form the soil in which plants grow; and these plants are the base of the food chain that supports most animals, including humans.

Figure 2.1
THE STATES OF WATER: ICE, WATER, AND VAPOR
One of the essential chemical substances needed for life, $H_2O$ occurs in three states at the Earth's surface: solid ice, liquid water, and gaseous water vapor. The state in which $H_2O$ occurs is determined by the temperature and pressure at any given time and place. Along the boundary lines between any two states, both states can coexist.

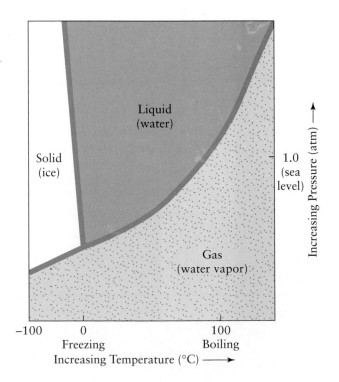

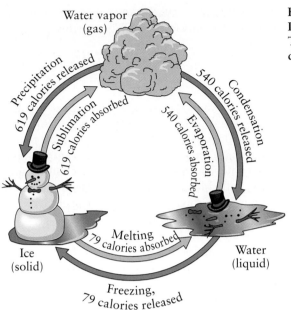

Figure 2.2
**LATENT HEAT**
The amount of heat absorbed by, or released from, one gram of $H_2O$ during a change of state.

The state of a substance is determined by temperature and pressure. When ice is heated, for example, a change of state occurs and the ice melts to become water. If the temperature is raised further, the water evaporates to become water vapor. By contrast, if water vapor is either compressed or cooled, it will condense, becoming either ice, as in snow or hail, or water, as in rain. Figure 2.1 shows the regions of temperature and pressure in which the three states of $H_2O$ are stable.

A change of state involves energy. The transition from ice to water, for example, requires the addition of heat energy; in the reverse case, when water freezes to ice, heat energy is released. The amount of heat energy released or absorbed per gram during a change of state is known as *latent heat* (from the Latin *latens*, meaning hidden, hence hidden heat). The latent heats corresponding to changes of state in $H_2O$ are shown in Figure 2.2. Latent heat plays an important part in the Earth's climate and thereby in the Earth system. For example, by evaporation of ocean water some of the Sun's heat energy is carried into the atmosphere as the latent heat of water vapor. When the vapor condenses to water droplets in clouds, or to rain drops or snowflakes, the latent heat is released and the atmosphere is warmed.

When we discuss the materials and processes that make and shape the Earth, it is important to be clear about the states of those materials. This may sound like a simple matter, but there are so many Earth processes in which materials change from one state to another—water freezing to ice, rock melting to a hot liquid that may flow out of the Earth as lava, and so on—that we must always be careful to specify the state of the material under discussion.

# SIMPLE SUBSTANCES

Regardless of the state of a substance, all matter on and in the Earth consists of one or more of the 90 chemical elements known to occur in the Earth. Even though you probably learned about the chemical elements in your high school chemistry course, it will be helpful to review the topic here because all of the chemical reactions that control our lives, and that make our planet habitable, depend on the ways in which the chemical elements interact. Without those interactions, there would be no planet Earth.

## Elements

If you were a chemist and were asked to analyze a sample of mineral or rock, your report would list the kinds and amounts of chemical elements present in the sample. The **elements** are the most fundamental substances into which matter can be separated and analyzed by ordinary chemical means. For example, table salt—the chemical compound sodium chloride, NaCl—is not an element because it can be separated into sodium (Na) and chlorine (Cl). But neither sodium nor chlorine can be further broken down by ordinary chemical means; therefore, they are both elements.

Every element is identified by a symbol, such as H for hydrogen and Si for silicon. Some of these symbols, such as that for hydrogen, come from the element's name in English. Other symbols come from other languages. For example, the symbol for iron is Fe, from the Latin *ferrum*; the symbol for copper is Cu, from the Latin *cuprum*, which in turn comes from the Greek *kyprios* (after Cyprus, an island in the Mediterranean Sea where copper was mined in antiquity); and the symbol for sodium is Na, from the Latin *natrium*. The chemical elements, their symbols, and their abundances in the Earth's crust are listed in Appendix B.

## Atoms and Isotopes

A piece of a pure element, even a tiny piece no bigger than the head of a pin, consists of a vast number of particles called *atoms*. An **atom** is the smallest individual particle that retains the distinctive properties of a given chemical element. Atoms are so tiny that they can be seen only with the most powerful microscopes, and even then the image is imperfect because individual atoms are only about $10^{-10}$m[1] in diameter (that is, 0.00000000010 m).

As you probably recall, atoms are built up from *protons,* which have positive electrical charges, *neutrons,* which are electrically neutral, and *electrons,* which have negative electrical charges (Fig. 2.3). An atom consists of a nucleus and a number of electrons. The nucleus contains all of the protons and neutrons; the electrons move rapidly around the nucleus in paths that cannot be traced in detail. The number of protons in an atom—the *atomic number*—is what gives an atom its special chemical characteristics. These characteristics, in turn, identify it as an atom of a specific chemical element. Atomic numbers increase from the lightest element, hydrogen, atomic number 1 (with one proton in the nucleus) through uranium, atomic number 92 (with 92 protons in the nucleus), to a synthetic element, meitnerium, atomic number 109 (with 109 protons in the nucleus), the heaviest element made so far.

Only 90 of the chemical elements are known to occur naturally on the Earth; 19 others have been made in the laboratory. Two elements, technetium (atomic number 43) and promethium (atomic number 61), probably did once exist on the Earth but have long since disappeared as a result of radioactive decay. Four other elements—astatine (atomic number 85), francium (atomic number 87), actinium (atomic number 89), and plutonium (atomic number 94)—have been detected in nature but only in vanishingly small amounts.

element (chemical) The most fundamental substance into which matter can be separated by chemical means.

? What's the difference between elements and atoms?

atom The smallest individual particle that retains the distinctive properties of a given chemical element.

---

[1]m is an abbreviation for meter, the unit of length in the metric system. One meter is equal in length to 1.094 yards or 3.281 feet.

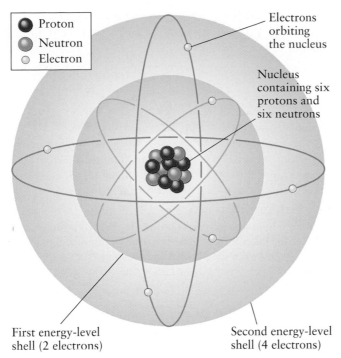

Proton
Neutron
Electron

Electrons
orbiting
the nucleus

Nucleus
containing six
protons and
six neutrons

First energy-level
shell (2 electrons)

Second energy-level
shell (4 electrons)

**Figure 2.3**
**INSIDE AN ATOM**
This is a schematic diagram of an atom of carbon-12 ($^{12}$C). At the center is the nucleus, containing protons and neutrons (six of each in the case of $^{12}$C). Protons have a positive electrical charge; neutrons have no charge. Six negatively charged electrons circle the nucleus in complex paths called orbitals, so the diagram is only schematic. Two electrons are in orbitals close to the nucleus; four are in more distant orbitals. The different groupings of orbitals are called energy-level shells. We chose carbon-12 for the illustration because it is one of the important atoms of which our bodies are made.

What role do neutrons play in the nucleus? Neutrons act like a glue that holds the nucleus together. The sum of the numbers of neutrons and protons in an atom is called the *mass number*. All atoms of a given chemical element have the same atomic number. But the atoms of an element can have different mass numbers because they can have different numbers of neutrons in their nuclei. Atoms with the same atomic number but different mass numbers are called **isotopes.** For example, there are three natural isotopes of carbon: carbon-12, carbon-13, and carbon-14. Each of the isotopes of carbon has 6 protons per atom and thus an atomic number of 6, but the three isotopes contain different numbers of neutrons: 6, 7, and 8, per atom, respectively.

**isotope** Atoms with the same atomic number but different mass numbers.

## Ions

The positive electrical charge of a proton is exactly equal, but opposite, to the negative electrical charge of an electron. In its ideal state, an atom has an equal number of protons and electrons and is electrically neutral. Chemical reactions involve the transfer of electrons between atoms, or the sharing of electrons between atoms.

An atom that has an overall positive or negative electrical charge because of the loss or addition of one or more electrons is called an **ion.** When the charge is positive, meaning that the atom has given up one or more electrons, the ion is called a *cation.* When the charge is negative, meaning that the atom has added

**ion** An atom that has an excess positive or negative electrical charge.

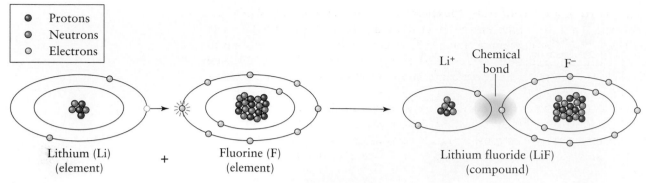

Protons
Neutrons
Electrons

Lithium (Li)
(element)

+

Fluorine (F)
(element)

Li$^+$    Chemical bond    F$^-$

Lithium fluoride (LiF)
(compound)

**Figure 2.4**
**HOW IONS AND COMPOUNDS FORM**
When an atom of lithium loses an electron, a positively charged lithium cation (Li$^+$) is the result. An atom of fluorine takes the extra electron and becomes a negatively charged anion (F$^-$). The attraction between the positive and negative charges bonds Li$^+$ to F$^-$, forming an ionic bond.

one or more electrons, it is called an *anion*. A convenient way to indicate ionic charges is to record them as superscripts. For example, Li$^+$ is the symbol for an atom of lithium that has given up an electron, while F$^-$ is a symbol for an atom of fluorine that has accepted an electron (Fig. 2.4).

# COMPOUNDS

**compound** (chemical) A combination of atoms of one or more elements with atoms of another element in a specific ratio.

**?** **Are the properties of chemical compounds the same as the properties of the elements of which it is made?**

**molecule** The smallest unit that has all the properties of a particular compound.

**bonding** The force that holds the atoms together in a chemical compound.

Chemical **compounds** form when atoms of one or more elements combine with atoms of another element in a specific ratio. For example, lithium and fluorine combine to form lithium fluoride (written LiF), a compound used in making ceramics and enamels. Writing the formula as LiF indicates that for every Li atom there is one F atom. Similarly, the compound H$_2$O forms when hydrogen combines with oxygen in the ratio of two atoms of hydrogen to one of oxygen. The formula of a compound is written by putting the element that tends to form cations first and the element that tends to form anions second. The relative numbers of atoms in a compound are indicated by subscripts. Thus, for water, we write H$_2$O.

## Molecules and Bonding

The properties of compounds are quite different from those of their constituent elements. For example, the elements sodium (Na) and chlorine (Cl) are highly toxic, but the compound NaCl (which occurs naturally as the mineral *halite*) is essential for human health. The smallest unit that has the properties of a compound is called a **molecule.** Do not confuse a molecule and an atom; the definitions are similar, but a molecular compound always consists of two or more atoms held together. The force that holds the atoms together in a compound is called **bonding.** There are several different types of bonding, and because bonding determines the physical and chemical properties of a compound, it is helpful to briefly review the subject.

### Bonding

Electrons move around the nucleus of an atom in complex, three-dimensional patterns called orbitals. Each orbital is characterized by a specific amount of energy, and electrons are identified by the energy levels of their orbitals. When several electrons have the same energy level, they are said to occupy the same *energy-level shell*. The maximum number of electrons that can occupy a given energy-level shell is fixed; shell 1, the lowest energy level, can only accommodate 2 orbital electrons; shell 2, however, can accommodate up to 8 orbital electrons, shell 3, 18, and shell 4, 32.

When an energy-level shell is filled with its quota of electrons, it is very stable, like an evenly loaded boat. To fill the shells and reach a stable configuration, atoms transfer or share electrons, and it is the transferring and sharing of electrons that forms bonds between atoms. There are four important kinds of bonding:

1. *Ionic bonding:* Electron transfers between atoms produce cations and anions, which can exist as free entities, but even so an electrostatic attraction draws these negatively and positively charged ions together, forming an *ionic bond* as shown in Figure 2.4. Compounds with ionic bonds tend to have moderate strength and moderate hardness. Table salt (NaCl) has ionic bonds. When you eat salt and it dissolves in your mouth, the NaCl separates into $Na^+$ and $Cl^-$ ions. The elements Na and Cl are toxic, but their ions in solution are not; it is the ions that create the familiar salty flavor.

2. *Covalent bonding:* Some atoms share electrons, and the force of sharing is called a *covalent bond*. Note that electron sharing does not produce ions. One important substance in which covalent bonds occur is diamond, a form of carbon. The highest energy-level shell of carbon has 4 electrons but requires 8 for maximum stability. Each carbon atom shares two electrons with four other carbon atoms as shown in Figure 2.5. Ele-

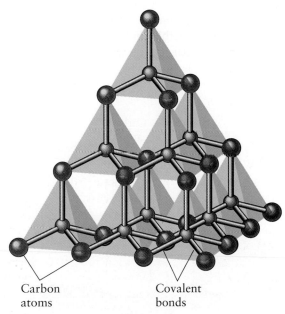

Carbon
atoms

Covalent
bonds

**Figure 2.5**
**BONDING IN DIAMOND**
This schematic diagram shows the three-dimensional geometric arrangement of carbon atoms in diamond. Each atom is surrounded by four others; this is because each carbon atom shares its four outer energy-level electrons with four other carbon atoms, so that all atoms have stable energy-level shells of eight electrons.

ments and compounds with covalent bonds tend to be strong and hard, and the sparkle that makes diamonds attractive gems is due to covalent bonds.

3. *Metallic bonding:* Metals have a special kind of bonding. In *metallic bonding,* the atoms are closely packed; electrons in the higher energy-level shells are shared between several atoms and are so loosely held that they can readily drift from one atom to another. The drifting electrons give metals their distinctive properties—for example, they are opaque, malleable, and good conductors of heat and electricity. Gold, copper, and silver are examples of naturally occurring metals with metallic bonds.

4. *Van der Waals bonding:* The last kind of bond arises as a weak secondary attraction between certain molecules formed by the transfer or sharing of electrons. The weak *Van der Waals bonding* plays an important role in certain minerals, of which graphite, another form of carbon, is an example. Graphite has a sheetlike structure in which each carbon atom has three nearest neighbors at the corners of an equilateral triangle (Fig. 2.6). The carbon atoms in the sheets form large, covalently bonded molecules that are strong and flexible. The graphite used in golf clubs and tennis rackets makes use of the strongly bonded sheets. Adjacent sheets of carbon are held together by weak Van der Waals bonds. The bonds are so weak they can be easily broken. That is why graphite feels slippery when you rub it between your fingers—the rubbing breaks the bonds and the sheets slide easily past each other. Talc, the mineral in talcum powder, has a sheet structure like that of graphite and also feels smooth and slippery because of the easily broken Van der Waals bonds.

Why have we spent all this time learning about simple substances, bonds, and compounds? It is because minerals—the main building blocks of Earth materials—occur in nature in the form of elements (like gold and diamond) or chemical compounds. To understand minerals, we needed first to understand something about how they are put together and why they have the properties they do. Now let's look more closely at the characteristics that help us to identify a specific material as a mineral.

> **?**
> **Why do diamond and graphite have such different properties?**

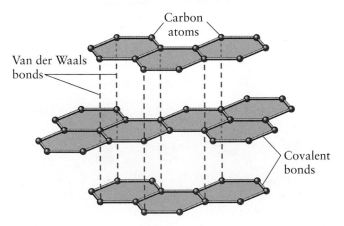

**Figure 2.6**
**BONDING IN GRAPHITE**
Geometric arrangement of carbon atoms in graphite. Bonding within the sheets is covalent; bonding between the sheets is Van der Waals.

# MINERALS

The word **mineral** has a specific connotation in science. To be classified as a mineral, a substance must meet five requirements:

1. It must be *naturally formed.*
2. It must be a *solid.*
3. It must have been *formed by inorganic processes.*
4. It must have a *specific chemical composition.*
5. It must have a *characteristic crystal structure.*

## Characteristics of Minerals

Let's briefly review these characteristics before going on to discuss the specific properties of minerals in more detail.

*Naturally Formed.* The requirement that minerals be naturally formed excludes any substance that is produced by artificial means, such as steel, plastic, or any of the laboratory-produced crystalline materials that have no natural equivalents. Technically, none of these substances is a mineral.

*Solid.* All liquids and gases—including naturally occurring ones such as oil and natural gas—are also excluded because minerals are solids. This requirement is based on the state of the material, not on its composition. For example, ice in a glacier is a mineral but water in a stream is not, even though both are made of the same chemical compound, $H_2O$.

*Formed by Inorganic Processes.* Materials such as leaves, which are derived from living organisms and contain organic compounds, are not minerals. Coal, for example, is not a mineral because it is derived from the remains of plant material and contains organic compounds. The teeth and bony parts of dead animals as well as the shells of sea creatures present a trickier case. The materials in bones, teeth, and shells are the same as those in minerals, but by definition, such materials are not minerals because they are formed by organic processes. However, when the bone or shell is fossilized, the original materials are usually replaced by minerals, an inorganic process called *mineralization*. If you were to perform a chemical analysis of a dinosaur bone, for example, you would find few or no organic compounds, only minerals formed by inorganic processes, even though the internal structure of the bone may have been preserved by mineralization.

*Specific Chemical Composition.* The requirement that a mineral must have a specific chemical composition has several implications. Most importantly, it means that minerals are either chemical elements (gold and diamond are examples) or chemical compounds in which atoms are present in specific ratios. Quartz, with the formula $SiO_2$, is an example of a chemical compound. The ratio of Si to O in quartz is always 1 to 2. Many minerals have much more complicated formulas than quartz; for example, the chemical formula of the mineral phlogopite, a common form of mica, is $KMg_3AlSi_3O_{10}(OH)_2$. Other minerals have even more complicated formulas, but in all cases the elements in compounds combine in specific ratios.

**mineral** A naturally formed, solid, inorganic substance with a specific chemical composition and a characteristic crystal structure.

**?** **Are bones made of minerals?**

A complication involving mineral compositions arises from a property called *atomic substitution*. Atoms of different elements that are similar in size and bonding properties can substitute for each other in a mineral. For example, $Mg^{2+}$ and $Fe^{2+}$ are so similar in size and electrical charge that extensive substitution of $Fe^{2+}$ for $Mg^{2+}$ can occur in the mineral olivine, $Mg_2SiO_4$. Atomic substitution does not change the combining ratios of the elements involved or the kinds of bonds involved. Atomic substitution is indicated in a chemical formula by putting brackets around the substituting elements and a comma between them; the formula for olivine, therefore, becomes $(Mg, Fe)_2SiO_4$.

The composition requirements for minerals serve to exclude materials whose composition varies within a range that cannot be expressed by an exact chemical formula. An example of such a material is glass, which is a mixture of many elements and can have a wide range of compositions.

*Characteristic Crystal Structure.* Glass—even naturally formed volcanic glass—is further excluded from being called a mineral by the requirement that a mineral must have a characteristic crystal structure. This term has to do with the arrangement of atoms in the material. Technically, glass is a frozen liquid. The atoms in liquids are randomly jumbled, while the atoms in minerals are organized in regular, repetitive geometric patterns, as shown in Figure 2.7. The geometric pattern of atoms in a mineral is referred to as the *crystal structure*. Because minerals have a crystal structure, they are said to be **crystalline**. Solids such as glass that lack a crystal structure are called *amorphous* solids (from the Greek for "without form"); they are not minerals. All minerals are crystalline, and the crystal structure of any mineral is a unique characteristic of that mineral.

All specimens of a given mineral have an identical crystal structure. Extremely sensitive, ultra-high-resolution microscopes enable scientists to look at the crystal structures of minerals and actually see the orderly arrangement of atoms in the mineral. As you can see from Figure 2.8, the atoms in a crystalline material resemble the regular, orderly rows in an egg carton.

**crystalline** Having an internal crystal structure, that is, a geometric pattern of atoms.

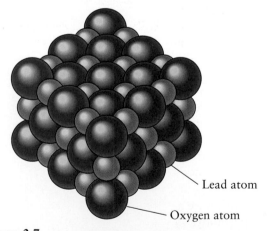

Lead atom

Oxygen atom

**Figure 2.7**
**ATOMS IN ORDERLY ARRANGEMENT**
The orderly arrangement of atoms in galena (PbS), the most common mineral containing lead. The geometric arrangement is repetitive throughout a mineral grain. The atoms are so small that a cube of galena 1 cm on its edge contains more than $10^{22}$ atoms each of lead and sulfur. ($10^{22}$ is 1 followed by 22 zeros!)

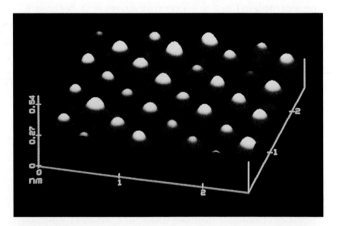

**Figure 2.8**
**SEEING ATOMS**
Atoms in galena (PbS), seen with a special, ultra-high-resolution device called a scanning-tunneling microscope. The packing of sulfur atoms (large) and lead atoms (small) is the same as the geometric arrangement of atoms shown in Figure 2.7.

**Figure 2.9**
**CRYSTAL FACES AND ANGLES**
The equivalent crystal faces on two specimens of quartz differ greatly in size, but the angles between the equivalent faces are the same on both specimens. Equivalent faces are identified by the same number on each specimen.

## Properties of Minerals

The properties of minerals are determined by their compositions and crystal structures. Once we know the properties of the various mineral species, we can often use a few simple tests to determine which species an unidentified mineral sample belongs to. It is usually not necessary to analyze an unidentified mineral sample chemically or to determine its crystal structure in order to discover its identity. The properties most often used to identify minerals are obvious ones, such as color, external shape (the crystal form or *habit*), and hardness. Less obvious properties, such as *luster* (the quality and intensity of light reflected from the mineral), *cleavage* (the tendency of the mineral to break in preferred directions), and *density* (i.e., the "heaviness" of the mineral), are also used to identify minerals.

### Crystal Form

The ancient Greeks were fascinated by ice. They were intrigued by the fact that needles of ice are six-sided and have smooth, planar surfaces. The Greeks called ice *krystallos*, and the Romans Latinized the word to *crystallum*. Eventually, the word **crystal** came to be applied to any solid body that has grown with planar surfaces. The planar surfaces that bound a crystal are called *crystal faces,* and the geometric arrangement of crystal faces is called the **crystal form.** During the seventeenth century, crystal form was a subject of intense study. Scientists discovered that crystal form could be used to identify minerals. But they were unable to explain certain features such as the wide variation in the relative sizes of crystal faces from one sample to another. Under some circumstances, a mineral species may grow a thin crystal; under others, the same mineral species may grow a fat crystal, as Figure 2.9 shows. It is apparent from the figure that the overall crystal size and the relative sizes of crystal faces are not the same for these two crystals of quartz: crystal size and the relative sizes of crystal faces are not definitive for any mineral.

The person who solved the mystery was a Danish physician, Nicolaus Steno.[2] In 1669 Steno demonstrated that an important property of a given mineral is not the size of its faces, but rather the angles between the faces. The angle between any designated pair of crystal faces of a given mineral species is constant, he wrote, and it is the same for all specimens of a mineral, regardless of

**?**  **How do geologists tell one mineral from another?**

**crystal** Any solid body with an internal crystal structure.

**crystal form** The geometric arrangement of planar faces that bound a crystal.

___
[2]Steno wrote in Latin and is best known by his Latinized name, Nicolaus Steno; his actual name was Nils Stensen.

overall shape or size. Steno's discovery that *interfacial angles* are constant is made clear by the numbering in Figure 2.9. Equivalent faces occur on both of the quartz crystals. The sets of faces are parallel, and the angle between any two equivalent faces is the same on each crystal.

Steno and other early scientists suspected that a mineral must have some kind of internal order that enables it to form crystals with constant interfacial angles. However, the particles on which that order depends—atoms—were too small for them to see, so they could only speculate. Proof that crystal form reflects internal order was finally achieved in 1912. In that year the German scientist Max von Lauë, using X rays, demonstrated that crystals must be made up of atoms packed in fixed geometric arrays, as shown in Figure 2.7.

Crystal faces can form only when mineral grains can grow freely in an open space. Because most mineral grains do not form in open, unobstructed spaces, crystals are uncommon in nature. Instead, most mineral grains grow in limited spaces where other mineral grains get in the way. As a result, most mineral grains are irregularly shaped. However, in both a crystal and an irregularly shaped grain of the same mineral, all the atoms present are packed in the same strict geometric pattern; that is, the crystals and the irregular grains have identical crystal structures, and both are crystalline. The term *crystal structure*, rather than crystal, is therefore used in the definition of a mineral.

## Habit and Cleavage

**habit** The distinctive shape of a particular mineral.

Some minerals grow such distinctively shaped grains that the grain shape—called the mineral's **habit**—can sometimes be used as an identification tool. For example, the mineral pyrite ($FeS_2$) is commonly found as a collection of intergrown cubes (Fig. 2.10). Figure 2.11 shows chrysotile asbestos, a variety of the mineral

**Figure 2.10**
**CUBES OF PYRITE**
A characteristic growth form of pyrite ($FeS_2$), known as "fool's gold," is a cube-shaped crystal with pronounced striations on the cube faces. The cubic crystals and striations on the sides are dictated by the internal structure of pyrite's atoms. The largest crystals in the photograph are 3 cm (about 1 in) on an edge.

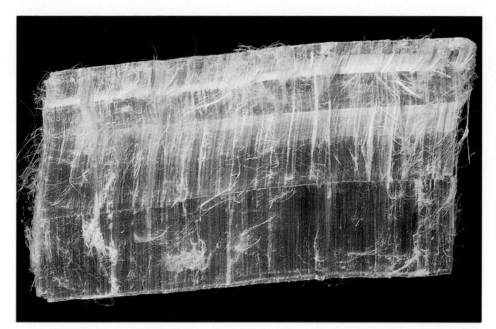

Figure 2.11
**FIBERS OF ASBESTOS**
Some minerals have distinctive growth habits, even though they do not develop well-formed crystal faces. The mineral chrysotile sometimes grows as fine, cotton-like threads that can be separated and woven into fireproof fabric. When the mineral occurs like this, it is called *asbestos*. Chrysotile is one of several different minerals that are mined and commercially processed to make asbestos fibers.

serpentine that characteristically takes the form of fine threads. Muscovite, a variety of mica, almost always grows as booklike stacks of clear, thin sheets (Fig. 2.12).

The tendency of certain minerals to break in preferred directions along brightly reflective planar surfaces is called **cleavage**. If you break such a mineral with a hammer or drop it on the floor so that it shatters, some of the broken fragments will be bounded by surfaces that are smooth and planar, and in that respect they are like crystal faces. This can also be seen in Figure 2.12. The "books" of muscovite cleave easily into thin sheets (the "pages" of the books) but do not cleave at all in any other direction. In certain cases, such as the halite (NaCl) fragments shown in Figure 2.13, several cleavage directions are present,

**cleavage** The tendency of a mineral to break in preferred directions.

Figure 2.12
**BOOKS OF MICA**
The perfect cleavage of muscovite (a mica) is shown by the thin, planar flakes into which this specimen is being split. The cleavage flakes suggest leaves of a book, a semblance embodied in the name "book of mica."

Figure 2.13
**MINERAL CLEAVAGE**
Because halite (NaCl) has three well-defined cleavage directions, it breaks into fragments bounded by three perpendicular faces. This cleavage pattern is controlled by the mineral's internal structure of atoms.

and all of the fragments are bounded by smooth planar surfaces. Don't confuse crystal faces and cleavage surfaces, even though the two often look alike. A cleavage surface is a breakage surface, whereas a crystal face is a growth surface.

The directions in which cleavage occurs are governed by the crystal structure: cleavage takes place along planes where the bonds between atoms are relatively weak as in the case of muscovite, or where there are fewer bonds per unit area, as in the case of diamond. Because cleavage directions are directly related to crystal structure, the angles between equivalent pairs of cleavage directions are the same for all grains of a given mineral. Thus, just as the interfacial angles of crystals are constant, so are the angles between cleavage planes. As we remarked earlier, crystals and therefore crystal faces are rare; however, almost every mineral grain you see in a rock shows one or more breakage surfaces. That is why cleavage is such a useful aid in the identification of minerals.

### Hardness

**hardness** A mineral's resistance to scratching.

The term **hardness** refers to a mineral's resistance to scratching. Hardness, like habit, crystal form, and cleavage, is governed by crystal structure and by the strength of the bonds between atoms. The stronger the bonds, the harder the mineral. Note the difference between *hardness* and *cleavage*: a mineral might be quite hard—that is, resistant to scratching—but still cleave easily along one or more surfaces.

**?** **If a mineral is very soft, does that mean it will break easily?**

Relative hardness values can be assigned by determining the ease or difficulty with which one mineral will scratch another. The *Mohs relative hardness scale* can be used to assign hardness values. The scale is divided into 10 steps, each marked by a common mineral (Table 2.1). Talc, the basic ingredient of most body ("talcum") powders, is the softest mineral known; Mohs assigned talc a value of 1 on the relative hardness scale. Diamond, the hardest mineral known, was given a value of 10. The 10 steps of the hardness scale do not represent equal intervals of hardness; the important feature of the scale is that any mineral on the scale will scratch all minerals below it. For convenience, we often test relative hardness by using a common object such as a penny or a penknife as the scratching instrument, or glass as the object to be scratched.

The ten minerals of Mohs scale, starting with the softest, talc, in the upper left-hand corner, and proceeding across two rows to diamond, the hardest mineral, in the lower right-hand corner.

**TABLE 2.1 The Mohs Scale\* of Relative Hardness of Minerals**

| | Relative Hardness Number | Reference Mineral | Hardness of Common Objects |
|---|---|---|---|
| Softest | 1 | Talc | |
| | 2 | Gypsum | Fingernail |
| | 3 | Calcite | |
| | 4 | Fluorite | |
| | 5 | Apatite | Copper penny |
| | 6 | Potassium feldspar | Pocketknife; glass |
| | 7 | Quartz | |
| | 8 | Topaz | |
| | 9 | Corundum | |
| Hardest | 10 | Diamond | |

\*Named for Friedrich Mohs, an Austrian mineralogist, who chose the 10 minerals of the scale.

| *A.* | *B.* | *C.* |

**Figure 2.14**
**MINERAL LUSTER**
Examples of three minerals with very different lusters. *A.* Quartz with a glassy luster. *B.* Sphalerite with a resinous luster—it resembles dried tree resin, such as amber. *C.* Talc with a pearly luster.

## Luster, Color, and Streak

The quality and intensity of light reflected from a mineral produce an effect called **luster.** Luster may be described as *metallic,* like that of a polished metal surface; *vitreous,* like that of glass; *resinous,* like that of resin; *pearly,* like that of pearl; or *greasy,* as if the surface were covered by a film of oil (Fig. 2.14). Two minerals with almost identical color can have quite different lusters.

The color of a mineral, though often striking, is not a reliable means of identification. A mineral's color is determined by several factors, but the main cause is chemical composition. Some elements can create strong color effects, even when they are present only as trace impurities (Fig. 2.15). For example, the min-

**luster** The quality and intensity of light reflected from a mineral.

**Figure 2.15**
**SAME MINERAL, DIFFERENT COLOR**
Two specimens of the same mineral, corundum ($Al_2O_3$), with distinctly different colors. The red crystal, ruby, is from Tanzania, about 2.5 cm (1 in) across. The blue crystal, sapphire, comes from Newton, New Jersey, and is about 3 cm (1.2 in) across.

Figure 2.16
MINERAL STREAK
This specimen of hematite is shiny, black, and metallic-looking, but it makes a reddish-colored smear on a porcelain streak plate. A red streak is a distinctive property of hematite.

eral corundum ($Al_2O_3$) is commonly white or grayish, but when small amounts of chromium are present as a result of atomic substitution of $Cr^{3+}$ for $Al^{3+}$, corundum is blood red and is given the gem name *ruby*. Similarly, when small amounts of iron and titanium are present, the corundum is deep blue, producing another prized gem, *sapphire*. Color can be very confusing in opaque minerals with metallic lusters. This is because the color is partly a property of the size of the mineral grains. One way to reduce error is to prepare a **streak,** a thin layer of powdered mineral made by rubbing a specimen on a nonglazed fragment of porcelain called a *streak plate*. The color of a streak is reliable because all the grains in the streak are very small and the effect of grain size is reduced. For example, hematite ($Fe_2O_3$) produces a red streak even though a specimen itself may look black and metallic (Fig. 2.16).

**streak** A thin layer of powdered mineral made by rubbing a specimen on a nonglazed fragment of porcelain.

### Density

Another important physical property of a mineral is how light or heavy it feels. We know that two equal-sized baskets have different weights when one is filled with feathers and the other with rocks. The property that causes this difference is **density,** the mass per unit volume. The unit of density is a gram per cubic centimeter ($1$ g/cm$^3$). Minerals with a high density, such as gold, have closely packed atoms. Minerals with a low density, such as ice, have loosely packed atoms.

**density** Mass per unit volume.

Density can be gauged by holding different minerals and comparing their weights. Metallic minerals generally feel heavy, whereas minerals with vitreous lusters tend to feel light. Each mineral can be placed at a specific location on a density scale. Gold has a density of $19.3$ g/cm$^3$ and feels very heavy. Many other minerals, such as galena (PbS) and magnetite ($Fe_3O_4$), which have densities of $7.5$ g/cm$^3$ and $5.2$ g/cm$^3$, respectively, also feel heavy by comparison with other minerals. Many common minerals have densities in the range of $2.5$–$3.0$ g/cm$^3$.

**?** If there are 90 chemical elements that make up minerals, then why are there only a small number of rock-forming minerals?

# THE COMMON MINERAL FAMILIES

Geologists have identified approximately 3500 mineral species. This number may seem large, but it is tiny compared to the large number of ways in which naturally occurring elements can be combined to form compounds. The reason

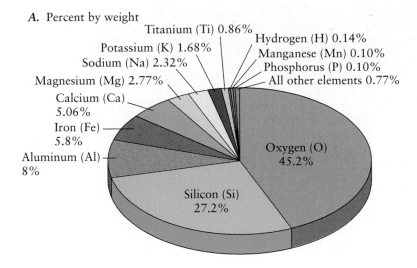

*A.* Percent by weight

Titanium (Ti) 0.86%
Potassium (K) 1.68%
Sodium (Na) 2.32%
Magnesium (Mg) 2.77%
Calcium (Ca) 5.06%
Iron (Fe) 5.8%
Aluminum (Al) 8%

Hydrogen (H) 0.14%
Manganese (Mn) 0.10%
Phosphorus (P) 0.10%
All other elements 0.77%

Oxygen (O) 45.2%
Silicon (Si) 27.2%

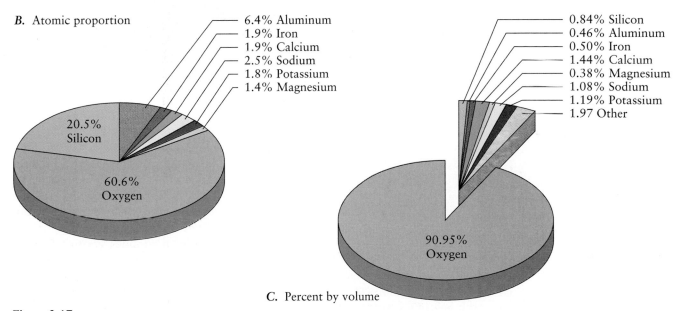

*B.* Atomic proportion

6.4% Aluminum
1.9% Iron
1.9% Calcium
2.5% Sodium
1.8% Potassium
1.4% Magnesium

20.5% Silicon
60.6% Oxygen

0.84% Silicon
0.46% Aluminum
0.50% Iron
1.44% Calcium
0.38% Magnesium
1.08% Sodium
1.19% Potassium
1.97 Other

90.95% Oxygen

*C.* Percent by volume

**Figure 2.17**
**ELEMENTS OF THE CRUST**
These pie charts show the relative proportions of the most abundant elements in the Earth's continental crust. *A.* Abundances by weight. *B.* Abundances by atomic proportions. *C.* Percentages of elements by volume. Oxygen makes up most of the volume because it has a large ionic radius. Oxygen occupies so much space that the crust is essentially a big oxygen mesh.

for the disparity becomes clear when we consider the relative abundance of the chemical elements. Only 12 elements—oxygen, silicon, aluminum, iron, calcium, magnesium, sodium, potassium, titanium, hydrogen, manganese, and phosphorus—occur in the Earth's crust in amounts greater than 0.1 percent of all the atoms present (Fig. 2.17A and B). Together, these 12 elements make up more than 99 percent of the crust's volume (Fig. 12.17C). The crust is constructed mostly of a limited number of relatively abundant mineral species, each of which

Look around you; wherever you are, almost everything you see is made in part or in whole from minerals and other materials dug from the Earth. Even the pages of this book are partly mineral because clay must be added to the wood pulp in order to make paper on which high-quality printing can be done—printing, incidentally, with inks that contain mineral pigments. In order to live the way we do, about 8 tons of new mineral supplies must be dug up, processed, and put to use each year for each and every one of us. Some of us, as in North America and Europe, use larger amounts; others, as in Kenya and India, use less, but the world average is about 8 tons per person per year. Considering that close to 6 billion people now live on the Earth, a truly enormous amount of material must be dug from the Earth each year.

Unfortunately, most minerals that are abundant in the Earth's crust have little or no commercial value or use except that they may be present in rocks crushed for road gravel or the preparation of concrete. Minerals that contain metals such as copper, zinc, tin, tungsten, silver, gold, and many others that are the raw materials of industry (*ore minerals*) tend to be rare and hard to find.

Can the ore minerals in the Earth's crust used by industry sustain both a growing population and a high standard of living for everyone? This difficult question has many experts worried. The minerals they worry most about are those used as sources of such important metals as lead, zinc, and copper. Without metals, we cannot make machines. Without machines, we cannot convert the chemical energy of coal and oil to useful mechanical energy. Without mechanical energy, the tractors that pull plows must grind to a halt; trains and trucks must stop running; and our whole industrial complex must become still and silent.

Experts have no sure way of telling how long the Earth's supplies of the rare but important ore minerals will last. Optimistic experts point to the success our technological society has enjoyed over the past two centuries as ever more remarkable discoveries have been made. If ore mineral supplies become limited, they suggest, we will find ways to get around the limits by recycling, substitution, and the discovery of new technologies. Many geologists have more pessimistic opinion. Technologically advanced societies have faced mineral resource limits in the past, they point out, but the solution has been to import new supplies from elsewhere rather than trying to develop substitutes or effective recycling measures.

England, for instance, was once a great supplier of metals (Fig. B2.1). Today the minerals are mined out, most of its mines are closed, and English industry

has at least one or more of the 12 abundant elements as an essential ingredient. Minerals containing scarcer elements occur only in small amounts and only under special circumstances.

As you can see from Figure 2.17, two elements—oxygen and silicon—make up more than 80 percent of the atoms in the Earth's crust and more than 90 percent by volume. Not only are there a lot of oxygen atoms in the crust, but also they take up a lot of space. Oxygen forms a simple anion, $O^{2-}$; compounds that contain the $O^{2-}$ anion are called *oxides*, or **oxide minerals.** Oxygen and silicon

**oxide mineral** A mineral that contains the $O^{2-}$ anion.

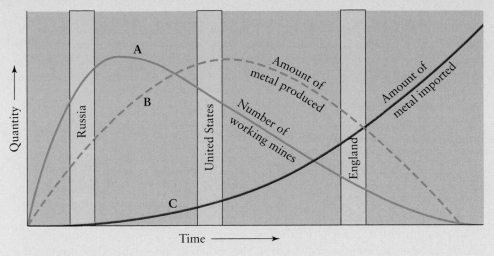

**Figure B2.1**
**A HISTORY OF MINING**
Changing metal production illustrated by the histories of three countries. The number of mines rises rapidly (curve A) but declines when the rate of mine exhaustion exceeds the discovery rate. The amount of metal produced also rises, then falls when mines become exhausted. Curve C represents the growing need to import metal when material production fails to meet needs. As time passes, the position of a country moves from left to right. In the late nineteenth century, Britain was about where the United States is today, at which time the United States was at about the present position of Russia.

runs on raw materials imported from abroad. The United States, too, was once self-sufficient in most minerals and an exporter of many. Slowly the situation has changed, so that now the United States is a net importer and has to rely on supplies from such countries as Australia, Chile, South Africa, and Canada. No country can supply its own mineral needs, and world trade is needed to make supplies available. But eventually the mines of Australia, Chile, and the rest of today's supply countries will be depleted of their minerals; where then will society turn?

The answer to the question just posed is not obvious, but it is one that must be answered in the foreseeable future. It is highly likely that within the lifetimes of the people who read this book, mineral limitations may occur. Which minerals, and therefore which metals, would first be in short supply is still an open question. How society will cope and respond, and when it will have to do so, are just two of the great social and scientific issues still to be solved.

together form a strongly bonded complex ion called a silicate anion $(SiO_4)^{4-}$. Minerals that contain the silicate anion are called *silicates*, or **silicate minerals.** These are the most abundant of all minerals; oxides are the second most abundant. There are other mineral groups based on different complex anions—for example, *carbonates* $(CO_3)^{2-}$, *sulfates* $(SO_4)^{2-}$, and *phosphates* $(PO_4)^{3-}$—but they are less abundant than the silicates and oxides.

A few silicate minerals, a few oxide minerals, calcium sulfate, and calcium carbonate are major constituents of the rocks that make up the bulk of the

**silicate mineral** A mineral that contains the silicate anion $(SiO_4)^{4-}$.

Earth's crust—an estimated 99 percent by volume. These common minerals, of which there are about 30, are the *rock-forming minerals*, so called because they are the main components of most common rocks. Rock-forming minerals are everywhere—not only in rocks, but also in soils and sediment. Rock-forming minerals are used in the construction of houses and roads; even the dust in the air we breathe contains grains of these minerals. The two most common rock-forming minerals are quartz and feldspar. There are several compositional varieties of feldspar—potassium, sodium, and calcium feldspars are the most abundant ones—so the term *feldspar mineral group* is a more accurate statement. Quartz and the feldspar mineral group together constitute about 75 percent of the volume of the Earth's crust. The bonding in most silicate minerals is a mixture of ionic and covalent. As a result, silicates tend to be hard, tough minerals.

Less common minerals are called *accessory minerals*. Many accessory minerals are present in common rocks but in such small amounts that they do not determine the properties of the rocks. Many accessory minerals are important sources of metals and chemicals for industry; the sulfide minerals sphalerite $(ZnS)$ and chalcopyrite $(CuFeS_2)$, for example, are the principal mineral sources of zinc and copper, respectively. The phosphate mineral apatite $[Ca_5(PO_4)_3(F,OH)]$ supplies all the phosphorus needed for agricultural fertilizers.

## Silicates: The Most Important Rock-Formers

The four oxygen atoms in a silicate anion are tightly bonded to the single silicon atom. As shown in Figure 2.18A and B, the four oxygen atoms are so placed that if their centers were connected by lines, those lines would outline a regular *tetrahedron*. The small ionically bonded silicon atom occupies the space between the oxygens at the center of the tetrahedron. The structures and properties of silicate

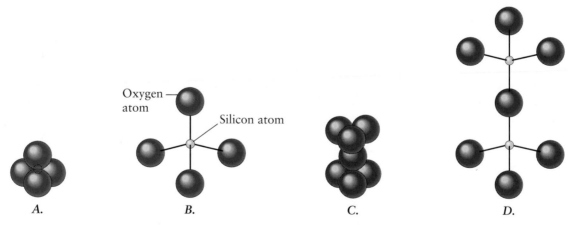

Oxygen atom

Silicon atom

*A.*     *B.*     *C.*     *D.*

**Figure 2.18**
**SILICATE LINKS**
The silicate tetrahedron has four large oxygen atoms at the corners and a small silicon atom at the center. *A.* The large spheres, which represent oxygens, form a tetrahedron, and the small sphere, which represents silicon, sits at the center of the tetrahedron. *B.* An expanded view of the tetrahedron using rods to indicate bonds between the silicon and oxygen atoms. *C.* Two tetrahedrons can link together by sharing an oxygen at one corner. *D.* Expanded view of two tetrahedrons sharing an oxygen.

minerals are determined by the way the $(SiO_4)^{4-}$ silicate tetrahedrons are bonded in the crystal structure.

Two silicate tetrahedrons can bond and form a still larger complex ion by sharing an oxygen atom at one apex, in a process called *polymerization* (Fig. 2.18C and D). By polymerization even larger complex ions, consisting of rings, chains, sheets, or three-dimensional frameworks of tetrahedrons, can form (Fig. 2.19). Different common cations (such as $Ca^{2+}$, $Al^{3+}$, $Mg^{2+}$, $Fe^{2+}$, and $Na^+$) can fit into the spaces or *interstices* between the polymerized silicate tetrahedrons. What determines the identity of a silicate mineral is a combination of (1) how the silicate tetrahedrons occur within the mineral (that is, whether they are single or polymerized); (2) which cations are present in the spaces; and (3) how the cations are distributed throughout the crystal structure. Figure 2.20 shows the common silicate mineral families and how the silicate tetrahedrons are polymerized. The significance of the different structures is the properties they give to each silicate mineral. For example, polymerization in mica produces a sheet and the pronounced cleavage of mica is parallel to the sheets, while in quartz there is a three-dimensional framework that is equally tough in all directions, so quartz is a hard, tough mineral that lacks cleavage.

## Oxides and Other Mineral Groups

The oxides are the second most abundant group of minerals in the Earth's crust. Oxides are compounds based on the $O^{2-}$ anion, which is bonded with the common cations in various arrangements. The oxide minerals of iron, magnetite $(Fe_3O_4)$ and hematite $(Fe_2O_3)$, are the two most common oxide minerals. Mag-

**?** In some parts of the world, animal bones are ground up for use as plant fertilizers. Can you suggest a reason why?

*A.* Single chain

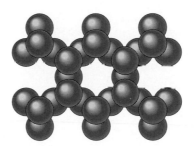

*B.* Double chain

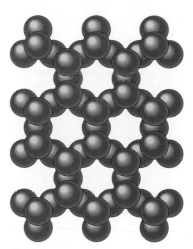

*C.* Sheet

Figure 2.19
**LARGE ANIONS FORMED
BY POLYMERIZATION**
By sharing oxygens, silicate tetrahedrons can form chains, sheets, and three-dimensional nets. Properties of silicate minerals, such as cleavage and habit, are determined to a great extent by the geometry of the linkages. *A.* A single chain in which each tetrahedron shares two oxygens with adjacent tetrahedrons. *B.* Double chain in which some tetrahedrons share two oxygens, some three. *C.* Sheet structure in which all tetrahedrons share three oxygens with other tetrahedrons.

| Silicate Structure | | Mineral/formula | Cleavage | Example of a specimen |
|---|---|---|---|---|
| | Single tetrahedron | Olivine $Mg_2SiO_4$ | None | |
| | Hexagonal ring | Beryl (Gem form is emerald) $Be_3Al_2Si_6O_{18}$ | One plane | |
| | Single chain | Pyroxene group $CaMg(SiO_3)_2$ (variety: diopside) | Two planes at 90° | |
| | Double chain | Amphibole group $Ca_2Mg_5(Si_4O_{11})_2(OH)_2$ (variety: tremolite) | Two planes at 120° | |
| | Sheet | Mica $KAl_2Si_3O_{10}(OH)_2$ (variety: muscovite) $K(Mg,Fe)_3AlSi_3O_{10}(OH)_2$ (variety: biotite) | One plane | |
| Too complex to draw. | Three-dimensional network | Feldspar $KAlSi_3O_8$ (variety: orthoclase) | Two directions at 90° | |
| | | Quartz $SiO_2$ | None | |

◀ Figure 2.20
**POLYMERIZATION AND THE COMMON SILICATE MINERALS**
This chart summarizes the ways in which silicate tetrahedrons can link together to form
the common silicate minerals. Typical examples of each type are listed and shown in the
small photographs. Note that there are many other minerals in each polymerization category—far too many to show on a diagram like this! Silicate linkages other than those
shown are also possible but do not occur in common minerals.

netite takes its name from the ancient Greek word *magnetis*, meaning "stone of
Magnesia," a town in Asia Minor. Magnetis had the power to attract iron particles; hence, it is the source of the word *magnet* as well as *magnetite*. The word
*hematite* is derived from the red color of the mineral in powdered form (see Fig.
2.16); the Greek word for red blood is *haima*. Magnetite and hematite are the
main minerals from which iron is derived. Other important oxides are uraninite
($UO_2$), the main source of uranium; cassiterite ($SnO_2$), from which tin is derived;
and rutile ($TiO_2$), a widely used starting material for making pigments used in
paints.

Another important mineral group is the *sulfides*, all of which contain $S^{2-}$ in
combination with different metal cations. Sulfide minerals typically are dense
and have a metallic luster. The two most common are pyrite ($FeS_2$) and
pyrrhotite ($FeS$). Many of the sulfides are referred to as *ore minerals* (Fig. 2.21),
which means that they are sought and processed for their valuable metal
content.

*A.*            *B.*            *C.*

Figure 2.21
**ORE MINERALS**
Examples of three common ore minerals. A. Galena ($PbS$), the main lead mineral, from
Joplin, Missouri. B. Chalcopyrite ($CuFeS_2$), an important copper mineral, from Ugo,
Japan. C. Sphalerite ($ZnS$), the principal zinc mineral, from Joplin, Missouri.

**Figure 2.22**
**COMMON CARBONATE MINERALS**
The two most common carbonate minerals, calcite ($CaCO_3$) on the left, and dolomite ($CaMgCO_3)_2$ on the right, have similar properties, such as the same three directions of cleavage by which the minerals break in rhomb-shaped fragments.

**rock** A naturally formed, coherent aggregate of minerals or other solid matter.

Like the silicates, oxides, and sulfides, each of the other mineral groups is based on combinations of cations with a particular anionic complex. The *carbonates* (Fig. 2.22) contain the $(CO_3)^{2-}$ anion, and *phosphates* are based on the $(PO_4)^{3-}$ anion. Each of these groups includes minerals that are important both in industry and, in certain cases, in biological processes. For example, the phosphate mineral apatite [$Ca_5(PO_4)_3(F,OH)$], is the principal source of phosphorus for fertilizers and is also an important substance in bones and teeth. Gypsum ($CaSO_4 \cdot 2H_2O$), a sulfate, is the raw material from which plaster is made. And the carbonate mineral calcite ($CaCO_3$), the principal mineral in limestone and marble, is also found in the shells of organisms such as molluscs.

Finally, some materials occur in nature as *native elements;* that is, they are not combined with other elements. Minerals that occur in this form include some metals, such as gold (Au) and silver (Ag), and some nonmetals, such as sulfur (S), graphite (C), and diamond (C). Note that diamond and graphite have the same composition—carbon—but different crystal structures, so they are different minerals. Diamond and graphite are *polymorphs* of carbon; polymorph means many (*poly*) forms (*morphs*).

# THE COMMON ROCK FAMILIES

Minerals are to rocks as letters are to words. **Rock** is a naturally formed aggregate of minerals together with other bits of solid material such as natural glass and organic matter. An important distinction between minerals and rocks is that rocks are *aggregates;* this means that rocks are *collections* of minerals (and sometimes other types of particles) stuck together. Rocks usually consist of several different types of minerals, but sometimes they are made of just one type of mineral. In any case, a rock will contain *many* grains of the constituent mineral or minerals.

For each of the different types of rocks there is a range of possible *mineral assemblages.* The mineral assemblage of a rock is the type and relative proportions of the minerals that constitute the rock. Different mineral assemblages are characteristic of rocks of different composition. Mineral assemblages also can reveal much about the geologic environment in which a particular rock formed, as we will see in subsequent chapters.

Rocks are grouped into three large families: *igneous, sedimentary,* and *metamorphic.* The three families are defined and distinguished from one another by the processes that form them. Let's briefly define each of the major rock families:

**igneous rock** A rock formed by the cooling and consolidation of magma.

**magma** Molten rock under the ground.

**lava** Magma that reaches the surface in a molten state.

*Igneous rocks.* **Igneous rocks** (named from the Latin *ignis,* meaning fire) are the first of the rock families. They are formed by the cooling and solidification of magma. **Magma** is molten rock when it is still under the ground (often containing crystals and dissolved gases in addition to the molten rock). If the magma reaches the surface and flows out in a molten state, it is called **lava.** As you will learn in chapter 6, some igneous rocks cool and crystallize deep under the ground; they are called *plutonic rocks.* Others reach the surface in a molten state and crystallize quickly; these are *volcanic rocks.* An extensive discussion of the kinds of igneous rocks is presented in chapter 6.

**sediment** Regolith particles that have been transported and then deposited.

*Sedimentary rocks.* The second major family is the sedimentary rock family. When particles in the regolith are transported in suspension by water, wind, or ice and then deposited, the deposit is called **sediment.** Sediment eventually becomes **sedimentary rock,** which is any rock formed by chemical precipitation from water at the Earth's surface, or by the cementation of sediment. A special kind of sediment is **soil,** which consists of loose particles that have been altered by biological processes to form a material that can support rooted plants. Soil-forming processes are discussed in detail in chapter 7, and the kinds of sedimentary rocks are examined in chapter 8.

*Metamorphic rocks.* The third rock family is **metamorphic rocks,** those whose original sedimentary or igneous form and mineral content have been changed as a result of high temperature, high pressure, or both. The term *metamorphic* comes from the Greek *meta,* meaning change, and *morphe,* meaning form—hence, change of form. How and why metamorphism occurs, and the kinds of metamorphic rocks that are formed, are discussed in detail in chapter 10.

## What Holds Rocks Together?

The minerals in some rocks are held together with great tenacity, whereas in others they are easily broken apart. The most tenacious rocks are igneous and metamorphic because both types contain intricately interlocked minerals. During the formation of igneous and metamorphic rocks, the growing minerals crowd against each other, filling all spaces and forming an intricate, three-dimensional jigsaw puzzle. A similar interlocking of grains holds together steel, ceramics, and bricks.

The forces that hold the grains of sedimentary rocks together are less obvious. Sediment is a loose aggregate of particles, and it must be transformed into sedimentary rock. The three ways sediment becomes sedimentary rock are:

1. *Compaction.* As mineral grains in a sediment are squeezed and compressed by the weight of overlying sediment, they become compacted into an interlocking network of grains.
2. *Deposition of a cement.* Water circulating slowly through the open spaces in a sediment deposits new materials such as calcite, quartz, and iron oxide, which cement the grains together.
3. *Recrystallization.* As sediment becomes deeply buried and the temperature rises, mineral grains begin to recrystallize, and the growing grains interlock and form strong aggregates. The process is the same as that which occurs when ice crystals in a snow pile recrystallize to form a compact mass of ice.

# WHAT'S AHEAD

You have now been introduced to the most important materials we find on the Earth in the solid state. Rocks are the words that tell the story of the Earth's long history; minerals are the letters that let us read part of that history. We say "part" because a lot of the Earth's history must be read from the sequence, or order, in which rocks were formed. The sequence of rock formation answers such questions as: How old is the Earth? How long did the dinosaurs last? When did the Rocky Mountains form? What were ancient environments like? Before we attempt to answer such questions, we need to find out how geologic time is determined. We will do this in the next chapter.

**sedimentary rock** A rock formed by precipitation from water at the surface, or by the cementation of sediment.

**soil** Sediment that has been altered by biological processes, so that it can support rooted plants.

**metamorphic rock** A rock that has changed as a result of high temperature, high pressure, or both.

## Chapter Highlights

**1.** In an ordinary environment, substances occur in three different states—solid, liquid, and gas. The Earth is unique in the solar system in that $H_2O$ occurs at the surface in all three of these states.

**2.** There are 90 naturally occurring chemical **elements.** The number of protons in the atomic nucleus of an atom is a unique property of an element, called the atomic number. The number of neutrons present can vary within limits. **Isotopes** are species of atoms of the same element with different numbers of neutrons.

**3.** Chemical **compounds** form when atoms share electrons, or when positive and negative **ions** (atoms that have lost or gained electrons through transfer) combine to form electrically neutral **molecules.**

**4.** There are four important kinds of **bonding** between atoms: ionic bonding, formed by electron transfer; covalent bonding, formed by electron sharing; metallic bonding, formed by sharing of electrons in higher energy-level shells ; and Van der Waals bonding, which are very weak and form between certain molecules formed by ionic or covalent bonds.

**5.** **Minerals** are naturally formed, inorganic solids, with specific chemical compositions and characteristic crystal structures. There are approximately 3500 minerals, but only a few (about 30) make up about 99 percent of the volume of the Earth's crust.

**6.** Minerals can be identified on the basis of common properties such as **habit, hardness, cleavage, density, luster,** and **streak,** as well as color.

**7.** **Silicate minerals** are the most abundant rock-forming minerals in the Earth's crust, and **oxide minerals** are the second most abundant. All common silicate minerals contain either the complex silicate anion $(SiO_4)^{4-}$ or larger anions formed by the polymerization of $(SiO_4)^{4-}$ into larger units by sharing oxygens. Other common mineral families are the carbonates, phosphates, sulfates, and sulfides.

**8.** **Rocks** are aggregates of minerals and (sometimes) other types of grains. There are three major rock families: **igneous, sedimentary,** and **metamorphic.** Igneous rocks form by the solidification of **magma;** sedimentary rocks form by the cementation of **sediment;** metamorphic rocks are sedimentary or igneous rocks whose original form and mineral content have been changed as a result of high temperature, high pressure, or both.

## ▶ The Language of Geology

- atom 32
- bonding 34
- cleavage 41
- compound (chemical) 34
- crystal 39
- crystal form 39
- crystalline 38
- density 44
- element (chemical) 32

- habit 40
- hardness 42
- igneous rock 52
- ion 33
- isotope 33
- lava 52
- luster 43
- magma 52
- metamorphic rock 53

- mineral 37
- molecule 34
- oxide mineral 46
- rock 52
- sediment 52
- sedimentary rock 53
- silicate mineral 47
- soil 53
- streak 44

## ▶ Questions for Review

1. Identify the state or states of H₂O stable at points A, B, and C, respectively, in the figure shown here.

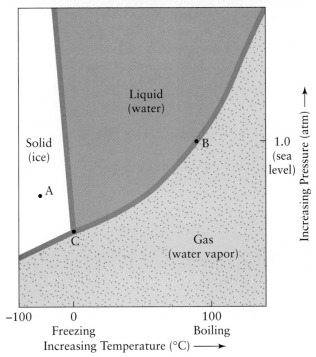

2. When chemists report the analyses of minerals, they list the kinds and amounts of chemical elements present. Why?
3. How does an atom differ from a molecule?
4. Uranium, atomic number 92, has three natural isotopes, with mass numbers 234, 235, and 238. Why are the mass numbers different?
5. How does an ionic bond differ from a covalent bond? Give an example of a compound with ionic bonding and one with covalent bonding.
6. What are the five requirements that must be satisfied if a substance is to be called a mineral?
7. Describe five properties that can be used to identify a mineral.
8. What is the difference between a cleavage surface and a crystal face?
9. Why do minerals differ in hardness?
10. How do rock-forming minerals differ from accessory minerals? Which are the two most common rock-forming minerals?
11. How does a rock differ from a mineral?
12. What are the major rock families and how do they form?

## ▶ Questions for Thought and Discussion

1. When astronauts brought back rock samples from the Moon, the minerals present were the same as those found on the Earth. Can you think of reasons why this might be so? Would you expect minerals on Mars and Venus also to be the same, or at least very similar, to those on the Earth?
2. When a volcano erupts, spewing forth a column of hot volcanic ash, the ash particles are tiny fragments of solidified magma that slowly fall down to the Earth's surface, forming a layer of sediment. Would a rock formed from cemented particles of volcanic ash be igneous or sedimentary? Can you think of other circumstances that might form rocks that are intermediate between two of the major rock families?
3. Which of the following materials are minerals, and why (or why not)? Water; beach sand; diamond; wood; vitamin pill; gold nugget; fishbone; emerald.
4. The minerals calcite and aragonite have the same formula (CaCO₃) but different crystal structures; are they polymorphs? The minerals halite (NaCl) and galena (PbS) have the same geometric patterns in their crystal structures but different compositions; are they polymorphs?

# · 3 ·

# EARTH IN TIME: THE ROCK RECORD AND GEOLOGIC TIME

**What's in a name?** The Jura are hills that separate France from Switzerland. Partly wooded, partly farmland, long inhabited, the Jura derive their name from *juria*, the Latin word for "forest."

The rocks of the Jura are fossiliferous limestones. They are famous for fossils of extinct sea creatures called *ammonites* that lived in coiled shells resembling the modern coiled nautilus. In the early nineteenth century, when European geologists started to arrange fossils in the sequence in which they had lived, fossils in the Jura were selected as the types characterizing certain ammonites, and rocks containing ammonites were selected as the examples of Jurassic sedimentary rocks, named after the Jura hills.

Eventually, geologists discovered that great marine reptiles had swum in the same seas as the ammonites, that dinosaurs had strode the land, and that ancient toothed birds had flown overhead. When Michael Crichton published his science fiction novel about the creation of a dinosaur world using genetic residues from fossil eggs, he called it "Jurassic Park." Steven Spielberg based one of the most successful movies of all time on the book, and the term *Jurassic* became part of the everyday vocabulary of people all over the world.

◆

### In this chapter we discuss

- The chronologic sequence of geologic events
- Geologic time, and how the geologic time scale was established
- Radiometric dating
- Using the magnetic record to date geologic events
- The age of the Earth

*Place Name of a Time Period*
These are the Jura Mountains in Switzerland. The Jurassic Period was named for the Jura because it was in these mountains that rocks of the period were first identified.

**?** How short is human history compared with the Earth's geologic history?

The oldest historical documents dealing with human activities come from places like Mesopotamia, Egypt, China, and Greece. They tell us about events that happened several thousand years ago. Those events seem incredibly distant, but compared to the billions of years of geologic time, they happened only a moment ago. We use the year as the time unit for both historic and geologic events, but many geologic events happened so long ago that the number of years becomes astronomically large. Appreciating the vast extent of geologic time is a real challenge.

The first attempts to determine the extent of geologic time were made about two centuries ago. Geologists speculated that if they measured the amount of sediment transported by streams, it might be possible to estimate the time needed to erode away a mountain range. These attempts were imprecise, but even so they indicated that the Earth is immensely ancient in comparison with recorded human history. One of the founders of geology, a Scot named James Hutton, was so impressed by the evidence that he wrote, in 1785, that for the Earth there is "no vestige of a beginning, no prospect of an end."

# RELATIVE AGE

**relative age** The age of a rock, fossil, or other geologic feature measured relative to another feature.

The geologists who followed Hutton realized that the Earth is very ancient, but they lacked a way to determine exactly how old. The only thing they could do was figure out the order in which past events had occurred; they could establish the **relative ages** of rock formations or other geologic features, which means that they could determine whether a particular formation or feature was older or younger than another. In other words, they were concerned with establishing the *chronological sequence* in which things had happened. Nineteenth-century geologists built a geologic time scale based on relative ages. In doing so, they took the essential first steps toward unraveling the Earth's geologic history. The relative time scale is derived from three basic principles of *stratigraphy,* as follows.

## Stratigraphy

**stratum** A distinct layer of sediment that accumulated at the Earth's surface.

**stratigraphy** The study of strata, that is, the study of rock layers and layering.

We call an individual layer of sediment a **stratum** (plural = **strata**), from the Latin word for layer. The layering that results from a pile of strata deposited one on top of another is called *stratification* (Fig. 3.1A), and the study of strata and stratification is **stratigraphy.** Stratigraphy is based on two principles used to determine relative ages. The first is the *principle of original horizontality,* which states that water-borne sediments[1] are deposited in horizontal strata. You can test this principle yourself, and as often as you do so it will always work. Shake up a bottle of muddy water so that all of the mud particles are suspended. Let the bottle stand and then examine the result—the mud will be deposited in a horizontal layer. The principle of original horizontality is important because it means that whenever we observe water-laid strata that are bent, twisted, or tilted so that they are no longer horizontal, we know that some tectonic force must have disturbed the strata after they were deposited (Fig. 3.1B).

---

[1]The word "sediment" comes from a Latin word meaning "settled."

A.

B.

The second of the principles on which stratigraphy is based is the *principle of stratigraphic superposition*, which states that in any undisturbed sequence of strata, each stratum is younger than the stratum below it and older than the stratum above. In other words, a stratigraphic sequence is like a pile of newspapers laid down day by day—it provides a record from the time the bottom (oldest) stratum was deposited to the time the uppermost (youngest) stratum was deposited. Newspapers, of course, have dates printed on them, so in a pile of papers it is possible to determine the exact age of a given newspaper. Because rock strata lack dates, using stratigraphy we can only assign relative ages to them. In Figure 3.2, for example, the horizontal strata at the top of the Grand Canyon are younger than the strata below them.

The third principle used in determining relative ages, the *principle of cross-cutting relationships*, states that a rock unit must always be older than any feature that cuts or disrupts it. If a rock unit is cut by a fracture, for example, the rock itself is older than the fracture that cuts across it, as shown in Figure 3.3. Sometimes igneous rocks are formed when magmas intrude and cut across previously existing sedimentary rock strata. In this case, too, the sedimentary rocks must be older than the igneous rocks that cut and disrupt the strata.

Figure 3.1
**PRINCIPLE OF ORIGINAL HORIZONTALITY**
These photos show undisturbed and disturbed rock strata. *A.* These horizontal strata exposed by erosion in the badlands of Theodore Roosevelt National Park, North Dakota have not been disturbed by tectonic activity since they were deposited. *B.* These strata on the island of Crete were horizontal when deposited, but over the centuries have been folded and distorted by tectonic activity.

Figure 3.2
**PRINCIPLE OF SUPERPOSITION**
These rock strata are exposed in the Grand Canyon of the Colorado River. The flat-lying strata, a little more than 1000 m thick (about two thirds of a mile), accumulated over 300 million years, were laid down on older strata that were tilted and tectonically deformed prior to deposition of the overlying strata.

Figure 3.3
**PRINCIPLE OF CROSS-CUT-
TING RELATIONSHIPS**
In accordance with the principle
of cross-cutting relationships,
these fractures are younger than
the rock strata they cut. The fractures are cutting through a sequence of sandstone strata in
Merseyside, United Kingdom.
The white sandstone strata are
about 0.3m (1 ft) thick.

**absolute age** The age of a rock in
years.

> **?** What happened during
> the time of a gap in the
> geologic record may be as
> interesting as what happened
> when sediments were
> accumulating. Why is this so?

**unconformity** A break or gap in a
stratigraphic sequence.

**fossil** The remains of plants and
animals that died and were preserved in the Earth's crust.

## Gaps in the Record

When nineteenth-century geologists had determined the relative ages of all the
strata that they could see, they tried to estimate the **absolute age**—that is, the
exact number of years since a given stratum had been deposited. They determined the time in years that it would take for a certain thickness of strata to accumulate. Then they multiplied that amount of time by the total thickness of
strata lying above the stratum in question. For example, if a geologist determined
that it would take 10 years for 1 inch of sediment to accumulate, and there were
100 yds (3600 in) of strata between the modern stratum at the top and the stratum in question, that stratum was estimated to be 36,000 years old (3600 in × 10
years).

At least three assumptions must be true for this method to work. First, the
rate of sedimentation must have been constant during deposition. Second, the
thickness of sediment must have been the same as the thickness of the sedimentary rock that eventually forms from it. Finally, all the strata must be *conformable*. This means that each layer must have been deposited on the one below
it without any interruptions—in other words, there must not be any depositional
gaps in the stratigraphic record. In a conformable pile of newspapers, every day's
paper would be present and in the correct order.

There are problems with each of the assumptions. Rates of sedimentation
vary greatly. Sediments are often compressed while they are turning into sedimentary rock, resulting in a much thinner stratum than was originally present.
The conformity assumption also fails; there are lots of gaps in the stratigraphic
record. Sometimes there are gaps because sedimentation simply stopped for a
time and previously deposited layers were removed due to erosion, as if some
newspapers were laid down and then removed for another purpose. A boundary
that represents a gap in the sedimentary record is called an **unconformity** (Fig.
3.4). When we find an unconformity, we have to conclude that a significant time
interval is not represented in the rock sequence. In sum, estimates of absolute
ages determined by early geologists from the thickness of stratigraphic sequences
were not very accurate.

## Fossils and Correlation

Many strata contain evidence of the remains of plants and animals that were incorporated into the sediment as it accumulated. **Fossils** are usually of things like

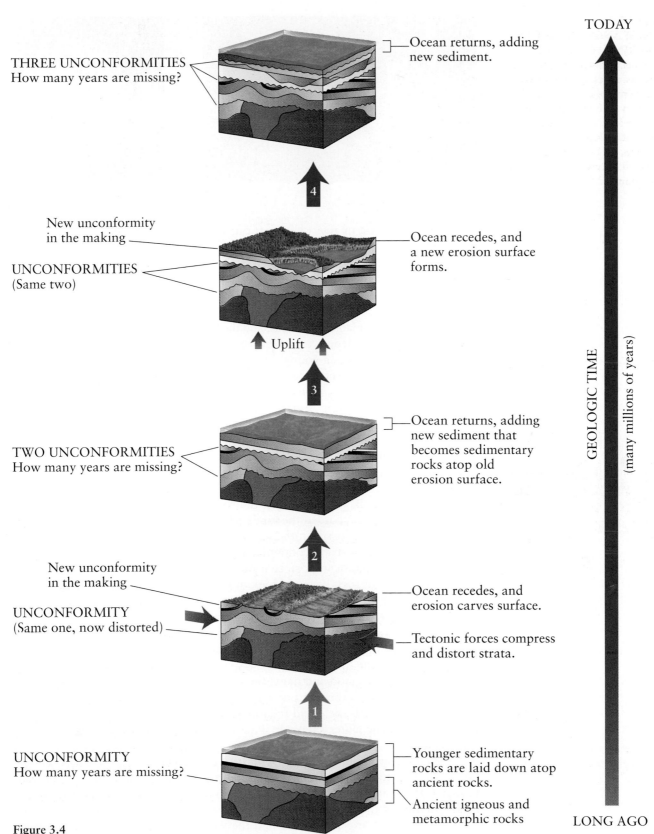

TODAY

THREE UNCONFORMITIES
How many years are missing?

Ocean returns, adding new sediment.

New unconformity in the making

UNCONFORMITIES
(Same two)

Ocean recedes, and a new erosion surface forms.

Uplift

TWO UNCONFORMITIES
How many years are missing?

Ocean returns, adding new sediment that becomes sedimentary rocks atop old erosion surface.

New unconformity in the making

UNCONFORMITY
(Same one, now distorted)

Ocean recedes, and erosion carves surface.

Tectonic forces compress and distort strata.

UNCONFORMITY
How many years are missing?

Younger sedimentary rocks are laid down atop ancient rocks.

Ancient igneous and metamorphic rocks

LONG AGO

GEOLOGIC TIME

(many millions of years)

Figure 3.4
**UNCONFORMITY: GAPS IN THE ROCK RECORD**
Like missing pages from a book, breaks in the continuous rock record frustrate geologists as they try to assemble a complete picture of the Earth's long history. This diagram shows how unconformities (gaps) develop in the rock record. (Read the diagram from the *bottom up*.)

Figure 3.5
**CLAMS: ANCIENT AND MODERN**
Both specimens in the photographs are clams: they lived in similar environments, but developed very different homes in which to live. On the left is a Cretaceous-age clam called Coralliochama, on the right, a much younger Pliocene-age clam of the Pecten family.

*A.*                                                                                                      *B.*

shells, bones, leaves, or twigs whose forms have been preserved in the stone, but in rare cases the imprints of soft animal tissues, like jelly fish, have also been preserved. Even the preserved tracks and footprints of animals—anything that records the former presence of life—are considered to be fossils. Many fossils found in geologically young strata look similar to plants and animals living today (Fig. 3.5A). The farther down in the stratigraphic sequence we go, the more likely we are to find fossils of extinct plants or animals, and the less familiar they seem (Fig. 3.5B).

One of the first people to investigate fossils in a serious way was a Dane, Nicolaus Steno,[2] who, as a young man studying in Italy, became interested in the origin of fossil sharks' teeth. He did not realize they were fossils when he started his study. In 1669 Steno published his ideas in a paper in which he stated the principles of stratigraphic superposition and original horizontality. He also stated that fossils were the remains of ancient life. His conclusions were ridiculed at the time. But a century later the fact that fossils are the remains of ancient plants and animals was widely accepted. The study of fossils, which are the record of ancient life on Earth, is known as **paleontology.**

**paleontology** The study of fossils and the record of ancient life on Earth.

At the same time that James Hutton was working in Scotland, a young surveyor, William Smith, was laying out the routes for canals in southern England. As Smith worked, he recorded the kinds of rock that had to be excavated, and in doing so he noted the fossils in the various strata. Smith made an important discovery: Each group of strata contained a specific assemblage of fossils. This was of some practical significance, since it meant that whenever Smith came across a new outcropping of rock he could look at the fossils and immediately say where the stratum belonged in the stratigraphic succession.

The stratigraphic ordering of fossil assemblages is known as the *faunal succession.* (Faunal means animals; succession means that new species succeed earlier ones as they evolve.) Smith used fossil assemblages to correlate strata in different places. In time he could look at a specimen of rock from any sedimentary layer in southern England and name the stratum from which it had come and, of course, the position of the stratum in the sequence (Fig. 3.6). For Smith this skill was entirely practical—he could reliably predict what kind of rock the canal excavators would encounter simply by looking at the fossils in the rock exposed at the surface.[3]

William Smith's practical discovery turned out to be of great scientific importance. Geologists soon demonstrated that the faunal succession in northern France is the same as the faunal succession found by Smith in southern England. By the middle of the nineteenth century, it had become clear that the faunal succession is the same everywhere around the world. What Smith's practical obser-

---

[2] Steno is the same person we learned about in chapter 2, who first determined that interfacial angles in crystals of all specimens of the same mineral are constant.

[3] As a result, Smith eventually gained the nickname "Strata."

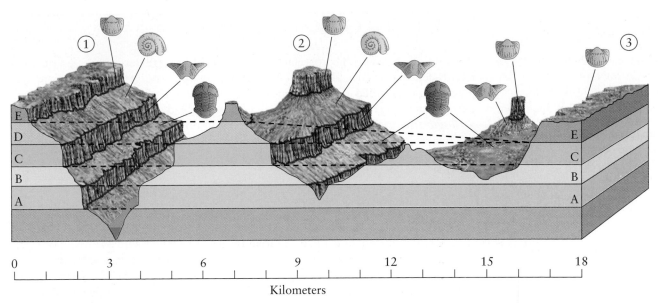

0       3       6       9       12      15      18
Kilometers

**Figure 3.6**
**FOSSILS AND CORRELATION**
Correlation of strata exposed at three localities (1, 2, and 3), many kilometers apart, on the basis of similarity of the fossils they contain. The fossils show that at Locality 3 stratum D is missing, because E directly overlies C. Why? Either D was never deposited there, or it was deposited and later removed before deposition of E. Either way, the boundary between C and E at Locality 3 is an unconformity.

vations led to was a means of worldwide **correlation**—that is, a method of equating the ages of strata from two or more different places.

During the second half of the twentieth century, another, quite unanticipated, means of stratigraphic correlation was discovered. The discovery involved unconformities. Remember that unconformities record gaps in a stratigraphic succession. In attempting to match strata deposited along the submerged margins of continents, geologists discovered that there had been times when the sea level had risen worldwide, so that on every continent sediment was deposited on parts of the continental margins that were formerly land. There were also times when the sea level dropped and exposed the continental margins so that sedimentation ceased and erosion occurred. In other words, a gap formed in the record, and an unconformity was created. The worldwide rises and falls of sea level, recorded in unconformities in the sediments deposited along the edges of continents, have given rise to *sequence stratigraphy*. Each *sequence* is a set of conformable strata (remember that conformable means without any gaps) bounded above and below by an unconformity (Fig. 3.7). Sequence stratigraphy is important in oil exploration because unconformities often serve as traps for oil.

**correlation** A method of equating relative ages in successions of strata from two or more different places.

**Figure 3.7**
This is a diagram of a group of sequences bounded above and below by unconformities. Within each sequence the strata are conformable.

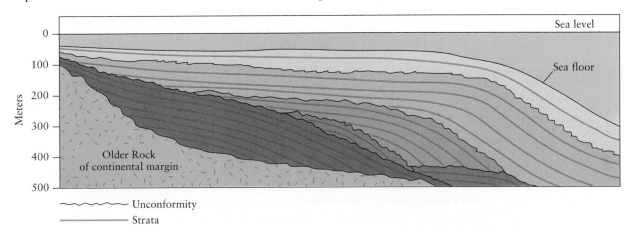

# THE GEOLOGIC COLUMN

**Geologic Column** The succession of all known strata, fitted together in relative chronological order.

Worldwide stratigraphic correlation was one of the greatest successes of nineteenth-century science. It meant that a gap in the stratigraphic record in one place could be filled using evidence from somewhere else. Through worldwide correlation, nineteenth-century geologists assembled the **Geologic Column,** or *geologic time scale* which is a composite diagram showing the succession of all known strata, fitted together in chronological order, on the basis of their fossils and other evidence of relative age (Fig. 3.8).

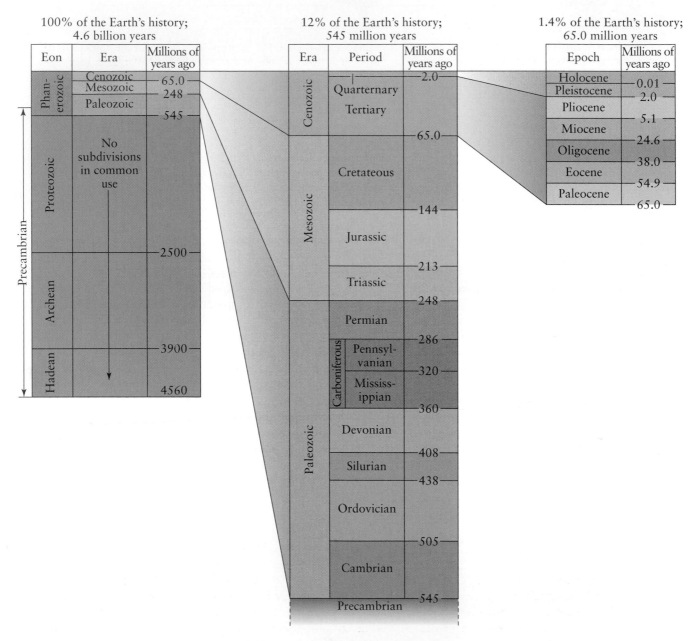

**Figure 3.8**
**THE GEOLOGIC COLUMN AND TIME SCALE**
The stratigraphic time scale, which we call the Geologic Column, puts all strata in chronological order. The absolute ages in millions of years were determined in the twentieth century, using radioactivity.

A.                                   B.                                   C.

**Figure 3.9**
**CHANGES THROUGH THE AGES**
Fossil evidence records dramatic changes in animal life on land, even though the environments in which the animals lived changed little. *A.* Ichthyostega, a primitive amphibian that lived both in and out of the water about 350 million years ago. Ichthyostega had features, such as a tail, that it inherited from fish, and features such as legs that made it possible to move around on land. *B.* A scene in the Cretaceous, about 100 million years ago. Two carnivorous dinosaurs have just killed a smaller animal and are beginning to feast on the body. *C.* The scene was very different in the Miocene, about 15 million years ago. Dinosaurs had died out and their places were taken by mammals. Horses and antelopes of various kinds, the direct ancestors of today's animals, were common.

## Eons and Eras

The geologic time scale has four major divisions, called **eons,** which have Greek names. They are the **Hadean** ("beneath the Earth"), **Archean** ("ancient"), **Proterozoic** ("early life"), and **Phanerozoic** ("visible life"). Some of the eons, which encompass hundreds of millions or even billions of years, are divided into shorter spans called **eras.** Eras are defined on the basis of fossil assemblages and therefore are most useful in studying the Phanerozoic Eon; fossils are absent or very rare in the rocks of the three earlier eons. The formal names for the three eras of the Phanerozoic Eon (also from Greek) are the **Paleozoic** ("ancient life"), **Mesozoic** ("middle life"), and **Cenozoic** ("recent life").

During the Paleozoic Era, the evolution of life progressed from marine *invertebrates* (animals without backbones) to fishes, amphibians, and reptiles (Fig. 3.9A). Early land plants appeared during the Paleozoic Era, and the oldest coal strata are Paleozoic. The Mesozoic Era saw the rise of dinosaurs, which were the dominant *vertebrates* (animals with backbones) on land for many millions of years (Fig. 3.9B). Mammals first appeared during the Mesozoic Era, but they did not become dominant until the dinosaurs disappeared at the end of the Mesozoic Era. Mammals have been dominant throughout the Cenozoic Era (Fig. 3.9C). The Mesozoic Era also witnessed the appearance of the first flowering plants. Grasses appeared during the Cenozoic Era and became important food for grazing mammals. All of these changes will be discussed in greater detail in chapter 15.

**eon** The four major time divisions of the geologic time scale.

**era** A unit of geologic time that is shorter than an eon and longer than a period.

## Periods and Epochs

The three eras of the Phanerozoic Eon are divided into shorter units called **periods.** These were defined through nearly a century of detailed work on fossil assemblages in Europe and North America. The naming of the periods was somewhat haphazard; some periods were named for the geographic locations where their strata were first studied, and some were named for the characteristics of their strata (Table 3.1). The Mesozoic Era, for example, is divided into three periods; the oldest is the Triassic Period, which is named for a threefold division of salt-bearing strata in Germany. The Jurassic Period is a time when dinosaurs grew to giant size; it is named for the Jura Mountains of France and Switzerland.

**period** A unit of geologic time that is shorter than an era and longer than an epoch.

**Precambrian** Geologic time prior to the beginning of the Phanerozoic Eon.

**epoch** A unit of geologic time that is shorter than a period.

The youngest unit is the Cretaceous Period, named for the Latin word for chalk (*creta*) after the chalk cliffs of southern England and France.

The earliest period of the Paleozoic Era, the Cambrian Period, is the time when animals with hard shells first appeared in the geologic record. Before that, all animals had soft bodies, and fossil evidence is rare (Fig. 3.10). Rocks formed during the Archean and Proterozoic Eons sometimes contain microscopic, soft-bodied fossils, but strata cannot be differentiated on the basis of the fossils they contain. For this reason, geologists often refer to the entire time span and all rocks deposited before the Cambrian Period as simply **Precambrian.**

Periods, we now know, lasted for tens of millions of years, so geologists split them into still smaller divisions called **epochs.** Epochs are most useful to specialists. The names of the epochs of the Tertiary and Quaternary Periods are somewhat more familiar than others because humans and their ancestors emerged during these epochs, and the names sometimes appear in the popular press. Names for the Tertiary and Quaternary Epochs were derived from studies of marine strata deposited in France and Italy; geologists who studied the strata subdivided them on the basis of the percentage of their fossils that are represented by still-living species (Table 3.1). Many plant and animal fossils found in Pliocene strata, for example, have still-living counterparts, but fossils in Eocene strata have few still-living counterparts.

**TABLE 3.1    Origin of Names for Periods of the Paleozoic, Mesozoic, and Cenozoic Eras, and the Epochs of the Quaternary and Tertiary Periods**

| Era | Period | Epoch | Origin of Name |
|---|---|---|---|
| Cenozoic | Quaternary | Holocene | Greek for wholly recent |
| | | Pleistocene | Greek for most recent |
| | Tertiary | Pliocene | Greek for more recent |
| | | Miocene | Greek for less recent |
| | | Oligocene | Greek for slightly recent |
| | | Eocene | Greek for dawn of the recent |
| | | Paleocene | Greek for early dawn of the recent |
| Mesozoic | Cretaceous | | Latin for chalk, after chalk cliffs of southern England and France |
| | Jurassic | Epoch Names Used Only By Specialists | Jura Mountains, Switzerland, and France |
| | Triassic | | Threefold division of rocks in Germany |
| Paleozoic | Permian | | Province of Perm, Russia |
| | Pennsylvanian | | State of Pennsylvania |
| | Mississippian | | Mississippi River |
| | Devonian | | Devonshire, County of Southwest England |
| | Silurian | | Silures, ancient Celtic tribe of Wales |
| | Ordovician | | Ordovices, ancient Celtic tribe of Wales |
| | Cambrian | | Cambria, Roman name for Wales |

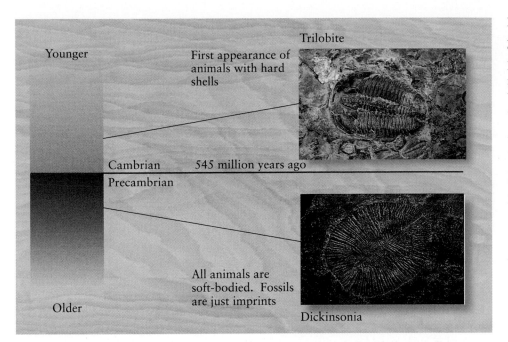

**Figure 3.10**
**ANCIENT ANIMALS**
The beginning of the Cambrian Period, 545 million years ago, marks the appearance of animals with hard shell such as the *trilobite Ptychoparia* that lived about 500 million years ago in what is today Pennsylvania. Prior to the Cambrian Period, animals were soft-bodied, and the rare fossils that have been found are just impressions left in sedimentary rocks. One such is the curious creature known as *Dickinsonia* from Ediacara, South Australia. *Dickinsonia* lived about 600 million years ago.

# ABSOLUTE AGE

The scientists who worked out the Geologic Column were tantalized by the challenge of absolute time. They wanted to know the age of the Earth, how fast mountain ranges rise, how long the Paleozoic Era lasted, and most challenging of all, how long humans have inhabited the Earth. To get the answers they sought, they needed a way to determine absolute age.

> **?** **What's the difference between absolute age and relative age?**

## Early Attempts

During the nineteenth century, many indirect attempts were made to solve the problem of absolute time. One method—estimating rates of sedimentation and multiplying by the thickness of stratigraphic sections—was mentioned earlier. Estimates for the age of the Earth, which was presumed to be equivalent to the amount of time needed for all sedimentary strata to form, ranged from 3 million to 1.5 billion years. We now know that these estimates fall far short of the Earth's actual age.

Because most sedimentary strata are deposited in the ocean, an indirect method of estimating the minimum age of the Earth would be to determine the age of the ocean. The first person to suggest this, in 1715, was Edmund Halley (for whom Halley's comet is named). Halley realized that when common rocks are eroded, the salts in them are transported to the sea by rivers. Why not measure the rate at which salts are added to river water and calculate the time needed to transport all the salts now present in the sea? Halley did not carry out his own suggestion, and it was not until 1889 that John Joly made the necessary measurements and calculations. Joly's estimate of the ocean's age was 90 million years. Unfortunately, this estimate was wrong too. Although salts are added to seawater by erosion and submarine volcanism, they are also removed by reactions between seawater and hot volcanic rocks on the seafloor and by the evaporation of seawater in isolated basins. What neither Halley nor Joly realized is that the ocean, like other parts of the Earth system, is an open system and that the composition of seawater is the result of a balance between the addition and removal of salts. This means that the salinity of seawater is approximately constant and has been so for hundreds of millions (even billions) of years.

The desire to determine absolute age became intense after the publication in 1859 of Charles Darwin's *On the Origin of Species*. Darwin understood that the evolution of new species by natural selection must be a very slow process that needed vast amounts of time. Opposing Darwin was Lord Kelvin, a leading physicist of his time. Kelvin attempted to calculate the amount of time the Earth has been a solid body. The Earth had come into existence as a very hot object, he argued. Once it had cooled sufficiently to form a solid outer crust, it was isolated from any heat source and would continue to cool as heat was conducted outward through the solid rock. By measuring the thermal properties of rock and estimating the present temperature of the Earth's interior, he calculated the amount of time required for the Earth to cool, by conduction, to its present state (Fig. 3.11). Kelvin's hypothesis that the Earth was initially molten and that it is cooling by conduction was in agreement with other hypotheses of the time, and his mathematical calculations were correct. The answer he obtained for the maximum value of the age of the Earth was about 20 million years. Darwin realized that this was too little time for natural selection to have occurred in the way he believed it had, so in later editions of his famous book he paid little attention to the absolute time problem.

Darwin died before geologists discovered that one aspect of Lord Kelvin's hypothesis was wrong. Kelvin assumed that no heat had been added to the Earth's interior since the Earth was formed. When he made his calculations, radioactivity was not known. But radioactivity continuously supplies heat to the Earth's interior. Instead of cooling rapidly, the Earth's interior is cooling so slowly that its temperature is nearly constant for periods as long as hundreds of millions or even billions of years. Kelvin, like Joly, had missed an important aspect of the Earth system, the balance between input and output of energy in a closed system—in this case, the input and output of heat from the Earth's interior.

What was needed to solve the problem of absolute geologic time was a way to measure events by some process that runs continuously, is not reversible, is not

**Figure 3.11**
**KELVIN'S ESTIMATE OF THE SOLID EARTH'S AGE**
Lord Kelvin made two assumptions: that the Earth was once a completely molten small star and that no heat was added once it formed. Once a solid rocky outer layer had formed, the Earth cooled by conduction. By measuring today's rate of heat loss, and knowing the heat conduction properties of rocks, Kelvin calculated how long ago the Earth first cooled sufficiently for solid rocks to form. His clever calculations proved to be incorrect because his assumptions were incorrect.

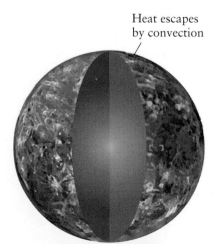

Heat escapes by convection

Earth is completely molten.

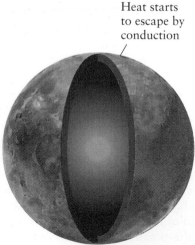

Heat starts to escape by conduction

Solid rock layer forms.
Heat escapes by conduction.
**KELVIN'S TIME ZERO**

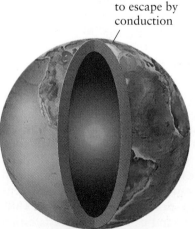

Heat continues to escape by conduction

Solid rock layer now thicker.
**TODAY**

influenced by such factors as chemical reactions and high temperatures, and leaves a continuous record without any gaps. In 1896 the discovery of radioactivity not only proved that Kelvin's assumptions were wrong, but by fortunate chance, also provided the breakthrough needed to measure absolute time.

## Radioactivity and Absolute Age

We learned in chapter 2 that the atomic number of a given chemical element— that is, the number of protons in the nucleus of each atom—is constant. We also learned that most elements have two or more *isotopes*, each of which has the same number of protons per atom but a different number of neutrons per atom. To put it another way, each isotope of an element has the same atomic number but a different mass number.

Most isotopes of the chemical elements that occur in the Earth have stable nuclei that do not change spontaneously. However, the nuclei of a few naturally occurring isotopes, such as carbon-14 and potassium-40, are unstable and subject to spontaneous change. The change happens by emission from the nucleus of protons and neutrons that are almost always grouped together as a pair of protons and a pair of neutrons, called an α-*particle* (alpha particle), hence α-*emission*; by emission from the nucleus of an electron, called a β-*particle* (beta particle), hence β-*emission*; or by capture into the nucleus of an orbiting electron, called β-*capture*. A change in the nucleus means, of course, a change in the atomic number, the mass number, or both. Any isotope that spontaneously undergoes such nuclear change is said to be *radioactive*. The process of change is referred to as *radioactive decay*, and the phenomenon is called **radioactivity**. An isotope that is undergoing radioactive decay is said to be a *parent*, and an isotope that results from radioactive decay is called a *daughter*. For example, carbon-14 decays to nitrogen-14 and uranium-238 to lead-206. Thus, carbon-14 and uranium-238 are parents, and nitrogen-14 and lead-206 are daughters.

Radioactive decay not only involves the release of particles from the nucleus, but also a penetrating form of radiation called gamma rays, and the production of heat. It was the heat produced by the decay of radioactive isotopes that made Lord Kelvin's estimates of the age of the Earth wrong.[4] The science of radioactivity is very complicated. What is important for our discussion is the fact that each radioactive isotope decays according to a distinct timetable. For this reason, the radioactivity of natural isotopes can serve as a clock, one that is built into rocks. Furthermore, because each radioactive isotope has its own rate of decay, a rock that contains several different radioactive isotopes has several built-in clocks that can be checked against each other.

### Rates of Decay

Many of the radioactive isotopes that were present when the Earth was formed have decayed away because they are species that decay relatively quickly. A few radioactive isotopes that decay very slowly are still present, however. Careful study in the laboratory has shown that, for nearly all radioactive isotopes, decay

?

**How does a radioactive "rock clock" work?**

**radioactivity** The process by which an element transforms spontaneously into another isotope of the same element or another element.

---

[4]Lord Rutherford (one of the scientists who discovered a lot about radioactivity) once gave a talk in which he spoke of Kelvin's errors in calculating the age of the Earth. Lord Kelvin attended the talk but reportedly slept through most of it. Rutherford was much younger and, therefore, less eminent than Kelvin. When Rutherford came to the part about heat sources within the Earth, one of Kelvin's eyes opened. Diplomatically, Rutherford stated that Kelvin's estimates of the age of the Earth had been based on calculations of cooling time, "provided no additional heat sources were ever found." The new heat source, of course, was radioactivity. This meant that Kelvin's estimates had been correct given the information available to him at the time. Kelvin happily went back to sleep.

Figure 3.12
**RADIOACTIVITY AND TIME**
These graphs illustrate the basic decay law of radioactivity. At time zero, a sample consists of 100 percent radioactive parent atoms. After one time unit, corresponding to the half-life of the material, 50 percent of the parent atoms will have decayed to daughter atoms. After two time units, 75 percent will have decayed. At any given time, the total number of daughter atoms plus remaining parent atoms equals the original number of parent atoms.

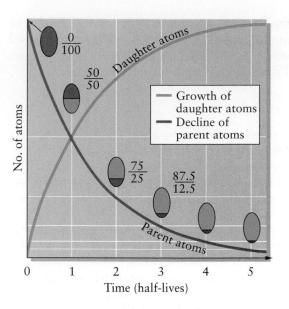

**half-life** The time needed for a radioactive material to lose half its radioactivity.

rates are unaffected by changes in the chemical and physical environment. Thus, the decay rate of a given isotope is the same in the mantle or in a magma as it is in a sedimentary rock. This is an important point because it means that rates of radioactive decay are not altered by geologic processes like erosion, metamorphism, or the melting of a rock to form magma.

In the least complicated process of radioactive decay, which happens to be the process of most utility in determining absolute ages, the number of radioactive parent atoms continuously decreases, while the number of nonradioactive daughter atoms continuously increases. More complicated processes of radioactive decay, which are of less utility for determining ages, involve daughter atoms that are also radioactive and so undergo further decay until a stable daughter is finally produced. Regardless of the complications involved in the decay scheme, all decay timetables follow the same basic law: The *proportion* of parent atoms that decay during each unit of time is always the same (see Fig. 3.12).[5] Proportion means a fraction, or percentage, not a whole number. The rate of radioactive decay is determined by the **half-life,** which is the amount of time needed for the number of parent atoms to be reduced by one-half. For example, let's say that the half-life of a radioactive isotope is 1 hour. If we started an experiment with 1,000,000 parent atoms and the daughter atoms are not radioactive, only 500,000 parent atoms would remain at the end of an hour, and 500,000 daughter atoms would have formed. At the end of the second hour, another half of the parent atoms would be gone, so there would be 250,000 parent atoms and 750,000 daughter atoms. After the third hour, another half of the parent atoms would have decayed, leaving 125,000 parent atoms and 875,000 daughter atoms. The proportion of parent atoms that decay during each time interval is 50 percent. (The same kind of law governs compound interest paid on a bank account.)

In Figure 3.12 the time units marked are half-lives. The units are of equal length, but at the end of each unit the number of parent atoms, and therefore the radioactivity of the sample, has decreased by exactly one-half. Figure 3.12 also shows that the growth of daughter atoms just matches the decline of parent

---

[5] The application of this law is valid only when the number of atoms is large enough (at least tens of thousands) so that statistical variations in the counting procedure are negligible.

atoms. When the number of remaining parent atoms is added to the number of daughter atoms, the result is the number of parent atoms in the original mineral sample. This fact is the key to the use of radioactivity to measure geologic time and determine absolute ages.

## Radiometric Dating: The Rock Clock

Radioactivity was discovered in 1896; the first estimates of the ages of rocks using radioactive decay were made in 1905. The long-hoped-for rock clock was finally available, and the results were, and continue to be, remarkable. Our ability to measure absolute age has revolutionized the way we think about the Earth.

A radioactive rock clock works as follows. When a new mineral grain has formed—for example, a grain of feldspar in a cooling lava—all the atoms in the grain have become locked into the crystal structure and removed from reaction with the environment outside the grain. In a sense, the atoms in the mineral grain are sealed in an atomic bottle. If some of the atoms are radioactive, and provided no daughter atoms were present in the mineral at the time of formation, we can determine how long ago the bottle was sealed—that is, how long ago the mineral grain formed—by measuring the number of remaining parent atoms and the number of daughter atoms. If the mineral grain was contaminated with some daughter atoms at the time of formation, an estimate of the amount of contamination must be made in order to know the fraction of daughter atoms that have come from the radioactive decay of parent atoms. Geologists have developed several ways to estimate the initial contamination of a sample by daughter atoms. Once that is done, and provided we know the half-life of the radioactive parent, it is a simple matter to calculate how long ago the mineral grain was formed. **Radiometric dating** is the use of naturally occurring radioactive isotopes to determine the time when minerals and rocks were formed.

Radiometric dating has been particularly useful for determining the ages of igneous rocks, since the mineral grains in an igneous rock form at the same time that the rock that contains them is formed. On the other hand, most sedimentary rocks consist largely of mineral grains that were formed long before the strata that contain them were deposited; dating such mineral grains will tell the age of the grain but not when the strata were deposited. The most frequently employed radioactive isotopes and their daughter isotopes are listed in Table 3.2.

**radiometric dating** The use of naturally occurring radioactive isotopes to determine the absolute age of minerals or rocks.

**TABLE 3.2    Some of the Principal Isotopes Used in Radioactive Dating**

| Parent | Decay System | Daughter | Half-life (years) | Effective Dating Range (years) | Materials That Can Be Dated |
|---|---|---|---|---|---|
| Uranium-238 | $\alpha + \beta$ emission | Lead-206 | 4.5 billion | 10 million to 4.6 billion | Uranium- |
| Uranium-235 | $\alpha + \beta$ emission | Lead-207 | 710 million | 10 million to 4.6 billion | bearing |
| Thorium-232 | $\alpha + \beta$ emission | Lead-208 | 14 billion | 10 million to 4.6 billion | minerals |
| Potassium-40 | $\beta$-capture $\beta$-emission | Argon-40 Calcium-40 | 1.3 billion 1.3 billion | 50,000 to 4.6 billion | Mica Amphibole Volcanic rock |
| Rubidium-87 | $\beta$-emission | Strontium-87 | 4.7 billion | 10 million to 4.6 billion | Mica Feldspar Igneous and metamorphic rock |
| Carbon-14 | $\beta$-emission | Nitrogen-14 | 5730 | 100 to 70,000 | Wood, charcoal, and other carbon-containing matter |

**?**

How can carbon–14 tell us the age of things that were alive in the past?

### Carbon-14 Dating

One of the frequently employed radioactive isotopes for organic materials, carbon-14, poses a special challenge because the daughter isotope, nitrogen-14, is the most abundant nitrogen isotope found in the Earth's atmosphere. This means that it is almost impossible to avoid atmospheric contamination of samples. Furthermore, nitrogen-14 formed by decay readily leaks out of most organic compounds, and so it is not possible to accurately measure the daughter atoms. Nonetheless, special circumstances enable us to use carbon–14 as a rock clock. Here is how it is done.

Carbon-14 is continuously created in the atmosphere as a result of cosmic radiation from outer space. The carbon-14, which has a half-life of 5730 years, mixes rapidly and uniformly with other carbon isotopes in the atmosphere. Therefore, carbon–14 is a constant fraction of the carbon in the atmosphere.

Plants use carbon from the atmosphere (as carbon dioxide) to build their substances by photosynthesis, and animals feed on plants, or on other animals in a food web that depends ultimately on plant tissue as the source of food. While an organism is alive, it will continuously take in carbon that has come, directly or indirectly, from the atmosphere. Thus, the carbon in most organisms will contain the same proportion of carbon-14 as is present in the carbon of the atmosphere. However, at death the intake of carbon-14 ends because photosynthesis, or feeding, ceases. The carbon-14 in dead tissues decreases continuously through radioactive decay. Therefore, to find the radiocarbon date of a sample all we need to do is determine the fraction of carbon–14 left in the sample. This can be done by measuring the level of radioactivity or, in a recent scientific development, by using a special device to measure the proportions of the three carbon isotopes (12, 13, and 14) in a sample.

Because of its relatively short half-life and its use in dating the remains of organisms (wood, charcoal, peat, bone, and shell material), carbon–14 can be used to determine the age of prehistoric human remains and recently extinct animals. It is therefore extremely important in archaeology. It is also of great value in dating recent geologic events, particularly the latest glacial age. For example, radiocarbon dates have been obtained for many samples of wood taken from trees killed by the last great ice sheet. They show that the ice reached its greatest extent in the Ohio-Indiana-Illinois region about 18,000 to 21,000 years ago. It is even possible to date young ice, such as that in the Greenland Ice Sheet. As the ice forms, bubbles of air are trapped. The carbon dioxide in the air bubbles can be released in a laboratory and dated, giving a date for the time of formation of the ice.

## Absolute Time and the Geologic Column

As geologists worked out the Geologic Column, they found many locations where layers of solidified lava and volcanic ash are interspersed with sedimentary strata. Through radiometric dating, it is possible to determine the absolute ages of the lavas and volcanic ash and thereby to bracket the ages of the sedimentary strata. An example of how this is accomplished is shown in Figure 3.13.

Through a combination of geologic relations and radiometric dating, twentieth-century scientists have been able to fit a scale of absolute time to the Geologic Column. The scale is continually being refined, so the numbers given in Figure 3.8 should be considered the best available now. Further work will make them more precise. It is a tribute to the work of geologists during the nineteenth century that the Geologic Column they established by the ordering of strata in respect to relative age has been fully confirmed by radiometric dating. It is interesting, too, to see just how wrong Lord Kelvin was in his estimate of the age of the Earth.

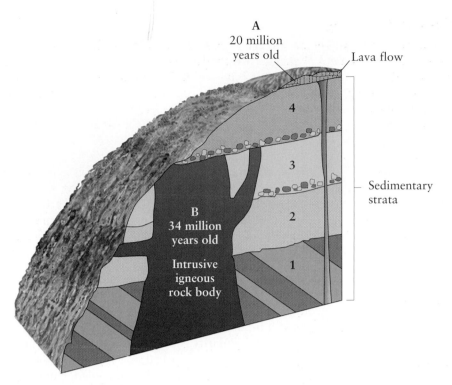

**A**
20 million
years old

Lava flow

4

3

Sedimentary
strata

**B**
34 million
years old

2

Intrusive
igneous
rock body

1

Figure 3.13
**RADIOMETRIC DATING AND THE GEOLOGIC COLUMN**
This idealized stratigraphic section shows how radiometric dating can be applied to the
Geologic Column. Ages determined for igneous rocks can be used to bracket the ages of
the sedimentary strata. Because it cuts across strata 1, 2, 3, and 4, the lava flow labeled A
must be younger than those strata. Because it cuts through 1, 2, and 3, igneous intrusion
B must be younger than 1, 2, and 3. A is 20 million years old, and B is 34 million years
old. Thus, the age of stratum 4 is bracketed by units A (dated by radiometric methods at
20 million years old) and B (dated at 34 million years old). Therefore, stratum 4 must be
intermediate in age between these two.

# THE MAGNETIC POLARITY DATING SCHEME

Time is so central to the study of the Earth that geologists are always seeking
new ways to estimate ages. One method involves magnetism.

The Earth has an invisible magnetic field that permeates everything. If a
small magnet is allowed to swing freely in the Earth's magnetic field, the magnet
will become oriented in the direction of the Earth's magnetic field; it will point
toward the Earth's magnetic north pole, and it will be inclined to the Earth's surface (Fig. 3.14). This is true for all places on the Earth: All free-swinging magnets
will point to the north magnetic pole. The reason the orientation of a magnet is
important is that certain rocks that contain magnetite and other iron-bearing
minerals are also permanent magnets as a result of the way they form and because the mineral grains act like free-swinging magnets.

In order to acquire permanent magnetism, a mineral must cool below a certain temperature—different for each mineral—called the *Curie point*. At temperatures above the Curie point, permanent magnetism is impossible. The Curie
point for magnetite is about 580°C (1076°F). Above that temperature, magnetite
is not a permanent magnet; below it, a magnetite grain is a permanent magnet.

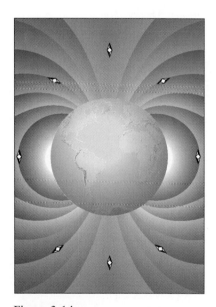

Figure 3.14
**THE EARTH'S
MAGNETIC FIELD**
A freely swinging magnetic needle
will align itself parallel to the
Earth's magnetic field. The needle
will point toward the magnetic
north pole but will change its angle
of inclination from horizontal at
the equator to vertical at the pole.

Figure 3.15
**LAVAS RECORD THE EARTH'S MAGNETIC FIELD**
Lavas retain a record of the polarity of the Earth's magnetic field at the instant they cool below the Curie point for magnetite. At Tjornes, in Iceland, a series of lava flows separated by glacial sediment demonstrate the reversibility of the magnetic field.

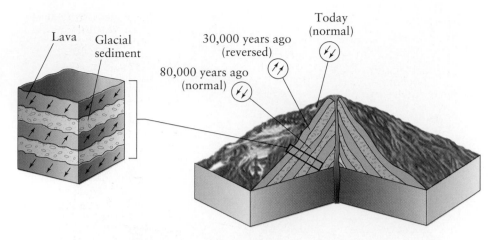

**polarity** The north-south directionality of the Earth's magnetic field.

**paleomagnetism** The magnetism taken on by a rock long ago.

**magnetic reversal** An event in which the Earth's magnetic polarity reverses itself.

Consider what happens when lava cools. All the mineral grains crystallize at temperatures above 700°C (1290°F)—well above the Curie point of magnetite. As the crystallized lava continues to cool, the temperature will drop below 580°C and all the magnetite grains will become tiny permanent magnets with the same polarity as the Earth's magnetic field. **Polarity** is the north-south directionality of the Earth's magnetic field. Grains of magnetite locked in a solidified lava, or any rock that has been heated above the Curie point and then cooled, cannot move and reorient themselves the way a freely swinging bar magnet can. As long as that lava or previously heated rock lasts, therefore, it will carry a record of the polarity of the Earth's magnetic field at the moment that it cooled through the Curie point of magnetite (Fig. 3.15).

Sedimentary rocks also acquire weak but permanent magnetism through the orientation of magnetic grains during sedimentation. As grains settle through ocean or lake water, or even as dust particles settle through the air, any magnetite particles present will act like freely swinging magnets and orient themselves parallel to the lines of force caused by the Earth's magnetic field. Once they are locked into a sediment, the grains turn the rock into a weak permanent magnet.

Both the polarity and the strength of a rock's magnetism can be measured with a *magnetometer*. When scientists measure the magnetism taken on by a rock long ago, they are measuring the rock's **paleomagnetism.** (*Paleo* is the same prefix we see in the word *paleontology*; it means "ancient.") Sometimes a rock's polarity is the same as the Earth's present polarity. This is called *normal* polarity. But sometimes it is *reversed* from the Earth's present polarity. How can this be?

It turns out that about every half-million years the Earth's magnetic field reverses its polarity, which means that the magnetic pole that had been near the Earth's north pole moves to a position near the Earth's south pole, and the magnetic pole that had been near the Earth's south pole moves to the north. Scientists don't fully understand how or why **magnetic reversals** happen. The important point here is that the reversal happens quickly by geologic time standards, and that the paleomagnetism of a rock retains or "remembers" the magnetic polarity of the Earth at the time that the rock solidified. This is called *remanent* (not "remnant") magnetism.

Studies of the paleomagnetism of solidified lava flows have revealed evidence of many magnetic reversals. As noted earlier, the ages of lavas can be determined by radiometric dating, and so, like the dating of strata from interlayered volcanic rocks, the dating of times of magnetic reversals in the past can be determined (Fig. 3.16). Through a combination of radiometric dating and magnetic polarity measurements, it has been possible to establish a time scale of magnetic polarity

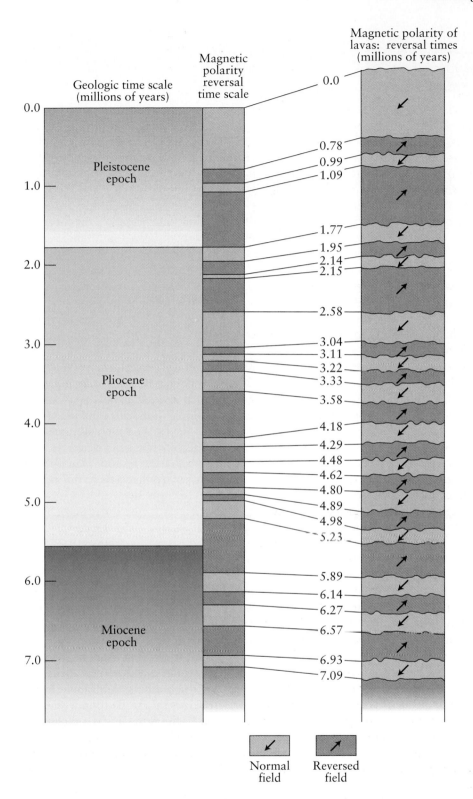

**Figure 3.16**
**MAGNETIC REVERSAL TIME SCALE**
Times of reversals of the magnetic polarity, as determined by radiometric dating of lavas,
back to the end of the Miocene Epoch. Periods of normal polarity, as today, and periods
of reversed polarity have been identified and dated using radiometric dating of lavas back
to the beginning of the Jurassic Period, about 200 million years ago.

Over the past 40 years, many fragments of fossil hominid bones have been found in southern and eastern Africa; details of human evolution are slowly emerging from studies of the fossils. The story is necessarily incomplete because many of the fossils have been found in ancient stream gravels where they were tumbled and battered by long-ago flood waters.

*Hominid* is a term for any upright-walking primate; we humans are hominids. The most ancient hominid fossils so far discovered are about 4 million years old. Adding to the complications of unraveling our origins is the fact that fossils of many different species of hominids have now been found and that at times in the past several different species of hominids were living at the same time. It is slowly becoming clear that the hominid genus *Homo* (which includes our own species, *Homo sapiens*) evolved a little more than 2 million years ago from an older genus, *Australopithecus*. To further our understanding, it is essential that the fragmentary fossil record be as precisely dated as possible. This has been accomplished through a combination of radiometric and magnetic polarity dates.

The Haddar region of northern Ethiopia in eastern Africa is a very productive place for finding fossils of ancestral hominids. The sediments that are interlayered with the fossil-bearing gravels give good magnetic signals (Fig. B3.1). The problem is to know where in the magnetic polarity time scale the Haddar reversals fall. Radiometric dates using the potassium–40 method were obtained on a lava flow and a layer of volcanic ash. The two radiometric dates determine the magnetic reversal ages unambiguously and indicate that early hominids lived in the region until about 3.15 million years ago; that is, they lived there during the Pliocene Epoch.

reversals dating back to the Jurassic Period. Still earlier reversals are a subject of ongoing research.

Correlation on the basis of magnetic reversals differs from other geologic correlation methods. One magnetic reversal looks just like any other in the rock record. When evidence of a magnetic reversal is found in a sequence of rocks, the problem is knowing which of the many reversals it actually is. Additional information is needed. When a continuous record of reversals can be found, starting with the present, it is simply a matter of counting backward. This is a technique sometimes used in the dating of oceanic crust, discussed in chapter 4.

Paleomagnetism in sedimentary rocks is the basis for a very important correlation technique. When fossils are present in a rock, an approximate age can be determined. Once they have an approximate age, geologists can determine a more precise age by comparing the magnetic reversals with the magnetic reversal time scale. Sediment cores recovered from the seafloor can be dated very accurately using a combination of fossil and magnetic reversal information. The correlations are so good that magnetic reversals can even be used to determine rates of sedimentation in the ocean.

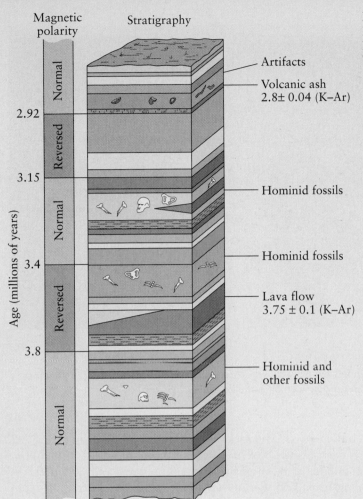

Magnetic polarity

Stratigraphy

Age (millions of years)

Normal

2.92

Reversed

3.15

Normal

3.4

Reversed

3.8

Normal

Artifacts

Volcanic ash
2.8± 0.04 (K–Ar)

Hominid fossils

Hominid fossils

Lava flow
3.75 ± 0.1 (K–Ar)

Hominid and
other fossils

**Figure B3.1**
**DATING HOMINID FOSSILS**
Example of two different dating techniques brought to bear on a geologic problem.

# PRECAMBRIAN TIME
# AND THE AGE OF THE EARTH

Throughout this book, examples of actual rates of geologic processes are mentioned. This would not be possible without the absolute dates obtained through magnetic and radiometric dating. In fact, more than any other contribution by geologists, the ability to determine absolute dates has changed the way humans think about the world. It has also changed the way we study the world because the ability to determine the rates of geologic processes is an essential step in studying the whole Earth as a system of interacting parts.

Now that we know how to determine the absolute ages of rocks, can we determine the age of the Earth? The earliest record, as indicated in Figure 3.8, comes from the great assemblage of metamorphic and igneous rocks that were formed during Precambrian time. Rocks that were formed before the Cambrian Period don't contain fossils with hard parts. Of the many radiometric dates obtained from Precambrian rocks, the youngest are around 545 million years and

Figure 3.17
THE EARTH'S OLDEST ROCKS
The Acastia gneiss in northern Canada was formed 4.0 billion years ago. It is the most ancient body of rock so far discovered on the Earth.

**? How has the age of the Earth been determined?**

the oldest about 4.0 billion years. The rocks of the Precambrian unit of the Geologic Column formed over a time span of at least 4.0 billion minus 545 million years, or about 3.46 billion years—a span of time about six and half times as long as the entire Phanerozoic Eon. Given that some igneous rocks from Canada (Fig. 3.17) are about 4 billion years old, the Earth must be even older.

No rocks older than 4.0 billion years that might provide radiometric dates have yet been found. The oldest radiometric data actually reported for the Earth is 4.1 billion years from an individual mineral grain from a sedimentary rock in Australia, so it is possible that igneous rocks older than 4.0 billion years may someday be located. How much older than 4.0 billion years might our planet be? There is strong evidence suggesting that the Earth was formed at the same time as the Moon, the other planets, and meteorites. Through radiometric dating, it has been possible to determine the ages of meteorites and of "Moon dust" brought back by astronauts. The ages of many of these objects cluster closely around 4.56 billion years. By inference, the formation of the Earth, and indeed of the Sun's entire planetary system, is believed to have occurred about 4.56 billion years ago.

Today, two centuries after Hutton, it is widely agreed that the Earth was formed about 4.56 billion years ago. When will it cease to exist? Astronomers tell us that billions of years in the future, the Sun will become what is called a red giant, that it will expand and wipe out the Earth. However, from evidence in the rocks on the Earth, Hutton is still correct: There is no prospect of an end to the Earth's existence. There are, of course, problems that arise because we humans are changing the Earth. From the point of view of an endangered animal, for example, an end may very well be in sight, and some writers have suggested that humans may eventually become endangered too. We don't agree with this hypothesis, even though we recognize that human activities do put many parts of the Earth system under great stress. Radiometric dating tells us that humans and their immediate ancestors have inhabited this planet for at least 2.5 million years. Perhaps by learning from our mistakes, we will be able to extend our stay on the planet at least as long into the future.

## WHAT'S AHEAD

In Part One you learned that the Earth can be studied as a large number of open systems that continually interact with each other and that these interactions make the Earth an active and dynamic planet. You also learned about the materials of the solid Earth—rocks and minerals—and about the evidence discovered by geologists that proves that the Earth is incredibly ancient.

Now we are ready to examine the inner workings of our dynamic planet, all of which are driven by the Earth's internal heat energy. Chapter 4 discusses why and how ocean basins and continents are continually being rearranged by the movement of lithosphere, a process called plate tectonics. In chapters 5 and 6 we will examine some of the consequences of plate tectonics, such as how, why, and where earthquakes and volcanic eruptions happen, and how, why and where great mountain ranges form.

## *Chapter Highlights*

**1.** Rock **strata** and **stratigraphy** provide a basis for reconstructing the Earth's history. There are two basic principles of stratigraphy: Strata are horizontal when they are deposited (principle of original horizontality); sedimentary strata accumulate in sequence, from the youngest on the bottom to the oldest on the top (principle of stratigraphic superposition).

**2.** Stratigraphic superposition and the principle of cross-cutting relationships (a rock stratum is always older than any other geologic feature that cuts across it, such as a fault) concern **relative age,** that is, the chronological sequence of geologic events and the age of geologic features relative to one another. **Absolute age,** on the other hand, is the exact time (in years) since the rock was formed or the stratum was deposited.

**3.** An **unconformity** is a break or gap in a stratigraphic sequence. It marks a period during which sedimentation ceased and erosion removed some of the previously laid strata.

**4. Correlation** of strata is the establishment of the time equivalence of strata in different places. Fossil assemblages have been the primary key to correlation of strata across long distances.

**5.** The **Geologic Column** is a composite diagram that shows the succession of all known strata, arranged in chronological order of formation, based on fossils and other age criteria.

**6.** Through measures of absolute age, it has been possible to determine specific times since given strata were formed and to thereby add times in years to the Geologic Column. The geologic time scale is divided into successively shorter time divisions: **eons** (the longest time unit), **eras, periods,** and **epochs. Fossils**  evidence of past life—are abundant only in strata of the youngest eon, the Phanerozoic, which were deposited during the past 545 million years.

**7.** The decay of **radioactive** isotopes of chemical elements provides a basis for **radiometric dating,** which give values for the absolute ages (age in years) of rock units and thus values for absolute dates of geologic events. The main radioactive isotopes used for radiometric dating, and their daughters, are potassium-40/argon-40, uranium-238/lead-206, uranium 235/lead-207, thorium 232/lead-208, rubidium-87/strontium-87, and carbon-14/nitrogen-14. Carbon-14 is used mostly for radiometric dating of organic materials. It is effective only for relatively young (less than 70,000 years old) materials because carbon-14 has a relatively short half-life.

**8.** Some iron-bearing minerals in igneous rocks, especially magnetite, lock in a record of the **polarity** of the Earth's magnetic field as they cool through the Curie point. The time sequence of **magnetic reversals,** as recorded in the **paleomagnetism** of igneous and sedimentary rocks, is a useful means of correlating strata and of estimating ages.

**9.** Most of the Earth's geologic history occurred during **Precambrian** time. The age of the Earth, estimated by radiometric dating of extraterrestrial materials, is 4.56 billion years.

## ▶ *The Language of Geology*

- absolute age 60
- Archean Eon 65
- Cenozoic Era 65
- correlation 63
- eon (geologic) 65
- epoch (geologic) 66
- era (geologic) 65
- fossil 60
- Geologic Column 64

- Hadean Eon 65
- half-life 70
- magnetic reversal 74
- Mesozoic Era 65
- paleomagnetism 74
- paleontology 62
- Paleozoic Era 65
- period (geologic) 65
- Phanerozoic Eon 65

- polarity (of magnetic field) 74
- Precambrian 66
- Proterozoic Eon 65
- radioactivity 69
- radiometric dating 71
- relative age 58
- stratigraphy 58
- stratum (plural = strata) 58
- unconformity 60

## ▶ *Questions for Review*

1. What is the difference between the relative ages and the absolute ages of a sequence of geologic events?

2. How do the principles of original horizontality and stratigraphic superposition help geologists unravel the geologic history of a region? Is that history known in absolute time or relative time?

3. What is the geologic significance of an unconformity? How many unconformities can you spot in the Grand Canyon section shown in Figure 3.2? See if you can establish the relative ages of all the rock strata in Figure 3.13 on the basis of the principles of superposition, original horizontality, and cross-cutting relationships.

4. How did William Smith make practical use of stratigraphic correlation?

5. What exactly is the Geologic Column? Is the Geologic Column the same in North America as it is in South America? in Australia? in Antarctica?

6. What is meant by the term faunal succession and what role did faunal succession play in working out the Geologic Column?

7. What are the three eras of the Phanerozoic Eon? What do their names mean?

8. Explain why Lord Kelvin's calculation of the age of the Earth was wrong.

9. If you start with 1,000,000 radioactive parent atoms and their half-life is 1 week, how many parent atoms will remain at the end of 4 weeks?

10. What features of radioactivity make it an ideal basis for determining absolute geologic time? Would the radiometric age of rocks on Mars and on Earth that were formed at exactly the same instant be the same? Why?

11. Why is radiometric dating possible for igneous rocks but not for sedimentary rocks?

12. How might you use carbon–14 dating to determine the date of the advance of the last great ice sheet in central North America? Explain how the same method might be used to determine the date of a major flood.

13. Answer the question posed at the beginning of the chapter: How would you determine how long ago the dinosaurs died out?

14. How is the irregular reversal of the Earth's magnetic field used in correlation of strata and solidified lava flows?

15. How was the age of the Earth determined?

# ▶ *Questions for Thought and Discussion*

1. If you were asked to make a geologic map of a place on the Moon based entirely on satellite photographs and then to determine the relative ages of the features on the map, how might you proceed? The kinds of geologic units present in the area to be mapped are lava flows and irregular blankets of pulverized rock scattered by meteorite impacts. This task has actually been carried out by geologists of the U.S. Geological Survey, so check in your library for examples of lunar geologic maps.

2. Check the area in which you live to see if there is an excavation—perhaps one associated with a new building or road repair. Visit the excavation and note the various layers, the paving (if the excavation is in a road), the soil below the surface, and whatever underlies the soil. Also note any pipes or wires exposed by the excavation. Determine the relative ages of the various features you see.

3. Suppose you are part of a group investigating some recently discovered hominid fossils in East Africa. A lava flow about 3 m (10 feet) above the fossil stratum yields a radiometric date of 2.8 million years, indicating that the fossils must be older than 2.8 million years. A few inches above and below the fossil stratum there are thin layers of volcanic ash. What method or methods would you use in trying to date the fossils more accurately than simply "older than 2.8 million years"?

For an interactive case study on the use of stratigraphy to study past climatic and environmental changes, visit our Web site.

# The
# Art
# of
# Geology

*William K. Hartmann, "Earth in Space."*

# The Cosmic Achoo!

I AM SITTING here 93 million miles from the sun on a rounded rock that is spinning at the rate of 1,000 miles an hour

and roaring through space to nobody-knows-where,

to keep a rendezvous with nobody-knows-what,

for nobody-knows-why,

and all around me whole continents are drifting rootlessly over the surface of the planet,

India ramming into the underbelly of Asia, America skidding off toward China by way of Alaska,

Antarctica slipping away from Africa at the rate of an inch per eon,

and my head pointing down into space with nothing between me and infinity but something called gravity, which I can't even understand and which you can't even buy anyplace so as to have some stored away for a gravityless day,

while off to the north of me the polar ice cap may, or may not,

be getting ready to send down oceanic mountains of ice that will bury everything from Bangor to Richmond in a ponderous white death,

and there, off to the east, the ocean if tearing away at the land and wrenching it into the sea bottom and coming back for more,

as if the ocean is determined to claim it all before the deadly swarms of killer bees,

which are moving relentlessly northward from South America,

can get here to take possession,

although it seems more likely that the protective ozone layer in the upper atmosphere may collapse first,

exposing us all, ocean, killer bees and me, too,

to the merciless spraying of deadly cosmic rays.

I AM SITTING here on this spinning, speeding rock surrounded by four billion people,

eight planets,

one awesome lot of galaxies,

hydrogen bombs enough to kill me 30 times over,

and mountains of handguns and frozen food,

and I am being swept along in the whole galaxy's insane dash toward the far wall of the universe,

across distances longer to traverse than Sunday afternoon on the New Jersey Turnpike,

so long, in fact, that when we get there I shall be at least 800,000 years old,

provided, of course, that the whole galaxy doesn't run into another speeding galaxy at some poorly marked universal intersection and turn us all into space garbage,

or that the sun doesn't burn out in the meantime,

or that some highly intelligent ferns from deepest space do not land from flying fern pots and cage me up in a greenhouse for scientific study.

SO, AS I SAY, I am sitting here with the continents moving, and killer bees coming, and the ocean eating away, and the ice cap poised, and the galaxy racing across the universe,

and the thermonuclear 30-times-over bombs stacked up around me,

and only the gravity holding me onto the rock,

which, if you saw it from Spica or Arcturus, you wouldn't even be able to see, since it is so minute that even from these relatively close stars it would look no bigger than an ant in the Sahara Desert as viewed from the top of the Empire State Building,

and as I sit here,

93 million miles from the sun,

I am feeling absolutely miserable,

and realize,

with self-pity and despair,

that I am

getting a cold.

—Russell Baker

# PART TWO

# THE DYNAMIC EARTH

**The Earth is a dynamic place,** a place of constant activity and change. Much of this activity originates in geophysical processes deep within the Earth, and these internal processes are the focus of Part Two. In the 1960s, the science of geology underwent a scientific revolution—a fundamental change in the way we view the functioning of the solid Earth. That revolution led to widespread acceptance of the theory of plate tectonics. This theory suggests that the outermost part of the Earth has broken into huge, rigid plates that shift around and interact with one another, causing earthquakes and other geophysical activity. There are still some details to be worked out, such as the exact nature of the driving mechanism for plate motion. But plate tectonics has provided a model that unifies much observational evidence about the Earth system. In the context of plate tectonics, much of what we know about the inside of the Earth from studies of earthquakes and volcanoes now makes sense.

▶ All of these topics and more are addressed in the chapters of Part Two:

Chapter 4: Plate Tectonics: A Unifying Theory

Chapter 5: Earthquakes and the Earth's Interior

Chapter 6: From the Earth's Interior: Volcanoes and Igneous Rocks

*Shaping the Landscape: From the Inside Out*
Trekkers with pack mules work their way up the mountains of the Cordillera Blanca in Peru. High mountain ranges like these are uplifted by great internal forces within the Earth.

# · 4 ·

# PLATE TECTONICS: A UNIFYING THEORY

**Mountains have awed, inspired, and challenged people** throughout history. Some mountains are revered and held sacred; others are objects of superstition and fear. In the eighteenth century, a traveler described the Alps as producing in him "an agreeable kind of horror." His words evoke the common sensation of being drawn to the mountains and yet, at the same time, feeling overwhelmed by them. Mountains confront us with the immense power and beauty of nature, and the relative insignificance of humanity.

Each great mountain range in the world has its own unique characteristics, but they all have one important feature in common: they were created by the intense tectonic forces characteristic of collisions between crustal plates. The Alps and the Himalayas, for example, formed as the result of grinding collisions between continents. If some day you are lucky enough to go trekking in the Himalayas, don't be surprised to find fossils of shelled organisms high in the mountains. They are the remains of seafloor sediments, uplifted by tectonic forces to form the highest peaks on the planet.

Popular science writer John McPhee expressed his awe of the mountain-building process when he wrote in his book *Basin and Range*, "If by some fiat I had to restrict all this writing to one sentence, this is the one I would choose: 'The summit of Mt. Everest is marine limestone.'" Plate tectonics has had many consequences for the inhabitants of this planet, but nowhere is the power of tectonic forces more evident than in the great mountainous regions of the world.

## In this chapter you will learn about

- How geologists turned years of isolated observations about geology into a coherent, unifying model of how the Earth system functions
- The basics of the theory of plate tectonics
- Which of the Earth's geologic features are explained by the plate tectonic model
- What kinds of questions about plate tectonics remain unanswered

*The Top of the World*
This is the peak of Ama Dablam (22,493 ft or 6856 m high) in Khumbu Himal, Nepal. The Himalayan Mountains are being uplifted as a result of the collision of the Indian subcontinent with the mainland of Asia.

Scientific revolutions challenge us to look at the world in a new way. They don't happen overnight, the way political revolutions often do. For two millennia it was almost universally believed that the Sun and other planets revolved around the Earth. It took hundreds of years to bring about the scientific revolution that led to the acceptance of the so-called Copernican model of the solar system, in which the Earth and the other planets are understood to revolve around the Sun. When this model was finally accepted, previous ideas about the solar system had to be adjusted to fit a new reality. Scientific revolutions can have as great an impact as political revolutions: They turn accepted ideas upside down. As we will see in this chapter, that is what happened in the case of the theory of plate tectonics.

## A REVOLUTION IN GEOLOGY

**How do scientific revolutions occur?**

In the 1960s, a revolution rippled through the staid old science of geology. That revolution was a result of many years of observations of how the Earth works and how the parts of the Earth system fit together. Like other scientific revolutions, this one didn't happen overnight. During the 1800s, people favored the idea that the Earth, originally a molten mass, had been cooling and contracting for centuries. Theorists pointed to mountain ranges full of folded rocks as expressions of the contraction and shrinkage of the Earth's interior. (If the crust didn't contract as much as the interior, it would fold and crumple like the loose, wrinkled skin of a dried prune.) However, while contraction did explain some features of the Earth's surface, it did not explain the shapes and positions of the continents. Nor did it explain features like great rift valleys, which appear to have been caused by stretching of the Earth's crust rather than by contraction.

At the beginning of the twentieth century, scientists discovered that the Earth's interior is heated by the decay of radioactive elements. This suggested that the Earth might not be cooling but rather heating up, and therefore expanding. A much smaller Earth could once have been covered mostly by continents. As the Earth expanded, the continents would crack into fragments, and eventually the cracks would grow into oceans. The theory of an expanding Earth does offer an explanation for the matching coastlines of Africa and South America, but it does not easily account for folded mountain ranges formed by compression. To get around the flaws in the expansion and contraction theories, geologists began to search for other theories to explain the shapes and positions of the continents, oceans, and mountain chains. By the middle of this century, however, all reasonable suggestions seemed to have been exhausted. The time was ripe for a totally new approach.

This approach turned out to be *plate tectonics*—the theory that the continents are carried along on huge slabs, or *plates,* of the Earth's outermost layer. In other places plates have been slowly converging, forming compressional features like huge mountain ranges. In other places plates have been moving apart, forming expansional features like the great rift valleys. The theory of plate tectonics provided, for the first time, a coherent, unified explanation for *all* of these features of the Earth's surface. It took years, but eventually the evidence convinced

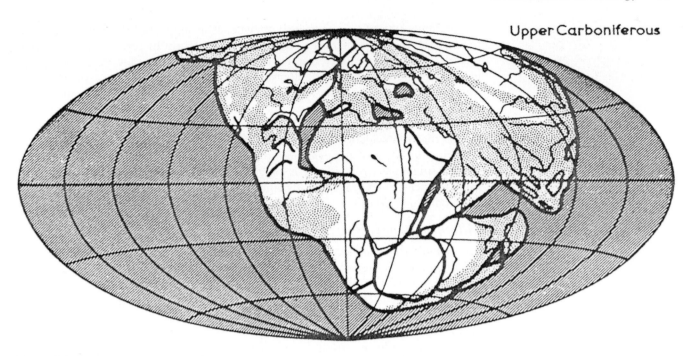

Upper Carboniferous

Figure 4.1
WEGENER'S PANGAEA
This is Alfred Wegener's 1915 map of Pangaea. It shows the positions of the continents during Late Carboniferous times (about 300 million years ago). Africa is shown at the center, in its present location. The shaded areas on the continents are shallow seas.

most scientists to accept the new theory of plate tectonics. In this chapter we tell the story of how the theory was conceived and developed and came to be accepted. The modern part of the story began in the early 1900s with a German meteorologist who had some controversial ideas about the shapes and positions of the continents.

## Continental Drift

In 1910, Alfred Wegener began lecturing and writing scientific papers about *continental drift*. The **continental drift** hypothesis[1] suggested that the continents have not always been in their present locations but instead have "drifted" and changed positions. Wegener's idea was that the continents had once been joined together in a single "supercontinent," which he called *Pangaea* (pronounced Pan-JEE-ah), meaning "all lands" (Fig. 4.1). He suggested that Pangaea had split into fragments like pieces of ice floating on a pond and that the continental fragments had slowly drifted to their present locations.[2]

**continental drift** The slow, lateral movement of continents across the surface of the Earth.

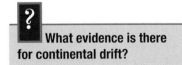

**What evidence is there for continental drift?**

---

[1] A *hypothesis* is a scientific concept or proposal that is a tentative explanation for a natural phenomenon. Before a hypothesis can become a theory, it must be supported by extensive experimentation and observation.

[2] Wegener was inspired by a trip to Greenland during which he observed ice floes breaking up and drifting apart. Actually, the idea of continental drift originated long before Wegener's time. As early as the 1600s, Leonardo da Vinci and Francis Bacon commented on the apparent fit of the continents. In the 1800s, geologists Eduard Suess and Antonio Snider both suggested that Africa, Australia, India, and South America had once been joined together. Suess called this supercontinent *Gondwanaland*. Like all scientists, Wegener developed his ideas by building on the observations of others.

Wegener presented a great deal of evidence in support of the continental drift hypothesis. Nevertheless, his proposal created a storm of protest in the international scientific community, a storm that continued for decades. Part of the problem was that contemporary geologists—particularly geophysicists—simply could not envision a reasonable mechanism whereby the continents could move around. (As you will see, the exact details of this mechanism still are not entirely understood.) Another part of the problem was that geologists had to be convinced that the evidence for the fit of the continents was truly conclusive. Let's look now at some of the evidence so that you can judge it for yourself. Notice that no single piece of evidence is conclusive on its own. It took the weight of all this evidence (and more) to finally convince geologists of the validity of the continental drift hypothesis.

## Matching Coastlines

If you look at a map of the world, it's obvious that the Atlantic coastlines of Africa and South America seem to match, almost like puzzle pieces. The southern coast of Australia similarly seems to match part of the coast of Antarctica, to its south; the same is true of some other continental coastlines. Is this apparent fit an accident, or does it truly support the hypothesis that the continents were once joined together?

To answer this question, we must first recognize that the edge of the land, that is, the shoreline, usually isn't the true edge of the continent (Fig. 4.2). Along a noncliffed shoreline (such as the Atlantic coasts of North America and Africa), the land usually slopes gently toward the sea. This gently sloping land, some of which may be above sea level and some below, is called the *continental shelf* or *platform*. At the edge of the continental shelf there is commonly a sharp drop-off to the steeper *continental slope*. At the bottom of the steep continental slope, the land begins to level off again. This more gently sloping land is called the *continental rise;* it marks the transition to the much flatter ocean floor, called the *abyssal plain.* (Recall from chapter 1 that continents consist primarily of granitic *continental crust,* whereas ocean floors consist primarily of basaltic *oceanic crust.* The place where the two types of crust meet is covered by sediment.) The actual configuration of the shoreline depends on several factors, including sea level, the presence or absence of cliffs, and the details of the topography of the continental shelf in any particular locality. Thus, parts of the continental shelf may (or may not) be underwater.

**Figure 4.2**
**WHERE IS THE TRUE EDGE OF A CONTINENT?**
Is the coastline—where the water meets the land—the true edge of a continent? To a geologist, the answer is "no." The true edge of a continent—the junction between continental and oceanic crust—usually occurs below sea level. We define the edge of a continent to be halfway down the continental slope.

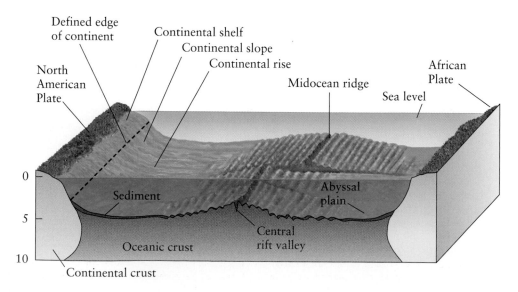

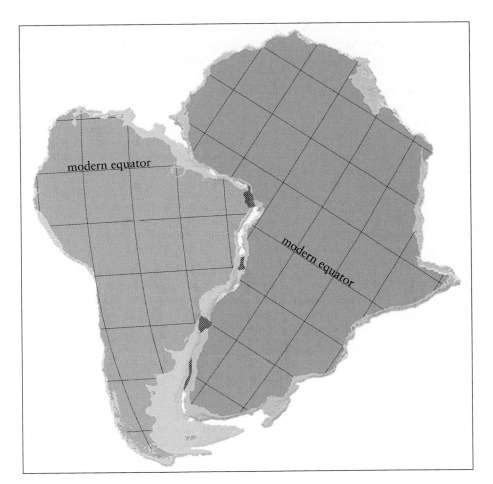

Figure 4.3
**HOW WELL DO
THE CONTINENTS FIT?**
This map shows the fit of
Africa and South America
along the "true edges" of the
continents. The dark areas
show overlap. The latitude and
longitude lines show how each
of the continents has been rotated from its present position
back to its position when the
continents were joined long
ago. This particular map was
constructed visually, but today
this kind of map is usually
made using computers.

So, how do we identify the true edge of a continent? Usually, the edge of a continent is defined as being halfway down the steep outer face of the continent, that is, the continental slope. When we try to fit the continents together, we should fit them along this line—the true edge of the continent—rather than along the present-day coastline. Figure 4.3 shows an example of such a fit. This example was constructed visually, but today this kind of map is usually drawn by computers programmed to find the best fit between the continents.[3] In the case of Africa and South America, the fit is remarkable; in the "best-fit" position, the average gap or overlap between the two continents is only 90 km (56 mi). Interestingly, the most significant overlapping areas consist of relatively large volumes of sedimentary or volcanic rocks that were formed *after* the time when the continents are thought to have split apart.

## Matching Geology

The exceptionally close fit between Africa and South America does suggest that they were once joined together. But if this is true, one would expect to find similar geologic features on both sides of the join. Indeed, such correlations provided some of the most compelling evidence presented by Wegener in support of the continental drift hypothesis. However, matching the geology of rocks on opposite sides of an ocean is more difficult than you might imagine. Rock-forming processes never cease. Some rocks were formed before the continents were

---

[3] Creating these reconstructions is not trivial. Part of the challenge is that the movement of the continents takes place not on a *flat* surface but on a *spherical* one—the surface of the Earth.

joined, some while they were joined, others during the splitting of the continents, and still others after they became separated. How can we tell which rock formations and geologic features are significant in trying to find a match between the continents?

A logical starting point is to see if the ages and orientations of similar rock types match up across the ocean. In Wegener's time the technique of radiometric dating was just being developed, so it was not easy to determine the exact age of a rock. But now we know that there is, indeed, some similarity in the ages of rocks and correlations in rock sequences across the oceans. As shown in Figure 4.4, the match is particularly good between rocks about 550 million years old in northeast Brazil and West Africa. This suggests that the two continents were joined together for some period of time prior to 550 million years ago.

Another thing to look for is continuity of geologic features such as mountain chains. Figure 4.5 shows a reconstruction of the continents at a time when they were joined together in the supercontinent Pangaea. Notice how mountain belts of similar ages seem to line up when the continents are moved back into this position. The oldest portions of the Appalachian Mountains, extending from the northeastern part of the United States through eastern Canada, match up with the Caledonides of Ireland, Britain, Greenland, and Scandinavia. A younger part of the Appalachians lines up with a belt of similar age in Africa and Europe. These and other correlations of bedrock features are strong evidence that the continents once were joined together.

## Glacial Deposits

Another geologic feature that matches up across continental joins is the deposits left by ancient ice sheets. These are similar to the deposits left by relatively recent

Older than 550 million years
About 550 million years old
Younger than 550 million yrs.

**Figure 4.4**
**DO THE ROCK AGES MATCH?**
When the continents are rotated back together, the ages of rock units generally match. They match particularly well between northeast Brazil and West Africa. This is evidence that the continents were once joined.

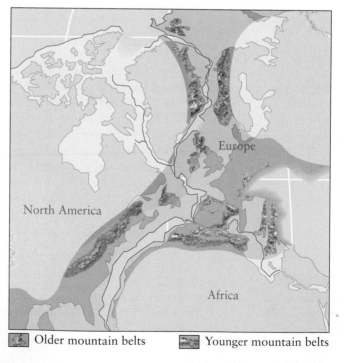

Older mountain belts       Younger mountain belts

**Figure 4.5**
**DO THE MOUNTAIN BELTS MATCH?**
When the continents are reconstructed into Pangaea, some mountain belts of similar ages match up. These correlations of bedrock features are further evidence that the continents were once joined together.

(Pleistocene) glaciations in Canada, Scandinavia, and northern United States, among other places (see chapter 12). In South America and Africa there are very thick glacial deposits of the same age (Permian-Carboniferous). The deposits match almost exactly when the continents are moved back together.

As glacial ice moves, it cuts grooves and scratches in underlying rocks and produces folds and wrinkles in soft sediments. These features left behind in glacial deposits provide evidence of the direction the ice was moving during the glaciation. When Africa and South America are moved back together, the direction of ice movement on both continents is consistent, radiating outward from the center of the former ice sheet. It's hard to imagine how such similar glacial features could have been created if the continents had not once been joined together. Africa and South America must have had similar climates during this period, colder than their present-day climates. This also suggests that they were not in their present equatorial locations. Figure 4.6 shows glacial deposits in a reconstruction of the southern continents in Pangaea, with grooves and scratches indicating ice movement outward in all directions from what was then the South Pole.

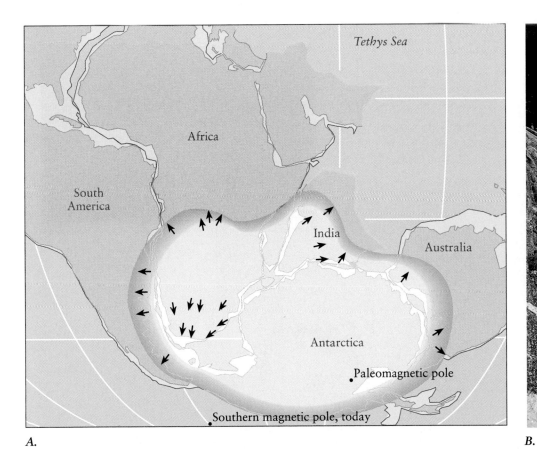

*A.*

*B.*

Figure 4.6
**DO THE GLACIAL DEPOSITS MATCH?**
In this reconstruction of the southern continents of Pangaea *A*, glacial deposits of Carboniferous age can be seen to match up remarkably well across the joins. Small arrows indicate the direction in which the ice was moving during the glaciation, as deduced from evidence like grooves and scratches in the bedrock (as shown in *B*), and folds and wrinkles in underlying soft sediments.

## Fossil Evidence

If Africa and South America were really joined together at one time, with the same climate and matching geologic features, then they also should have had the same plants and animals. To check this hypothesis, Wegener turned to the fossil record. It revealed that communities of plants and animals appear to have evolved together until the time of the splitting apart of Pangaea, after which they evolved separately.

In seeking support for the continental drift hypothesis, Wegener pointed to specific fossil species found in matching areas across the continental joins. One example he used was an ancient fern, *Glossopteris,* whose fossil remains have been found in southern Africa, South America, Australia, India, and Antarctica (Fig. 4.7). Could the seeds of this plant have been carried by wind or water from one location to another? Probably not. The seeds of *Glossopteris* were large and heavy and could not have been carried very far by wind or water currents. This fern flourished in a cold climate; it would not have thrived in the warm present-day climates of the continents where its fossil remains are found. This, too, is consistent with the idea that these continents were once joined together with similar, polar climates.

There are other examples as well. The fossil remains of *Mesosaurus,* a small reptile from the Permian Period, are found both in southern Brazil and in South

Figure 4.7
**DO THE FOSSIL PLANTS MATCH?**
This map shows the locations of fossils of *Glossopteris,* a fern of Carboniferous age (shown in the small photograph *B*). The present-day distribution of *Glossopteris* fossils suggests that these continents were once joined together.

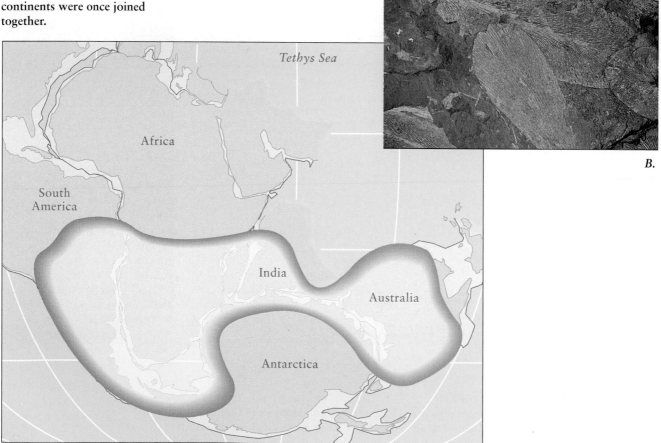

*B.*

*A.*

Figure 4.8
**DO THE FOSSIL ANIMALS MATCH?**
This photograph shows part of the fossilized skeleton of *Mesosaurus,* a Permian-aged reptile found in South Africa. Similar fossils have been found in Brazil. Although *Mesosaurus* could swim, it is thought to have been too small (less than half a meter, or about 2 feet) to have swum all the way across the ocean.

Africa (Fig. 4.8). The types of rocks in which the fossils are found are very similar. *Mesosaurus* did swim but was too small (about half a meter long, less than 2 ft) to swim all the way across the ocean. Fossil remains of specific types of earthworms also occur in areas that are now widely separated. How could they possibly have migrated across the oceans? The land masses in which they lived must once have been connected.

## Apparent Polar Wandering Paths

Wegener and his few supporters gathered more and more evidence in support of continental drift, but many scientists remained unconvinced. Wegener died in 1930 without seeing a resolution to the debate, which continued after his death. A turning point occurred in the 1950s, through the study of *paleomagnetism.*

Recall from chapter 3 that when magma cools and solidifies into rock, it becomes magnetized and takes on the prevailing *polarity*—the north-south directionality—of the Earth's magnetic field (Fig. 4.9). The paleomagnetic signature of a rock also provides useful information about the location of the Earth's poles. Just as a free-swinging magnet today will point toward today's magnetic north

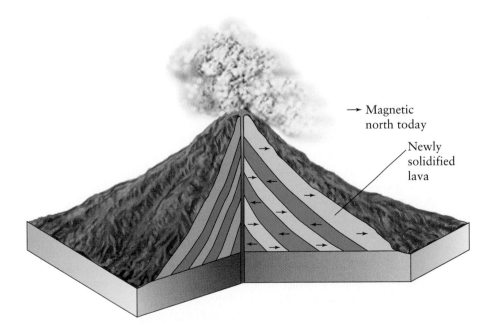

→ Magnetic north today

Newly solidified lava

Figure 4.9
**MAGNETIC FIELD REVERSALS RECORDED IN ROCK**
When molten rock solidifies into hard rock, as each of these lava flows did, it retains the polarity of the Earth's magnetic field at the time. In this drawing, the recent volcanic rocks in the top layer have the present polarity, while the older layers record the direction of the magnetic field—sometimes normal, sometimes reversed—that existed at the time that the rocks cooled and solidified.

Figure 4.10
WANDERING POLES
The curves on this map trace the apparent path followed by the north magnetic pole through the past 600 million years. The numbers are millions of years before the present. The paths don't actually show movement of the magnetic pole itself, but rather movement of the continents *relative* to the pole. The apparent polar wandering path for Europe is different from the path determined from measurements made in North America. If the continents are reassembled into a single supercontinent, the two paths coincide, indicating that Europe and North America moved as one continent during this period.

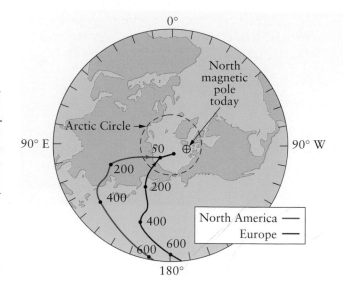

pole, so too does a rock's paleomagnetism act as a pointer toward the Earth's magnetic north pole at the time of rock formation.

Paleomagnetism provides another useful piece of information about the Earth's magnetic poles at the time of rock formation. The angle or *magnetic inclination* of a freely swinging bar magnet varies with latitude, because the bar magnet always points directly to the pole (refer to Fig. 3.14). For example, near the equator a magnet will lie relatively flat (horizontal). The closer it gets to the poles, the steeper will be its angle from the horizontal (that is, its inclination). When it is located exactly at the pole, the bar magnet will be vertical, pointing directly to the pole—it will be at the highest possible angle from the horizontal (90°). Because this angle varies systematically with latitude, it is always possible to figure out how far away you are from the pole, on the basis of magnetic inclination. The same is true of a rock's *paleomagnetic inclination,* which thus provides a record of the geographic latitude where the rock formed.

In the 1950s, geophysicists studying paleomagnetic pole positions found evidence suggesting that the Earth's magnetic poles had wandered all over the globe for at least the past several hundred million years. They plotted the pathways of the poles on maps and referred to the phenomenon as *apparent polar wandering* (Fig. 4.10). Geophysicists were puzzled by this evidence because they knew that the Earth's magnetic poles and its axis of rotation are always close together. When it was discovered that the path of apparent polar wandering measured in North America differed from that measured in Europe, geophysicists were even more puzzled. They knew that it was extremely unlikely that the magnetic poles had moved. Instead they concluded, somewhat reluctantly, that it must have been the continents themselves that had moved, carrying their magnetic rocks with them. Thus, the apparent polar wandering path of a continent, determined from the paleomagnetism of rocks of different ages, provides a historical record of the movement of that continent relative to the positions of the magnetic poles.

Note from Figure 4.10 that the apparent polar wandering paths of Europe and North America are separate from 600 million years ago to about 50 million years ago. You might also notice that the shapes of the paths look rather similar. In fact, if you rotate Europe and North America from their present positions and reassemble them as a single continent, as if the Atlantic Ocean weren't there, an interesting thing happens: the two paths overlap and fit together exactly. This indicates that Europe and North America were moving together as a single continent during, the several hundred million years represented by these paths. This was—and still is—a very powerful argument for the existence, long ago, of a supercontinent that eventually split into separately moving pieces.

## The Missing Clue: Paleomagnetic Bands and the Spreading Sea Floor

This new understanding of apparent polar wandering helped revive the hypothesis of continental drift. But many scientists were still holding out for a final piece of evidence that would demonstrate conclusively that a supercontinent had actually split apart and seas had flowed into the widening rift. Specifically, they were trying to envision a mechanism whereby the crust could actually split open. Evidence concerning that mechanism finally appeared but not until the early 1960s—three decades after Wegener's death. The clue was found by scientists who made a crucial discovery while studying the paleomagnetic properties of Atlantic seafloor rocks.

When oceanographers surveyed the floor of the Atlantic Ocean with magnetometers, they were astonished to find that parts of the sea floor consist of magnetized rocks with alternating bands of normal and reversed polarities (Fig. 4.11). The bands are hundreds of kilometers long. More important, they are ex-

> **?** **What piece of evidence finally convinced geologists that continental drift had occurred?**

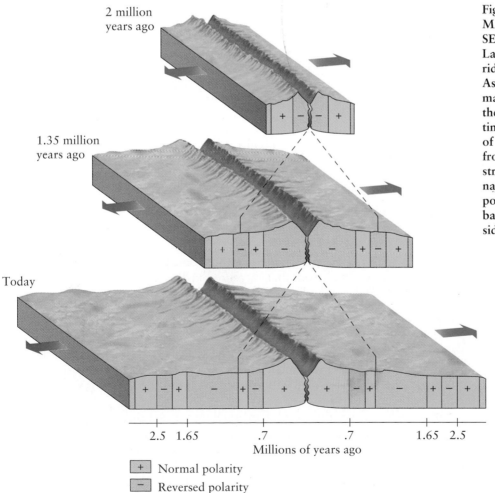

**Figure 4.11**
**MAGNETIC BANDS ON THE SEA FLOOR**
Lava extruded along a midocean ridge forms new oceanic crust. As the lava cools, it becomes magnetized with the polarity of the Earth's magnetic field at that time. As the plates on either side of the midocean ridge move apart from one another, successive strips of oceanic crust have alternating normal and reversed polarities. The resulting magnetic bands are symmetrical on either side of the midocean ridge.

2 million years ago

1.35 million years ago

Today

2.5   1.65   .7   .7   1.65   2.5
Millions of years ago

[+] Normal polarity
[−] Reversed polarity

**?** How can patterns of magnetized rock prove that seafloor spreading has occurred?

**seafloor spreading** The processes through which the sea floor splits and moves apart along a mid-ocean ridge, and new oceanic crust forms along the ridge.

actly symmetrical on either side of a centerline that corresponds to the crest of the ridge running down the middle of the Atlantic Ocean. In other words, if you could fold the sea floor in half along the midocean ridge, the bands on either side would match exactly.

The symmetrical patterns of magnetic reversals discovered in seafloor rocks mystified scientists at first. Then several groups of geophysicists, working independently, came up with the same explanation. They proposed that the sea floor had split apart along the midocean ridge and that the rocks on either side were moving away from one another. As the rocks spread apart, molten material from the mantle below welled up into the crack, solidifying into new volcanic rocks on the seafloor. When the molten rock solidified, it took on the magnetic field polarity of the Earth at that time. Over time the spreading sea floor functioned like a conveyor belt, carrying the newly magnetized bands of rock away from the centerline of the ridge in either direction. This process came to be known as **seafloor spreading.**

Geologists also have demonstrated that the ages of seafloor rocks increase with distance from the ridge. The youngest rocks are found along the centerline ridge, where new molten material wells up (Fig. 4.12). When magnetic reversals occur, they are recorded in the newly formed rocks along the midocean ridge. The result is the formation of symmetrical bands of volcanic rock with alternating magnetic polarities. This final piece of evidence convinced the great majority of geologists that seafloor spreading indeed occurs. As it turned out, geophysicists—those who had most vigorously opposed Wegener's ideas—ultimately provided the paleomagnetic evidence that supported continental drift.

# THE PLATE TECTONIC MODEL

By the early 1960s, the evidence was in and the scientific jury was, for the most part, convinced: continental drift had really occurred. However, it remained to put all of this together into a coherent model. This model is the theory of plate tectonics. (Now it is called a *theory* instead of an idea or *hypothesis*. Theories are supported by scientific evidence, whereas hypotheses are educated guesses.) Let's briefly summarize the theory of plate tectonics and then find out what geologic features this model successfully explains.

## Plate Tectonics in a Nutshell

**lithosphere** The crust and the uppermost part of the mantle.

The outermost part of the Earth—the **lithosphere,** consisting of the crust and the uppermost part of the mantle—is thin, cool, and strong compared to the rock below it. The rest of the mantle is very hot and therefore relatively malleable (like putty), even though it is composed of solid rock. The part of the mantle immediately beneath the lithosphere is particularly weak because it is close to the temperature at which it begins to melt. This weak layer underneath the lithosphere is called the *asthenosphere.* (We will learn more about the asthenosphere in chapter 5.)

If you place a very thin, cool, hard shell on top of hot, weak material that is moving around, what will happen? Predictably, the thin shell will break. That is exactly the state of the Earth's lithosphere: it consists of many large fragments, which we call *plates.* Today there are six large plates, each extending for several thousands of kilometers, and a number of smaller ones (Fig. 4.13). The plates are

**?** What will happen if a cold, thin, brittle layer is placed on top of hot material that is moving around?

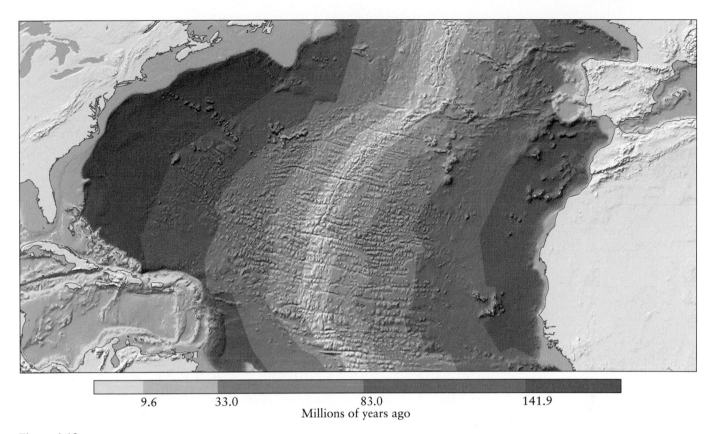

9.6          33.0                     83.0                   141.9
Millions of years ago

**Figure 4.12**
**HOW OLD ARE SEAFLOOR ROCKS?**
This drawing shows the ages of magnetically banded rocks on either side of the midocean ridge in the Atlantic Ocean. The numbers show the ages of the bands of rock in millions of years before the present. The youngest bands are along the mid-ocean ridge, and the oldest are far away from the ridge.

in a condition called *isostasy*, which means that they are essentially "floating" on the asthenosphere, like blocks of wood floating on water.

Think again about thin, brittle fragments floating on top of hot, mobile material. You might expect movement in the underlying material to cause the fragments to shift about. Again, that is exactly what happens to the Earth's lithospheric plates. As thermal movement occurs in the hot mantle, the plates move about, interacting with one another. Such movements involve complicated events, both seen and unseen, which are described by the term *tectonics* (from the Greek word *tekton*, meaning "carpenter" or "builder"). Tectonics is the study of the movement and deformation of the lithosphere. The branch of tectonics that deals with the processes by which the lithospheric plates move around and interact with one another is **plate tectonics.**

**plate tectonics** The horizontal movement and interactions of large fragments of lithosphere at the surface of the Earth.

## Plate Margins

Lithospheric plates move as individual units, and interactions between them occur mainly along their edges. Probably the most important aspect of this interaction, therefore, is the nature of plate margins. There are three fundamental ways in which plates can interact: they can move away from each other or *diverge;* they can move toward each other or *converge;* or they can slide past each

**?** **What happens when lithospheric plates interact?**

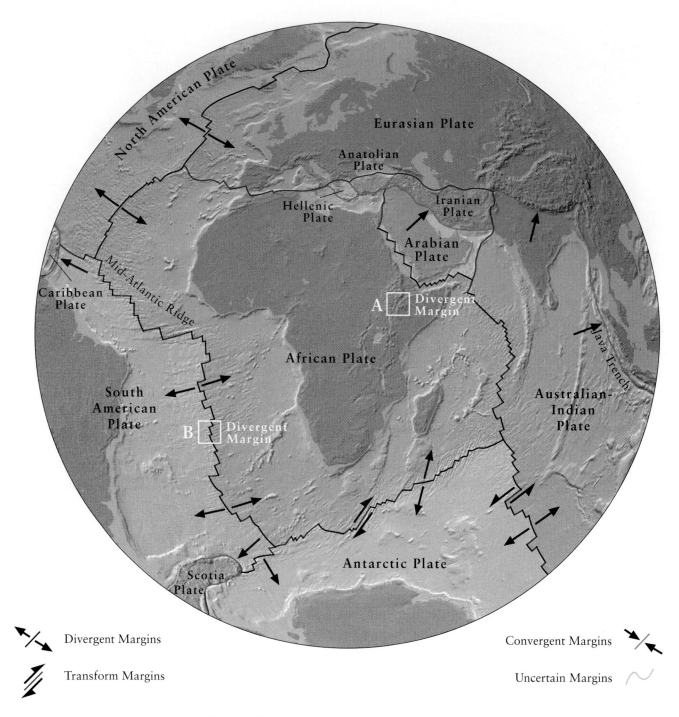

**Figure 4.13**
**THE EARTH'S LITHOSPHERIC PLATES**
Six large plates and a number of smaller ones comprise the Earth's surface. They are moving very slowly in the directions shown by the arrows. The labels *A*, *B*, *C*, *D*, and *E* correspond to the different types of plate margins shown in Figure 4.14.

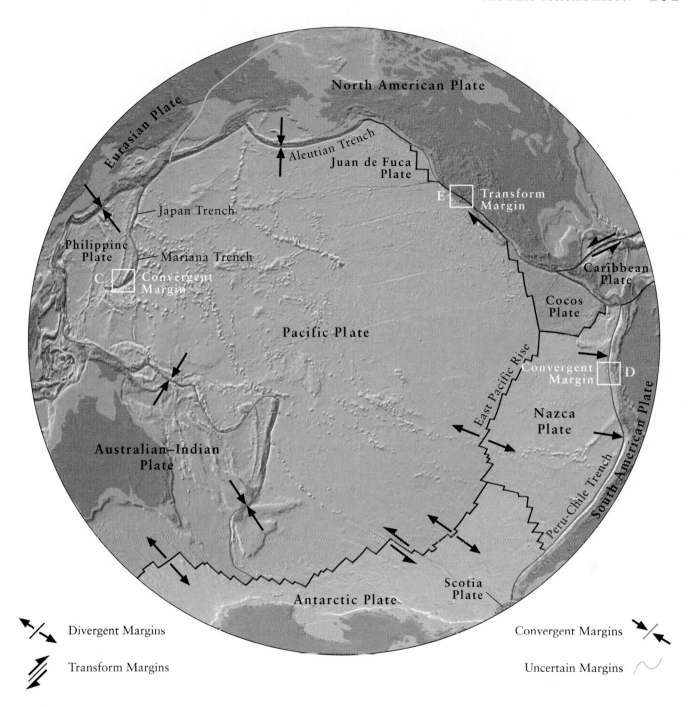

North American Plate

Eurasian Plate

Aleutian Trench

Juan de Fuca Plate

E  Transform Margin

Japan Trench

Philippine Plate

Mariana Trench

C  Convergent Margin

Caribbean Plate

Cocos Plate

Pacific Plate

East Pacific Rise

Convergent Margin  D

Nazca Plate

Australian–Indian Plate

South American Plate

Peru-Chile Trench

Antarctic Plate

Scotia Plate

Divergent Margins

Transform Margins

Convergent Margins

Uncertain Margins

other along a large fracture called a *transform fault*. Consequently, we can identify three kinds of plate margins (Fig. 4.14):

**divergent margin** A boundary along which two plates move apart from one another.

1. **Divergent margins** are also called *rifting* or *spreading centers* because they are fractures in the lithosphere where two plates move apart. Divergent margins can occur in either continental or oceanic crust. Where continental crust is splitting apart, the result is a great *rift valley* like those in East Africa, where the African Plate is being stretched and torn apart (Fig. 4.14A). Eventually, a new ocean may form in the widening continental rift; a modern example is the Red Sea. Where oceanic crust is splitting apart, the result is a midocean ridge and seafloor spreading, as described earlier (Fig. 4.14B).

**convergent margin** A boundary along which two plates come together.

**subduction zone** A boundary along which one lithospheric plate descends into the mantle beneath under another plate.

2. **Convergent margins** occur where two plates move toward each other. There are three types of convergent margins: ocean–ocean, ocean–continent, and continent–continent. Whenever oceanic crust is involved in a convergent margin, one plate of oceanic crust will sink beneath the other plate; this is called a **subduction zone**. Subduction zones are marked by very deep oceanic trenches and lines of volcanoes, as in Indonesia (ocean–ocean subduction zone) or the Andes (ocean–continent subduction zone) (Fig. 4.14C). When one continent meets another continent along a convergent margin, they collide and crumple up, forming huge mountain ranges like the Himalayas; this is called a **collision zone** (Fig. 4.14D).

**collision zone** Where one continent meets another continent along a convergent plate margin.

**transform fault margin** A fracture in the lithosphere where two plates slide past each other.

3. **Transform fault margins** are fractures in the lithosphere where two plates slide past each other, grinding and abrading their edges as they do so. A modern example is the San Andreas fault in California, where the Pacific Plate is moving north-northwest past the North American Plate (Fig. 4.14E).

Through the tectonic cycle, a balance is maintained. New material is added to the crust by volcanism along divergent margins, while subduction consumes an equal amount of crustal material along convergent margins. All these types of plate interaction are occurring today, just as they have occurred throughout much of the Earth's history. We don't often notice plate motion because lithospheric plates move very slowly—usually between 1 and 10 cm (0.4 to 4 in) per year. But the scars and remnants of ancient plate interactions are preserved in the rock record.

## *Earthquakes and Plate Margins*

**How do earthquakes reveal the nature of plates and plate boundaries?**

Earthquakes and volcanoes, which occur primarily along plate margins, are the most obvious manifestation of active plate interaction (Fig. 4.15). By studying these phenomena, particularly earthquakes, geologists have been able to decipher the shapes of the plates and other important characteristics of plate margins.

Earthquakes occur where huge blocks of rock are grinding past each other. (This topic is discussed in greater detail in chapter 5.) Tectonic motions produce directional pressure, which causes rocks on either side of a large fracture to move past each other. A fracture in the crust along which movement has occurred is called a **fault**. The movement is rarely smooth; usually the blocks stick because of friction, which slows down their movement. Eventually, the friction is overcome and the blocks slip abruptly, releasing pent-up energy with a huge "snap"—an earthquake. The actual location beneath the Earth's surface where the earthquake begins is the *focus* (plural *foci*). The spot on the surface directly above the focus, used to identify the map location of the earthquake, is the *epicenter* (Fig. 4.16).

**fault** A fracture in the crust along which movement has occurred.

# Plate Margins

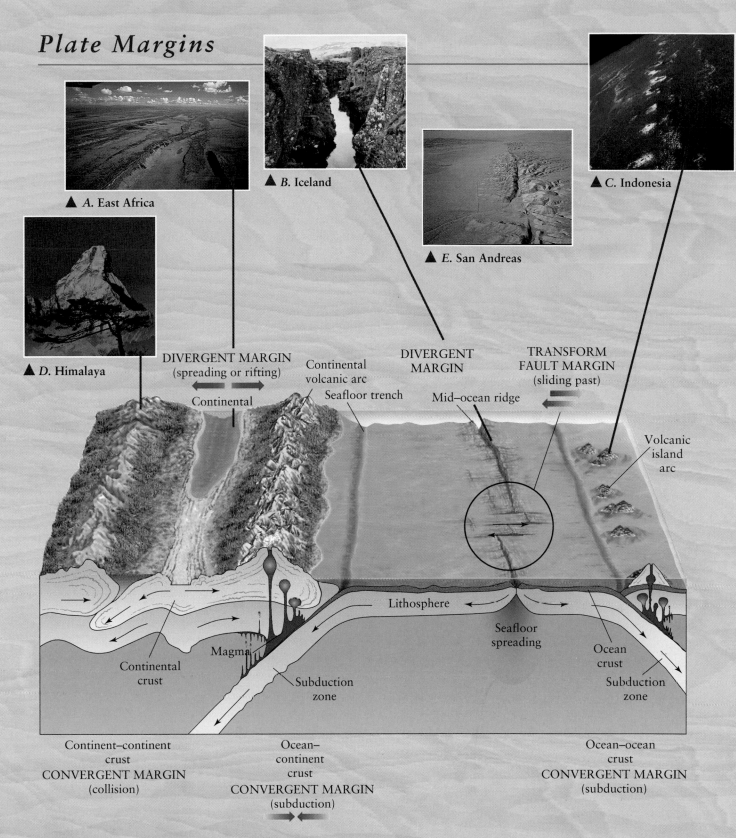

▲ A. East Africa

▲ B. Iceland

▲ C. Indonesia

▲ D. Himalaya

▲ E. San Andreas

DIVERGENT MARGIN
(spreading or rifting)

Continental

Continental
volcanic arc

Seafloor trench

DIVERGENT
MARGIN

Mid–ocean ridge

TRANSFORM
FAULT MARGIN
(sliding past)

Volcanic
island
arc

Lithosphere

Magma

Seafloor
spreading

Ocean
crust

Continental
crust

Subduction
zone

Subduction
zone

Continent–continent
crust
CONVERGENT MARGIN
(collision)

Ocean–
continent
crust
CONVERGENT MARGIN
(subduction)

Ocean–ocean
crust
CONVERGENT MARGIN
(subduction)

**Figure 4.14**
**PLATE MARGINS**
The different types of plate margins–divergent, convergent, and transform–are shown here. On continents, divergent (spreading or rifting) margins create rift valleys, as in A, which shows the Great Rift Valley in East Africa. Divergent margins in ocean basins create midocean ridges; in photo B, the Mid-Atlantic Ridge is shown cutting across Iceland. Ocean-ocean and ocean-continent convergent margins create subduction zones marked by deep oceanic trenches, strong earthquakes, and strings of volcanoes and volcanic islands (C). Continent-continent collisions are marked by great, folded mountain ranges (D). Transform fault margins, such as the San Andreas Fault (shown in photograph E), are zones of sliding-past motion; they are sometimes marked by a long, linear valley.

As shown in Figure 4.15, concentrated earthquake activity occurs at plate margins. In fact, geologists have deduced the shapes and outlines of the plates by examining the locations of earthquakes. More important, different types of earthquakes occur at different types of plate margins. For example, earthquakes that occur along divergent margins are usually fairly weak and have shallow foci (Fig. 4.17A). This is consistent with the suggestion that molten material lies just under the surface along divergent margins like midocean ridges, because earthquakes can occur only in rocks that are cold and brittle enough to break.

Other types of plate margins also have characteristic earthquake activity. Transform fault margins like the San Andreas fault usually have earthquakes that occur at shallow to intermediate depths (Fig. 4.17B). These earthquakes can be

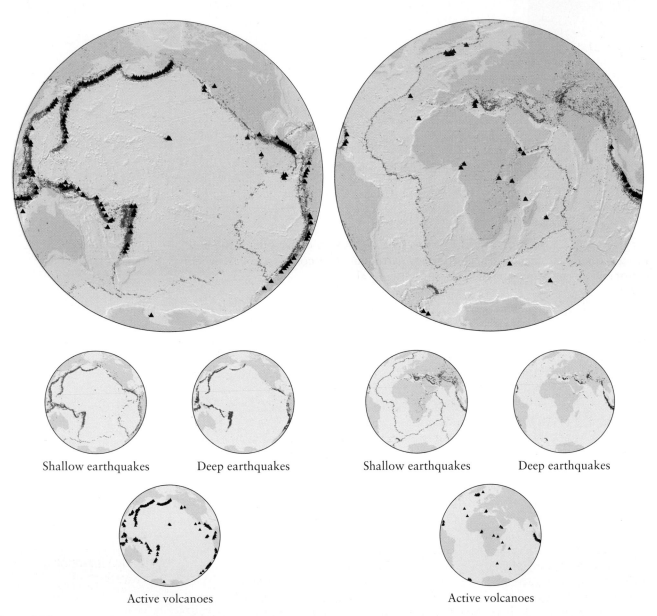

**Figure 4.15**
**EARTHQUAKES AND**
**VOLCANOES OUTLINE PLATE MARGINS**
The locations of earthquakes and volcanoes have enabled scientists to determine the shapes and boundaries of the Earth's lithospheric plates.

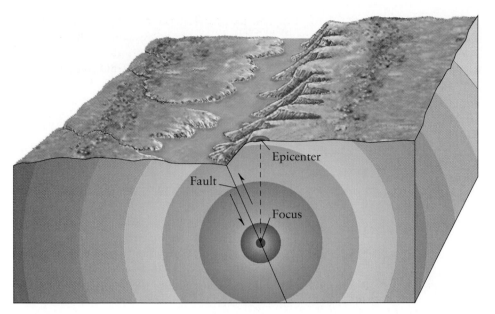

**Figure 4.16**
**EARTHQUAKE FOCUS AND EPICENTER**
The focus of an earthquake is the actual site of first movement along a fault and the center of energy release, which usually occurs at some depth. The epicenter is the point on the Earth's surface directly above the focus; it is used to specify the map location of an earthquake.

very powerful, consistent with the fact that the blocks on either side of the fault are grinding past each other. Deep-focus earthquakes in collision zones, where continents crumple into great mountain ranges, also can be very powerful (Fig. 4.17C).

In subduction zones, the entire surface along which one oceanic plate moves downward relative to the other plate is marked by powerful earthquakes. These earthquakes usually have shallow foci near the oceanic trench, where subduction begins. The earthquake foci become deeper and deeper along the descending edge of the subducting oceanic plate (Fig. 4.17D). If you look back at Figure 4.15, you will see that the different-colored dots indicate earthquake foci of different depths. In some areas there are lines of shallow earthquakes, corresponding to the locations of deep oceanic trenches. Moving away from the trench in the direction of subduction, we find that the earthquake foci become deeper and deeper (Fig. 4.18). These zones of shallow- to deep-focus earthquakes, called *Benioff zones,* first alerted scientists to the presence and geometry of subduction zones.

## The Search for a Mechanism

Although virtually all geologists accept the theory of plate tectonics, some questions remain. Among these questions are: What, exactly, causes plate motion? How does the mantle interact with the crust? What makes subduction occur? Scientists have a basic understanding of these processes, but the details have not been completely worked out. We do know that thermal movement in the mantle is at least partly responsible for the movement of lithospheric plates. We also know that movement in the mantle is caused by the release of heat from inside the Earth. Let's examine the Earth's heat-releasing processes and consider how they might cause plate motion.

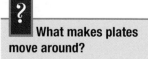
**What makes plates move around?**

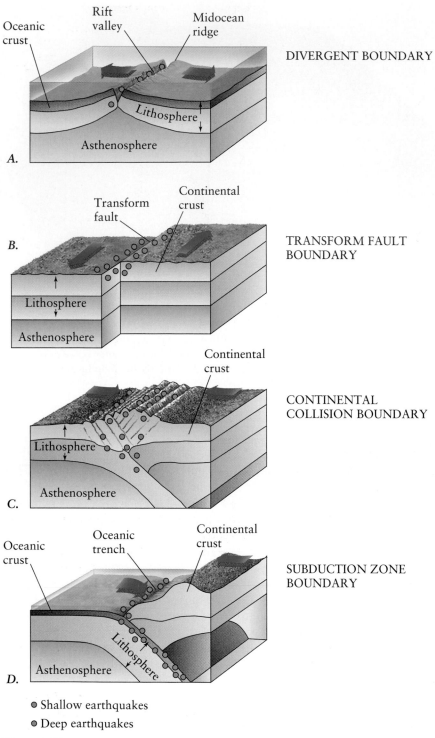

**DIVERGENT BOUNDARY**

**TRANSFORM FAULT BOUNDARY**

**CONTINENTAL COLLISION BOUNDARY**

**SUBDUCTION ZONE BOUNDARY**

- Shallow earthquakes
- Deep earthquakes

**Figure 4.17**
**EARTHQUAKES AND PLATE MARGINS**
The dots on this map indicate earthquake foci.

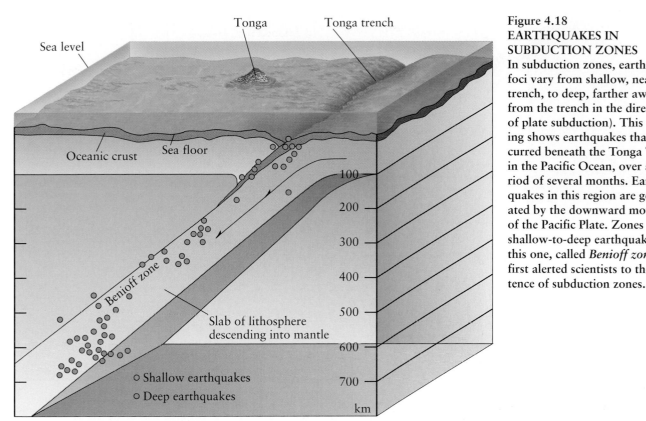

Figure 4.18
**EARTHQUAKES IN SUBDUCTION ZONES**
In subduction zones, earthquake foci vary from shallow, near the trench, to deep, farther away from the trench in the direction of plate subduction). This drawing shows earthquakes that occurred beneath the Tonga Trench in the Pacific Ocean, over a period of several months. Earthquakes in this region are generated by the downward movement of the Pacific Plate. Zones of shallow-to-deep earthquakes like this one, called *Benioff zones*, first alerted scientists to the existence of subduction zones.

## The Earth's Internal Heat

The temperature inside the Earth is high—about 5000°C (more than 9000°F) in the core. Some of this heat is left over from the Earth's beginnings, and some is constantly being generated by the decay of radioactive elements inside the Earth. If this heat were not released into outer space, the Earth would eventually become so hot that its entire interior would melt.

Some of the Earth's internal heat makes its way slowly to the surface through the process of *conduction*, in which heat energy is passed from one atom to the next. However, conduction is a slow way to transfer heat. It is faster and more efficient for an entire packet of hot material to be transported from the Earth's interior to the surface. This is similar to what happens when a fluid boils on a stovetop. If you watch a fluid such as water or spaghetti sauce as it boils, you will see that it turns over and over as packets of hot material rise from the bottom of the pot to the top (Fig. 4.19A). When it reaches the surface, the hot fluid loses its heat, moves sideways as it cools, and is then swept back down to the bottom of the pot, where it is reheated. The circuitous motion of heated material from bottom to top and back is called a *convection cell,* and this mechanism of heat transfer is termed **convection.**

## Convection as a Driving Force

Even though the Earth's mantle is composed mostly of solid rock, it is so hot that it, too, releases heat through convection. Rock deep in the mantle heats up and expands, becoming buoyant. As a result, the rock moves toward the surface—very, very slowly—in huge convection cells of solid rock (Fig. 4.19B). Near the surface (primarily within the asthenosphere), the hot rock moves laterally while losing some of its heat. The lateral movement of hot rock in the asthenosphere is

**?**  **How is heat released from inside the Earth?**

**convection**  The process whereby hot, less dense materials rise upward and are replaced by cold, more dense, downward-flowing material.

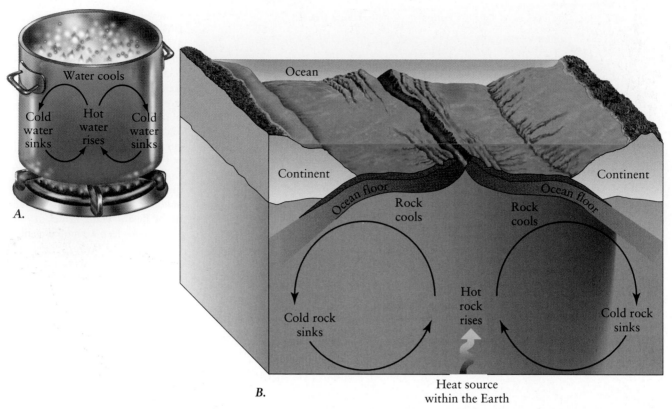

**Figure 4.19**
**MANTLE CONVECTION AND PLATE MOTION**
Convection may shape the Earth's surface by slowly moving its crustal plates. In *A,* you see convection at work on a small scale, in a saucepan of water. The heated water expands, becoming less dense and rising as a result. As it rises, it cools, flows sideways along the top surface, and eventually sinks. At the bottom of the saucepan it is reheated and cycles through the convection cell again. *B* shows convection as it is thought to occur in the Earth. Although much slower than convection in a saucepan, the principle is the same. Hot rock rises slowly and plastically from deep inside the Earth, cools, flows sideways, and sinks. The rising hot rock and sideways flow are believed to be part of the driving mechanism for plate motion. They create surface features like deep ocean trenches and midocean ridges.

thought to be the primary cause of movement of lithospheric plates. Cool rock is dense (heavier than hot rock); therefore, it tends to sink back into the deeper parts of the mantle. This convection cycle provides an efficient way for the Earth to rid itself of some of its internal heat. Convection and the lateral movement of plates near the surface create some of the most distinctive geologic and topographic features of the Earth's surface: the deep trenches where plates subduct into the mantle; the midocean ridges and continental rift valleys where plates split apart; and the high, folded and crumpled mountain chains.

Convection in the mantle is not nearly as simple as convection in a pot on a stovetop. Some of the most challenging unanswered questions about plate tectonics have to do with the exact nature of this process. Does the whole mantle convect as a unit, or is the top part of the mantle convecting separately from the bottom? In subduction zones, are lithospheric plates dragged down into the mantle or do they sink under their own weight? What are the shape and distribution of convection cells in the mantle? We know, for example, that sometimes hot rocks don't travel in neatly packaged cells. Instead, they may rise in long, thin blobs called **mantle plumes.** (The islands of Hawaii were formed atop such a

**mantle plume** A long, thin body of hot rock rising through the mantle.

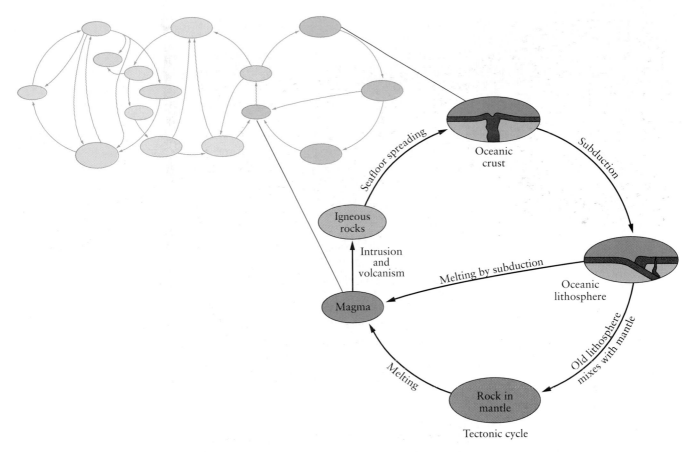

**Figure 4.20**
**THE TECTONIC CYCLE**
Internal Earth processes are highlighted on this diagram of the tectonic cycle. The chapters of Part Two focus principally on these processes. Compare this diagram to Figure 1.18.

mantle plume; see *Geology Around Us*.) Do such plumes originate in the mantle, or perhaps even deeper in the Earth? Scientists seek the answers to these and other important questions through observations of the natural world and through computer modeling.

Recall that in chapter 1 we introduced the concept of interacting cycles in the Earth system (Fig. 1.19). In chapter 4 and the subsequent chapters of Part Two, we focus primarily on internal Earth processes. These processes are summerized in the righthand cycle of our Earth system diagram, as shown here in Figure 4.20. You can refer to this diagram, and to Figure 1.19, when you want to review these processes and their relationship to other parts of the Earth system.

# WHAT'S AHEAD

Plate tectonics affects all life on Earth. Sometimes it does so in subtle ways—by influencing climate through the distribution of continents and ocean basins, for example. Other times it does so in obvious ways, such as through the effects of major earthquakes or volcanic eruptions. We describe plate tectonics as a "unifying" theory of how the Earth system works. By this we mean that the plate tectonics model brings together many diverse observations of the Earth's geologic features and unifies them into a single, reasonably straightforward "story." As a

**?**
**What geologic features are explained by plate tectonics that were not adequately explained by previous theories?**

# The Hawaiian Islands: A Record of Plate Motion

An ancient Hawaiian legend tells of the goddess **Pele** and her older sister Namakaokahai. They fought, and the scars of that battle are still visible in the Pacific Ocean today, in the form of the Hawaiian Islands. Pele eventually settled in the summit crater of the volcano Kilauea, from which she still vents her anger periodically. Her hair is visible as filamentous strands of volcanic glass, and her tears rain down as tiny glass beads (Fig. B4.1). Pele's temper is most clearly manifested when she argues with one of her lovers. As the legends relate:

> The marriage of Pele, goddess of Earth and fire, and Kamapua'a, god of water, was short and violent. In a rage, she routed him from her crater of fire and chased him with streams of lava into the sea.

The Hawaiian Islands are a vivid reminder of the awesome power of the Earth's volcanic activity. They also provide valuable information about the rate and direction of movement of the Pacific Plate. If you examine the ages of volcanic rocks from the Hawaiian Islands (Fig. B4.2), you will discover that they vary systematically. The youngest rocks (ranging from rocks that are forming today to rocks 0.8 million years of age) are found on the "big island," Hawaii. The rocks on the next island, Maui, are a bit older, and those on Molokai and Oahu are older still. The oldest rocks (approximately 5.5 million years old) are found on Kauai, located at the northwest end of the chain of islands. Interestingly, if you look beneath the ocean surface just past Kauai to the northwest, you will find that the chain of islands appears to continue for quite a distance. (The "islands" in this part of the chain are not quite tall enough to poke up above sea level, so they are called *seamounts* instead of islands.)

**Figure B4.1**
**PELE'S TEARS AND HAIR**
These tiny needles are not straw, but hair-like filaments of volcanic glass that were formed when magma from Hawaii's Kilauea volcano cooked very quickly. They are said to represent the hair of the goddess Pele. Tiny beads of volcanic glass represent Pele's fallen tears.

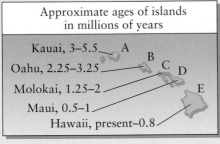

Approximate ages of islands
in millions of years

Kauai, 3–5.5    A
Oahu, 2.25–3.25
Molokai, 1.25–2
Maui, 0.5–1
Hawaii, present–0.8

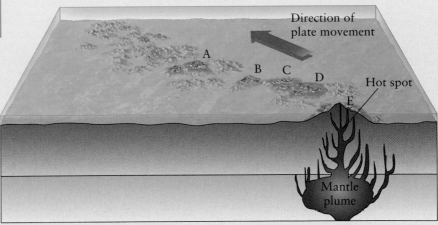

**Figure B4.2**
**HOT SPOT IN HAWAII**
The Hawaiian Island chain of volcanoes formed above a deep-seated source of hot material within the mantle, is called a plume. The hot spot over the plume has remained stationary for at least 70 million years. Meanwhile the Pacific Plate has moved over it, carrying with it the old volcanic landforms built over the hot spot. The Hawaiian volcanoes are progressively older with increasing distance from the hot spot. The currently active volcanoes, Mauna Loa and Kilauea, and Loihi (a submarine volcano) are now located on top of the hot spot. Eventually they, too, will be carried off with the moving plate, and a new volcano will be built over the hot spot.

The explanation for these observations is that the island of Hawaii is currently sitting on top of a long, thin plume of hot material rising from deep within the mantle. This is called a *mantle plume*. Wherever there is an active mantle plume, volcanic activity occurs at the *hot spot* above it. This is the origin of the volcanic activity at Kilauea and Mauna Loa, the volcanoes that are currently active on the island of Hawaii. Mantle plumes and the hot spots they create are thought to be rooted in place within the mantle. This means that lithospheric plates are moving over them. Although Kilauea and Mauna Loa are now sitting on top of a hot spot, the motion of the Pacific Plate will eventually carry these volcanoes off the hot spot, toward the northwest. Their volcanic activity will cease. This has happened to each of the islands and seamounts in the Hawaiian Island chain. Each was volcanically active during the period of time when it sat on top of the hot spot, and each, in turn, has been carried off toward the northwest with the rest of the Pacific Plate. By looking at the progression of ages in the Hawaiian Island chain of islands and seamounts, geologists have been able to determine how long this mantle plume has been rooted in its place (about 70 million years), the direction of movement of the Pacific Plate (mostly north-northwest), and how fast the plate has been moving (about 10 cm/yr, or 4 in/yr). As the island of Hawaii moves off the hot spot, we can expect new volcanic structures to form in its place. Indeed, a new volcano, named Loihi, is already beginning to form in underwater eruptions off the southeast coast of Hawaii.

model, plate tectonics is truly representative of the Earth system science approach, because it illustrates how solid-Earth processes are integrated with and fundamentally affect all other parts of the Earth system.

We will spend much of the rest of this book examining different rock-forming and rock-modifying processes in the context of plate tectonics and from a systems perspective. Geologists are far from answering all questions about plate tectonics, and certainly this model hasn't answered all of our questions

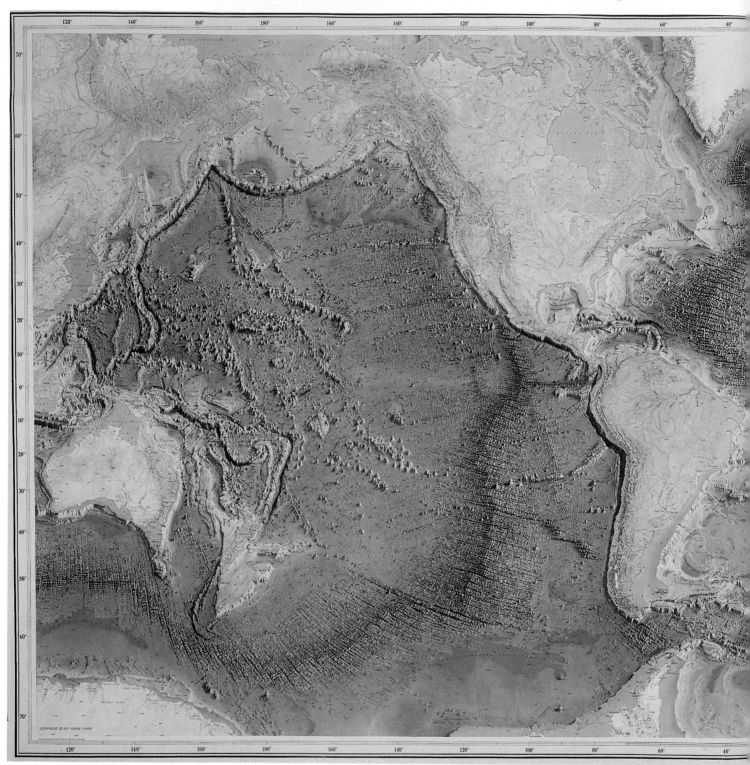

about how the Earth system works. But plate tectonics does explain many of our observations about the Earth, as you will see in upcoming chapters. For example, the plate tectonics model explains how and why great mountain chains form where they do. It explains rift valleys, deep ocean trenches, midocean ridges, and other aspects of the topography of the ocean floor (Fig. 4.21). It explains the locations of earthquakes, and it even tells us the specific characteristics of earthquakes, such as the depths of their foci in Benioff zones. As you will see in chap-

**Figure 4.21**
**TOPOGRAPHY OF CONTINENTS AND SEA FLOOR**
If we could drain all the water from the oceans, we would see the sea floor as vast, flat areas, long chains of underwater mountains, and deep trenches like those east of Australia. Note the difference in elevation between the continents and the oceans, and the various features that are explained by the plate tectonics model, including the midocean ridges, deep oceanic trenches, and high mountain ranges.

ter 6, plate tectonics explains the distribution of different types of volcanoes around the world; the shapes of continents and ocean basins; the formation of metamorphic rocks (chapter 10); and the locations of huge, sediment-filled basins (chapter 11). Before we go on to consider these topics, chapter 5 offers a closer look at the inside of the Earth and the techniques scientists use to learn about the Earth's deep interior.

## Chapter Highlights

**1.** A revolution in geology began almost 100 years ago with the suggestion that the continents had not always been in their present positions; this became known as the **continental drift** hypothesis. By the 1960s, enough evidence had been amassed to convince the majority of scientists that continental drift had actually occurred.

**2.** The evidence for continental drift included the fit between the coastlines of adjacent continents; the match of bedrock geology and continuity of geologic features across former continental joins; the match of glacial deposits across former continental joins; and distinctive fossil assemblages in areas that are now widely separated by ocean basins.

**3.** The debate about continental drift received a boost from paleomagnetic evidence indicating that the continents had moved relative to the position of the magnetic north pole. Even more convincing was the discovery that Europe and North America had apparently moved as a single continent for a period of at least several hundred million years, prior to splitting apart.

**4.** The piece of evidence that finally tipped the scale in favor of continental drift was the discovery of bands of magnetized rock on the sea floor with alternating normal and reversed polarities, aligned symmetrically on either side of the midocean ridges. The only plausible explanation for this discovery was that **seafloor spreading** had occurred.

**5.** After continental drift was generally accepted, it remained to put everything together into a simple, coherent model. That model is **plate tectonics**—the theory that the **lithosphere** has fragmented into several large plates, which are moved and jostled by movement in the underlying hot mantle. Because they are part of the lithospheric plates, the continents move along with them.

**6.** One of the most important aspects of plate tectonics is the nature of interactions along plate margins.

There are three types of plate margins: (1) **divergent margins,** also called rifts or spreading centers, which are fractures in the lithosphere where two plates move apart, causing midocean ridges and seafloor spreading in the oceanic setting and great rift valleys in the continental setting; (2) **convergent margins,** where two plates move toward each other, forming **subduction zones** and deep oceanic trenches whenever oceanic crust is involved, and **collision zones** when only continental crust is involved; and (3) **transform fault margins,** which are fractures in the lithosphere where two plates slide past each other.

**7.** Many earthquakes and active volcanoes are located along plate margins. Studies of the locations and characteristics of earthquakes have enabled scientists to determine the shapes of lithospheric plates and other characteristics of plate margins. For example, the varying depths of earthquakes along Benioff zones alerted scientists to the presence of subduction zones.

**8.** An important unanswered question about plate tectonics is what drives plate motion. The release of heat from the interior of the Earth in huge **convection** cells is at least partly responsible for the movement of lithospheric plates across the Earth's surface. But this is a complex process, and the details of mantle convection, plate motion, and mantle–crust interaction are not yet completely understood. For example, it is known that hot rocks from the mantle sometimes rise to the surface as long, thin **mantle plumes** rather than always traveling in neat convection cells.

**9.** Plate tectonics is a "unifying theory" because it is supported by the weight of geologic evidence, is accepted by the majority of scientists, and offers a coherent explanation for so many of the geologic features of the Earth system. However, there are still many questions about plate tectonics that are unresolved.

## ► *The Language of Geology*

- collision zone 102
- continental drift 89
- convection 107
- convergent margin 102

- divergent margin 102
- fault 102
- lithosphere 98
- mantle plume 108

- plate tectonics 99
- seafloor spreading 98
- subduction zone 102
- transform fault margin 102

## ► *Questions for Review*

1. What was Pangaea?
2. Describe the main lines of evidence that supported Wegener's hypothesis of continental drift.
3. Explain why a shoreline is not always the same as the true edge of a continent.
4. What is apparent polar wandering, and why is it such a powerful line of evidence in support of continental drift?
5. Describe how paleomagnetic bands develop in rocks at midocean ridges.
6. What are the three main types of plate margins?

Describe each type and the landforms associated with it.
7. How hot is the Earth's core?
8. How does convection act to release heat from the Earth? How does it cause plate motion?
9. What is a fault?
10. How are mantle plumes and hot spots used to measure plate motion?
11. What is a Benioff zone, and what is the relationship between Benioff zones and subduction zones?
12. How fast do lithospheric plates move?

## ► *Questions for Thought and Discussion*

1. What was wrong with the "contracting Earth" and "expanding Earth" theories that preceded the theory of plate tectonics?
2. Why was the discovery of paleomagnetic bands on the Atlantic sea floor such an important turning point in the development and eventual acceptance of the theory of plate tectonics?

3. What are some of the questions about plate tectonics that remain unanswered today? Research your answer.
4. Why do we call plate tectonics a "unifying theory"?

## *Virtual Internship: You Be the Geologist*

Your GiA CD-ROM contains a virtual internship through which you can learn more about volcanoes and volcanic eruptions.

For an interactive case study on the East African Rift and plate tectonics, visit our Web site.

# · 5 ·

# EARTHQUAKES AND THE EARTH'S INTERIOR

It's 5:00 A.M. and dawn is breaking, but you're sound asleep in a warm, comfortable bed. Suddenly, you're hurled to the floor. The bed is rattling and the floor is heaving ominously. You hear a cracking sound in the walls as you stumble blindly. You reach the door and just hang on. Thirty seconds later, the nightmare stops as abruptly as it began. Everything is still as you peer outside into the gray light of day. The streetlights are out, and some of the wires are crackling and sparking. You see smoke and dust rising and piles of rubble along the street. You smell gas, and you hear screams and sirens. In those few seconds your life has changed. You have lost your faith in the solidity of the Earth beneath your feet.

This imaginary scenario and others like it become reality for the hundreds, often thousands of people each year who experience a damaging earthquake. Over the past few decades, geologists have greatly improved their understanding of the causes, frequency, and distribution of earthquakes. From each major event we learn more about earthquakes and how they cause damage. Improvements in the structural engineering of buildings and highways have helped people in earthquake-prone areas to prepare for such occurrences. Yet true preparedness would require scientists to issue a prediction and early warning; unfortunately, this capability remains tantalizingly out of reach.

◆

### In this chapter you will learn about

- The internal structure and composition of the Earth
- The techniques used by geologists to study the Earth's interior
- What causes earthquakes and how geologists measure them
- How geologists are working toward the elusive goal of earthquake prediction
- How earthquakes reveal important information about the Earth's interior

*Earthquakes: Bad News, Good Data*
Although earthquakes can cause devastation, they also enable scientists to learn more about the interior of the Earth. This highway collapsed during a major earthquake in Kobe, Japan, in 1995.

*How could you find out what's deep inside the Earth?*

*. . . This chapter will give you some ideas.*

How can we live on this planet and not know everything about its interior? The obvious answer is that the inside of the Earth is not very accessible. We know a lot about the Earth's surface because we can move around on it and collect samples for study. But the Earth's interior is mostly beyond our reach. We *do* know that the Earth—like the other terrestrial planets—is a *differentiated* body, which means that early in its history it separated chemically into a *core, mantle,* and *crust.* We know that the chemical composition of the mantle and core is different from that of the crust. We also know that the rocks of the mantle and core have very different *physical* properties from those of the crust. But what exactly are the characteristics of those deeper layers, and how were they discovered? How did scientists come to realize that what happens deep inside the Earth influences what happens on the surface? In this chapter we will try to answer these questions.

**If you wanted to study the interior of the Earth, how would you go about getting a sample of it?**

# LOOKING INTO THE EARTH

Basically, there are two ways of studying something scientifically. You can study by *direct sampling;* that is, you can examine samples that you can hold in your hand and analyze in a laboratory. Or you can study something indirectly or *remotely,* using techniques that enable you to collect and analyze information without actually coming into contact with the object you are studying. Photography is an example of a remote technique: the camera collects information (reflected light) about an object without actually touching it. Medical techniques such as X rays allow doctors to study the inside of the body without having direct access to it.

Geoscientists use both direct and remote techniques to study the inside of the Earth. The remote techniques are the most important, because most of the Earth's interior is not accessible for direct sampling. Let's look briefly at the techniques used by geoscientists to study the inside of the Earth.

**Why can't geoscientists just drill a deep hole to get samples from inside the Earth?**

## Direct Sampling

How would you go about obtaining samples from the Earth's interior? We already know quite a lot about the crust, so your main goal would be to get samples of the deeper layers—the mantle and the core.

### Drilling Through the Crust

One thing you might try is to drill a very deep hole and collect samples from the bottom of the hole. Indeed, scientists have tried this approach (Fig. 5.1). The deepest hole ever drilled (in the Kola Peninsula of Russia) reached a depth of almost 12 km (about 7.5 mi). Recall from chapter 1 that the Earth's crust varies from an average thickness of 8 km (5 mi) for oceanic crust to an average of 45 km (28 mi) for continental crust. A 12-km hole sounds just about right for sampling the top part of the mantle—or does it?

Figure 5.1
**DRILLING THROUGH THE CRUST**
Scientists have acquired much useful data by drilling holes deep into the crust of the Earth. These cylindrical cores of rock, which were drilled from oceanic crust by the Deep Sea Drilling Program, have been split in half for analysis.

There are problems with using drilling as a means of sampling the Earth's deeper layers. One problem is that areas where the crust is thin tend to have high heat flow. In other words, if you try to drill a hole through thin oceanic crust all the way to the mantle, you will quickly encounter temperatures that will destroy your drilling equipment. Another problem with oceanic crust is that it's deep under water, which makes drilling more difficult.[1] The only place where the rocks are both accessible and cool enough for drilling to proceed to great depths is where the crust is very thick—that is, on the continents. The hole in the Kola Peninsula went through almost 12 km of thick continental crust and never even came *close* to reaching the mantle. Thus, while drilling has yielded much interesting and useful information about the composition of the crust, the distribution of heat, the flow of fluids through the crust, and the way materials change with depth, it isn't useful for sampling below the level of the crust—at least not with today's drilling technologies.

### Mantle Magmas and Xenoliths

There is another way to get samples from deep within the Earth: We can wait for them to be delivered to the surface by natural geologic processes. Molten rock, or *magma,* is formed deep in the Earth in areas where the temperature is higher than the temperatures at which rocks begin to melt. The characteristics of magmas are influenced by the nature of their source regions, as we will learn in chap-

---

[1] Drilling on the sea floor is difficult but not impossible. The Deep Sea Drilling Project has yielded much useful information about the nature of oceanic crust.

# Messengers From Far Below

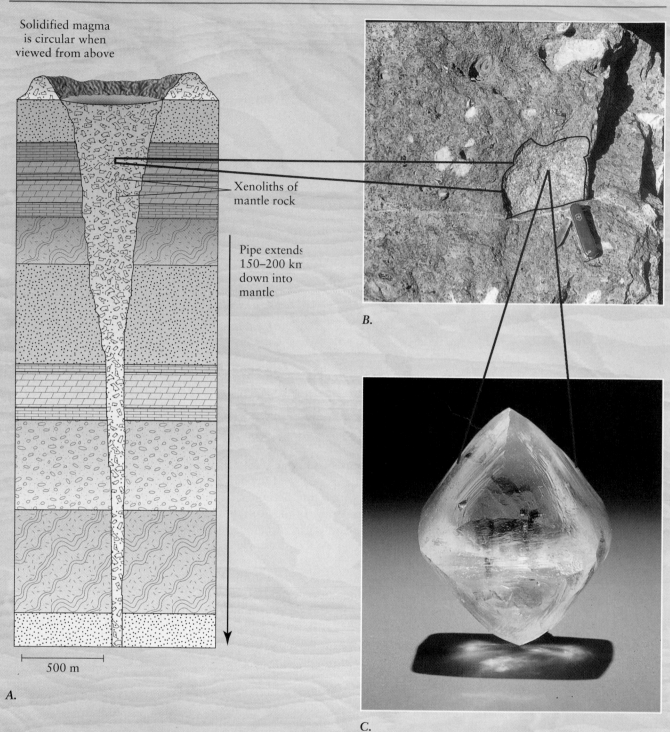

Solidified magma is circular when viewed from above

Xenoliths of mantle rock

Pipe extends 150–200 km down into mantle

500 m

A.

B.

C.

Figure 5.2
**MESSENGERS FROM THE MANTLE**
Magma is generated deep in the mantle. As the magma ascends, fragments of mantle rock can be ripped off by the rising magma and carried to the surface as xenoliths. *A.* One type of rock that commonly contains mantle xenoliths is kimberlite, which occurs in long, thin, pipelike bodies extending from deep in the mantle up to the surface. *B.* This angular fragment is a mantle xenolith in a kimberlite from a kimberlite pipe in South Africa. By studying mantle xenoliths, scientists can learn more about the rocks deep within the Earth. *C.* Like all natural diamonds, this one was formed deep within the Earth under conditions of very great pressure. This is the Oppenheimer diamond, a rare yellow stone the size of a small egg (4.5 cm, or almost 2 in long). The stone shows the eight-sided (*octahedral*) form that is characteristic of natural diamonds—it hasn't been cut by a jeweler.

ter 6. By studying the characteristics of magmas that originate at great depth, scientists can learn about the "environment" in their source regions—that is, the temperature, pressure, and composition of the mantle. Furthermore, when magma rises toward the surface, fragments of the surrounding rock are sometimes ripped off and carried along. We call these fragments **xenoliths,** from the Greek words *xenos* (foreigner) and *lithos* (stone).

**xenolith** A fragment of rock that is carried to the surface by magma.

An unusual type of rock called *kimberlite* forms from magma that originates at depths of 100 to 300 km (almost 200 mi)—deeper than even the thickest portions of the Earth's crust. Kimberlite magmas rise quickly and violently to the surface through long, thin volcanic vents called *pipes* (Fig. 5.2A). It is common for kimberlite magmas to rip fragments of mantle rock from their source regions, carrying the fragments to the surface as xenoliths (Fig. 5.2B). Such occurrences provide a perfect opportunity to study samples from deep within the Earth. Mantle xenoliths may contain minerals, such as diamonds, that can only form under conditions of very high pressure and temperature (Fig. 5.2C), correlating to depth ranges of 100 to 300 km.

## Remote Study of the Earth's Interior

Because access to the Earth's interior is so limited, scientists must rely on indirect techniques to study it. Techniques from a variety of disciplines are used to probe the inside of the Earth without actually coming into contact with it.

**?** **How can we study the Earth's interior without actually getting samples of it?**

### *Astronomy*

The same astronomical techniques that are used to study the interiors of other planets can be used to study the Earth's interior. The first step is to determine the planet's *mass*. This can be done by observing the planet's gravitational influence on other planets and satellites and applying the physical laws that govern planetary motion. Second, we need to know the *diameter* of the planet. Today this is done by satellites and global positioning systems. But the problem of determining the Earth's diameter was more difficult for early geoscientists, who had to calculate it painstakingly from careful surveying measurements. Knowing the dimensions of the planet, we find it relatively simple to figure out the *volume* of the body as long as we know its *shape*—another problem that seems trivial now but was very challenging for early geoscientists.[2] Once we know the mass and volume of the planet, we can figure out its average *density* (density equals mass divided by volume).

These kinds of measurements tell us about the size and mass of the planet, but what do they reveal about its interior? One of the most important things they can tell us is whether or not material is distributed evenly throughout the planet. On the Earth, the rocks at the surface are very light (low-density) compared to the planet as a whole. The surface rocks have an average density of about 2.8 g/cm³, whereas the Earth's overall density is 5.5 g/cm³. (For comparison, water has a density of 1.000 g/cm³ at 4°C [39°F].) For the planet as a whole to have such a high density, there must be a concentration of denser material inside the planet.

### *Geochemistry and the Meteorite Analogy*

The use of chemical techniques to study Earth materials and processes is called **geochemistry.** Sometimes geochemists conduct experiments in which rocks and

**geochemistry** The use of chemical techniques to study Earth materials and processes.

---

[2]Although we think of the Earth as a sphere, in fact it is not exactly spherical. Because it is rotating, the pull of gravity is stronger at its poles than at the equator, so the Earth bulges out at the equator. The science of measuring the exact shape and size of the Earth is called *geodesy*.

A.

B.

**Figure 5.3**
**GEOCHEMISTRY LAB**
In this geochemistry laboratory, a laser microscope *A* is used to study samples of rocks and minerals under the extreme temperatures and pressures characteristic of the mantle. In *B* (greatly magnified), a sample of the mineral pyroxene is being heated by lasers and squeezed at enormous pressure between the points of two diamonds, in an instrument called a diamond anvil.

**geophysics** The application of physics to the study of the Earth.

**gravity anomaly** A measurement of gravity that differs from the expected value.

minerals are subjected to extreme pressures and temperatures (Fig. 5.3). In this way, they can learn what types of chemical compounds are formed and what kinds of processes may occur under the conditions found at great depths in the Earth.

Another geochemical technique involves *meteorites,* samples of extraterrestrial material that have fallen to Earth. Most meteorites were formed at about the same time and, in some cases, the same part of the solar system as the Earth. Some meteorites are *primitive*—that is, they have never been affected by melting or differentiation since the time of their formation. The Earth, in contrast, has been greatly changed by differentiation and other geologic processes, and its chemical elements have been moved around and redistributed. If we examine primitive meteorites, we can get an idea of what the composition of the Earth, as a whole, must have been like prior to the redistribution of its elements. Other meteorites—the *irons, stony-irons,* and some kinds of *stony meteorites*—are thought to be fragments of planetary objects that differentiated into core, mantle, and crust, just as the Earth did. It is highly suggestive that a core with the composition of a typical iron meteorite—that is, a core composed mostly of iron and nickel metal—would provide just the right concentration of dense material at the center of the Earth to yield the correct overall density of 5.5 g/cm$^3$.

## Geophysics

The application of physics to the study of the Earth is called **geophysics.** One geophysical technique used to study the Earth's interior involves measuring variations in the Earth's gravity. *Gravity* is the attractional force that causes a downward pull on any object at the Earth's surface. Variations in gravity are measured with a *gravimeter,* which is basically a weight suspended from a spring attached to sensitive measuring devices (Fig. 5.4). The stronger the pull of gravity, the more the weight is pulled down and the more it stretches the spring. The stretching of the spring thus provides a measure of the local pull of gravity.

How can gravity measurements tell us about the material inside the Earth? If the rock between the gravimeter and the center of the Earth were the same everywhere, the force of gravity would be the same for every point on the Earth's surface (with adjustments for variations in topography). In fact, gravity measurements reveal significant variations, called **gravity anomalies,** caused by underlying bodies of rock with differing densities.[3] Because the pull of gravity de-

---

[3]Gravity *feels* the same to us everywhere on the Earth because the variations are so small that we cannot detect them without sensitive instruments.

pends on mass, a heavy (dense) mass will cause a greater pull than a light (less dense) mass (Fig. 5.5). A greater-than-average pull is called a *positive anomaly,* while a less-than-average pull is called a *negative anomaly.* By making gravity measurements over the Earth's entire surface, we can get a good idea of the distribution of dense and less dense materials underground. One of the most important things this type of measurement has revealed is the deep-seated structure of mountain chains, discussed in greater detail in chapter 11.

Measurements of variations in magnetism can also yield important information about the Earth's interior. Recall from chapter 3 that the Earth is surrounded by a magnetic field, similar to the much smaller magnetic field that surrounds a bar magnet. Therefore, there must be a source of magnetism inside the Earth. We could think of the Earth as having a huge dipole magnet (a bar magnet with north and south poles) at its center, offset slightly from the geographic north and south poles. The problem with this analogy is that bar magnets lose their magnetism at temperatures above about 500°C (or 900°F), and the temperature deep inside the Earth is *much* higher than this.

If there isn't a bar magnet at the center of the Earth, then how does the Earth generate its magnetic field? The most widely accepted explanation is the *dynamo* hypothesis. A simple experiment shows that the movement of a conducting material through the magnetic field of a bar magnet will generate electricity. This can be demonstrated, for example, by rotating a copper disk through the magnetic field of a bar magnet. If the bar magnet is replaced by a coil of wire, the electric current will continue to sustain a magnetic field in the coil. As long as the conducting metal disk keeps rotating, the electric current will be generated; this is called a *self-exciting dynamo.* Through a mechanism that is similar but much more complex, the movement of an electrically conducting liquid inside a planet could generate a magnetic field. In general, it is assumed that planets with strong magnetic fields have cores that are liquid (because liquids can move freely) and metallic or otherwise conducting. In the case of the Earth, it is believed that the core is composed mainly of metallic iron and that movement within the liquid portion of the iron core generates the magnetic field. Note that this is consistent with information gained through the study of meteorites.

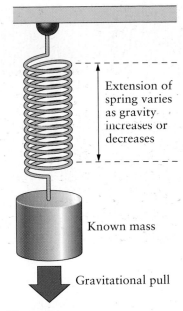

**Figure 5.4**

**A GRAVIMETER MEASURES GRAVITY**

A gravimeter is a heavy mass of metal suspended on a sensitive spring. Gravity varies slightly from place to place, depending on the rocks below. As the pull of gravity varies, the mass in the gravimeter exerts a greater or lesser pull on the spring, extending the spring to a greater or lesser extent. The mass of metal and the spring are contained in a vacuum along with sensitive measuring devices to record the variations in spring extension.

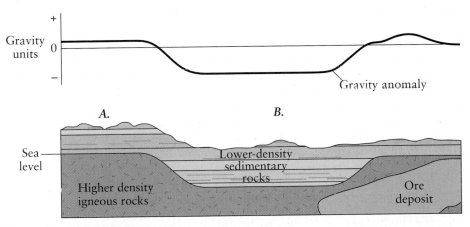

**Figure 5.5**

**GRAVITY ANOMALIES**

*A.* These areas have positive gravity anomalies, with gravity slightly greater than average. A mass suspended from a spring is pulled downward more strongly over a body of dense material, such as an ore deposit or a mountain, than in adjacent areas. *B.* This area has a negative gravity anomaly. The mass in the gravimeter is pulled down less strongly over low-density materials.

## GEOLOGY AROUND US

### All Shook Up:
### A Postcard from California

*In this excerpt from the April 1994 issue of the magazine* Saturday Night, *Ian Brown has written eloquently of his family's experience of a major earthquake in California.*

**We heard the earthquake first,** as we always do: all the metal drawer handles in the house jingling against their backplates, as if a large truck had argued by. The other times, earthquakes stopped there: the drawer handles clicked, the house gave a shimmy, it was gone. We thought: earthquake! Then we proceeded with our lives as if nothing had happened, our mortality barely dented. This one was different. . . . This earthquake was five seconds old and already it was the worst one I had experienced in four years in California (Fig. B5.1).

I had to find a doorway. Standing in a doorway is what experts on television advise in the event of an earthquake. The theory is that the doorframe prevents the ceiling from falling directly on one's head. In the days after the earthquake, I learned that door frames are now considered unsafe earthquake havens: if the shaking is violent enough, you can smash your head against the doorframe and

**Figure B5.1**
**RATTLING AND ROLLING IN CALIFORNIA**
This apartment building crumpled and collapsed during the 1994 earthquake in Northridge, California.

Probably the single most important geophysical technique used to learn about the Earth's interior is the study of *seismic waves,* that is, vibrational energy waves generated by earthquakes. Through measurements of earthquake vibrations, we can find out about parts of the Earth that are inaccessible to direct study. When used in this way, earthquake waves are like the X rays a doctor uses to study the inside of a human body. By examining how these waves travel through the internal layers of the Earth, we can learn a lot about the physical characteristics (density and state of matter) of those layers. Let's begin by examining earthquakes and what causes them. Then we will be able to evaluate the information seismic waves provide about the interior of the Earth.

be knocked unconscious. This might not be such a bad thing. Another reason a doorway can be hazardous, according to an earthquake documentary that has been playing day and night in L.A. since that terrifying morning, is that the fridge could come bouncing through from another room. But as with all things seismological, no one knows for sure. Preparing for earthquakes is almost beside the point, because they're like plane crashes: in the end, it doesn't matter where you sit. . . .

. . . By the time I made the doorframe, ten seconds had passed. I could hear my wife in the next doorframe over, and so I made my way there. I don't know what I was thinking: perhaps that it made sense for all of us to be in the same place. I remember thinking: *my, how clear my mind is at this moment!* . . . We stood together in a doorway, and the Earth shook beneath us for twenty seconds longer. We looked at each other and did the only thing there was to do: we waited. Other dangers have the courtesy to announce themselves: one can avoid a flood, evacuate a fire zone, stay inside when it snows. But no one can avoid the Earth, not when it is an animal chasing you down your street into a place you thought was your own and therefore safe; not when it is a giant that has picked up your home and is shaking it for loose change.

The worst part, though, was the sound, as if a vast sheet of cellophane was being screwed into a ball. This was the sound of the wooden joints in our house twisting and wowing. Two- and three-foot cracks spread out from the doorways like antlers. Our livingroom wall looked like a road map of northern Ontario.

Then it stopped. In the silence I realized how noisy an earthquake is, how terrifying the sound is all by itself. All I could hear now was dogs barking and the whoops of car alarms. "That was a big one," I said, like an idiot. I was surprised to find I was trembling. . . . [Then] the first of more than a thousand aftershocks greeted us. . . . I lit some candles (a potentially disastrous mistake, as a fallen chimney across the street had broken an underground gas main). We cleaned up, and felt more aftershocks. The kitchen was strewn with glass and sesame seeds. Finally we blew out all the candles and sat down at the kitchen table and turned on the transistor radio. . . .

. . . At the kitchen table in our apartment in Los Angeles, my wife and my baby and I sat as still as we could, as if we were trying to appreciate stillness now that we knew what it was not to have it, and we listened. . . . I like to think we'll remember the morning of the earthquake, and all that broken glass in the kitchen, and sesame seeds everywhere, and the way the cat disappeared for hours, and Hayley (who was still a baby) pointing and cooing at all the fallen books and plates, and the smell of those damn candles that we shouldn't have lit, and a hundred other details of that cool shaken night when we all knew, without a doubt, that we loved each other.

# EARTHQUAKES

We learned in chapter 4 that earthquakes tend to occur in very specific tectonic settings, primarily along active plate boundaries. In fact, the locations of plate boundaries have been identified mainly on the basis of the localization of earthquakes along linear zones. This close connection between earthquakes and plate tectonics activity suggests that the mechanisms that cause earthquakes must be closely related to those that drive plate motion. But what, exactly, causes earthquakes?

**?** **What is the connection between earthquakes and plate tectonics?**

Figure 5.6
**EVIDENCE OF FAULT MOTION**
This orange grove is planted across the San Andreas fault in Southern California. Movement along the fault has displaced the rows of trees significantly in only a few decades. The displacement is not in the direction it first appears; the plate in the background has moved from left to right relative to the trees in the foreground. The arrows show the direction of movement along the fault.

## What Causes Earthquakes?

Most earthquakes are caused by the sudden movement of strained blocks of the Earth's crust. Tectonic forces produce *stress,* or directional pressure, which causes large blocks of rock on either side of a fault to fracture and move past each other. Typically, the movement of blocks along a fault is not a smooth, continuous slippage. Instead, strain energy builds up slowly as plate motion continues, until friction between the two blocks is overcome. Then the blocks on either side of the fault slip abruptly. The stresses persist, and the cycle of slow buildup followed by abrupt movement repeats itself many times.

Although movement along a large fault may eventually total many kilometers, this distance is the sum of numerous small, sudden slips. Each of those slips will cause an earthquake and, if the movement occurs near the Earth's surface, disrupt and displace surface features. Figure 5.6 is an example of horizontal movement that occurred along a fault in California, displacing the neat rows of trees in the orange grove. Vertical movements also occur along faults. The largest abrupt vertical displacement ever observed occurred in 1899 at Yakutat Bay, Alaska, when a stretch of the Alaskan shore was suddenly lifted 15 m (almost 50 ft) above sea level during a major earthquake. Sometimes movement along a fault is abrupt, as in Yakutat Bay, but in other cases the fault blocks creep past each other slowly but continuously (Fig. 5.7).

### The Elastic Rebound Theory

**elastic rebound theory** The theory that continuing stress along a fault results in a buildup of stored energy, which is released suddenly when the fault blocks move past one another, resulting in an earthquake.

The most widely accepted explanation of the origins of earthquakes is the **elastic rebound theory.** It is based on the mechanics of *elastic deformation* of rocks: reversible changes in the volume or shape of a rock that is subjected to stress. When the stress is removed, an elastically deformed material snaps back to its original size and shape. You can demonstrate the storage of energy in an elastically deformed material with a steel spring or a long metal ruler (Fig. 5.8). When you compress the spring or bend the ruler across your knee, the material undergoes strain in the form of elastic deformation. When you suddenly release the spring or ruler, it bounces back to its original shape, releasing the built-up energy. The elastic rebound theory states that energy can be stored in bodies of rock when they are subjected to stress along a fault. Eventually the increasing stress along the fault is sufficient to overcome the friction between the blocks. The

**Figure 5.7**
**SLOW MOVEMENT ON THE SAN ANDREAS FAULT**
Like Figure 5.6, this photograph shows evidence of the movement of plates on either side of the San Andreas fault in Carrizo Plains, California. In this case, slow movement has detached the headwaters of a stream (upper right) from the downstream portion (lower left). The fault runs almost horizontally across the center of the photo. The land to the rear is moving to the right relative to the land in the foreground. Because motion on the fault is slow and continual, the two halves of the stream system remain in contact as water flows down the depression caused by the fault.

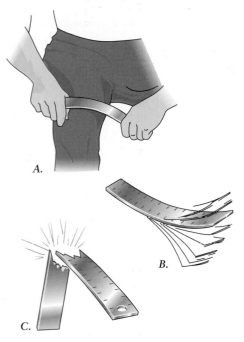

**Figure 5.8**
**ELASTIC DEFORMATION OF COMMON MATERIALS**
*A.* An elastically deformed material, such as a bent metal ruler, will spring back to its original shape *B* when released. If the material is pushed beyond the limits of its strength, it may break *C*, but the two halves will each snap back to their original shape (straight, in the case of the ruler). This is an analogy for the elastic rebound theory of the origin of earthquakes.

blocks slip, the stored energy is suddenly released in the form of an earthquake, and the rocks rebound to assume their original shapes.

The first evidence to support the elastic rebound theory came from studies of the San Andreas fault. During long-term field observations beginning in 1874, scientists from the U.S. Coast and Geodetic Survey determined the precise positions of many points both adjacent to and distant from the fault. As time passed, movement of the points revealed that the crust was slowly being bent. Near San Francisco, however, the fault was locked and did not slip. On April 18, 1906, the two sides of this locked fault shifted abruptly (Fig. 5.9). The elastically stored energy was released as the blocks on either side of the fault moved and the bent crust snapped to its new position, creating a violent earthquake. Subsequent repetition of the survey measurements revealed that the bending of the crust had disappeared.

## How Earthquakes Are Studied

**seismology** The scientific study of earthquakes and seismic waves.

**seismograph** An instrument that detects and records vibrations of the Earth resulting from earthquakes.

The study of earthquakes is known as **seismology,** from the ancient Greek word for earthquake, *seismos*. Scientists who study earthquakes are called *seismologists*, and the devices used to record the shocks and vibrations caused by earthquakes are **seismographs.**

### Seismographs

A seismograph must stand firmly attached to the Earth's vibrating surface, and it will therefore vibrate along with that surface. This means that there is no fixed frame of reference for making measurements. The problem is the same one that a sailor in a small boat faces when attempting to measure waves at sea. Because the boat moves up and down with each wave, there is no "platform" for measuring the height of the waves. To overcome this problem, most seismographs use *inertia*, the resistance of a large stationary mass to sudden movement.

If you suspend a heavy mass, such as a block of iron, from a light spring and suddenly lift the upper end of the spring, you will notice that the block remains almost stationary because of inertia, while the spring stretches upward (Fig. 5.10). To measure vertical ground movement, a heavy mass is supported by a spring and the spring is connected to a support, which in turn is connected to the ground. When the ground vibrates, the spring expands and contracts, but the mass remains almost stationary. The distance between the ground and the mass can be used to sense movement of the ground surface (Fig. 5.11A). Horizontal movement can be measured by suspending a heavy mass from a string to make a

**?** How can seismographs measure the Earth's vibrations when they themselves are sitting on a vibrating surface?

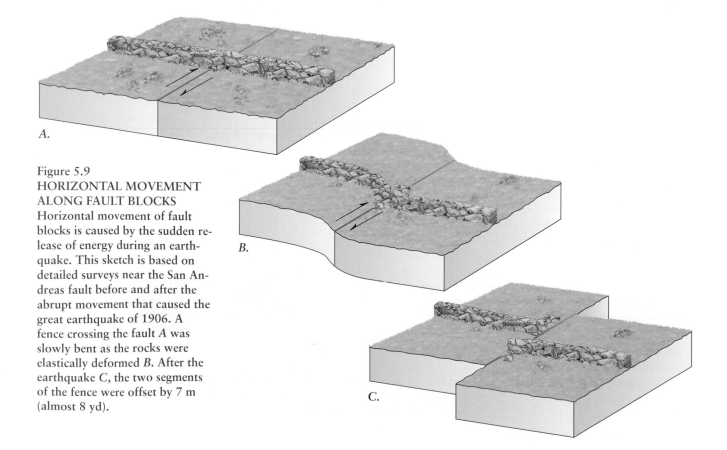

Figure 5.9
HORIZONTAL MOVEMENT ALONG FAULT BLOCKS Horizontal movement of fault blocks is caused by the sudden release of energy during an earthquake. This sketch is based on detailed surveys near the San Andreas fault before and after the abrupt movement that caused the great earthquake of 1906. A fence crossing the fault *A* was slowly bent as the rocks were elastically deformed *B*. After the earthquake *C*, the two segments of the fence were offset by 7 m (almost 8 yd).

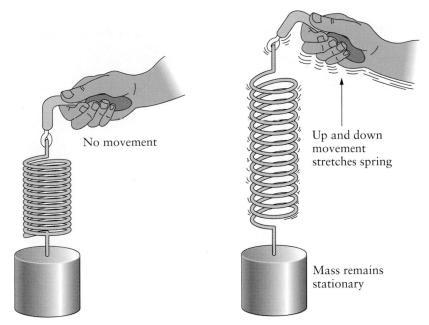

**Figure 5.10**
**THE PRINCIPLE OF INERTIA**
Inertia is the resistance of a stationary mass to sudden movement. Inertia is the reason a car cannot instantly go from parked to full highway speed; energy and time are required to get it up to speed. Seismographs use the principle of inertia, as shown here. In this example, the hand moves up and down, but the mass remains nearly motionless because of inertia.

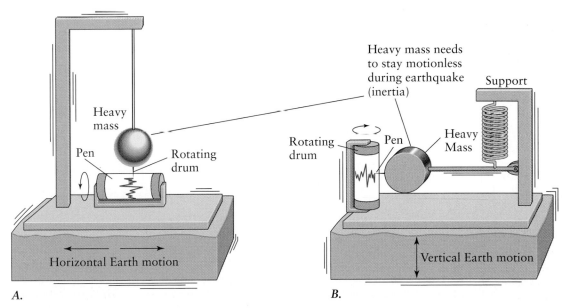

**Figure 5.11**
**SEISMOGRAPHS**
Seismographs measure vibrations caused by earthquakes. *A.* A seismograph for measuring vertical movements uses the inertia of a mass attached to a spring. *B.* A seismograph for measuring horizontal movements uses the inertia of a mass suspended from a pendulum.

Figure 5.12
### TRAVEL PATHS OF SEISMIC BODY WAVES

The energy released during an earthquake travels outward through the Earth from its source (the focus of the earthquake). If the Earth were uniform throughout, the waves would travel in straight lines. But in the real Earth, seismic waves travel faster at greater depths in the Earth. That is, the velocity of seismic waves increases with depth, which causes their paths to curve. (As we will see, the increase in velocity is not completely smooth, as suggested in this diagram; seismic velocities vary as the waves pass from one layer to another within the Earth.)

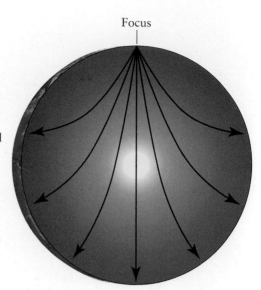

Focus

pendulum (Fig. 5.11B). Because of inertia, the mass does not keep up with the motion of the ground, and the difference between the movement of the pendulum and that of the ground serves as a measure of ground motion.

These simple machines are the basis for the seismograph. Modern seismographs are incredibly sensitive because the movements they detect are measured optically and amplified electronically. Vibrational movements as tiny as one hundred millionth ($10^{-8}$) of a centimeter can be detected. Indeed, many instruments are so sensitive that they can detect the ground depression caused by a moving automobile several blocks away.

## Seismic Waves

Most of the energy released by an earthquake is transmitted to other parts of the Earth. (Another considerable portion of the energy goes to the heating of rock along the fault.) The released energy travels outward in the form of vibrational waves from the earthquake's source, the **focus** (Fig. 5.12). These waves, called **seismic waves,** are elastic disturbances. The rocks through which they pass return to their original shapes after the waves have passed; the waves therefore must be measured while the rock is still vibrating. For this reason, a network of continuously recording seismograph stations has been installed around the world. When an earthquake occurs, the seismic waves are recorded by many seismographs and instantly evaluated by computers. Records obtained in this way are called **seismograms.**

There are two main types of seismic waves. **Body waves** travel outward from the point of origin and have the capacity to travel through the interior of the Earth. **Surface waves,** on the other hand, are guided by and restricted to the Earth's surface.

### Body Waves.
**Compressional waves,** one of two types of body wave, consist of alternating pulses of compression and expansion acting in the direction in which the wave is traveling (Fig. 5.13A). Sound waves are also compressional waves. When a sound wave passes through the air, it does so by alternating compression and expansion of the air. Compressional waves can pass through solids, liquids, and gases. They have the greatest velocity of all seismic waves—6 km/s (almost 4 mi per second) is a typical value for the velocity of compressional waves through the uppermost portion of the crust—and they are the first waves to be recorded after an earthquake. They are therefore called **P** (for *primary*) **waves.**

**Shear waves,** the other type of body wave, travel through materials in an alternating series of sidewise movements (Fig. 5.13B). Shearing involves changing

**focus** The point within the Earth where an earthquake's energy is first released.

**seismic wave** A vibrational energy wave that travels outward from an earthquake's source.

**seismogram** The record made by a seismograph.

**body wave** Any seismic wave that travels through the interior of the Earth.

**surface wave** A seismic wave that travels along the surface of the Earth.

**compressional wave** A seismic body wave consisting of alternating pulses of compression and expansion.

**P wave** Primary wave, a compressional seismic body wave.

**shear wave** A seismic body wave that travels in shearing movements perpendicular to the direction in which the wave is traveling.

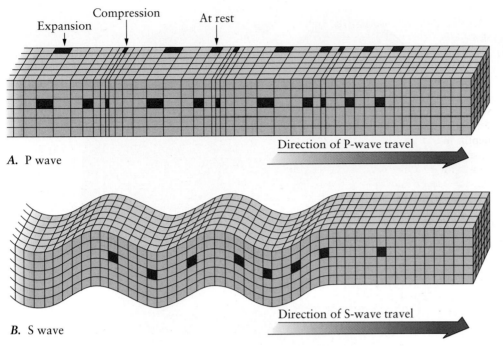

Expansion · Compression · At rest

*A.* P wave

Direction of P-wave travel

*B.* S wave

Direction of S-wave travel

Figure 5.13
**SEISMIC BODY WAVES:
P AND S**
*A.* A P wave alternately compresses and expands the rock as the wave passes through. These squares illustrate the alternating compression and expansion. As waves pass through, a square will repeatedly expand to a long rectangle, return to a square, contract to a short, squashed-looking rectangle, and so on. *B.* S waves cause a shearing motion, like that of a rope being shaken up and down. In this case, as the waves pass through, the squares move up and down perpendicular to the direction of S-wave travel. The square doesn't expand or contract but changes its shape, starting as a square, then becoming a distorted parallelogram, and then changing back to a square.

the shape of an object. Solids have elastic characteristics that provide a restoring force for recovery from shearing; liquids and gases lack these characteristics. Therefore, shear waves cannot be transmitted through liquids or gases. This is very important, as we will soon see. A typical velocity for shear waves in the upper crust is 3.5 km/s (more than 2 mi per second). Shear waves are slower than P waves, so they reach a seismograph some time after the arrival of P waves from the same earthquake. For this reason, they are called **S** (for *secondary*) **waves**.

**S wave** Secondary wave, a seismic body wave that travels with a shearing motion.

*Surface Waves.* Surface waves travel along or near the surface of the Earth, like waves along the surface of a body of water. They travel more slowly than P and S waves, and they pass around the Earth rather than through it. Thus, surface waves are the last to be detected by a seismograph. Figure 5.14 shows a typical seismogram, in which the P wave's arrival is seen first, followed by the arrival of the S wave and finally by that of the surface waves. Surface waves are

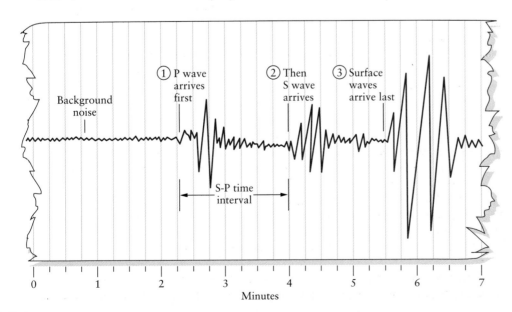

Background noise

① P wave arrives first

② Then S wave arrives

③ Surface waves arrive last

S-P time interval

Minutes

Figure 5.14
**SEISMOGRAM OF A TYPICAL EARTHQUAKE**
This typical seismogram illustrates the different travel times for P waves, S waves, and surface waves. The P waves and S waves leave the point of origin (the focus) at the same instant. The faster-moving P waves reach the seismograph first, followed some time later by the slower-moving S waves. The surface waves, which travel the long way around on the Earth's surface, arrive even later.

Figure 5.15
**FINDING AN EARTHQUAKE EPICENTER BY TRIANGULATION**
Waves from an earthquake arrive at three different seismic stations. The time difference between the first arrival of the P waves and S waves depends on the distance of the station from the epicenter; the greater the difference in their arrival times, the farther the waves have traveled. The distance from each station to the epicenter can be calculated from this difference in arrival times. Then three circles are drawn on a map. Each circle has a seismic station at its center and a radius equal to the distance from that station to the epicenter. The point where the three circles meet is the epicenter of the earthquake.

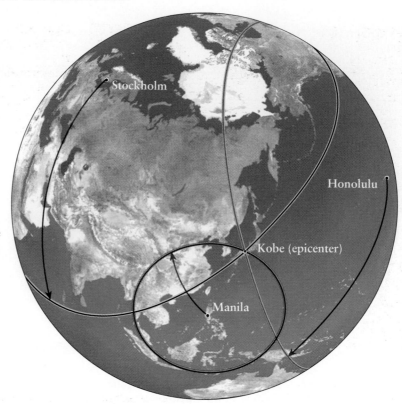

**epicenter** The point on the Earth's surface directly above an earthquake's focus.

important for planners and builders because they cause much of the ground shaking that causes damage to buildings and infrastructure (roads, pipes, sewers, etc.) during large earthquakes.

## Locating Earthquakes

If an earthquake's body waves have been recorded by three or more seismographs, the location of its **epicenter**—the point on the Earth's surface directly above the earthquake's focus—can be determined through simple graphical calculations. The first step is to find out how far each seismograph is from the epicenter. This is done by comparing the arrival times of P and S waves as revealed by a seismogram (Fig. 5.14). The greater the distance traveled by the waves, the greater the difference between their arrival times. In other words, the farther the waves travel, the more the S waves lag behind the P waves. Thus, the lag time can be used to calculate the distance traveled.

After determining the distance from the seismograph to the epicenter, the seismologist draws a circle on a map, with the seismic station at the center of the circle. The radius of the circle is the distance from the seismograph to the epicenter. It has to be a circle because the seismologist only knows the *distance* to the epicenter—usually there is no easy way to know the *direction* from which the waves have come. When similar information is plotted for three or more seismographs, the location of the epicenter can be determined (Fig. 5.15). This process is called the *Geiger method of triangulation*.

## Measuring Earthquakes

Several different scales are used to quantify the strength or *magnitude* of an earthquake. The most widely used of these are the *Richter magnitude*, the *modified Mercalli intensity*, and the *moment magnitude*.

***The Richter Magnitude Scale.*** Seismologists have developed a way to estimate the energy released by a quake by measuring the recorded heights, or *am-*

Figure 5.16
ALASKA "GOOD FRIDAY" EARTHQUAKE,
1964
An extremely powerful earthquake—one of the
largest ever recorded—struck Alaska on March 27,
1964. This photo shows the damage along part of
the main thoroughfare in Anchorage. These cars,
parked along the side of the street, wound up several meters below street level.

*plitudes*, of the resulting seismic waves. The **Richter magnitude scale,** named
after seismologist Charles Richter, who developed it in 1935, is calculated from
the amplitudes of P and S waves recorded on a seismogram. The Richter scale is
*logarithmic*, meaning that each increase of one on the Richter scale corresponds
to a tenfold increase in the amplitude of the wave signal. Thus, a magnitude 6
earthquake has an amplitude ten times larger than that of a magnitude 5 quake.
A magnitude 7 earthquake has an amplitude 100 times larger than that of a magnitude 5 quake (10 × 10). Richter magnitudes also are corrected for distance
from the epicenter. Thus, the magnitude calculated for a given earthquake is the
same whether you are standing at the epicenter, where the effects of the earthquake would be felt strongest, or thousands of kilometers away, where the effects
would be felt less intensely.

   Each step in the Richter scale corresponds to a tenfold increase in the *amplitude* of the seismic wave signal—that is, the actual amount of ground shaking
that occurred. However, the amount of *energy released* is closer to a thirtyfold increase for each step in the scale. Thus, a magnitude 6 earthquake may release
900 times as much energy as a magnitude 4 quake (a thirtyfold increase for each
step in the scale, or 30 × 30 = 900). A magnitude 8 earthquake may release almost a million times as much energy as a magnitude 4 quake (30 × 30 × 30 × 30
= 810,000). So even though large earthquakes happen infrequently, a single very
large quake can release as much stored energy as many thousands of smaller
ones. The largest earthquakes recorded to date have had Richter magnitudes of
about 8.6. Many of these occurred in subduction zones, including the Chilean
earthquake of 1960 (magnitude 9.1) and the Alaskan "Good Friday" earthquake
of 1964 (magnitude 8.6) (Fig. 5.16).

### The Mercalli Intensity Scale.
The **Mercalli intensity scale,** developed in 1902
by an Italian geologist and later modified, is based on descriptions of the vibrations that people felt, saw, and heard, and on the extent of damage to buildings
during earthquakes. The scale ranges from I (not felt except under favorable circumstances) to XII (waves seen on ground surface, practically all works of construction destroyed or severely damaged). The approximate correspondence

**Richter magnitude scale** A scale
of earthquake intensity based on
the recorded heights, or *amplitudes,* of the seismic waves.

**Mercalli intensity scale** A scale of
earthquake intensity, based on
damage to buildings and descriptions of the vibrations experienced during earthquakes.

**TABLE 5.1**    Earthquake Magnitudes and Frequencies of Occurrence, with Characteristic Damaging Effects

| Richter Magnitude | Number per Year | Modified Mercalli Intensity Scale[a] | Characteristic Effects of Shocks in Populated Areas |
|---|---|---|---|
| <3.4 | 800,000 | I | Recorded only by seismographs |
| 3.5–4.2 | 30,000 | II and III | Felt by some people who are indoors |
| 4.3–4.8 | 4,800 | IV | Felt by many people; windows rattle |
| 4.9–5.4 | 1,400 | V | Felt by everyone; dishes break, doors swing |
| 5.5–6.1 | 500 | VI and VII | Slight building damage; plaster cracks, bricks fall |
| 6.2–6.9 | 100 | VII and IX | Much building damage; chimneys fall; houses move on foundations |
| 7.0–7.3 | 15 | X | Serious damage, bridges twisted, walls fractured; many masonry buildings collapse |
| 7.4–7.9 | 4 | XI | Great damage; most buildings collapse |
| >8.0 | One every 5–10 years | XII | Total damage, waves seen on ground surface, objects thrown in the air |

[a]Mercalli numbers are determined by the amount of damage to structures and the degree to which ground motions are felt. These depend on the magnitude of the earthquake, the distance of the observer from the epicenter, and whether an observer is in or out of doors. There is not an exact correspondence between the Mercalli intensity and the Richter magnitude.

among Mercalli intensity near the epicenter, Richter magnitude, and the frequency of occurrence of earthquakes of different sizes are shown in Table 5.1.

The Mercalli intensity of an earthquake varies with distance from the epicenter. A single earthquake could have a Mercalli intensity of IX or X near the epicenter where the intensity is greatest, whereas a few hundred kilometers away its intensity could be only I or II. The Mercalli scale is useful in the study of earthquakes that occurred before the development of modern seismic equipment. For example, the exact magnitudes of a series of devastating earthquakes that struck New Madrid, Missouri, in 1811–1812 are unknown. However, historical eyewitness accounts indicate that all three earthquakes had Mercalli intensities of XI near the epicenter (Fig. 5.17). Combining this with information about the nature of the bedrock and how far away the effects were felt, researchers estimate that the quakes had Richter magnitudes ranging from 7.3 to 7.8. The Mercalli intensity of an earthquake is controlled by many factors ranging from soil types to construction codes; therefore, no clear equivalency can be established between the Richter magnitude and Mercalli intensities of earthquakes.

***The Moment Magnitude.*** The **moment magnitude** is an expression of the strength of an earthquake based on field observations of the area over which the fault ruptured, and the average amount of ground displacement. Faults sometimes rupture over an area several kilometers to several hundred kilometers long. Richter scale calculations are based on the concept that earthquake foci are points. Therefore, the Richter scale is best suited as a measure of the strength of an earthquake in which energy is released from a relatively small volume of rock. In contrast, the calculation of seismic moment takes account of the fact that energy may be released from a large area. Seismic moment also accounts for variations in the physical characteristics of Earth materials, which can affect the efficiency with which seismic waves are transmitted. Therefore, the moment magnitude may provide a better measure of the actual strength of an earthquake.

**moment magnitude** A measure of earthquake strength, based on the rupture size, rock properties, and amount of ground displacement along the fault.

## Earthquake Hazards and Prediction

Each year many hundreds of thousands of earthquakes occur around the world. Fortunately, only a few are large enough, or close enough to major population centers, to cause loss of life. A great deal of research focuses on earthquake prediction and hazard assessment. The hope is that through such research seismologists will be able to improve their forecasting ability to the point at which effective and accurate early warnings can be issued. Let's look briefly at the hazards associated with earthquakes and at efforts to predict earthquakes.

## Earthquake Hazards

Earthquakes can cause total devastation in a matter of seconds. The most disastrous quake in history was the one that occurred in Shaanxi Province, China, in 1556, in which an estimated 830,000 people died. The worst earthquake disaster of the twentieth century also occurred in China. At 3:42 A.M. on July 28, 1976, while most of the 1 million inhabitants of T'ang Shan were asleep, a 7.8 magnitude quake leveled the city. Hardly a building was left standing, and the few that withstood the first earthquake were destroyed by a second one (magnitude 7.1) that struck at 6:45 P.M. the same day. When the wreckage was cleared, 240,000 people were dead (unofficial accounts were much higher, rivaling the loss of life from the 1556 Shaanxi earthquake). In all, 18 earthquakes are known to have caused 50,000 or more deaths apiece (Table 5.2).

A variety of hazards are associated with earthquakes. Ground motion is usually the most significant cause of damage. In the most intense earthquakes (magnitude > 8.0), the surface of the ground can sometimes be observed moving in

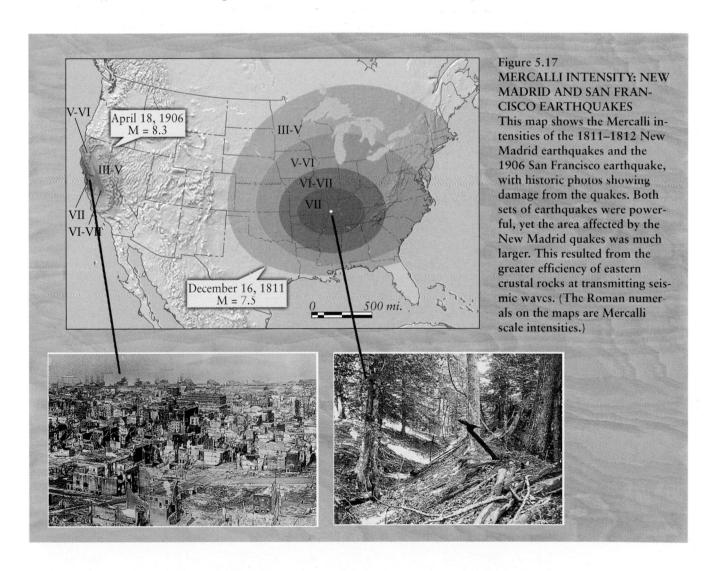

**Figure 5.17**
**MERCALLI INTENSITY: NEW MADRID AND SAN FRANCISCO EARTHQUAKES**
This map shows the Mercalli intensities of the 1811–1812 New Madrid earthquakes and the 1906 San Francisco earthquake, with historic photos showing damage from the quakes. Both sets of earthquakes were powerful, yet the area affected by the New Madrid quakes was much larger. This resulted from the greater efficiency of eastern crustal rocks at transmitting seismic waves. (The Roman numerals on the maps are Mercalli scale intensities.)

**TABLE 5.2** Earthquakes Occurring During the Past 800 Years That Have Caused 50,000 or More Deaths

| Place | Year | Estimated Number of Deaths |
|---|---|---|
| Silicia, Turkey | 1268 | 60,000 |
| Chihli, China | 1290 | 100,000 |
| Naples, Italy | 1456 | 60,000 |
| Shaanxi, China | 1556 | 830,000 |
| Shemaka, USSR | 1667 | 80,000 |
| Naples, Italy | 1693 | 93,000 |
| Catalina, Italy | 1693 | 60,000 |
| Beijing, China | 1731 | 100,000 |
| Calcutta, India | 1737 | 300,000 |
| Lisbon, Portugal | 1755 | 60,000 |
| Calabria, Italy | 1783 | 50,000 |
| Messina, Italy | 1908 | 160,000 |
| Gansu, China | 1920 | 180,000 |
| Tokyo and Yokohama, Japan | 1923 | 143,000 |
| Gansu, China | 1932 | 70,000 |
| Quetta, Pakistan | 1935 | 60,000 |
| T'ang Shan, China | 1976 | 240,000 |
| Iran | 1990 | 52,000 |

waves. Proper design of buildings and other structures can do much to prevent damage from ground motion, but in a very strong earthquake even the best-designed buildings may collapse. Sometimes large cracks and fissures open in the ground (Fig. 5.18). Where a fault breaks the ground surface, buildings can be split, roads disrupted, and anything that lies on or across the fault broken apart. To make matters worse, major earthquakes are often followed by numerous smaller (usually) earthquakes called *aftershocks*. Aftershocks triggered by large earthquakes may be quite far from the original epicenter. The Landers (near Los Angeles) earthquake of 1992 (magnitude 7.3) triggered major aftershocks at 14 locations, some of them hundreds of kilometers from the original epicenter.

In regions with steep slopes, earthquake vibrations may cause cliff collapses, landslides, and other rapid downslope movements of Earth material. In 1970, for example, a magnitude 7.75 earthquake triggered a devastating landslide in Yungay, Peru, in which at least 18,000 people were killed (Fig. 5.19). The sudden shaking of water-saturated sediment can turn seemingly solid ground into a quicksand-like material. This process, called *liquefaction*, was a major cause of damage during the earthquake that destroyed much of Anchorage, Alaska, in 1964. In the same year earthquake-triggered liquefaction and uneven ground settling caused apartment houses to sink and collapse in Niigata, Japan (Fig. 5.20).

Fire is a secondary effect that can pose an even greater hazard than ground motion. Ground motion displaces stoves, breaks gas lines, and loosens electrical wires, often causing fires. Because ground motion also breaks water mains, no water may be available to put out the fires. In the earthquake that struck San Francisco in 1906, most of the damage was caused by fires that destroyed 521 city blocks in three days. For many years the quake was referred to as "The Great Fire." In 1989 fires caused by the Loma Prieta earthquake again ravaged the Marina district of downtown San Francisco (Fig. 5.21).

# Earthquake Hazards

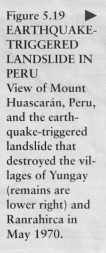

**Figure 5.19** ►
EARTHQUAKE-TRIGGERED LANDSLIDE IN PERU
View of Mount Huascarán, Peru, and the earthquake-triggered landslide that destroyed the villages of Yungay (remains are lower right) and Ranrahirca in May 1970.

▲
**Figure 5.18**
EARTHQUAKES CAN OPEN LARGE FISSURES
Fissures opened in the ground in Santa Cruz, California, as a result of the Loma Prieta earthquake, October 21, 1989.

▲
**Figure 5.20**
EARTHQUAKE-TRIGGERED LIQUEFACTION IN JAPAN
In 1964 liquefaction and uneven ground settling resulting from an earthquake caused apartment buildings to sink and collapse in Niigata, Japan. Many of the buildings were not structurally damaged; they simply keeled over onto their sides. Apartment dwellers retrieved their belongings by rolling wheelbarrows up the walls and lowering themselves through the windows.

▲
**Figure 5.21**
FIRES FROM THE "WORLD SERIES" EARTHQUAKE
These fires in the Marina District of San Francisco were caused by gas lines that broke as a result of the Loma Prieta "World Series" earthquake in 1989.

Another secondary effect of earthquakes is *seismic sea waves,* or *tsunami* (sometimes mistakenly called tidal waves). Submarine earthquakes are the main cause of these waves, which are especially destructive around the Pacific Ocean rim. A well-known example is the tsunami generated by a severe submarine earthquake near Unimak Island, Alaska, in 1946. The wave traveled across the Pacific Ocean at a velocity of 800 km/h (almost 500 mi/hr), striking Hilo, Hawaii, about 4.5 hours later. When it hit Hilo, the wave had a crest 18 m (60 ft) higher than normal high tide. It demolished nearly 500 houses, damaged 1000 more, and killed 159 people. More recently, five destructive Pacific rim tsunami events (Nicaragua, 1992; Flores Island, Indonesia, 1992; Hokkaido, Japan, 1993; southeast Java and Bali, Indonesia, 1994; and Papua New Guinea, 1998) killed a total of approximately 2000 people.

## Earthquake Prediction

Charles Richter once said, "Only fools, charlatans, and liars predict earthquakes." Today, seismologists attempt to predict earthquakes by using sensitive instruments to monitor seismically active zones. It still is not possible to predict the exact magnitude and time of occurrence of an earthquake. However, scientists' understanding about seismic mechanisms and the tectonic settings in which earthquakes occur has improved greatly since Richter's time, and advances in modern seismology may yet prove him wrong.

There are two aspects to the problem of earthquake prediction. *Long-term forecasting* involves predicting a large earthquake years or even decades in advance of its occurrence. *Short-term prediction* would ideally be the precise prediction of the time, magnitude, and location of an earthquake event, providing an opportunity for authorities to issue an *early warning*.

**Long-Term Forecasting** Long-term earthquake forecasting is based mainly on our understanding of the tectonic cycle and the geologic settings in which earthquakes occur. In places where earthquakes are known to occur repeatedly, such as along plate boundaries, it is sometimes possible to detect a regular pattern in the recurrence intervals of large quakes. To do so, seismologists require information about seismic activity going back farther than historical records. This information is provided by **paleoseismology,** the study of prehistoric earthquakes.

**paleoseismology** The study of prehistoric earthquakes.

The primary goal of paleoseismology is to search the stratigraphic record for evidence of major earthquakes and, if possible, to discern the intervals between them (Fig. 5.22). Evidence may include vertical displacement of sedimentary layers, as shown in Figure 5.22, indicators of liquefaction, or horizontal displacement of geologic features. If the pattern of recurrence suggests intervals of, say, a century between major quakes, it may be possible to predict approximately when a large quake is due to happen next in that location. Such studies have identified a number of *seismic gaps* around the Pacific rim. These are places along a fault where a large earthquake has not occurred for a long time, even though tectonic movement is still active. Seismic gaps are considered by some geophysicists to be the places most likely to experience large earthquakes.

Long-term earthquake forecasting has met with reasonable success. Seismologists know where the most hazardous areas are. They can accurately calculate the probability that a large earthquake will occur in a particular area within a given period. And they have a theory of earthquake generation that successfully unites their predictions and observations in the context of plate tectonic theory.

**Short-Term Prediction and Early Warning.** Unfortunately, the short-term prediction of earthquakes has been mostly unsuccessful. Attempts at short-term prediction are based on observations of *precursor phenomena,* that is, anomalous occurrences that may serve as early-warning signs of earthquake activity.

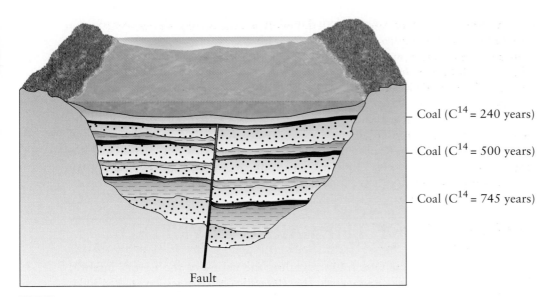

Coal (C$^{14}$ = 240 years)

Coal (C$^{14}$ = 500 years)

Coal (C$^{14}$ = 745 years)

Fault

 Sand layer

 Clay layer

Organic (carbon-containing) layer

**Figure 5.22**
**PALEOSEISMOLOGY: EVIDENCE OF ANCIENT QUAKES**
Layers of sediment were offset by ancient earthquakes. The organic-rich layers of sediment have yielded C-14 dates, which help scientists pinpoint the ages of the ancient quakes.

Most research on short-term prediction involves monitoring changes in the properties of the rocks, such as magnetism and electrical conductivity. Even simple observations, such as the level of water in wells or the amount of radon (a soil gas) in well water, might indicate changes in the properties of the underlying rocks. Strange animal behavior, glowing auras, and unusual radio waves have all been reported as precursors near the epicenters of large earthquakes.

Tilting or bulging of the ground and slow rises and falls in elevation are among the most reliable indications that strain energy is building up. Small cracks and fractures that develop in severely strained rock can cause swarms of tiny earthquakes—*foreshocks*—that may be a clue that a big quake is coming. The most famous successful earthquake prediction, made by Chinese scientists in 1975, was based on slow tilting of the land surface, fluctuations in the magnetic field, and the numerous foreshocks that preceded a magnitude 7.3 quake that struck the town of Haicheng (Fig. 5.23). Half the city was destroyed, but because

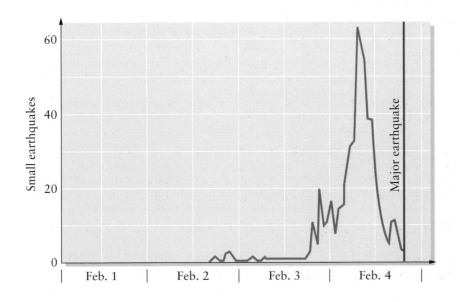

**Figure 5.23**
**FORESHOCKS USED TO PREDICT EARTHQUAKE IN CHINA**
Frequency of foreshocks before the 1975 earthquake (M 7.3) at Haicheng, People's Republic of China. Foreshock activity peaked just before the main event.

authorities had evacuated more than a million people before the quake, only a few hundred were killed.

Short-term prediction of earthquakes has been less successful than long-term forecasting, partly because the processes associated with earthquakes are hidden under the ground. An even greater problem, however, is that earthquakes are highly inconsistent in terms of their precursor phenomena. Sometimes they give no discernible warning signs at all. Less than two years after the prediction of the Haicheng earthquake, Chinese geophysicists were dealt a major blow when the devastating 1976 T'ang Shan earthquake struck with no apparent precursory activity. Short-term prediction and early warning of earthquakes remain elusive goals for seismologists.

# THE INSIDE OF THE EARTH REVISITED

Earthquakes are extremely useful in revealing the nature of the Earth's interior. In this section we briefly examine the characteristics of seismic waves that make them useful in studying the inside of the Earth. We then summarize what we know about the Earth's interior from all of the sources of information discussed in this chapter.

## How Earthquakes Reveal the Inside of the Earth

Seismograms from seismic stations around the world provide records of waves that have traveled along many different paths. By examining these records, it is possible to determine how the properties of rocks change with increasing depth below the Earth's surface. The following characteristics of seismic waves are particularly useful:

> **?**  S waves cannot be transmitted through liquids. Why is this so important in the study of the Earth's interior?

1. Seismic waves travel outward in all directions from the focus of an earthquake. By examining what happens to the waves while they are passing through the Earth—how they slow down, speed up, or *refract* (bend)—scientists can get a picture of the Earth's interior. The pathways traveled by body waves are curved, as shown in Figure 5.12, because the waves travel at greater velocities with increasing depth in the Earth.

2. P waves generally travel faster than S waves, but the exact velocity of a seismic wave depends on the physical properties of the material it is traveling through. The velocities of seismic waves in different materials can be determined by laboratory experiments. For example, seismic waves travel at greater velocities through more compact materials than through less compact materials. (That's why they travel more quickly at greater depths—materials deep in the Earth are more compact than materials near the surface.) Seismic waves also travel more quickly through cold materials than through hot materials.

3. P waves can be transmitted through solids and liquids, but S waves cannot be transmitted through liquids.

> **?**  What happens when seismic waves cross the boundaries between different types of rock?

### Seismic Refraction

Imagine what happens when a seismic wave traveling through a homogeneous material encounters a sharp boundary with a material that has different physical properties. Let's say that the boundary separates a rock unit of low seismic velocity from a rock unit of high seismic velocity. As it crosses the boundary, the wave will speed up and its path will bend sharply, as in Figure 5.24A. This bending is called **refraction**. The same process causes light waves to bend when they pass from air to water (Fig. 5.24B).

**refraction** The bending of a wave as it passes from one material into another material of differing density.

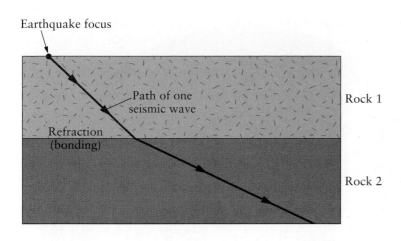

A.                                                                                                      B.

**Figure 5.24**
**REFRACTION: BENDING THE WAVES**
*A.* When seismic waves cross a boundary between rocks with differing physical properties, their paths bend. This is called seismic refraction. *B.* Refraction also happens to light waves when they pass from water to air, making the straw in this photograph look bent.

Refraction of seismic waves reveals the presence of layers of rock with differing physical properties within the Earth. For example, Figure 5.25 shows P-wave paths from an earthquake through the Earth. The paths are smoothly curved until the waves encounter the boundary between the mantle and the core. At that point the paths are bent—that is, refracted—quite sharply. The reason for this is that the properties of the outer part of the core are very different from those of the mantle. The P waves from the earthquake begin to travel much more slowly as soon as they hit the core. The refraction causes a *P-wave shadow zone,* an area on the Earth's surface where the P waves do not arrive as expected. A seismic station situated in the shadow zone would not record any P waves from this earthquake.

A similar thing happens with S waves (Fig. 5.25). However, when the S waves encounter the boundary between the mantle and the core they are not refracted; they are blocked altogether. This creates a large *S-wave shadow zone* on the side of the Earth opposite the earthquake's epicenter. Recall that S waves cannot be transmitted through liquids. From this we can deduce that the outer core of the Earth must be a liquid, which blocks the passage of S waves through the center of the Earth.

## Seismic Tomography and Discontinuities

As more sophisticated seismic equipment was developed, scientists began to make more detailed observations of this type, and they discovered more boundaries and layers within the Earth. Today seismologists use seismic waves to probe the Earth's interior in much the same way that doctors use X rays and CAT scans. In CAT (*computer-aided tomography*) scanning, a series of X rays along successive planes can be used to create a three-dimensional picture of the inside of the body. Similarly, *seismic tomography* allows seismologists to superimpose many two-dimensional seismic "snapshots," like the one shown in Figure 5.25, to create a three-dimensional image of the inside of the Earth.

We now know that there are many different layers inside the Earth, creating numerous distinct boundaries where the velocities of seismic waves change

**Figure 5.25**
**P-WAVE REFRACTION AND S-WAVE SHADOW ZONE**
P waves are strongly bent (refracted) when they encounter the boundary between the mantle and the core. The refraction occurs because the velocity of P waves is much slower in the outer part of the core than in the mantle. S waves are blocked by the outer core, creating a large S-wave shadow zone on the other side of the Earth. Seismic station A would detect P waves, S waves, and surface waves from an earthquake. Seismic station B is located within both the P- and S-wave shadow zones, and would detect surface waves but no P or S waves). Seismic station C, in the S-wave shadow zone, would detect refracted P waves and surface waves, but no S waves.

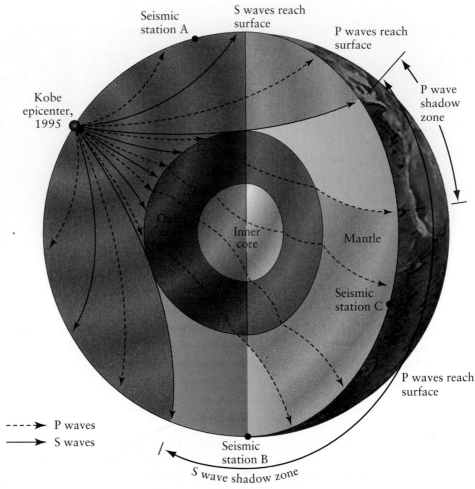

**seismic discontinuity** A boundary inside the Earth, where the velocities of seismic waves change abruptly.

abruptly. These are called **seismic discontinuities**. In some cases, seismic discontinuities represent true boundaries between layers of different chemical composition. In other cases, discontinuities may represent boundaries between rocks of the same chemical composition but different physical properties. For example, the contrast in physical properties between cold, hard, sinking lithospheric plate and the surrounding hot mantle could cause a discontinuity in seismic velocities. The transition from a low-pressure crystal structure to a denser, high-pressure crystal structure could also cause a detectable seismic discontinuity. Detailed seismic studies allow scientists to map the locations of seismic discontinuities, the distribution of hot and cold masses, and the distribution of dense and less dense materials inside the Earth.

## Summary: Layers of the Earth

Using information from the sources discussed in this chapter, especially seismic studies and other geophysical methods, scientists have arrived at a very detailed understanding of the inside of the Earth. Let's complete our study of the Earth's interior with a brief tour, starting with the crust and working down to the innermost layers. As we proceed, keep in mind that some boundaries inside the Earth separate layers with differing composition, while others separate layers with the same composition but differing physical properties.

### The Crust

**crust** The outermost compositional layer of the Earth.

The outermost layer of the Earth is the **crust**. As discussed previously, the thickness of the crust varies greatly, from an average of 8 km for oceanic crust to an

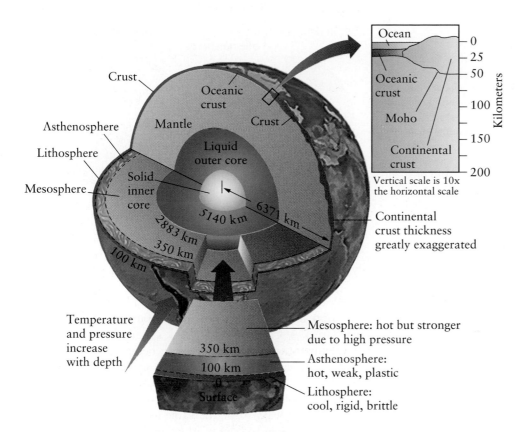

**Figure 5.26**
**INSIDE THE EARTH**
This diagram shows the Earth's internal structure. The upper part of the cutaway shows compositional layers and layers with differing rock properties. Note that the boundaries between zones of rock that differ in strength—such as the rigid lithosphere and the underlying, more plastic asthenosphere—do not coincide with compositional boundaries.

average of 45 km for continental crust (Fig. 5.26). Even at its thickest (about 70 km, or about 44 mi), the crust is extremely thin compared to the Earth as a whole—it's like a thin, brittle eggshell, or about the same relative thickness as the glass of a lightbulb.

The composition of the crust also varies from place to place. About 95 percent of the crust is igneous rock, or metamorphic rock derived from igneous rock. In general, the rocks of the crust are lighter (less dense) than the material that makes up the Earth's interior; the crust is composed of material that has "floated" to the top during planetary differentiation.[4] *Continental crust* is made mostly of *granitic* rocks, which are very low in density. *Oceanic crust,* as discussed in chapter 1, is made of *basalt,* which is denser than granitic rocks. The very outermost layer of much of the crust is a thin veneer of sediment and sedimentary rock, both on land and on the ocean floor. Thus, although the crust is composed primarily of igneous rock, most of what we actually see at the surface is sedimentary (Fig. 5.27).

The boundary that separates the crust from the mantle was the first internal boundary to be discovered using seismic techniques. This boundary was named the *Mohorovičić discontinuity* after the seismologist who discovered it in 1909, but it is usually called the **Moho.** The Moho is an example of a boundary between two layers of rock with differing compositions (Fig. 5.26).

**Moho** The boundary between the crust and the mantle.

### The Mantle

The **mantle** extends from the Moho to the core (Fig. 5.26). About 80 percent of the volume of the Earth is contained in the mantle. Geologists believe that the mantle consists mainly of iron- and magnesium-silicate minerals. The upper part

**mantle** The middle compositional layer of the Earth, between the core and the crust.

---

[4] As you learned in Chapter 1, the core segregated during a major period of planetary differentiation early in the Earth's history. However, new oceanic and continental crust is still forming today; thus, the process of planetary differentiation is ongoing.

Figure 5.27
**IGNEOUS CRUST, SEDIMENTARY SURFACE**
These diagrams show the relative amounts of sedimentary and igneous rocks in the Earth's crust and at the surface. (For the purpose of this diagram, metamorphic rocks are considered to be either igneous or sedimentary, depending on their origin.) In the Earth's crust (*A*), most of the rocks (95%) are igneous, with a thin covering (5%) of sedimentary rock. But if we look only at the surface of the Earth (*B*), most of the rocks we see are sedimentary (75%). This is because erosion, weathering, and deposition are constantly at work, breaking down the rocks of the Earth's outer shell.

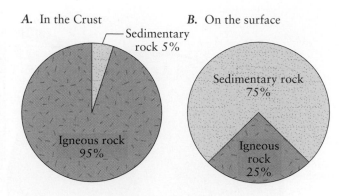

**asthenosphere** A weak layer within the mantle, just below the lithosphere.

**lithosphere** The tough, rocky, outermost part of the Earth, comprising the crust and the uppermost part of the mantle.

**mesosphere** The mantle between the bottom of the asthenosphere and the core-mantle boundary.

of the mantle has a composition similar to that of the rock *peridotite,* an igneous rock that consists mainly of the iron-magnesium silicate minerals olivine and pyroxene.

Seismic studies have revealed layering within the mantle. Extending from about 100 to 350 km (60 to more than 200 mi) below the surface is a layer called the **asthenosphere,** from the Greek words meaning "weak sphere." In this zone, the rocks are very near the temperatures at which melting begins, so they have little strength. Within the asthenosphere is a discontinuous zone in which P and S waves slow down markedly (Fig. 5.28). This is called the *low-velocity zone,* and it is believed to result from the presence of small amounts of melted rock. The composition of the asthenosphere appears to be the same as that of the mantle just above and below it. Thus, the asthenosphere is an example of a layer whose distinctiveness is based on its physical properties, not on its composition.

The outermost 100 km (60 mi) of the Earth, which includes the crust and the part of the mantle just above the asthenosphere, is called the **lithosphere** (Fig. 5.26). The rocks of the lithosphere are cooler, more rigid, and much stronger than the rocks of the asthenosphere, which are weak and easily deformed, like butter or warm tar. In plate tectonic theory, it is the entire lithosphere (not just the crust) that forms the plates. The movement of these plates is facilitated by the presence of the underlying weaker rocks of the asthenosphere.

The rest of the mantle, from the bottom of the asthenosphere (at about 350 km, or a little more than 200 mi depth) down to the core–mantle boundary, is the **mesosphere** (Fig. 5.26). Although temperatures in the mesosphere are very high, the rocks are strong because they are so highly compressed. There is additional layering within the mesosphere, revealed by the presence of seismic discontinuities. One important discontinuity occurs at about 400 km (250 mi) and another at about 670 km (415 mi) below the surface (Fig. 5.28). These discontinuities reveal the existence of boundaries or transitions within the mantle, as discussed earlier, but the nature of these boundaries is not well understood. They may be compositional boundaries (i.e., changes from one type of rock to another). Or they may result from changes in physical properties. For example, when the mineral olivine is squeezed at a pressure equal to that found at a depth of 400 km, the atoms rearrange themselves into a mineral structure that is more compact but still has the same chemical composition—that is, a high-pressure *polymorph* of olivine. (Polymorphs are discussed in chapter 2.) It may be this change to a more compact form that causes the 400-km seismic discontinuity.

It is important to remember that the mantle is basically solid rock, except for the small pockets of melt in the low-velocity zone. We know that the mantle must be solid because both P waves and S waves can travel through it. As you know from chapter 4, however, the mantle is undergoing convection, even

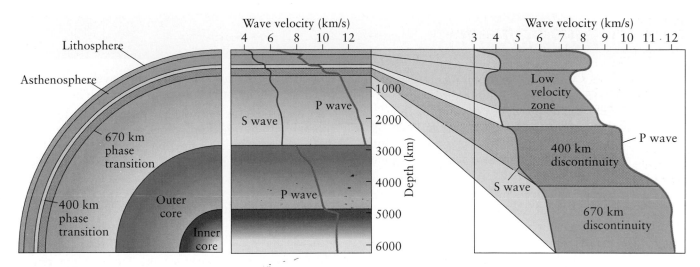

**Figure 5.28**
**SEISMIC DISCONTINUITIES IN THE MANTLE**
Changes in seismic wave velocity occur at the boundaries between the crust and the mantle (the Moho) and between the mantle and the core, owing to changes in composition. Another change occurs at a depth of 100 km (about 60 mi) and corresponds to the lithosphere–asthenosphere boundary. Changes also occur at depths of 400 km (250 mi) and 670 km (415 mi).

though it is solid. The pressures and temperatures deep within the Earth are so high that even solid rock can flow, although it flows very, very slowly.

### The Core

At a depth of 2883 km (1787 mi), another change in seismic velocity occurs (Figs. 5.26 and 5.28). This is the core–mantle boundary, which represents a change in both the composition and the physical properties of the rocks. The **core**, the innermost of the Earth's compositional layers, is the densest part of the Earth. It consists of material that "sank" to the center during the process of planetary differentiation. Geologists believe that the Earth's core is composed primarily of iron–nickel metal.

**core** The innermost compositional layer of the Earth.

As mentioned earlier, the S-wave shadow zone (and other sources of information) tells us that the *outer core,* from 2883 km to 5140 km (3187 mi) depth, must be a liquid. Within this is the *inner core,* where the pressure is so great that iron must be solid in spite of the very high temperature. (We know this from high-pressure experiments on iron.) The main difference between the inner and outer core is not one of composition; the compositions are believed to be virtually the same. Instead, the difference is in the physical states: one is a solid, the other a liquid. Note that the inner, solid core must have formed later than the core itself, because core formation occurred when the metallic iron of the core was molten. As heat escapes from the core and works its way to the surface, the core crystallizes. The solid inner core must be growing larger, but very slowly.

# WHAT'S AHEAD

Now you have learned some things about the interior of the Earth, its properties, and how it works. Heat escaping from the Earth's interior is the major driving force for plate tectonics, so what happens far below affects the surface of the Earth profoundly. In the next chapter we will look at how these internal processes influence and shape the Earth's surface through magmatic activity and volcanism.

## Chapter Highlights

**1.** Both direct and remote techniques are used to study the interior of the Earth. One direct method of study is drilling, which provides information about the distribution of rocks and heat within the crust. Another direct method of study is to examine **xenoliths** brought to the surface by volcanic rocks that originate at great depth within the Earth.

**2.** Basic astronomical techniques—determining the mass, diameter, volume, and average density of the Earth—tell us about the planet as a whole and about the distribution of material inside the planet. The study of meteorites also helps us make educated guesses about the composition of the Earth.

**3.** **Geochemistry** helps scientists study minerals that form at very high pressures and temperatures within the Earth. The techniques of **geophysics,** including studies of the Earth's gravity, magnetism, and seismicity, are used to learn about the Earth's interior.

**4.** Earthquakes are caused by sudden movements of blocks of the Earth's crust. As crustal blocks move past each other, they store elastic strain energy, bending and deforming until the stored energy overcomes the frictional forces between the blocks. When the blocks suddenly slip, the energy is released in the form of an earthquake, and the rocks snap back to their original form. This is the **elastic rebound theory.**

**5.** The energy of an earthquake is transmitted through the Earth in the form of **seismic waves. Surface waves** travel along the surface, whereas **body waves** travel through the Earth. There are two types of body waves: **compressional (P)** and **shear (S) waves. Seismographs** are used to measure the passage of seismic waves through rocks.

**6.** Earthquake hazards include ground motion, surface rupturing, landslides, fires, and tsunami. Using the techniques of **paleoseismology,** seismologists have been fairly successful at long-term prediction of earthquake hazards. However, it is not yet possible to predict the exact time, location, and magnitude of an earthquake.

**7.** Scientists can get a picture of the inside of the Earth by examining what happens to seismic waves while they are passing through the Earth. Certain characteristics of seismic waves are useful in this regard. They travel outward in all directions from the **focus** of an earthquake; their velocities depend on the physical properties of the material they are traveling through; and S waves cannot be transmitted through liquids.

**8.** When seismic waves pass from one material to another with different properties, their pathways bend; the bending is called **refraction.** Seismic wave refraction indicates the presence of **seismic discontinuities,** which can be caused by layering or changes in properties within the Earth. For example, refraction at the seismic discontinuity of the core–mantle boundary creates P-wave and S-wave shadow zones that reveal the presence of the Earth's core.

**9.** The outermost layer of the Earth is the **crust,** which varies in thickness from an average of 8 km (5 mi) for oceanic crust to an average of 45 km (28 mi) for continental crust. Continental crust is made up mostly of low-density granitic rocks, whereas oceanic crust is composed of higher-density basalt. The boundary between the lower crust and the mantle is called the **Moho** (short for Mohorovičić discontinuity).

**10.** The **mantle,** which extends from the Moho to the core, has a composition similar to that of the rock peridotite. The **lithosphere** comprises the outermost 100 km of the Earth, including the crust and the upper mantle just above the asthenosphere. The rocks of the lithosphere are cooler, stronger, and much more rigid than the underlying rocks of the asthenosphere.

**11.** In the **asthenosphere,** which extends from about 100 to 350 km (60 to more than 200 mi) below the surface, the rocks are weak and easily deformed because they are very near the temperature at which melting begins. Within the asthenosphere is a zone where there are small quantities of melted rock. The melted rock causes seismic waves to slow down markedly when they pass through this zone, so it is called the low-velocity zone.

**12.** The **mesosphere** extends from the bottom of the asthenosphere to the core–mantle boundary. Within the mesosphere there is additional layering at 400 km (250 mi) and 670 km (415 mi), as revealed by seismic discontinuities.

**13.** The **core,** the innermost of the Earth's compositional layers, is the densest part of the Earth. It is composed primarily of iron–nickel metal. The outer core is liquid. It surrounds the inner core, which is solid in spite of very high temperatures.

# ▶ *The Language of Geology*

- asthenosphere 144
- body wave 130
- compressional wave 130
- core 145
- crust 142
- elastic rebound theory 126
- epicenter 132
- focus 130

- geochemistry 121
- geophysics 122
- gravity anomaly 122
- lithosphere 144
- mantle 143
- Mercalli intensity scale 143
- mesosphere 144

- Moho 143
- moment magnitude 134
- P wave 130
- paleoseismology 138
- refraction 141
- Richter magnitude scale 133
- S wave 131

- seismic discontinuities 142
- seismic wave 130
- seismogram 130
- seismograph 128
- seismology 128
- shear wave 130
- surface wave 130
- xenolith 121

# ▶ *Questions for Review*

1. What is the density of the Earth? When we compare the density of the whole Earth to the density of surface rocks, what does it tell us about the properties of materials inside the Earth?
2. Why is the study of meteorites useful and applicable to the study of the Earth's interior?
3. What is a gravity anomaly? What kind of information can gravity anomalies provide about the Earth's interior?
4. What is the elastic rebound theory, and how does it work?
5. What are the main differences between P and S waves?
6. What techniques do seismologists use to predict earthquakes?
7. What are the three main scales that are used to measure the sizes of earthquakes?
8. Explain how refraction and seismic discontinuities reveal the presence of layering within the Earth.
9. What is the Moho? Is it a compositional boundary?
10. Draw a cross section showing the inside of the Earth, with the different layers and their thicknesses clearly labeled.

# ▶ *Questions for Thought and Discussion*

1. Why is short-term prediction of earthquakes so much less successful than long-term prediction? Why do you think seismologists are extremely cautious about making predictions? Do you think it will ever be possible to predict earthquakes accurately? Research your answer.
2. If you were asked to determine the exact shape and size of the Earth, how would you go about it?
3. Some of the boundaries inside the Earth represent transitions between layers with differing compositions, whereas others represent transitions between layers with differing physical states. Discuss this in greater detail.
4. Summarize the techniques, direct and indirect, that scientists use to study the Earth's interior.
5. Which of the techniques used to study the interior of the Earth do you think could also be used on other planets? Which cannot, and why?

For an interactive case study on earthquakes, visit our Web site.

# · 6 ·

# FROM THE EARTH'S INTERIOR: VOLCANOES AND IGNEOUS ROCKS

In January 1992, a series of earthquakes shook the lush tropical Caribbean island of Montserrat. The earthquakes continued, and puffs of steam and ash began to emanate from the Soufrière Hills volcano. Montserrat is no stranger to volcanic activity. It is a volcanic island, part of a chain that includes Mont Pelée, the scene of one of the worst volcanic disasters of twentieth century, in which 20,000 people were killed.

After hundreds of years of quiescence, however, the awakening of the volcano took residents by surprise. The U.S. Geological Survey sent volcanologists to Montserrat as part of the Volcano Disaster Assistance Program. In August 1995, after several years of intermittent seismic and volcanic activity, a large eruption occurred, darkening the town of Plymouth and blanketing it with ash. The first evacuation of the southern part of the island took place that month. Volcanic events and earthquake activity continued through 1996 and 1997, and the first direct casualties were reported in June 1997: 7 confirmed dead, 19 missing.

Only the northern third of Montserrat is now habitable. The island's remaining residents have started on the long road to recovery, but it won't be easy. The mainstays of Montserrat's economy—fishing, agriculture, and ecotourism—have all been damaged. The economic, social, and ecological recovery of the island and its resources will take careful planning. Unfortunately, volcanic activity continues intermittently; Soufrière Hills doesn't seem quite ready to settle down for its next long rest.

#### In this chapter you will learn about

- The conditions under which rocks and minerals melt, and the characteristics of molten rock
- How a few simple melting and crystallization processes can lead to the formation of a wide variety of igneous rock types
- How magmas cool and crystallize underground
- Why some volcanoes erupt explosively
- Why volcanoes occur where they do
- How volcanic hazards affect human interests

*Molten fire*
Glowing hot molten rock and lava spew forth from this volcano in Costa Rica.

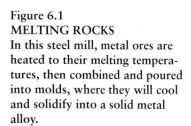

**? How hot do rocks have to be before they will melt?**

If a rock becomes very, very hot, it may reach its melting temperature, that is, the temperature at which it begins to change from a solid to a liquid state. You can take a rock into a specially equipped laboratory, heat it up, and watch it turn into a liquid (Fig. 6.1). The process is similar to heating ice to its melting temperature (0°C, or 32°F) and watching it turn into water, except that rocks melt at much higher temperatures than ice. Also, for reasons that will be revealed later in the chapter, rocks typically melt over a range of temperatures rather than at one specific temperature like ice. Most common rocks do not begin to melt until they reach a temperature of 800°C (almost 1500°F) or more.

The fact that rocks can melt, and that the resulting material is similar to the molten material that is erupted from volcanoes, was demonstrated almost 200 years ago by Sir James Hall, an experimentalist who worked with James Hutton. We continue to study the processes involved in rock melting because it helps us understand why different types of volcanoes behave the way they do. As you will see, it also helps us interpret the geologic processes involved in the formation of the wide variety of igneous rocks that make up the crust and mantle of the Earth.

## WHY DO ROCKS MELT?

The melting of rocks inside the Earth is not as simple as the melting of ice. One complication is that rocks are composed of many different minerals, each with its own characteristic melting temperature. This means that some components of a rock may melt at a lower temperature than others. Pressure and the presence of water in the rock's surroundings can also affect melting. Let's briefly examine these factors and then learn some things about the products of rock melting: *magma* and *lava*.

Figure 6.1
**MELTING ROCKS**
In this steel mill, metal ores are heated to their melting temperatures, then combined and poured into molds, where they will cool and solidify into a solid metal alloy.

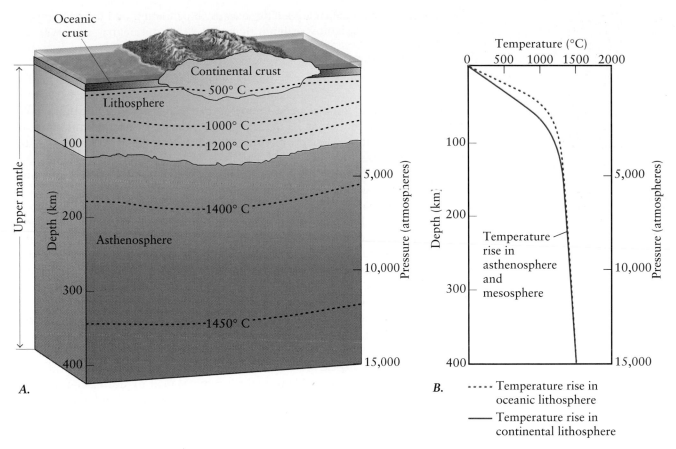

**Figure 6.2**
**GEOTHERMAL GRADIENT**
*A.* Temperature increases with depth in the Earth, as shown here. The dashed lines are *isotherms,* lines of equal temperature. Note that temperature increases more slowly with depth under the continents than under the oceans, where it gets quite hot at a shallow depth. *B.* This shows the same information as in (*A*) but in graph form. The Earth's surface is at the top, so depth (and corresponding pressure) increases downward. Temperature increases from left to right. The blue curve shows how temperature increases with depth under the oceans. The red curve shows how temperature increases more slowly with depth under the continents.

## Heat and Pressure Inside the Earth

The interior of the Earth is hot. If you go into a mine and measure rock temperatures, you will find that the deeper you go, the higher is the temperature. This relationship between depth and temperature is shown in Figure 6.2. The usual way to draw this diagram is with the Earth's surface at the top, temperature increasing toward the right, and both depth and pressure increasing toward the bottom.

### Geothermal Gradient

The rate at which temperature increases with depth in the Earth, shown by the red and blue curves in Figure 6.2B, is called the **geothermal gradient.** The red curve on the diagram shows how temperature increases with depth beneath the continents. You can see that temperature rises quite rapidly with depth along this curve, reaching 1000°C (more than 1800°F) at a depth of just over 100 km (62 mi). The temperature increase under the oceans, shown by the blue curve, is even more precipitous, reaching 1000°C at a shallower depth. In other words, the ge-

**geothermal gradient** The rate at which temperature increases with depth in the Earth.

othermal gradient is steeper under the oceans (blue curve), and hotter temperatures are reached at a shallower depth than under the continents (red curve).

## Effects of Pressure

**Why isn't the interior of the Earth entirely molten?**

This brings up an interesting question. As we have noted, common rocks usually melt at temperatures above 800°C (almost 1500°F), depending on the exact composition of the rock. In fact, measurements made at erupting volcanoes show that most lavas are molten at 1000°C (Fig. 6.3). Yet we can see from Figure 6.2 that temperatures in the Earth's mantle are well over 1000°C. So why isn't the mantle entirely molten?

The answer is that pressure also influences melting temperatures. As pressure increases, the temperature at which a rock or mineral melts also rises (Fig. 6.4A). For example, the feldspar mineral albite ($NaAlSi_3O_8$), a common rock-forming mineral, melts at 1104°C (2019°F) at the Earth's surface, where the pressure is low. However, at a depth of 100 km, where the pressure is 35,000 times greater, the melting temperature of albite is 1440°C (2624°F). Therefore, whether a particular rock or mineral melts at a given depth depends on both the temperature and the pressure at that depth. The mantle is mostly solid because the pressures are so great that the rocks do not melt.

Here is another interesting question: If most minerals and rocks melt at *higher* temperatures when they are under pressure, do they melt at *lower* temperatures when the pressure is removed? The answer, of course, is yes. You can make a mineral or a rock melt by increasing the temperature sufficiently at a given pressure. It is also possible to make a mineral melt by decreasing the pressure sufficiently at a given temperature. This is also illustrated in Figure 6.4A. If a rock from deep inside the Earth moves up toward the surface where pressures are much lower, it may melt—even if its temperature hasn't increased. This process is called *decompression melting*.

## Effects of Water

The effect of pressure on melting is straightforward—provided that the mineral is dry. When water (or water vapor) is present in the rock or mineral, however, another effect is important. At a given pressure, a wet mineral will melt at a lower temperature than a dry mineral of the same composition (Fig. 6.4B). The effect is similar to that of salt on ice. Anyone who lives in a cold climate knows that salt can melt the ice on an icy road. This happens because a mixture of salt and ice

**Figure 6.3**
**TAKING THE VOLCANO'S TEMPERATURE**
This geologist, wearing a protective suit, is measuring the temperature of lava erupting from Mauna Loa volcano, Hawaii. The temperature of molten rock can be measured from a distance, with some accuracy, on the basis of its color: bright orange, yellow, and white indicate the hottest regions, while deeper red, brown, and black colors indicate cooler regions.

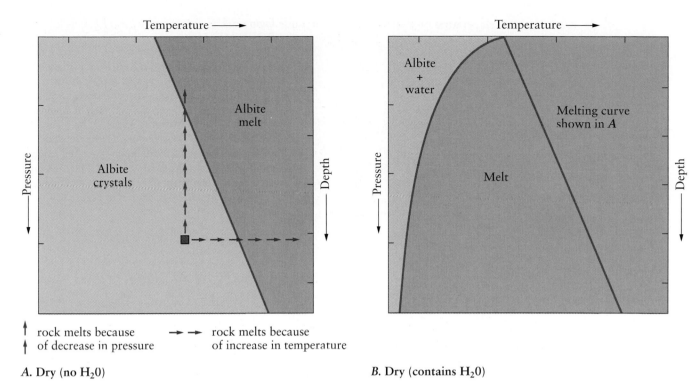

Figure 6.4
**EFFECTS OF PRESSURE AND WATER ON MELTING**
*A.* The melting temperature of a dry mineral will increase if pressure is increased, as shown here for the mineral albite. If you increase the temperature of a mineral sufficiently at a given pressure, it will surpass its melting point. This is shown by the blue arrow. Alternatively, if you decrease the pressure on a mineral sufficiently at a given temperature, it will surpass its melting point. This is shown by the red arrow. *B.* The melting temperature of a wet mineral typically gets lower and lower as pressure increases. This is the opposite of what happens to dry minerals.

has a lower melting temperature than pure ice. In the same way, a mineral and-water mixture generally has a lower melting temperature than the pure mineral.

The effect of water on the melting temperatures of most rocks and minerals is opposite to the effect of pressure: In general, melting temperatures *increase* with increasing pressure but *decrease* with increasing concentration of water in the rock. What's more, the effect of water becomes more pronounced as pressure increases. In other words, if you increase the pressure on a wet rock or mineral, it will melt at lower and lower temperatures—exactly the opposite of what happens with a dry rock or mineral. As we will see in chapter 11, this effect can be very important in environments such as subduction zones beneath the ocean, where water is carried down into the mantle by subducting oceanic plates.

## Partial Melting

Most common rocks are composed of several different mineral components. For this reason, a rock typically does not melt at a particular temperature the way a single mineral does. Instead, as shown in Figure 6.5 and 6.6, most rocks melt over a range of temperatures. For most common rocks, the temperature at which melting begins—called the *eutectic* temperature—is lower than the melting temperatures of the constituent minerals. In other words, the melting temperature of the mixture (the rock) is lower than the melting temperature of the individual ingredients or components of the mixture (the minerals). An analogy is the ice-and-salt mixture mentioned earlier; the mixture begins to melt at a temperature lower

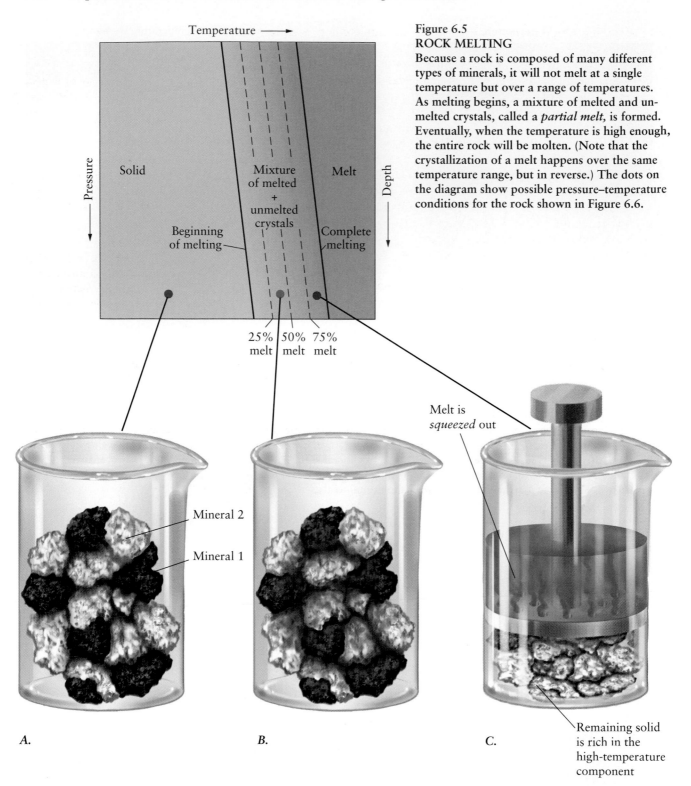

Figure 6.5
ROCK MELTING
Because a rock is composed of many different types of minerals, it will not melt at a single temperature but over a range of temperatures. As melting begins, a mixture of melted and unmelted crystals, called a *partial melt,* is formed. Eventually, when the temperature is high enough, the entire rock will be molten. (Note that the crystallization of a melt happens over the same temperature range, but in reverse.) The dots on the diagram show possible pressure–temperature conditions for the rock shown in Figure 6.6.

Figure 6.6
FRACTIONATION BY PARTIAL MELTING
A. This rock in a laboratory container consists of a mixture of two minerals. B. As temperature increases to the eutectic temperature, the mixture begins to melt, producing a partial melt. C. If the liquid produced by partial melting is squeezed out and removed, the solid residue that is left behind will be rich in the components that have not yet begun to melt. The process of partially melting a rock and separating it into components with different compositions is called *fractionation* or *fractional melting.*

than the melting temperatures of either of the individual components of the mixture, ice and salt.

If the temperature continues to increase after a rock has begun to melt, there will be some melted material and some unmelted material—a **partial melt.** It is also called a *fractional melt,* because only a fraction or portion of the rock has melted. Eventually, when the temperature is high enough, the entire rock will melt. Imagine a rock that is a mixture of two minerals (Fig. 6.6A). As the rock heats up to its eutectic temperature, melting begins (Fig. 6.6B) and a partial melt is formed. If the temperature were to rise high enough, the entire rock would melt. But what if something happened in this rock-melting scenario, such as compression caused by the movement of lithospheric plates, so that the liquid portion of the partial melt were squeezed out? The liquid portion would consist mostly of components that melted at a low temperature, whereas the solid left behind would consist mostly of components with a higher melting temperature (Fig. 6.6C). We would be left with a melt and a solid, each with a different composition (and, in fact, each with a composition different from that of the starting mixture). This process, whereby a melt is separated from residual solid material during the course of melting, is called **fractionation** (or *fractional melting*).

## Magma and Lava

Molten rock underground is called **magma.** Magma may contain dissolved gases and mineral grains suspended in the molten rock. Magma is usually less dense than the solid rock that surrounds it. Therefore, it is buoyant and tends to rise toward the Earth's surface. If the magma makes it all the way to the surface, it may emerge as **lava** and flow over the land. We can't study magma directly, but by observing and studying eruptions of lava we can draw three important conclusions concerning magma:

1.  Magma has a *range of compositions* in which silica ($SiO_2$) is usually dominant.
2.  Magma is characterized by *high temperatures.*
3.  Magma has the properties of a liquid, including the *ability to flow.*

Let's examine each of these observations in greater detail.

*Composition.* The bulk chemical composition of magma is dominated by the most abundant elements in the Earth—silicon (Si), aluminum (Al), iron (Fe), calcium (Ca), magnesium (Mg), sodium (Na), potassium (K), hydrogen (H), and oxygen (O). As we learned in chapter 2, $O^{2-}$ is the most abundant anion. Therefore it is usual to express the compositional variations of magmas in terms of oxides, such as $SiO_2$, $Al_2O_3$, $CaO$, and $H_2O$. The most abundant component of common magmas is silica, $SiO_2$, which usually accounts for 45 to 75 percent by weight. Because of the predominance of silica in such magmas, we call them *silicate magmas.* In addition, small amounts of dissolved gas (usually 0.2 to 3% by weight) are found in most magmas. The principal magmatic gases are water vapor and carbon dioxide. Despite their low abundance, these chemically active gases can strongly influence the properties of the magma.

*Temperature.* As discussed earlier, the temperatures at which rocks melt are very high. Sometimes a magma or a lava is *superheated*—that is, its temperature is even higher than the temperature at which the rock became completely molten. During eruptions of volcanoes such as Kilauea in Hawaii and Mount Vesuvius in Italy, magma temperatures have been recorded as ranging from 800° to 1200°C (1472° to 2192°F). Experiments using synthetic magmas in the laboratory sug-

**partial melt** A mixture of molten material and solid material.

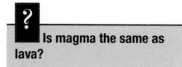
**Is magma the same as lava?**

**fractionation** A process in which a melt becomes separated from the remaining solid material during the course of melting.

**magma** Molten rock under the ground.

**lava** Magma that reaches the surface in a molten state.

gest that under some conditions magma temperatures may be as high as 1400°C (2552°F).

*Viscosity.* All magmas are liquids and have the ability to flow, although some are very stiff and therefore flow very slowly. The degree to which a substance offers resistance to flow is its **viscosity;** the more viscous a substance, the less fluid it is. Water is an example of a very fluid, runny, low-viscosity liquid. Tar—especially on a cold day—is an example of a high-viscosity liquid that does not flow easily. Dramatic videos of lava flowing rapidly down the sides of a volcano prove that some magmas can be very fluid (Fig. 6.7). Lava moving down a steep slope on Mauna Loa in Hawaii has been clocked at 64 km/h (40 mi/h), indicating a very low viscosity.

> **viscosity** The degree to which a substance offers resistance to flow.

The viscosity of a magma or a lava depends on its composition. The $SiO_4^{4-}$ anions that form the foundation for most common rock-forming minerals are also present in magmas. In magmas, as in minerals, the anions link together, or *polymerize,* by sharing oxygens. Unlike the anions in silicate minerals, however, those in magma form irregularly shaped groupings. As the polymerized groupings become larger, the magma becomes more viscous, that is, more resistant to flow, and behaves increasingly like a solid. The higher the silica content of the magma, the larger the polymerized groups. For this reason, high-silica magma tends to be more viscous than low-silica magma.

Temperature and gas content also affect the viscosity of magmas. The higher the temperature and, in general, the higher the gas content, the lower the viscosity and the more readily the magma flows. A very hot magma that erupts from a volcano may flow readily, but as it cools it becomes more viscous and eventually stops flowing. In Figure 6.8 the smooth, ropy-surfaced material, called *pahoehoe* (a Hawaiian word), was formed from a hot, fluid lava with a lot of dissolved gas. The rubbly, rough-looking material was formed from a cooler lava with a higher viscosity. Scientists call this rough, blocky lava *aa* (another Hawaiian word).

So far we've learned how and why rocks melt and how the melting process influences the properties of magmas. Next we consider what happens when magmas cool and crystallize.

# COOLING AND CRYSTALLIZATION

> **igneous rock** A rock that solidified from a magma or lava.
>
> **crystallization** The process whereby crystals form and grow in a cooling magma (or lava).

Rocks that have crystallized from a molten state are **igneous rocks.** When magmas cool and mineral grains start to form, the process is called **crystallization.** Why do geologists study crystallization? Just as melting influences the properties of magmas, cooling and crystallization influence the properties of igneous rocks. To understand how igneous rocks form, we must study these processes. For example, the *rate* of cooling and crystallization determines how large the crystals in the rock will grow and affects other aspects of the rock's overall appearance or *texture.* The *composition* of the magma is also important; it is reflected in the final rock composition and *mineral assemblage.* Let's first examine the factors that affect crystallization and then learn more about the process of crystallization.

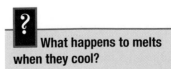

**What happens to melts when they cool?**

## Rate of Cooling and Crystallization

> **texture** The size, shape, and arrangement of mineral grains that give a rock its overall appearance.

Rate of cooling is the main factor that distinguishes the two main classes of igneous rocks from one another. *Volcanic rocks* solidify quickly at or near the Earth's surface, whereas *plutonic rocks* cool and crystallize more slowly, deep underground. How fast a magma cools will determine whether crystals have a chance to grow in the rock and, if so, how large they will be. In other words, the rate of cooling and crystallization is very important in determining the **texture** of

◀ Figure 6.7
**FREE-FLOWING LAVA**
This stream of low-viscosity (runny) lava moving smoothly and quickly away from an eruptive vent demonstrates how fluid and free flowing lava can be. The temperature of the lava is about 1100°C (about 2000°F). The eruption occurred in Hawaii in 1983.

Figure 6.8 ▶
**PAHOEHOE AND AA FLOWS**
The flow of lava is controlled by viscosity. Two strikingly different lava flows are visible in this scene from Kilauea volcano in Hawaii. They have the same composition but different gas contents, which caused them to have different viscosities when they were molten. The lower flow, on which the geologist is standing, is a pahoehoe flow formed from a rapidly flowing, low-viscosity lava such as that shown in Figure 6.7. The upper flow (the one being sampled by the geologist), which was very viscous and slow moving, is an aa flow that erupted from kilauea in 1989. The pahoehoe flow erupted 40 years earlier.

the resulting rock—that is, the size, shape, and arrangement of mineral grains that give a rock its overall appearance.

## Volcanic Rocks and Their Textures

As you have learned, magmas are formed deep underground and then rise toward the Earth's surface. Sometimes they make it all the way to the surface, pouring out as lava over the land or underwater (Fig. 6.9). When lavas and other

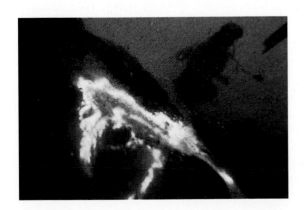

Figure 6.9
**UNDERWATER ERUPTION**
Hardening lava forms pillow-shaped masses in this underwater eruption off the coast of Hawaii.

# Volcanic Rock Textures

Figure 6.10

◄ *A.* GLASSY TEXTURE: OBSIDIAN
This photograph shows volcanic glass, or *obsidian,* from the Jemez Mountains in New Mexico. The curved ridges are typical of the fracture pattern in glass broken by a sharp blow. This specimen is 10 cm (4 in) across and is glass of rhyolitic composition.

▲ *B.* PORPHYRITIC TEXTURE: PHENOCRYSTS IN GROUNDMASS
Magmas often contain suspended crystals. If the magma rises to the surface, it may cool and crystallize very quickly. In this way, a very fine-grained rock will form, with some relatively large crystals scattered throughout. The large crystals are *phenocrysts,* the very fine-grained material surrounding them is the *groundmass,* and the texture of the rock as a whole is called *porphyritic.* This particular photograph shows a porphyritic rock with feldspar phenocrysts from Oslo, Norway.

◄ C. VESICULAR TEXTURE: GAS BUBBLES CAPTURED
The holes in this rock were left by small bubbles of gas in the magma. This is called *vesicular texture,* and the holes are called *vesicles.* The specimen is 4.5 cm (almost 2 in) across.

**volcanic rock** An igneous rock formed from lava and other volcanic materials that cooled and solidified at or near the surface of the Earth.

**volcanic glass** A volcanic rock that solidified too rapidly for crystals to grow from the lava.

volcanic materials cool and solidify, they become **volcanic rocks.** They are also called *extrusive* rocks because the magmas they came from were extruded at the Earth's surface.

Lavas that erupt from volcanoes cool and solidify very quickly, because the temperature of the air (or water) is much lower than that of the lava. Sometimes lava cools so quickly that recognizable crystals do not form. The resulting noncrystalline rock has a *glassy* texture and is called **volcanic glass,** or *obsidian* (Fig. 6.10A). Even when crystals do form in solidifying lavas, the crystals are typically very small—they simply don't have enough time to grow very big before the rock has completely solidified. Thus, volcanic rock is commonly very fine-grained, a texture that is termed *aphanitic.* Usually it is impossible to see the individual crystals in aphanitic rocks without a microscope (Fig. 6.11).

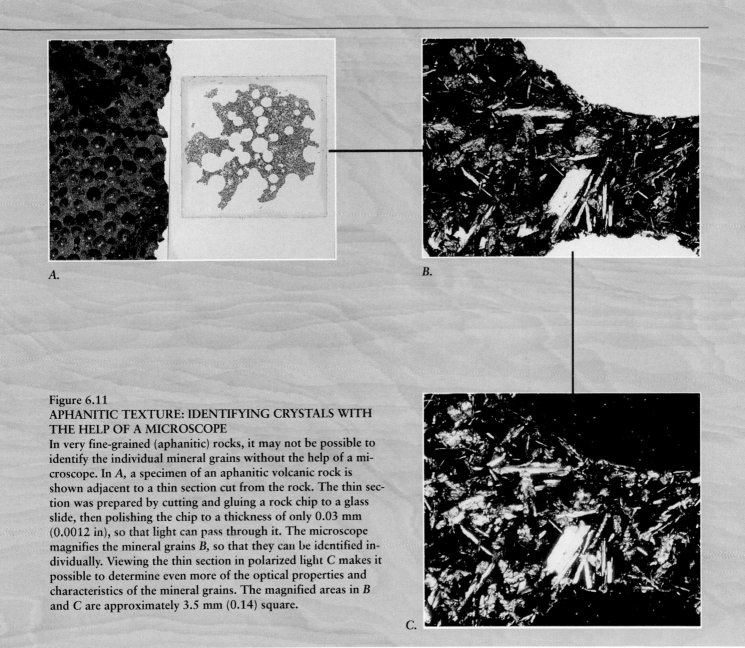

A.

B.

**Figure 6.11**
**APHANITIC TEXTURE: IDENTIFYING CRYSTALS WITH THE HELP OF A MICROSCOPE**
In very fine-grained (aphanitic) rocks, it may not be possible to identify the individual mineral grains without the help of a microscope. In *A*, a specimen of an aphanitic volcanic rock is shown adjacent to a thin section cut from the rock. The thin section was prepared by cutting and gluing a rock chip to a glass slide, then polishing the chip to a thickness of only 0.03 mm (0.0012 in), so that light can pass through it. The microscope magnifies the mineral grains *B*, so that they can be identified individually. Viewing the thin section in polarized light *C* makes it possible to determine even more of the optical properties and characteristics of the mineral grains. The magnified areas in *B* and *C* are approximately 3.5 mm (0.14) square.

C.

Other textures also occur in volcanic rocks. If a magma contains suspended crystals that are extruded along with the lava, the resulting rock will consist of large crystals surrounded by aphanitic rock. The large grains are called *phenocrysts*. The fine-grained part of the rock that encloses the large grains is called the *groundmass,* and the rock texture is called *porphyritic* (Fig. 6.10B). Porphyritic texture is very common in volcanic rocks. Dissolved gases, too, can have an impact on the texture of volcanic rock. As lava cools, its viscosity increases and it becomes increasingly difficult for gas bubbles to escape. When the lava finally solidifies into rock, the bubbles that were expelled last become trapped. This leaves bubble holes, called *vesicles,* in the rock (Fig. 6.10C). If the magma is highly charged with gas, a frothy mass of bubbles may form, resulting in a spongy-textured, low-density rock called *pumice.*

# Plutonic Rock Textures

Figure 6.12

*A.* PLUTONIC ROCKS: LARGE CRYSTALS
Large crystals of quartz (light gray) and feldspars (pink and white) are clearly visible and identifiable in this plutonic rock, a granite from Death Valley National Park in California. The pink feldspar crystals in this rock are about 1 cm (half an inch) long.

▼

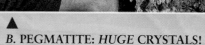

▲
*B.* PEGMATITE: *HUGE* CRYSTALS!
The crystals in *pegmatites* grow to be very large, usually because of the presence of fluids while the rock is crystallizing. In this outcrop of pegmatite from the Black Hills, North Dakota, very large crystals of mica (dark), feldspar (pink), and quartz (clear) are visible.

## Plutonic Rocks and Their Textures

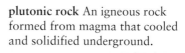

**plutonic rock** An igneous rock formed from magma that cooled and solidified underground.

Some magmas never make it to the surface. They remain deep underground, cooling and solidifying very slowly. Rocks that form in this way are called **plutonic rocks** (after Pluto, the Greek god of the underworld). They are also called *intrusive* rocks because the magmas they came from intruded into the surrounding rocks.

Unlike the minerals in volcanic rocks, those in plutonic rocks usually have time to form recognizable crystals. Thus, compared with that of volcanic rocks, the texture of plutonic rocks is typically coarse-grained, or *phaneritic* (Fig. 6.12A). In phaneritic rocks the individual mineral grains can be seen without a microscope. A plutonic rock that contains unusually large mineral grains (larger than 2 cm or about 1 in) is called a *pegmatite*. Sometimes individual mineral grains in pegmatites are huge—several meters or more in length (Fig. 6.12B). Pegmatites typically form in the last stage of crystallization of a plutonic rock body. At this stage, there may be a buildup of gases, including water vapor, in the remaining fraction of magma. The presence of vapor facilitates the growth of large crystals by allowing chemical components to migrate very quickly to the growing crystal surfaces.[1]

---

[1] It is a common misconception that the crystals in pegmatites are large because they grow very slowly. In fact, the rate of cooling can be the same for pegmatites as for the rest of the magma body (or even faster). The crystals grow to large size because of the rapid movement (called *diffusion*) of chemical components to their growing surfaces by means of *vapor transport*.

## Chemical Composition

The main volcanic and plutonic rock types are derived from silicate magmas that range from 45 to 75 percent silica by weight (Table 6.1).[2] The chemical composition of a magma determines which minerals will crystallize from the melt. The types and amounts of minerals in a rock constitute its *mineral assemblage*. When magma of a given composition solidifies, the resulting mineral assemblage can be the same for both extrusive and intrusive rocks of that composition. The main difference is in the texture of the rock. Depending on the exact conditions of crystallization, a single magma composition might yield a fine-grained volcanic rock or a coarse-grained plutonic rock, each containing the same mineral assemblage.

Most common igneous rocks are composed of one or more of six minerals or mineral groups: quartz, feldspar, mica, amphibole, pyroxene, and olivine. These are the most abundant rock-forming minerals. Quartz, feldspar, and some micas are light-colored minerals; the others are dark-colored. Rocks that contain large amounts of light-colored minerals are usually light in color themselves and relatively rich in silica; they are termed *felsic* (from "feldspar" and "silica"). Rocks that contain large amounts of dark-colored minerals are usually dark in color and relatively low in silica; they are termed *mafic* (from "magnesium" and "ferric," or iron-rich, reflecting the common chemical constituents of dark minerals).

## Composition of Volcanic Rocks

The three main types of volcanic rock are derived from the common silicate magmas.

1. **Basalt** is the dominant rock in oceanic crust. In fact, it is the most common volcanic rock in the solar system, as far as we know. Basalt is a dark gray, dark green, or black, fine-grained volcanic rock with low silica content (approximately 45 to 50% $SiO_2$ by weight). The mafic minerals olivine and pyroxene may make up more than 50 percent of the volume of the rock; the rest is usually plagioclase feldspar.

   **basalt** A mafic volcanic rock with low silica content, consisting primarily of plagioclase feldspar, pyroxene, and olivine.

2. **Andesite** is an intermediate-silica volcanic rock (approximately 60% $SiO_2$ by weight), with lots of feldspar and some amphibole or pyroxene. It is named for the Andes, the major volcanic mountain system of western South America. Andesite is usually gray, purple, or dark green.

   **andesite** An intermediate-silica volcanic rock, consisting primarily of abundant feldspar with some mafic minerals.

3. **Rhyolite** represents the felsic, high-silica end of the scale (approximately 70 to 75% $SiO_2$ by weight), consisting largely of quartz and feldspars. The volcanic rock shown in Figure 6.10A is a rhyolite. Another high-silica volcanic rock, *dacite*, is similar to rhyolite but contains plagioclase feldspar, instead of the alkali feldspar found in rhyolites. Because rhyolite and dacite tend to be extremely fine-grained or even glassy, it is difficult to tell them apart without a microscope. Rhyolite and dacite are usually pale in color, ranging from nearly white to shades of gray, yellow, red, or purple.

   **rhyolite** A felsic, high-silica volcanic rock, often very fine-grained or glassy in texture.

## Composition of Plutonic Rocks

For each of the common volcanic rocks described in the preceding section there is a plutonic rock that has the same bulk composition (and sometimes the same mineral assemblage) but a different texture (Table 6.1).

---

[2] You may hear low-silica rocks and magmas referred to as *basic* and high-silica rocks and magmas referred to as *acidic*. These are old-fashioned terms that are still in use even though they no longer reflect our understanding of the chemistry of rocks and magmas.

**TABLE 6.1    Volcanic and Plutonic Equivalents**

| Silica Content of Magma | Approx. Melting Temperature of Rock | Viscosity of Magma | Color of Mineral Assemblage | Hand Samples of | |
|---|---|---|---|---|---|
| | | | | Resulting Plutonic Rocks | Resulting Volcanic Rocks |
| High (≈70–75%) | Low (≈800°C) | High (not runny) | Felsic (light-colored) | Granite | Rhyolite |
| Intermediate (≈60%) | Medium (≈1000°C) | Medium | Intermediate | Diorite | Andesite |
| Low (≈45–50%) | High (≈1200°C) | Low (runny) | Mafic (dark-colored) | Gabbro | Basalt |

1. **Gabbro** is the plutonic rock with the same composition as the volcanic rock basalt. Gabbro is a low-silica, relatively coarse-grained, dark-colored rock, with large amounts of olivine, pyroxene, and plagioclase feldspar. An even more mafic rock called *peridotite* (which consists almost entirely of the minerals pyroxene and olivine) is an important constituent of the mantle.

2. **Diorite** is an intermediate-silica rock, the plutonic equivalent of andesite. The felsic, high-silica end of the scale is represented by

3. **Granite,** which is compositionally equivalent to rhyolite. As with volcanic rocks, we distinguish two important felsic rock types on the basis of the feldspars they contain: Granites contain alkali feldspar, whereas plagioclase is the dominant feldspar in *granodiorites*. Thus, granodiorite is the plutonic equivalent of the volcanic rock dacite. Rocks of granitic and granodioritic composition are common in the continental crust, especially in the cores of mountain ranges.

# Fractional Crystallization

The magmas and igneous rocks just discussed are the most common types, but hundreds of different kinds of igneous rocks are found on the Earth. Most are rare, but the fact that they exist suggests an important point: A single magma of a given composition can crystallize into many different kinds of igneous rock. This is called **magmatic differentiation.**[3] How does magmatic differentiation occur? In other words, how can all these different rock types form from a limited range of magma compositions? A particularly important mechanism is the process of *fractional crystallization.*

Recall how fractionation occurs through partial melting: A newly forming melt is squeezed out, leaving behind a solid residue with a different composition. The same thing can happen—but in reverse—during crystallization. Like melting, crystallization occurs over a range of temperatures. A solidifying magma forms several different minerals, and those minerals start to crystallize from the melt at different temperatures. The temperature range across which crystallization occurs is the reverse of the melting range discussed previously (Fig. 6.5).

What would happen if the newly formed crystals were separated from the remaining melt? Crystals can become separated from melt in a variety of ways. For example, it might happen if the crystallizing magma began to move as a result of compression. If magma is squeezed through small openings, such as pore spaces or fractures, crystals may clog the openings so that only the liquid can pass through (Fig. 6.13A); this is called *filter pressing.* Crystal-melt separation can also happen if newly formed crystals are denser (heavier) than the liquid from which they are crystallizing. In that case, the crystals may sink to the bottom of the magma body, forming a solid layer of crystals covered by melt that has a different composition; this is called *crystal settling* (Fig. 6.13B). Crystal-melt separation can also occur if the newly formed crystals are lighter (less dense) than the liquid, in which case they might float to the top; this is called *crystal flotation* (Fig. 6.13C). Finally, a newly forming crystal may grow a protective rim around its outer edges, sheltering the original core of the crystal from any further contact with the melt; this is called *crystal zonation* or *mantling* (Fig. 6.13D).

**gabbro** A low-silica, mafic igneous rock, the plutonic equivalent of basalt.

**diorite** An intermediate-silica igneous rock, the plutonic equivalent of andesite.

**granite** A high-silica, felsic igneous rock, the plutonic equivalent of rhyolite, consisting primarily of quartz, feldspar, and mica.

**?** **Why are there so many different kinds of igneous rocks?**

**magmatic differentiation** The formation of many different kinds of rock from a single magma.

---

[3] Compare this with the use of the term *differentiation* in chapter 1, to describe the process whereby a homogeneous planet segregates into layers of differing composition.

Figure 6.13
**SEPARATING CRYSTALS FROM MELTS**
Magmatic differentiation—the separation of crystals from melts—can occur during crystallization. There are several ways in which newly formed crystals can be separated from melt. *A.* If pressure is placed on the magma, the liquid portion may squeeze out through small openings while the crystals remain behind. *B.* If the newly formed crystals are denser (heavier) than the liquid, they may sink; this is called crystal settling. *C.* If the newly formed crystals are less dense (lighter) than the liquid, they may float; this is called crystal flotation. *D.* Sometimes a newly forming crystal grows a protective rim or a series of rims that shelter the core of the crystal from any further contact with the melt; this is called crystal zonation or mantling.

*A.* Filter pressing

Crystals clog opening

① Crystals and melt

② Melt is pushed through and separated from crystals

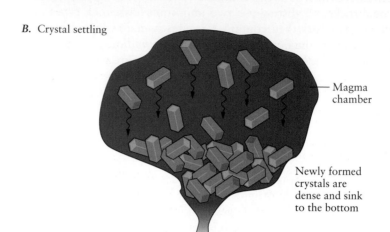

*B.* Crystal settling

Magma chamber

Newly formed crystals are dense and sink to the bottom

*C.* Crystal flotation

Newly formed crystals are light and float to the top

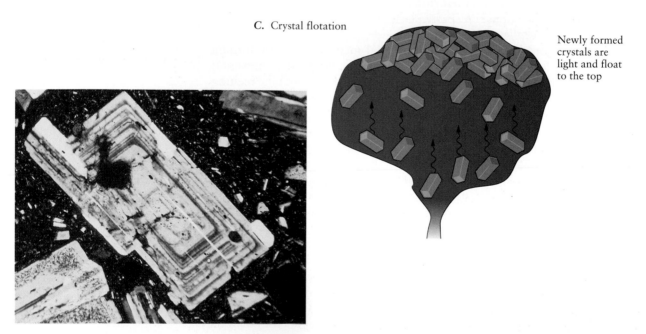

*D.* Crystal zonation or mantling

**fractional crystallization** A process in which crystals and liquids become separated from one another during crystallization.

No matter how it happens, when crystals and liquids are separated from one another during crystallization, the process is termed **fractional crystallization.** Through fractional crystallization a single magma can be separated into a liquid and a solid with different compositions. You can see that different combinations of magma composition, rates of cooling, conditions of crystallization, and fractional crystallization can lead to many, many variations in the resulting rock types. This is magmatic differentiation, and it is the main reason we find so many different types of igneous rocks on the Earth.

### Bowen's Reaction Series

It was a Canadian-born scientist, N. L. Bowen, who first established the importance of magmatic differentiation by fractional crystallization. Because magma of basaltic composition is far more common than either rhyolitic or andesitic magma, he suggested that all other magmas may be derived from basaltic magma through fractional crystallization. At least in theory, he argued, a single magma could crystallize into basalt, rhyolite, and a whole series of intermediate rock compositions, as a result of fractional crystallization. The sequence of rock compositions and mineral assemblages that would form during such a process is known as *Bowen's Reaction Series,* which is discussed in greater detail in Appendix D. We now know that such extreme differentiation rarely happens. Different magma types do not all evolve from basaltic magma; rather, they form as a result of partial melting in different geologic environments (which we will learn about in chapter 11). Nevertheless, fractional crystallization is an extremely important process. It contributes to the production of a wide range of rock compositions from a relatively limited number of magma compositions.

So far we have examined melting and its influence on the properties of magma, as well as crystallization and its influence on the properties of rock. We now turn to some of the processes involved in the formation of specific plutonic and volcanic rocks.

> **?** **Can two different rocks crystallize from the same melt?**

# PLUTONS AND PLUTONISM

All bodies of intrusive igneous rock, regardless of shape or size, are called **plutons.** The magma that forms a pluton did not originate at the location where we find the pluton. Rather, it intruded into the surrounding rock, which is termed the *country rock.* The common types of plutons are illustrated in Figure 6.14.

**pluton** An intrusive igneous rock body.

## Batholiths and Stocks

Plutons are given special names depending on their shapes and sizes. A **batholith** is the largest kind of pluton (the name comes from the Greek roots *bathy-* and *lithos,* meaning "deep rock"). It is an irregularly shaped igneous body that cuts across the layering of the rock into which it intrudes. Some batholiths exceed 1000 km (620 mi) in length and 250 km (155 mi) in width. The largest in North America is the Coast Range Batholith of British Columbia and northern Washington, which is about 1500 km (more than 900 mi) long (Fig. 6.14B). **Stocks** are also irregularly shaped plutons, but their maximum dimension at the surface is 10 km (about 6 mi). A stock may be associated with a batholith, as shown in Figure 6.14A, or it may even be the top of a partly eroded batholith.

Where it is possible to see them, the walls of batholiths tend to be nearly vertical. This early observation led to a commonly held perception that most batholiths extend downward to the base of the Earth's crust. However, geophysical measurements and studies of very deeply eroded bodies of igneous rock suggest that this perception is incorrect. Most batholiths seem to be only 20 to 30 km (12 to 19 mi) thick, which is relatively thin compared to their great lengths and widths.

**batholith** A large, irregularly shaped igneous body that cuts across the layering of the rock into which it intrudes.

**stock** An irregularly shaped pluton, smaller than a batholith.

# Ancient Plutons Exposed by Erosion

Figure 6.14

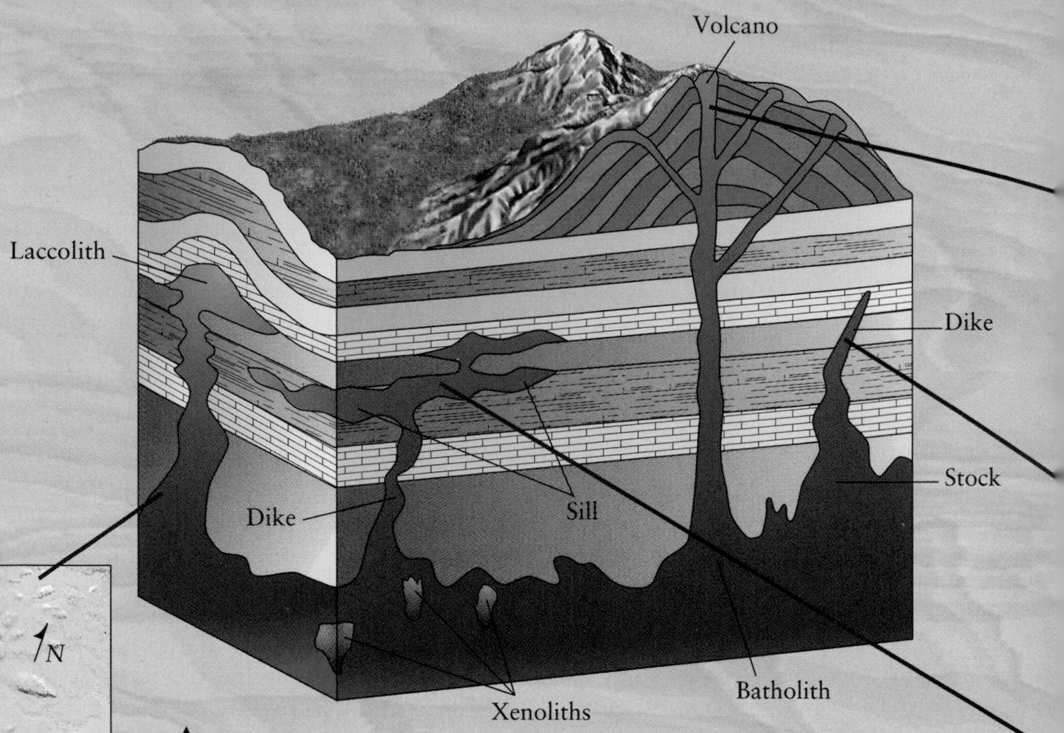

**A. PLUTONS**
This diagrammatic section of part of the crust shows the various forms taken by plutonic rocks. Many plutons were once connected to volcanoes, and there is a close relationship between intrusive and extrusive rocks.

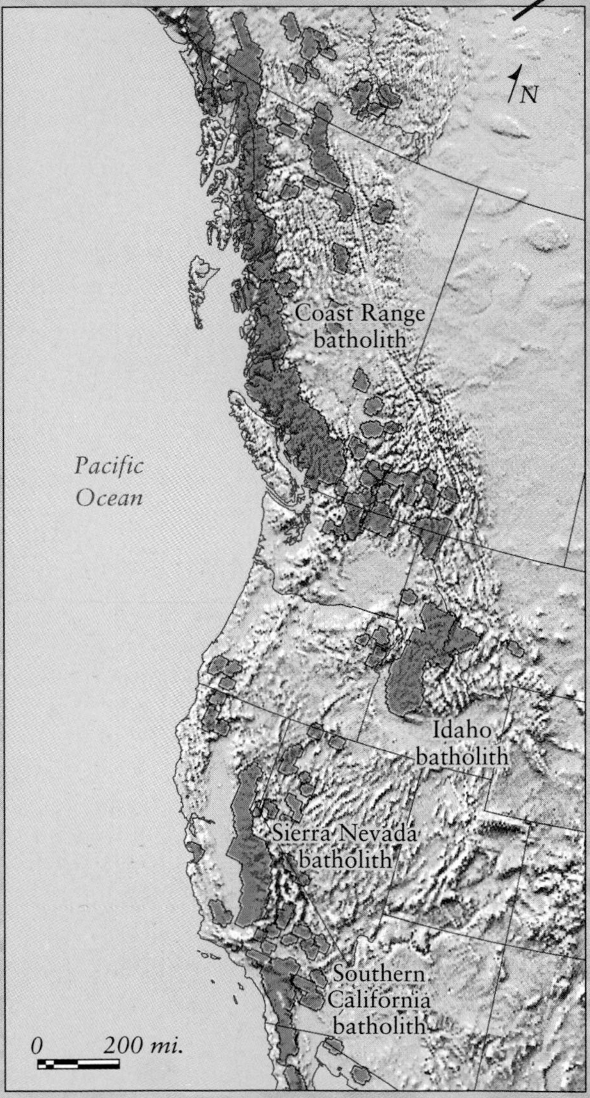

**B. BATHOLITHS**
The Idaho, Sierra Nevada, and Southern California batholiths, the largest batholiths in the United States, are dwarfed by the Coast Range Batholith in Southern Alaska, British Columbia, and Washington. Each of these giant batholiths formed from magma generated by the partial melting of continental crust.

**C. VOLCANIC NECK IN AN ▶ OLD VOLCANO**
This volcanic neck is the eroded remnant of an ancient volcano in the Hoggar Mountains of Algeria.

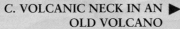

**◀ *D.* DIKE**
A dike of gabbro cuts across horizontally layered sedimentary rocks in Hance Rapid, Grand Canyon National Park, Arizona.

**_E._ SILL ▶**
A sill of gabbro (dark brown) is intruded parallel to layering of sedimentary rocks above and below, in Big Bend National Park, Texas.

Most stocks and batholiths are granitic or granodioritic. Many do not consist of just one type of rock but instead contain zones of different composition. Batholiths are not formed by magma squeezing into fractures (the way some smaller intrusive igneous bodies are formed), because no fractures are that large. They are probably formed by extensive partial melting of the lower continental crust. Despite their huge size, the magma bodies that form batholiths do move upward, pushing overlying rocks out of the way. The magma dislodges fragments of overlying rock in a process known as *stoping*. The dislodged fragments are usually denser than the rising magma, so they sink. As they sink, some are partly dissolved and absorbed by the magma, and others may sink to the floor of the magma chamber. Any fragment of rock that is enclosed in a solidified magma is known as a *xenolith*, like the mantle xenoliths discussed in chapter 5.

## Dikes and Sills

**dike** A sheetlike body of igneous rock that cuts across layering or contacts in the rock into which it intrudes.

**sill** A tabular body of intrusive igneous rock, parallel to the layering of the rocks into which it intrudes.

An obvious indication of past igneous activity is a **dike,** a sheet-like body of igneous rock that cuts across the layering of the rock into which it intrudes (Fig. 6.14D). A dike forms when magma squeezes into a cross-cutting fracture and then solidifies. Tabular bodies of intrusive igneous rock that are parallel to the layering of the rocks into which they intrude are called **sills** (Fig. 6.14E). A variation of a sill is a *laccolith,* an igneous body that has been intruded parallel to the layering of the country rock, which has been pushed up to form a mushroom-shaped dome as a result. These intrusive forms may occur together as part of a network of plutonic bodies, as shown in Figure 6.14A.

Dikes and sills can range from very small to very large. For example, the Great Dyke in Zimbabwe is a mass of gabbro nearly 500 km (310 mi) long and 8 km (5 mi) wide. It fills what must once have been a huge fracture in the crust. A large and well-known sill-like mass, also gabbroic, can be seen in the Palisades, the cliffs that line the Hudson River opposite New York City. The Palisades Intrusive Sheet is about 300 m (almost 1000 ft) thick. It formed from multiple charges of magma intruded between layers of sedimentary rock about 200 million years ago. The sheet is visible today because tectonic forces raised that portion of the crust upward, and the covering sedimentary rocks were then largely removed by erosion.

As Figure 6.14A shows, there is a close link between plutonic rocks and volcanoes. Beneath every volcano is a complex network of channels and chambers through which magma reaches the surface. When a volcano stops erupting, the magma in the channels solidifies into plutonic rocks of various types. Some of these rocks take the form of dikes or sills. *Volcanic pipes* are the remnants of pipe-like channels that originally fed magma to the volcanic vent. A pipe that has been stripped of its surrounding rock by erosion is called a *volcanic neck;* an example is shown in Figure 6.14C.

**?** **Are magmas under the ground connected to volcanoes at the surface?**

# VOLCANOES AND VOLCANISM

**volcano** A vent through which lava, pyroclastic material, and gases are erupted.

A **volcano** is a vent through which magma, solid rock debris, and gases are erupted. The term *volcano* comes from Vulcan, the Roman god of fire. For most people, the thought of a volcano conjures up visions of rivers of glowing lava pouring out over the landscape. But other types of materials, such as gases and fragments of rock, can also emerge from volcanoes. There are different types of volcanic eruptions, each characterized by different volcanic materials and landforms.

# Eruptions, Landforms, and Materials

**?** **Why do volcanoes have different shapes?**

Volcanic eruptions are difficult to classify. Most eruptions change as they happen, either gradually (from month to month or year to year) or abruptly (from one day or even one hour to the next). However, we can distinguish among several types of volcanic eruptions on the basis of style of eruptive activity and type of landform built up by the volcano. These are summarized and illustrated in Figure 6.15. Eruptions that come from a central vent are categorized mainly on the basis of the explosivity of the eruption. Other kinds of eruptions, such as those that come from long cracks or *fissures* in the crust, are also summarized in the figure.

## Characteristics of Nonexplosive Eruptions

Most people regard any volcanic eruption as hazardous and view volcanoes as dangerous places that should be avoided. However, geologists have discovered that some volcanoes are comparatively quiet. Nonexplosive eruptions, such as those we can witness in Hawaii, are relatively safe compared to explosive eruptions like those at Mount St. Helens in Washington (1980), Mount Pinatubo in the Philippines (1991), or Soufrière Hills, Montserrat (1997), all of which caused substantial destruction and loss of life.

The differences between nonexplosive and explosive eruptions are largely a function of the viscosity and dissolved gas content of the magma. Recall that low-silica, basaltic magmas have relatively low viscosities; in other words, they are very runny. They tend to flow easily out of a volcanic vent, like the free-flowing lava shown in Figure 6.7. Successive basaltic lava flows may eventually build up to form a broad, flat volcano with gently sloping sides. Because it resembles a shield lying on its back, this type of volcanic landform is called a **shield volcano**. A classic example is the Hawaiian volcano Mauna Kea (Fig. 6.15A). Although Mauna Kea rises only about 4.2 km (2.6 mi) above sea level, its true height when measured from the sea floor is 10.2 km (more than 6.3 mi). This exceeds the height of Mount Everest (9.1 km, or 5.6 mi), making Mauna Kea the tallest mountain on the Earth.

**shield volcano** A broad, flat volcano with gently sloping sides, built of successive fluid lava flows.

Some basaltic lava reaches the surface via elongate fractures or *fissures*. Eruptions of this type are called *fissure eruptions*. Low-viscosity lavas emerging from fissures on land tend to spread widely and may create vast, flat lava plains called *basalt plateaus* or *flood basalts* (Fig. 6.15B). The Deccan Traps in India resulted from an extensive fissure eruption of this type. The Roza flow, a great sheet of solidified basaltic lava in eastern Washington State, can be traced over an area of 22,000 km$^2$ (almost 8500 mi$^2$) and has a volume of 650 km$^3$ (156 mi$^3$). Centralized volcanoes like the Hawaiian shield volcanoes Mauna Loa and Mauna Kea commonly display some fissure activity as well (Fig. 6.15C). Volcanic activity at seafloor spreading centers such as the Mid-Atlantic Ridge is often related to extensive fissuring (Fig. 6.15D).

## Characteristics of Explosive Eruptions

When little or no dissolved gas is present, a magma will erupt as a lava flow. If dissolved gas is present, however, it must escape somehow. Gases dissolved in a low-viscosity basaltic magma may cause the lava to bubble and fountain quite dramatically, especially at the beginning of an eruption (Fig. 6.15E). After the fountaining has died down, most basaltic eruptions are relatively quiet and are characterized by extensive lava flows.

The higher the viscosity of the magma, however, the more difficult it is for the gas to form bubbles and escape. In andesitic and rhyolitic magmas, more than in basaltic magmas, gas bubbles are held back by the viscosity of the fluid.

# Volcanoes and Eruptions

A. This is Mauna Kea, a 4.2-km-high shield volcano in Hawaii, as seen from Mauna Loa. Note the gentle slopes formed by highly fluid basaltic lava flows.

B. These nearly flat-lying basaltic lava flows are typical of plateau or flood basalts. These flows form part of the Columbia River Plateau near Idaho Falls, Idaho.

C. This is an Aerial view of a fissure eruption, Mauna Loa, Hawaii. Basaltic lava is erupting from a series of parallel fissures.

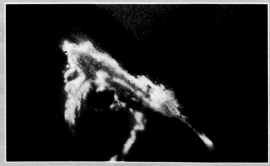

D. Lava oozes out, forming pillows of basalt in an underwater eruption.

**LESS VIOLENT ERUPTIONS**

Figure 6.15
This figure illustrates the main types of volcanic eruptions and landforms associated with each type of volcano, with some examples.

A. Hawaiian
- gently sloping shield volcanoes
- fluid, low-silica, basaltic magma
- nonexplosive fountaining
- lava flows, lava tubes, spatter cones
- examples: Kilauea (Hawaii); Mauna Loa (Hawaii)

B. Plateau (Flood) Basalts
- voluminous flows of very fluid, low-silica, basaltic magma
- extensive sheets of nearly flat flows
- broad, coalescing shields
- examples: Columbia River Plateau (Washington); Deccan Traps (India)

C. Fissure Eruptions
- radial fractures and lines of vents
- magma and eruption style variable
- may be related to a central vent
- examples: Laki (Iceland); Yellowstone Plateau (Wyoming)

D. Submarine Volcanism
- often related to seafloor spreading along midocean rifts
- pillow lavas and glassy flows
- fluid, low-silica, basaltic magma
- examples: Mid-Atlantic Ridge; East Pacific Rise; Hawaiian-Emperor seamount chain

*E.* Incandescent lava fountains from this spatter cone at Cerro Azul in the Galápagos Islands.

**E. Strombolian**
- spatter cones and cinder cones
- mild explosive eruptions; fumaroles
- incandescent bombs, ash, and lapilli
- magma-type variable
- examples: Stromboli (Italy); Parícutin (Mexico); Kivu (Zaïre)

*F.* Glowing lava and ash cascade down Mount Mayón in the Philippines, 1993. Ash from the eruption cloud blanketed the town.

**F. Peléan and Vulcanian**
- cones
- hot flows of blocks and ash; glowing avalanches
- viscous, silica-rich, gas-rich magma
- violent, destructive eruptions
- examples: Mont Pelée (Martinique); Vulcan (Italy); Mount Mayón (Philippines)

*G.* This photo shows steam-blast, or *phreatic* eruption of the volcanic island, Surtsey, Iceland.

**G. Steam-blast (Phreatic)**
- steep-sided domes, collapse caldera
- rising magma comes into contact with seawater or groundwater
- violent, explosive eruptions, glowing avalanches
- powerful blasts of rock ± magma ± steam
- often viscous, silica-rich magma
- examples: Surtsey (Iceland); Krakatau (Indonesia)

*H.* This is the plinian ash column from the May 1980 eruption of Mount St. Helens. At least 63 people died as a result of this eruption.

**H. Plinian**
- stratovolcanoes
- viscous, gas-rich, silica-rich magma
- voluminous pyroclastics
- exceptionally powerful blast from eruptive column
- examples: Vesuvius (Italy); Krakatau (Indonesia); Mount Pinatubo (Philippines); Mount St. Helens (Washington)

**MORE VIOLENT ERUPTIONS**

*A.*

*B.*

*C.*

Figure 6.16
TEPHRA
*A.* Large volcanic bombs up to half a meter long cover the surface of Haleakala volcano in Maui, Hawaii. *B.* Intermediate-sized tephra called lapilli cover the Kau Desert, also in Hawaii. *C.* Volcanic ash, the smallest tephra, blankets a farm in Oregon after the eruption of Mount St. Helens in 1980.

pyroclastic rock A rock formed from the consolidation of rock fragments ejected during a volcanic eruption.

As viscous magmas move up toward the Earth's surface, the decrease in pressure eventually allows the dissolved gas to come out of solution and expand to form bubbles. When the gas finally escapes, it usually does so explosively, blasting small bits of magma and other fragmental material in all directions.

A fragment of rock ejected during a volcanic eruption is called a *pyroclast* (from the Greek words *pyro,* meaning "fire," and *klastos,* meaning "broken"; hence, hot, broken fragments). Pyroclasts are also referred to as *tephra,* from the Greek word for ash. Tephra of different sizes are illustrated in Figure 6.16. Volcanic *bombs* are the coarsest, ranging up to the size of a small car, and volcanic *ash* is the finest type of tephra. Volcanic ash is not the same as the ash that forms from the burning of wood, which typically contains a lot of carbon; rather, volcanic ash consists largely of particles of volcanic glass, which are so small they may be individually distinguishable only under a microscope.

Rocks formed from the consolidation of pyroclasts are called **pyroclastic rocks.** The names of pyroclastic rocks are keyed to the size of the tephra they are composed of. Pyroclastic rocks are called *agglomerate* when the tephra particles are large and *tuff* when the particles are small. Loose tephra particles can be converted into pyroclastic rock by cementing with another mineral or by the welding together of hot, glassy particles.

Pyroclastic material can erupt in different ways. If the mixture of hot pyroclastic material and gases is denser than air, it may rush down the flanks of the volcano in a *pyroclastic flow.* The presence of gases in the mixture allows the pyroclastic material to move downslope very rapidly, the same way that air facilitates the downslope rush of snow in an avalanche. Pyroclastic flows are also called *glowing avalanches* or *nuées ardentes*. They are extremely dangerous because of their intense heat and the great speed at which they surge down the volcano's slopes (Figure 6.15F). If seawater or groundwater mixes with hot magma, a powerful *steam blast* or *phreatic eruption* may occur (Fig. 6.15G).

The most violent type of pyroclastic eruption is an *eruption column* (also called a *Plinian column*), a turbulent mixture of hot gas and tephra that rises explosively into the cooler air above the vent (Fig. 6.15H). Eruption columns may reach heights as great as 45 km (28 mi). Eventually the dark, ominous-looking eruption column spreads out to form a mushroom cloud similar to that created by nuclear bomb explosions. As the cloud drifts with the wind, particles of debris fall out and accumulate on the ground as tephra deposits. During exceptionally explosive eruptions, pyroclastic material may spread over distances of thousands of kilometers. Sometimes, instead of emerging from the top of the volcano, an explosive column may burst out from the side. This happened in May 1980 during part of the modern eruption of Mount St. Helens. An eruption of this type is called a *lateral blast*. (Mount St. Helens' 1980 eruption had both a lateral blast and a vertical Plinian column phase.)

In contrast to nonexplosive volcanoes, explosive volcanoes tend to build volcanic landforms with steep sides. **Stratovolcanoes,** also called *composite volcanoes,* are composed of solidified lava flows interlayered with pyroclastic material. The steep, snow-capped peaks of classic stratovolcanoes like Mount Fuji in Japan (Fig. 6.17) are among the most beautiful sights on the Earth. The majority of andesitic and rhyolitic volcanoes in continental and subduction zone settings are stratovolcanoes. In general, however, stratovolcanoes are much smaller than basaltic shield volcanoes like Mauna Loa and Mauna Kea. A very steep-sided volcanic structure composed entirely of loose pyroclastic material is called a *spatter cone* or *tephra cone;* examples are seen in Figures 6.15E and 6.18.

**stratovolcano** A volcano composed of solidified lava flows interlayered with pyroclastic material.

**Figure 6.17**
**MOUNT FUJI: CLASSIC STRATOVOLCANO**
Mount Fuji, Japan, a snow-clad giant towering over the countryside, displays the classic steep-sided profile of a stratovolcano.

Figure 6.18
**CRATER LAKE**
Beautiful Crater Lake in Oregon occupies a caldera that crowns the summit of a once-lofty strato-volcano called Mount Mazama. Wizard Island, a small tephra cone in the middle of the lake, was formed after the collapse that created the caldera. Crater Lake is the deepest lake in the United States.

## Other Volcanic Features

**Craters and Calderas.** Near the summit of most volcanoes is a *crater*, a funnel-shaped depression from which gas, tephra, and lava are ejected. Some volcanoes have a much larger depression known as a *caldera*, a roughly circular, steep-walled basin that may be several kilometers or more in diameter. Calderas can form explosively as a result of the partial emptying of a magma chamber. Rapid ejection of magma or tephra during an eruption can leave the magma chamber empty or partly empty, causing the unsupported roof of the chamber to collapse under its own weight. Crater Lake in Oregon occupies a caldera 8 km (5 mi) in diameter that formed after a great eruption about 6600 years ago by a volcano we now call Mount Mazama (Fig. 6.18). Tephra deposits from that enormous eruption can still be seen in Crater Lake National Park and over a vast area of the northwestern United States and southwestern Canada.

**Resurgent Domes.** A volcano does not necessarily become inactive after a major eruption. If magma begins to enter the chamber again, it may lift the collapsed floor of the caldera or crater and form a *resurgent dome*. Small lava flows and tephra cones may also build up inside the caldera. Wizard Island in Crater Lake is a tephra cone that was formed in this way. The caldera of Mount St. Helens contains a dome of sticky, viscous lava that has been growing since the explosive eruption of May 1980.

**Geysers, Fumaroles, and Thermal Springs.** A vent or fracture from which volcanic gases are emitted is referred to as a *fumarole*. The gases tend to be mostly water vapor, but other gases, such as evil-smelling sulfur, may be present as well. When volcanism finally ceases, the rock in an old magma chamber may remain hot for hundreds of thousands of years. Groundwater that comes into contact with the hot rock is heated, creating *thermal springs;* many such springs have become famous health spas. A thermal spring with a natural system of plumbing that causes intermittent eruptions of water and steam is a *geyser*. The name comes from the Icelandic word *geysir*, "to gush" (Fig. 6.19).

**?** **Why do so many people live near active volcanoes?**

## Hazards and Prediction

Since A.D. 1800, there have been 19 volcanic eruptions in which a thousand or more people have died (Fig. 6.20). Like other natural hazards, volcanic eruptions have *primary effects*, those directly caused by the eruption; *secondary effects,*

Figure 6.19
**THE GREAT GEYSIR**
The Great Geysir, Iceland, from which all geysers take their name.

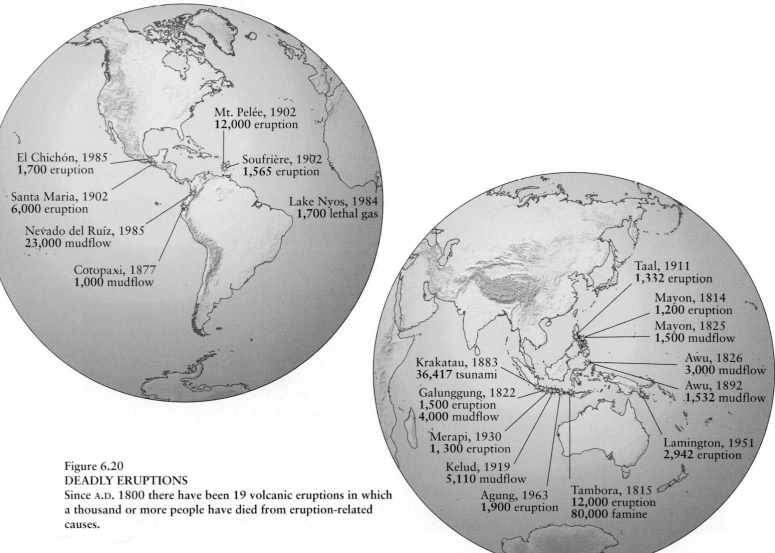

Mt. Pelée, 1902
**12,000** eruption

Soufrière, 1902
**1,565** eruption

El Chichón, 1985
**1,700** eruption

Santa Maria, 1902
**6,000** eruption

Lake Nyos, 1984
**1,700** lethal gas

Nevado del Ruíz, 1985
**23,000** mudflow

Cotopaxi, 1877
**1,000** mudflow

Taal, 1911
**1,332** eruption

Mayon, 1814
**1,200** eruption

Mayon, 1825
**1,500** mudflow

Awu, 1826
**3,000** mudflow

Awu, 1892
**1,532** mudflow

Krakatau, 1883
**36,417** tsunami

Galunggung, 1822
**1,500** eruption
**4,000** mudflow

Lamington, 1951
**2,942** eruption

Merapi, 1930
**1, 300** eruption

Kelud, 1919
**5,110** mudflow

Tambora, 1815
**12,000** eruption
**80,000** famine

Agung, 1963
**1,900** eruption

Figure 6.20
**DEADLY ERUPTIONS**
Since A.D. 1800 there have been 19 volcanic eruptions in which a thousand or more people have died from eruption-related causes.

those indirectly triggered by the eruption; and *tertiary effects,* the long-term or even permanent after-effects of a major eruption. Most volcanic hazards cannot be controlled, but their negative impacts can be mitigated by effective prediction methods. It's also worth remembering that not all the effects of volcanism are harmful—some are beneficial.

## Primary Effects

Most volcanoes produce at least some lava flows. Because people are usually able to outrun them, lava flows typically cause more property damage than deaths or injuries. In Hawaii, where Kilauea has erupted almost continuously for more than a decade, homes, cars, roads, and forests have been buried by lava flows, but no lives have been lost (Fig. 6.21). It is sometimes possible to control a flow, at least partially, by building retaining walls or chilling the front of the flow with a water spray. These approaches are designed to divert the flow from threatened property.

Many of the primary hazards of volcanism are directly related to the effects of violent eruptions, particularly pyroclastic activity. Unlike slowly moving lava flows, hot, rapidly moving pyroclastic flows and lateral blasts may overwhelm people before they can run away. The most destructive pyroclastic flow in this century (in terms of loss of life) occurred on the island of Martinique in 1902. An avalanche of hot ash rushed down the flanks of Mont Pelée at a speed of more than 160 km/h (99 mi/h), killing 29,000 people. In A.D. 79, many of the citizens of Pompeii and Herculaneum were buried under hot pyroclastic material from an eruption of nearby Mount Vesuvius (Fig. 6.22). Pliny the Elder was killed during the eruption, which is why this type of eruption is called a Plinian column. Many of the victims were killed by poisonous volcanic gases, and their bodies were later buried by pyroclastic material. In 1986, at least 1700 people and 3000 cattle lost their lives when poisonous gas was emitted from a volcano at Lake Nyos, Cameroon (see Box).

## Secondary Effects

Secondary effects, those triggered by or related to volcanic activity but not directly caused by an eruption, include fires, which are often caused by lava flows;

**Figure 6.21**
**HOUSES THREATENED BY LAVA FLOWS**
An advancing tongue of basaltic lava sets fire to a house in Kalapana, Hawaii, during an eruption of Kilauea in June 1989. Flames at the edge of the flow are due to burning lawn grass.

Figure 6.22
VICTIMS OF MOUNT VESUVIUS
Scientists made plaster casts of the bodies of these citizens of Pompeii, Italy, who were killed during the eruption of Mount Vesuvius in A.D. 79. Death was caused by poisonous gases; then the bodies were buried by pyroclastic material. Over the centuries the bodies decayed, but casts of their body shapes were preserved in the tephra.

Figure 6.23
HOT VOLCANIC MUDFLOW TURNS DEADLY
This mudflow was formed when volcanic ash from a small eruption of Nevado del Ruíz mixed with melted snow. The thick, hot mudflow inundated the town of Armero, Colombia, in 1985. More than 23,000 people were killed.

flooding, which may happen if a river channel is blocked or a crater lake bursts; *tsunami,* a destructive type of ocean wave that is also caused by submarine earthquakes; and *volcanic tremors,* a type of seismic activity that is associated with volcanism.

Pyroclastic material can cause hazardous effects even after an eruption has ceased. Rain or meltwater from snow at the volcano's summit can loosen volcanic ash piled on a steep slope and start a deadly mudflow, referred to as a *lahar.* In 1985, a small eruption of Nevado del Ruíz in Colombia melted part of the snow cap on the volcano's summit. Mudflows were formed when the meltwater mixed with volcanic ash. Massive lahars moved swiftly down river valleys on the flanks of the volcano, killing at least 23,000 people (Fig. 6.23). A related phenomenon is a volcanic *debris avalanche,* in which many different types of materials—mud, blocks of pyroclastic material, trees, and so on—are mixed together. Much of the damage from the 1980 eruption of Mount St. Helens was caused by a devastating debris avalanche.

## Tertiary and Beneficial Effects

Volcanic activity can change a landscape. River channels may be blocked and the flow of water diverted. Mountainous terrains may be drastically altered; in the 1980 eruption of Mount St. Helens, for example, the entire top and side of the mountain were blown away. New land can be formed, like the black sand beaches of Hawaii, which are made of dark pyroclastic material. Volcanic eruptions can also alter the chemistry of the atmosphere. In fact, the Earth's atmosphere and oceans originated from the outgassing of volatile material through volcanoes. The atmospheric effects of major eruptions can include salty, toxic, or acidic precipitation; spectacular sunsets; extended periods of darkness; and

**Two small lakes in a remote part of Cameroon,** a small country in central Africa (Fig. B6.1), made international news in the mid-1980s when lethal clouds of carbon dioxide ($CO_2$) gas from deep beneath the surface of the lakes escaped into the surrounding atmosphere, killing animal and human populations far downwind. The first gas discharge, which occurred at Lake Monoun in 1984, asphyxiated 37 people. The second, which occurred at Lake Nyos in 1986, released a highly concentrated cloud of $CO_2$ that killed more than 1700 people. The two events have similarities other than location: both occurred at night during the rainy season; both involved volcanic crater lakes; and both are likely to recur without some type of technologic intervention.

Immediately after the disasters scientists began monitoring the lakes in an attempt to understand what had triggered these catastrophic incidents. They identified several factors that caused instability in the lakes and they produced models of the conditions that probably preceded the degassing. The researchers found that Lake Nyos had a huge reservoir of $CO_2$ stored deep beneath unstable, density-layered lake waters. When it was disturbed by some event—wind at the surface, a landslide, an earthquake, or a minor eruption—the stratified column of water turned over, allowing approximately 100 million m$^3$ (more than 3.5 billion ft$^3$) of $CO_2$ to bubble to the surface in just 2 hours (Figs. B6.2 and B6.3).

It soon became apparent that despite the release of huge amounts of $CO_2$ during the overturns, vast amounts of gas remain dissolved at great depths in the lakes, and more is being added to each year from natural sources. On the basis of

**Figure B6.1**
**PATH OF DEATH**
These maps show the locations of Cameroon and Lake Nyos and the path taken by the cloud of carbon dioxide that emerged from the lake.

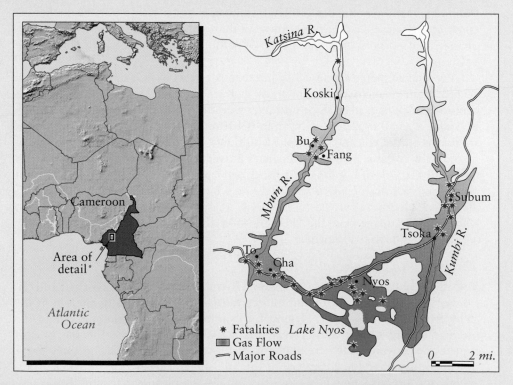

*Fatalities   Lake Nyos
◼ Gas Flow
━ Major Roads

0 ⸻ 2 mi.

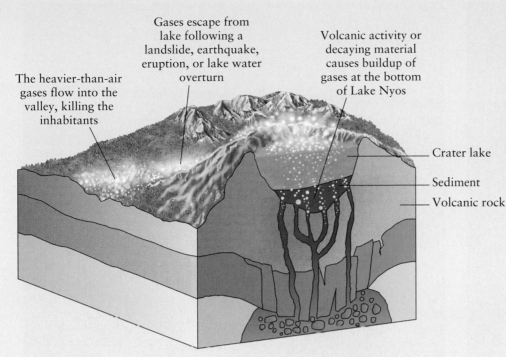

The heavier-than-air gases flow into the valley, killing the inhabitants

Gases escape from lake following a landslide, earthquake, eruption, or lake water overturn

Volcanic activity or decaying material causes buildup of gases at the bottom of Lake Nyos

Crater lake

Sediment

Volcanic rock

**Figure B6.2**
**LETHAL GAS ESCAPE**
A lethal cloud of $CO_2$ bubbled up from the bottom of Lake Nyos after the stratification of the lake water was disturbed. The gas, which is invisible and heavier than air, flowed over the natural dam surrounding the lake and down the adjacent valley.

6 years of records of dissolved gases, temperature, density, and recharge rates in each lake, scientists predicted that Lake Monoun was in danger of a violent degassing event within 10 years. Lake Nyos, which is much larger and deeper, would build up its $CO_2$ to dangerous concentrations within about 20 years.

Gas-rich volcanic crater lakes are known to occur in other localities as well, including Japan, Democratic Republic of the Congo (Zaïre), and Indonesia. In theory, the hazards associated with such lakes can be alleviated by controlled degassing. This technique involves installing a subsurface network of pipes to reduce the concentrations of gas in deep water. Gas-charged water is pumped to the surface, where it releases gases to the air in nontoxic amounts. In Lake Monoun a prototype system of this type was tested successfully in 1992. It used three pipes, each 14 cm (5.5 in) in diameter, placed deep beneath the surface of the water. Calculations showed that this procedure would restore $CO_2$ concentrations to stable levels in 2 to 3 years.

However, Lake Nyos presents a more complicated situation because of its greater size and depth. The challenge is to control the flow of partially degassed water without upsetting the stratification of the lake water. One proposed solution is to divert water to a storage reservoir and then reinject it into the lake at the level of its natural buoyancy. The goal would be to establish a balance between the degassed water flowing into the lake and the bottom water removed from the lake. But unless the system is precisely calculated and perfectly designed, there is a danger that it will initiate the very situation the scientists are trying to prevent.

**Figure B6.3**
**THE AFTERMATH**
In addition to the loss of human life, many cattle were killed by asphyxiation as a result of the gas discharge from Lake Nyos.

Figure 6.24
RICH VOLCANIC SOILS
Seemingly tranquil, lush fields grow on rich volcanic soils around an active volcano in the Philippines. Although an active volcano can be deadly during an eruption, the volcanic ash deposits fertilize and enrich the soil.

Figure 6.25
MONITORING VOLCANOES BY SATELLITE
This is a satellite image showing the area around Mount Vesuvius and the Bay of Naples, Italy. Vesuvius, an active volcano, is the circular structure, center right. Recent lava flows show up bright red in this image, which recorded infrared radiation (heat). Older lavas and volcanic ash show up as yellows and oranges. The dark blue and purple region at the head of the bay is the city of Naples. Left of Naples, near the center of the image, is a cluster of smaller volcanoes called the Flegreian Fields. Changes in the temperature of the ground surface can be monitored by comparing successive satellite images, because infrared imagery is sensitive to the temperature of ground materials.

global cooling resulting from the blockage of solar radiation by aerosols and fine pyroclastic material. The 1815 eruption of Mount Tambora in Indonesia caused three days of total darkness as far as 500 km (300 mi) from the volcano. The following year was so cool that it was called "the year without a summer"; average global temperatures fell more than 1°C below normal, and there were widespread crop failures.

Not all the impacts of volcanism are negative, and it is no accident that many people live near active volcanoes. Periodic volcanic eruptions replenish the mineral content of soils, ensuring continued fertility (Fig. 6.24). Volcanism provides geothermal energy and is linked with the formation of some types of mineral deposits. Volcanoes also provide some of the most magnificent scenery on this planet.

## Predicting Eruptions

It isn't possible to stop volcanic eruptions, but sometimes they can be predicted. The first step in prediction is to identify a volcano as active, dormant, or extinct.

An *active* volcano is one that has erupted within recorded history; a *dormant* volcano has not erupted in recent history; and an *extinct* volcano shows no signs of activity and is deeply eroded. Volcanoes can change from one of these categories to another; for example, Mount Pinatubo in the Philippines has been dormant for several hundred years prior to its eruption in 1991, when it became active once again.

Another important step in prediction is to identify the volcano's past eruptive style. Mount Pinatubo is surrounded by thick deposits of pyroclastic material, a sign that the volcano erupted violently in the past and might do so again. The composition and tectonic setting of the volcano are also important. For example, andesitic and rhyolitic continental and subduction zone volcanoes are more likely to erupt explosively than are basaltic volcanoes and fissure eruptions.

Scientists can monitor a potentially dangerous volcano for signs of increasing activity. Some of the warning signs that have proven useful are changes in the shape or elevation of the ground, such as bulging, swelling, or doming caused by inflation of the underlying magma chamber; changes in the temperature of crater lakes, well water, or hot springs; changes in heat output at the surface; changes in the amount or composition of gases emitted from the volcanic vent; and sudden increases in local seismic activity. Some monitoring, such as the monitoring of seismic activity, is done with local instruments placed directly on the flanks of the volcano. Other monitoring, such as monitoring gas emissions or the temperature of the ground surface, can be done by means of remote sensing from satellites (Fig. 6.25).

# WHAT'S AHEAD

As you have discovered in this chapter, learning about the different processes involved in rock melting and crystallization can help us understand the formation of some of the many different types of plutonic and volcanic rocks on the Earth. The chapters in this part of the book have focused principally on internal processes such as magmatic activity, earthquakes, and plate tectonics. In the next part of the book, we turn to the crustal part of the rock cycle, in which weathering, erosion, rock deformation, and metamorphism act to modify and break down the products of igneous activity and tectonic uplift. Then, in chapter 11, we will bring all of this information together, with a look at all the common rock types in the context of the plate tectonic environment in which they form.

## Chapter Highlights

**1.** The **geothermal gradient** is the increase of temperature with increasing depth in the Earth. The gradient is steeper—that is, higher temperatures are reached at shallower depths—beneath the oceans than beneath the continents.

**2.** The mantle is mostly solid, even though temperatures in the mantle are very high. This is because the melting temperatures of most rocks and minerals increase as pressure increases—as long as they are dry. Water lowers the melting temperatures of most rocks

and minerals, and this effect is stronger with increasing pressure.

**3.** When a rock reaches its eutectic temperature, it begins to melt. As temperature rises, more of the rock melts, creating a mixture of melted and unmelted crystals known as a **partial melt**. Eventually, when the temperature is high enough, the entire rock melts.

**4.** Any geologic process that causes melts and solids to become separated is called **fractionation**. Fractionation (or fractional melting) can occur during partial melting if the newly formed melt is squeezed out of the crystalline residue, yielding a solid and a melt of different composition.

**5.** Molten rock underground is called **magma.** Magma, which may contain dissolved gases and suspended mineral grains, is buoyant and tends to rise toward the Earth's surface. Magma that reaches the Earth's surface and flows over the land is called **lava.** Magmas and lavas have a range of compositions, usually dominated by $SiO_2$ (silica).

**6.** All magmas and lavas are fluids. Some are very runny, whereas others flow only very slowly. A fluid's resistance to flow is its **viscosity.** In general, the higher the silica content of a magma, the higher its viscosity. The higher its temperature and gas content, the lower its viscosity.

**7.** Cooling and crystallization influence the properties of rocks. **Volcanic rocks** are generally glassy or fine-grained because they cool very quickly, whereas **plutonic rocks** are generally coarse-grained because they cool slowly enough for large crystals to form.

**8.** The composition of a melt influences its final mineral assemblage. If a mineral assemblage is dominated by light-colored minerals, such as quartz, feldspar, or muscovite, it is felsic; if it is dominated by dark-colored minerals, such as olivine, pyroxene, amphibole, or biotite, it is mafic.

**9.** The three main types of volcanic rocks are **basalt, andesite,** and **rhyolite.** For each of these common volcanic rocks there is a corresponding plutonic rock with the same composition and often the same mineral assemblage but a different texture. **Gabbro** is the plutonic equivalent of basalt; **diorite** is equivalent to andesite; and **granite** is equivalent to rhyolite.

**10.** A range of igneous rock types can form from a single magma through **magmatic differentiation. Fractional crystallization,** in which newly formed crystals become separated from melt, is an important process in producing the great diversity of igneous rock types observed on the Earth. Crystals can separate from melt during crystallization as a result of filter pressing, crystal settling, crystal flotation, or crystal zonation.

**11.** Bodies of igneous rock that have cooled and solidified underground are called **plutons. Batholiths** are the largest plutons, sometimes exceeding 1000 km in length. **Stocks** are similar to batholiths but smaller. Stocks and batholiths are generally granitic in composition and are probably formed by extensive melting of the lower continental crust. **Dikes,** which are formed from magma intruded into fractures that cross-cut the layering of the surrounding rock, and **sills,** which are formed from magma intruded between the layers of the country rock, are common types of minor plutons.

**12.** Nonexplosive eruptions are characterized by relatively quiet, low-viscosity lava flows of basaltic composition. Volcanoes that erupt in this way tend to build broad, gently sloping **shield volcanoes** or plateau basalts. Explosive eruptions are characterized by the violent ejection of large quantities of **pyroclastic** material. Volcanoes of this type are mostly andesitic and rhyolitic, in continental or subduction zone settings. They tend to build steep-sided **stratovolcanoes.**

**13.** Primary volcanic hazards include lava flows, explosive pyroclastic eruptions, and emissions of poisonous gas. Secondary effects include volcanic mudflows (lahars), debris avalanches, floods, fires, tsunamis, and volcanic tremors. Tertiary effects—some of which are beneficial—include changes in the landscape, changes in atmospheric chemistry, global cooling, fertilization of soils, the creation of new land, the formation of mineral deposits, and geothermal energy.

**14.** Predictions are based on identifying the past eruptive behavior of a volcano and carefully monitoring the volcano for signs of renewed activity.

# ▶ *The Language of Geology*

- andesite 161
- basalt 161
- batholith 165
- crystallization 156
- dike 168
- diorite 163
- fractional crystallization 164
- fractionation (fractional melting) 155
- gabbro 163
- geothermal gradient 151

- granite 163
- igneous rock 156
- lava 155
- magma 155
- magmatic differentiation 163
- partial melt 155
- pluton 165
- plutonic rock 160
- pyroclastic rock 172
- rhyolite 161

- shield volcano 168
- sill 168
- stock 165
- stratovolcano 173
- texture 156
- viscosity 156
- volcanic glass 158
- volcanic rock 158
- volcano 168

## ▶ *Questions for Review*

1. Referring to Figure 6.2, at what depth under the continents would you expect to find a temperature of 1500°C? What would be the temperature at a depth of 200 km below the oceans?

2. Describe the effects of pressure and water on the melting temperatures of minerals.

3. From what you know about the relationship between the viscosity and the silica content of a magma, which of the common magma types—basalt, andesite, or rhyolite—would have the lowest viscosity and which would have the highest?

4. What is fractional melting? Use Figure 6.5 to describe the conditions under which a rock of a given composition would be (a) completely solid, (b) partially molten, and (c) completely molten.

5. What is fractional crystallization? What are some of the physical circumstances that can lead to fractional crystallization?

6. How does the cooling rate affect the texture of a rock?

7. Construct a table showing the three main types of magma, the volcanic rocks formed from them, and their plutonic equivalents. Which types of rock are felsic? Which are mafic?

8. Summarize the main types of plutons on the basis of their sizes and shapes.

9. Summarize the common primary, secondary, and tertiary effects of volcanic eruptions. Which of these effects are commonly associated with shield volcanoes, and which are more commonly associated with stratovolcanoes?

10. What techniques are used by volcanologists to predict volcanic eruptions?

## ▶ *Questions for Thought and Discussion*

1. Why is the geothermal gradient steeper under the oceans than under the continents?

2. How can the mantle be mostly solid, even though its temperature is higher than the melting temperature of most rocks?

3. Think about the types of geologic environments in which melting occurs. How might water be introduced into these environments? How would water affect the melting process?

4. The slopes of active volcanoes tend to be populated. How many reasons can you think of for living near a volcano?

5. Why is it so difficult to predict volcanic eruptions accurately? Research your answer.

## *Virtual Internship: You Be the Geologist*

On your *Geosciences in Action* CD-ROM, you will become a geologist as you assess volcanic hazards on the imaginary island of Paradiso.

For an interactive case study on volcanism in the East African Rift, visit our Web site.

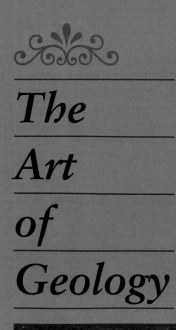

# The
# Art
# of
# Geology

...Oft the teeming earth
Is with a kind of colic pinch'd and vex'd
By the imprisoning of unruly wind
Within her womb; which, for enlargement striving
Shakes the old beldam earth, and topples down
Steeples and moss-grown towers...

—*William Shakespeare*
**First Part of King Henry IV**

We rode on a sea of mountains and jungles, sinking in rubble and drowning in the foam of wood and rock. The Earth was boiling under our feet…making bells ring, the towers, spires, temples, palaces, houses, and even the humblest huts fall; it would not forgive either one for being high or the other for being low.

———————

*—From an eyewitness account
of the 1773 earthquake in Guatemala*

And the mount of Olives shall cleave in the midst thereof toward the east and toward the west, and there shall be a very great valley; and half of the mountain shall remove toward the north, and half of it toward the south.

———————

*—Zechariah 14:4*

*Katsushika Hokusai, "Fuji in Clear Water."*

185

# PART THREE

# THE CHANGING EARTH

The Earth's surface is a meeting place; it is the interface between the geophysical activity of plate motion, seismicity, rock deformation, and volcanism within the Earth and the constant, churning activity of the biosphere, atmo-sphere and hydrosphere outside the Earth. It is the most active and dynamic part of the Earth system. When rocks are uplifted by internal processes, they are attacked and worn down by the external processes of weathering and erosion. Wind and water constantly modify the surface, cutting away material here, depositing material there, sculpting and creating the familiar landscapes that surround us. The heat and pressure that result from plate motion and mountain-building in the tectonic cycle can also cause the minerals in crustal rocks to change, or metamorphose.

The chapters of Part Two focused primarily on the internal part of the rock cycle, that is, on geologic processes that originate deep within the Earth. Now we turn our attention to the ways in which these internal processes affect crustal rocks and what happens to rocks in the crust when they are uplifted. Plate tectonics provide a context within which both the internal and external parts of the rock cycle can be understood as parts of a unified, integrated system. In the last chapter of this part we will look once again at the common rocks and rock-forming processes in the context of the tectonic environments in which they occur.

> ▶ All of these topics and more are addressed in the chapters of Part Three, which are

*Badlands*
Moonrise over the Badlands National Park, South Dakota. Erosion has exposed the horizontal layering in the sedimentary rocks of the Badlands.

# WEATHERING AND EROSION

**Nanga Parbat, a mountain in the Himalayas,** is 8125 m (26,658 ft) high. Thousands of meters below, the Indus River flows through a deep canyon that slices through the side of the mountain. Such pronounced relief means that slopes are steep and landslides are frequent. One landslide caused a historic disaster.

In June 1841 a large Sikh army was camped at the mouth of the canyon near Attock, Pakistan, 400 km (250 mi) downstream from Nanga Parbat. Suddenly, in midafternoon, a huge wall of muddy water rushed out of the canyon and overwhelmed the entire encampment. A rescue worker described the scene: "It was a horrible mess of foul water, carcasses of soldiers, peasants, war-steeds, camels, prostitutes, tents, mules, asses, trees and household furniture, in short every item of existence jumbled together in one flood of ruin."

The cause of the disaster was a massive landslide off the slopes of Nanga Parbat which dammed the river and formed a lake 32 km (20 mi) long and 150 m (500 ft) deep. When the rising water crested the landslide dam, it quickly cut through the unconsolidated rubble and released the deadly flood.

Tectonic forces are lifting the Himalayas at a rate of almost 5000 m (16,000 ft) every million years—it is the fastest rising place on Earth. But the erosive processes of the rock cycle are tearing the mountains down about as rapidly as they are rising. The tectonic cycle and the rock cycle are dynamically balanced in this geologically active area.

◆

### In this chapter you will learn about

- How rock breaks down mechanically and chemically
- The factors that control the rate of rock disintegration
- How soils form
- Erosion by water, wind, and ice

*A Dangerous River*

This is the Indus River where it cuts through the Himalaya, in Pakistan. The scene is not far from the site of the landslide that caused the major disaster in 1841, discussed above. The carved stones in the foreground are Buddhist prayer stones.

**weathering** The chemical and physical breakdown of rock exposed to air, moisture, and organic matter.

**regolith** The loose layer of broken rock and mineral fragments that covers the Earth's surface.

**erosion** The wearing away of bedrock and transport of loosened particles.

Without regular maintenance, even the sturdiest house will succumb to decay caused by natural processes—water damage, freezing and thawing, rusting, attacks by fungus and mold. Rock is affected by weather in similar ways. No rock on or near the Earth's surface is immune to **weathering**, the chemical and physical breakdown of rock exposed to air, moisture, and organic matter. The rate at which the rock breaks down varies with the climate and rock type, but all rocks are susceptible to weathering.

The result of weathering is a loose layer of broken rock and mineral fragments on the Earth's surface. This is called the **regolith**, from the Greek *rhegos*, meaning "blanket," and *lithos*, meaning "stone." Fragments in the regolith range in size from microscopic to many meters across, but all have been formed by chemical and physical breakdown of rock, or they have been loosened and freed from the solid bedrock below. Eventually, as larger particles break down, the uppermost layer of the regolith starts to support rooted plants, in which case it is called *soil*. Because regolith (including soil) is loose, it can be readily moved around by ice, water, or wind. We use the term **erosion** to describe the combined processes of transport of regolith and the direct wearing away of bedrock by transported particles.

## THE ROCK CYCLE

When mountain ranges rise up as a result of plate tectonics or when lava and volcanic ash erupt from volcanoes, the newly exposed or newly formed rocks are quickly attacked by water, wind, and ice. These constantly modify the Earth's surface, cutting away material here, depositing material there, and in the process creating the landscapes we see around us. Through weathering and erosion the atmosphere, hydrosphere, and biosphere continually react with and change the surface of the solid Earth and are thereby closely linked to the rock cycle. As discussed in chapter 1, the *rock cycle* describes all the processes by which rock is formed, modified, transported, decomposed, and reformed as a result of the Earth's internal and external processes.

Figure 7.1 is a diagram of the geologic cycle, and by now it should be familiar because it appeared previously as Figures 1.19 and 4.20. The geologic cycle, as shown here, has three interconnected parts: the hydrologic cycle, on the lefthand side of Figure 7.1, the rock cycle, in the middle, and the tectonic cycle, on the righthand side of the diagram. Chapters 4 through 6 looked at the tectonic cycle. In chapters 7 through 11 we will address various aspects of the rock cycle.

Cycles are continuous and have no beginning or end. The rock cycle is like a merry-go-round, endlessly turning, powered by the Earth's internal heat energy and by incoming energy from the Sun. In order to discuss the rock cycle, we have to jump in somewhere, so let's start at the top of the diagram with the uplift of continental crust and the exposure of crustal rocks. Exposed crustal rocks are vulnerable to weathering and are transformed into regolith by processes discussed in this chapter. The combined processes of erosion then transport the regolith and eventually deposit it as sediment. After deposition, the sediment is buried and compacted, eventually becoming sedimentary rock; these processes are discussed in chapter 8. The processes of plate motion and crustal uplift lead to rock deformation, discussed in chapter 9. Deeper burial turns sedimentary

**?** Continental crust is recycled through the rock cycle. How is oceanic crust recycled? (Refer back to chapter 4 for the answer.)

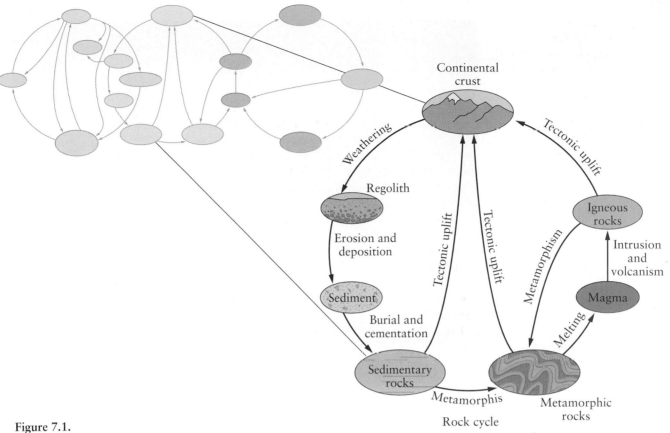

**Figure 7.1.**
**THE ROCK CYCLE**
The rock cycle traces the processes whereby materials within and on top of the Earth's crust are weathered, transported, deposited, and deformed. Note that rocks don't necessarily follow the path that leads around the outside of the circle; they can follow any of the "short circuits" from one stage to another.

rock into metamorphic rock (chapter 10), and even deeper burial may cause some of the metamorphic rock to melt, forming magma from which new igneous rock will form (chapter 6). As a consequence of the rock cycle, material of the continental crust is constantly being recycled. At any stage in the cycle, tectonic processes might elevate the crust and expose the rocks to weathering and erosion. Thus, the rock cycle and the tectonic cycle are closely connected.

The rock cycle is dominated by external processes and is driven partly by solar energy, whereas the tectonic cycle is dominated by internal processes and is driven by geothermal energy. But neither is independent of the other. The Earth's surface is dynamic and changeable because of these cycles and the ways the processes of the cycles interact. The rock cycle, the tectonic cycle, and the hydrologic cycle, and the interactions between the three cycles are essential parts of the Earth system. We have chosen to begin our examination of the rock cycle by looking at the process of weathering.

# WEATHERING: HOW ROCK DISINTEGRATES

Weathering takes place throughout the zone in which materials of the lithosphere, hydrosphere, atmosphere, and biosphere can mix. This zone extends downward below the Earth's surface as far as air, water, and microscopic organisms can readily penetrate, and can range from 1 m (3.3 ft) to hundreds of meters

Figure 7.2
**FROM ROCK TO SOIL**
This soil developed by weathering of a sedimentary rock. Rock disintegrates as rainwater and air penetrate through fractures and react with the minerals present. This photograph was taken in South Africa.

in depth. Rock in the mixing zone usually contains numerous fractures, cracks, and pores (small spaces between mineral grains) through which water, air, and organisms can enter. Given enough time, they can produce major changes in the rock (Fig. 7.2).

Notice in Figure 7.2 that the regolith near the surface is loose and consists mainly of soil. It grades downward to partly altered rock. Weathering clearly proceeds from the surface downward. When we look carefully at the breakdown processes involved in weathering, we find they are of two kinds. In **mechanical weathering** the rock is physically broken up, but there is no change in its mineral content or chemical composition. **Chemical weathering** is the dissolving in rainwater, together with the decomposition of rocks and minerals as a result of chemical and biochemical reactions that produce new minerals that are stable at the Earth's surface. Although mechanical weathering is distinct from chemical weathering, the two processes almost always occur together and their effects are sometimes difficult to separate.

**mechanical weathering** The breakdown of solid rock into fragments by physical processes, with no change in chemical composition.

**chemical weathering** The decomposition of rocks and minerals as a result of chemical and biochemical reactions.

> **?** **By what means can air and water penetrate the crust of the Earth?**

**joint** A fracture in a rock, along which no appreciable movement has occurred.

## Mechanical Weathering

Most rocks are subject to a kind of mechanical fracturing that plays an important role in weathering. Rocks in the upper half of the crust are brittle, and like any other brittle material they break at the weak spots when they are twisted, squeezed, or stretched by tectonic forces. The timing and origin of such tectonic forces are not always obvious, but they leave their imprints in the form of **joints**, which are fractures in a rock along which no appreciable movement has occurred (Fig. 7.3). One common way joints form is through erosion. Rock masses buried deep beneath the surface are subject to enormous pressures by the weight of the overlying rock. As erosion removes material from the surface, the weight of the overlying rock, and hence the pressures, are reduced. The rock responds by expanding upward. As the rock expands, it fractures and forms joints. Joints are the main passageways through which rainwater, air, and small organisms enter the rock, leading to mechanical and chemical weathering.

Mechanical weathering takes place in four main ways—through freezing of water, crystallization of salt crystals, heating by forest fires, and penetration by plant roots. By far the most important type of mechanical weathering involves the freezing of water.

Water is an unusual substance. When most liquids freeze, the volume of the resulting material decreases. When water freezes to ice, there is an *increase* in volume of about 9 percent. That is why ice floats. It also explains why, if you put a full, capped bottle of water in the freezer, the bottle will eventually break. Wherever temperatures fluctuate around the freezing point for part of the year, water in the ground will freeze and thaw. The increase in volume resulting from each freeze causes high stresses and eventually shatters rocks. This process—the freezing of ice in a confined opening within a rock, causing the rock to be forced apart—is known as **frost wedging** (Fig. 7.4).

Water moving slowly through rock fractures will dissolve soluble material, which may later precipitate out of solution to form salt crystals of various kinds. The force exerted by growing crystals, either within rock cavities or along grain boundaries, can be large enough to cause disruption. Mechanical weathering by crystal growth occurs mostly in desert regions, where salts such as calcium carbonate and calcium sulfate are precipitated from groundwater as a result of evaporation.

Fire can also be very effective in causing rocks to break. Anyone who has seen a rock shatter explosively in a campfire knows this. Because rock is a poor conductor of heat, an intense fire may rapidly heat only a thin outer shell of the rock. The thin shell expands and breaks away as a *spall*. Forest and brush fires contribute significantly to mechanical weathering through spalling from boulders and exposed bedrock (Fig. 7.5).

When a tree grows in a crack, its roots grow too, eventually widening the crack and wedging apart the bedrock (Fig. 7.6). In much the same way, roots disrupt stone walls, sidewalks, and even buildings. Large trees swaying in the wind can also cause fractures to widen. When the trees are blown over, they can disrupt the rock still further Although it is difficult to measure, the total amount of rock breakage caused by plants must be very large.

A final form of rock breakage, for which there is no generally accepted explanation, is *sheet jointing*. This is the peeling off of large, curved slabs of rock from the surface of uniformly textured igneous rock (Fig. 7.7). The breakage may be due to pressure release, as with other joints, or it may be caused by a combination of forces that contribute to mechanical weathering.

**frost wedging** The freezing of ice in a confined opening within a rock, causing the rock to be forced apart.

## Chemical Weathering

Chemical weathering is caused by water that is slightly acidic. As raindrops form and fall through the air, they dissolve carbon dioxide. Rainwater thus is a weak solution of carbonic acid ($H_2CO_3$)—a weak version of Perrier water! When weakly acidified rainwater sinks into the regolith and becomes soil water, it may dissolve more carbon dioxide from decaying organic matter, becoming more strongly acidified. Another way that rainwater can become more strongly acidified is by interacting with *anthropogenic* (human-generated) sulfur and nitrogen compounds released into the atmosphere. This produces a phenomenon called *acid rain*. Human-caused acid rain is stronger than natural, weakly acid rain and causes accelerated weathering.

The term *hydrolysis* describes any chemical reaction in which water is a participant.

One hydrolysis reaction that is of special importance in chemical weathering is *ion exchange*. All acids, including carbonic acid and the acids in acid rain, form hydrogen ions ($H^+$), the smallest of all ions when they are dissolved. Ions form in solution in the same way ions in minerals do, by giving up or accepting electrons (see chapter 2). The difference is that ions in minerals are tightly

**?  In what way is rainwater like Perrier water?**

# Mechanical Weathering

◄ Figure 7.3
JOINTS
Two well-developed sets of vertical joints and a horizontal bedding plane cause this red sandstone in Brecon Beacons, Wales, to break into roughly rectangular blocks.

Figure 7.4
FROST WEDGING
This high-mountain granite boulder in the Sierra Nevada of California has been split by repeated freezing and thawing of water that penetrated along joints.
▼

◄ Figure 7.5
HEAT SPALLING
A fast-moving forest fire in the pine forests of Yellowstone National Park, Wyoming, caused mechanical weathering of a boulder of igneous rock. Fresh, light-colored spalls, which flaked off as heat caused the outermost part of the rock to expand rapidly, litter the fire-blackened ground.

◄ Figure 7.6
ROOT WEDGING
A Ponderosa pine tree that began growing in a crack on this outcrop of bedrock has caused a large flake of rock to break away, exposing the tree's expanding root system. The rock is granite.

Figure 7.7 ▲
EXFOLIATION
The photo shows exfoliation of granite boulders (the Devil's Marbles) in central Australia. Thin, sheet-like spalls flake off the rock as it weathers, gradually causing the boulders to become rounder.

bonded and fixed in a crystal lattice, whereas ions in solutions can move about randomly and cause chemical reactions. In ion exchange, hydrogen ions in solution enter and alter a mineral by displacing larger, positively charged ions such as potassium ($K^+$), sodium ($Na^+$), and magnesium ($Mg^{2+}$). The displaced ions go into solution and flow away in the groundwater. Eventually, the ions released by ion exchange reach the sea; they are a major source of sea salts. The removal of these ions leaves behind a relatively insoluble residue of clay minerals (Fig. 7.8).

Another hydrolysis reaction is solution. Through solution, some minerals can be completely dissolved and removed without leaving a residue. The most soluble of the rock-forming minerals are the two carbonate minerals, calcite and dolomite (Fig. 7.9). Soluble substances dissolved from rocks, together with ions displaced during ion exchange, are present in all groundwater and surface water. Sometimes the concentrations of these substances are so high that the water has an unpleasant taste.

Although it is not the same as hydrolysis, a very important process of chemical weathering that commonly accompanies hydrolysis is *oxidation*. Iron and manganese can be chemically altered by oxidation. Rust is a product of chemical weathering that results from the oxidation of iron. Iron and manganese are present in many rock-forming minerals. When such minerals are chemically weathered, the iron and manganese are released and immediately oxidized. Iron forms an insoluble yellowish mineral called limonite, and manganese forms an insoluble black mineral called pyrolusite. Limonite may later be dehydrated (meaning that it loses water) to form hematite, a brick-red-colored mineral that gives a distinctive red color to tropical soils (Fig. 7.10).

> **?** What type of chemical weathering causes the underside of your family car to rust?

## Factors That Influence Weathering

From the perspective of human lifespans, the chemical weathering of rocks happens very slowly. For example, granite and other hard bedrock surfaces in New England, Canada, and Scandinavia still display polish and fine grooves made by glaciers of the last ice age, which ended more than 10,000 years ago. In such regions, which have cool climates, it takes hundreds of thousands of years for a regolith like that shown in Figure 7.2 to develop. In regions that have not been repeatedly glaciated and have been continuously exposed to weathering, the zone of weathering often extends to depths of many tens of meters. Through the use of various dating techniques, it has been estimated that deep tropical weathering of 500 m (about a third of a mile) or more requires many millions and possibly several tens of millions of years.

Many factors influence how susceptible a rock is to chemical and mechanical weathering and how fast weathering will proceed. The most important factors are the kind of rock, the presence of openings such as joints, the steepness of slope, climate (especially temperature and rainfall), vegetative cover, and the activity of burrowing animals (Table 7.1). Let's look briefly at each of these factors.

*Resistance to Weathering.* Because specific minerals react to weathering processes in different ways, the type of rock subjected to weathering is clearly important. Quartz, one of the most common rock-forming minerals, is also one of the most resistant to chemical weathering. This is so because quartz dissolves very slowly and is not affected by ion exchange, which affects almost every other silicate mineral. Rocks that are rich in quartz, such as quartzite (a metamorphic rock consisting largely of quartz), therefore are affected by chemical weathering only very slowly.

*Rock Structure.* Weathering is also strongly influenced by the number of openings in the rock, especially the closeness of joints. Even a rock that consists en-

# Chemical Weathering

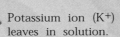

Water and hydrogen ion (H+) enter from solution.

Potassium ion (K+) leaves in solution.

▲ A. Grain of feldspar attacked by acid rainwater

Unaltered feldspar

Alteration products

▲ B. Feldspar partly altered to a mixture of clay and quartz

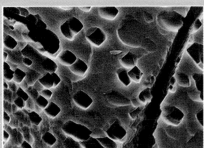

C.

▶

Figure 7.8
**ION EXCHANGE**
Chemical weathering by ion exchange happens when hydrogen ions (H+) in weakly acid rainfall displace cations such as potassium (K+) and form new minerals in the process. *A.* Potassium feldspar is altered to a mixture of clay and quartz *B* as a result of ion exchange. Dotted lines show the direction of cleavage. Note that alteration penetrates parallel to cleavage. *C.* Scanning electron microscope photo of a feldspar grain that has been altered by ion exchange. The alteration products have been removed in order to reveal how acid attack is controlled by the cleavage. The image is 0.05 mm (0.002 in) across.

◀ Figure 7.9
**SOLUTION OF CALCITE**
Because marble is composed of soluble calcite, this marble tombstone standing in a New England cemetery since the early nineteenth century shows the effectiveness of rainfall in dissolving the calcite. Over the years, the rock surface has been slowly dissolved, making the once sharply chiseled inscriptions barely legible.

Figure 7.10
**OXIDATION**
This is an oxidized soil in Hawaii. The bright red color is due to the oxidation of iron to form hematite. Sugar cane is being grown on the soil.
▼

**TABLE 7.1 Factors that influence weathering**

| | Rate of Weathering Slow ————————————————→ Fast | | |
|---|---|---|---|
| Mineral resistance to chemical weathering | High (e.g., quartz) | Intermediate (e.g., mica, feldspar) | Low (e.g., calcite, olivine) |
| Frequency of joints | Few joints (meters apart) | Intermediate (0.5-1.0 meters apart) | Many (centimeters apart) |
| Depth of regolith | Zero | Shallow | Deep |
| Steepness of slope | Steep | Moderate | Gentle |
| Vegetation | Sparse | Moderate | Dense |
| Temperature | Cold (average about 5°C) | Temperate (average about 15°C) | Warm (average about 25°C) |
| Rainfall | Low (<40 cm/y) | Intermediate (40–130 cm/y) | High (>130 cm/y) |
| Burrowing animals | Rare | Frequent | Abundant |

tirely of quartz may break down rapidly if it contains closely spaced joints that are susceptible to frost wedging. On the other hand, rock with very widely spaced joints may weather so slowly that it will stand up as a prominent peak in an otherwise deeply eroded landscape (Fig. 7.11). In places where a deep regolith layer already exists, weathering can proceed at a greater rate because regolith is capable of holding the major agent of weathering, water.

*Steepness of Slope.* A mineral grain that has been loosened by weathering may be washed downslope by the next rainfall. On a steep slope, the solid products of weathering are quickly moved away, continually exposing fresh bedrock. As a result, regolith is very thin or even absent on steep slopes, and weathered rock seldom extends far below the surface. On gentle slopes, the products of weathering are not as easily washed away, so they accumulate and form deep deposits. Slopes that are cleared of vegetation may also erode more rapidly than vegetated slopes. Removal of vegetation can occur naturally—through landslides, for example—or as a result of activities such as clearcut logging, slash-and-burn agriculture, and grazing. This can lead to increased rates of weathering and erosion of newly exposed regolith.

*Climate.* High temperatures and abundant rainfall promote chemical reactions. It is hardly surprising that chemical weathering is more intense and extends to greater depths in a warm, wet tropical climate than in a cold, dry Arctic climate (Fig. 7.12). In moist tropical lands, such as Central America, Southeast Asia, and parts of West Africa, the effects of chemical weathering can be detected to depths of 100 m (110 yd) and more. As Figure 7.10 shows, the regolith in these areas tends to be bright red because it contains hematite. By contrast, in cold, dry regions such as Greenland and Antarctica, chemical weathering proceeds very slowly, mechanical weathering is relatively rapid, and evidence of weathering disappears at shallow depths. The scene shown in Figure 7.4 is an example of such a

Figure 7.11
**STRIKING ROCK PILLAR**
Sugarloaf Mountain in Potafogo Bay, Rio de Janiero, Brazil is a large, unjointed mass of granite. Because rainwater cannot penetrate Sugarloaf along joints, the rock weathers away very slowly and as a result stands as a striking erosion residual.

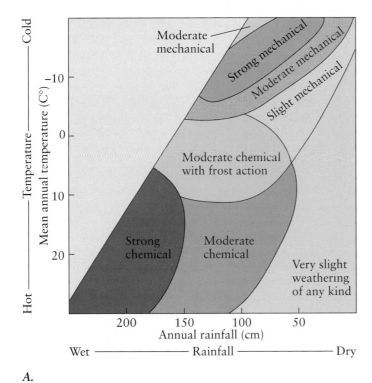

A.

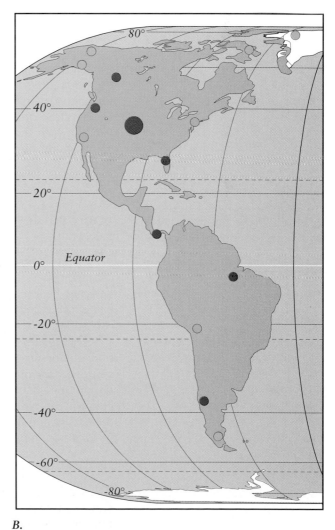

B.

Figure 7.12
**CLIMATE AND WEATHERING**
*A.* Climate (a combination of temperature and rainfall) strongly influences weathering processes. Mechanical weathering dominates in areas of low rainfall and cold temperatures. High temperature and high rainfall favor chemical weathering. *B.* Places in the Americas where different regimes of climate, and therefore different types of weathering, predominate.

region; the surface, which was covered with ice and free of rubble only a few thousand years ago, is now covered by fragments dislodged by frost wedging.

*Burrowing Animals.* Burrowing animals such as worms, ants, and rodents bring partly decayed rock particles to the surface where they are more fully exposed to chemical weathering. More than 100 years ago, Charles Darwin calculated that earthworms bring particles to the surface at a rate in excess of 2.5 kg/m$^2$ (10 tons/acre) per year. After a study in the Amazon River basin, geologist J. C. Branner wrote that the regolith there looked "as if it had been literally turned inside out by the burrowing of ants and termites." Burrowing animals do not break rock down directly, but the volume of rock fragments moved by them over millions of years and exposed to increased rates of chemical weathering must be truly enormous.

# SOIL

As we learned previously, **soil** is the uppermost part of the regolith, the portion that can support rooted plant life. Examine a bit of soil with a microscope or magnifying glass and you will see that it contains a lot of stuff besides bits of rock and mineral. You will find fragments of **humus**, which is partially decayed organic matter, and possibly some tiny insects and worms. If your microscope is very strong, you might even discover bacteria living on the humus.

Soil is a complex medium in which all parts interact and play important roles. Humus retains some of the chemical nutrients released by decaying organisms and by the chemical weathering of minerals. Humus is critical to *soil fertility,* the ability of a soil to provide nutrients needed by growing plants such as phosphorus, nitrogen, and potassium. All of the processes that involve living organisms and other soil constituents produce a continuous cycling of plant nutrients between the regolith and the biosphere. With its partly mineral, partly organic composition, soil forms a bridge between the lithosphere and the biosphere. Like liquid water, soil is one of the features that makes the Earth unique among the planets.

## Soil Profiles

Soil evolves gradually. When fully developed it consists of a succession of zones, or **soil horizons**, each of which has distinct physical, chemical, and biological characteristics. The sequence of soil horizons from the surface down to the underlying bedrock, is a **soil profile** (Fig. 7.13). Soil profiles vary considerably, being influenced by such factors as climate, topography, and rock type. However, certain kinds of horizons are common to many profiles.

The uppermost horizon in many soil profiles, the *O horizon,* is an accumulation of organic matter. Below it lies the *A horizon,* which is typically dark in color because of the humus present. The term *topsoil* is essentially a synonym for the combined O and A horizons. An *E horizon,* which is sometimes present below A, is typically grayish in color because it contains little humus and the mineral grains do not have dark coatings of iron and manganese hydroxides. E horizons are most common in acidic soils in evergreen forests.

The *B horizon* underlies the A horizon (or E, if one is present). B horizons are brownish or reddish in color because of the presence of iron hydroxides that have been transported downward from the horizons above. The *C horizon,* consisting of parent rock material in various stages of weathering, is deepest. Oxidation of iron in the parent rock gives the C horizon a yellowish brown color.

**What's the difference between *soil* and *regolith?***

**soil** The uppermost part of the regolith, the portion that can support rooted plant life.

**humus** Partially decayed organic matter in soil.

**What is topsoil?**

**soil horizon** One of a succession of zones or layers within a soil profile, each of which has distinct physical, chemical, and biological characteristics.

**soil profile** The sequence of soil horizons from the surface down to the underlying bedrock.

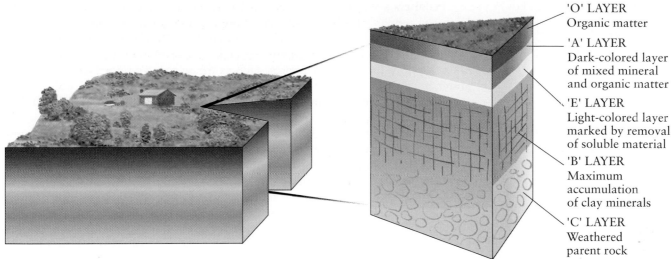

'O' LAYER
Organic matter

'A' LAYER
Dark-colored layer
of mixed mineral
and organic matter

'E' LAYER
Light-colored layer
marked by removal
of soluble material

'B' LAYER
Maximum
accumulation
of clay minerals

'C' LAYER
Weathered
parent rock

**Figure 7.13**
**SOIL PROFILE**
This is a typical sequence of soil horizons as they might appear in the soil profile of a soil that developed in a cool, moist climate. The uppermost layers, within reach of plant roots, are commonly termed the topsoil.

## Classification of Soils

Because the kind of soil that develops in a given place depends on many variables, it is hardly surprising that classifications used by soil scientists are very complicated. The classification scheme used in the United States and many other countries is a hierarchical one headed by 11 orders. Each order is distinguished by easily recognizable characteristics such as the presence or absence of well-developed horizons; accumulation of aluminum, iron, clay, or carbonate minerals in the B horizon; or high acidity and organic content. The 11 soil orders are divided into suborders and then, in increasing detail, into groups, subgroups, families, series, and types. Despite the complexity of modern soil classifications, the major differences between soils can readily be understood in terms of a simpler classification scheme that involves only three large families (Fig. 7.14): *pedalfers*, *pedocals*, and *laterites*.

**Figure 7.14**
**CLIMATE AND SOILS**
Examples of soil horizons that have developed under different climatic and vegetation conditions.

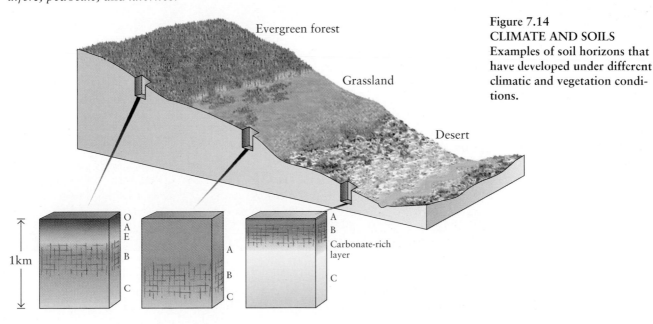

Evergreen forest

Grassland

Desert

1km

O
A
E
B
C

A
B
C

A
B
Carbonate-rich
layer
C

**pedalfer** A soil that is rich in clays, plus aluminum and iron hydroxides.

**Pedalfers** are named from the Greek *pedon*, meaning "soil," and *al* and *fer*, from the names *aluminum* (meaning bitter, in reference to a bitter-tasting aluminum salt,) and the Latin *ferrum* (meaning "iron"). They are rich in clays, as well as aluminum and iron hydroxides. Pedalfers develop in regions with moderate rainfall and temperate climates, such as the eastern and northwestern United States and much of Europe. They have well-developed A, B, and C horizons, and—with suitable vegetation—a distinct E horizon. They do not contain carbonates. Pedalfers make excellent agricultural soils (Fig. 7.15).

**pedocal** A soil that is rich in calcium carbonate and other soluble minerals.

**Pedocals** are rich in calcium carbonate and sometimes in other soluble minerals such as gypsum. They derive their name from *pedon* and the Latin *calx*, meaning "calcium." Pedocals are common in the southwestern United States and northern Mexico, many parts of Australia, and other places with dry climates. Under such conditions soil water tends to evaporate, causing the precipitation of soluble salts such as calcium carbonate, near the base of the B horizon. Such precipitates can form hard layers in the soil profile, sometimes called *hardpan* or *caliche* (Fig. 7.16). Pedocals tend to be less fertile than pedalfers, but with sufficient water for irrigation, they can be extremely fertile.

**laterite** A soil composed largely of insoluble substances such as iron and aluminum oxides and hydroxides.

**Laterites** are soils of equatorial and tropical lands that have both high rainfall and high temperature. Laterites are soils that have been strongly *leached*, which means that most of the original substances have been removed in solution. Laterites are composed largely of insoluble iron and aluminum oxides and hy-

Figure 7.15
RICH AGRICULTURE ON A PEDALFER
Vegetable crops are growing on agriculturally rich pedalfer in southern Ontario, Canada.

Figure 7.16
NEW MEXICO CALICHE
This pedocal soil profile in semiarid central New Mexico includes a layer of caliche. The white caliche forms a prominent horizon between the yellowish-brown C horizon (below) and the reddish-brown B horizon (above). The dark brown layer at the very top is the A horizon.

droxides; they get their name from the Latin *latere,* meaning "brick," because when they are exposed to the Sun's warmth, the iron hydroxides present dehydrate and form a hard, iron oxide-rich brick. Laterite bricks are used for building in many tropical countries (Fig. 7.17). Laterites are not good agricultural soils inasmuch as nearly every nutrient has been stripped from them. They can sometimes support lush tropical forests, however; a thin surface layer of humus is very effective in recycling the small store of soil nutrients back into the vegetation. When tropical forests are cleared, the humus oxidizes, the nutrients wash away, the soil becomes hard, and fertility drops dramatically.

Figure 7.17
QUARRYING LATERITE BRICKS IN INDIA
Laterite is formed by the natural hardening of deeply weathered, iron-rich soils in tropical climates. In this scene from southern India, bricks of laterite are being quarried for building stone.

**Our food comes from the soil;** we must remember that soil is a precious commodity that we can't afford to waste. It takes centuries or millennia for soil to develop. The amount of time required varies, but in all cases it is a multiple of human life spans. Historic observations at Glacier Bay, Alaska, show that it has taken about 150 years for 10 cm (4 in) of soil to develop on glacial sediment. Glacier Bay has a temperate climate and high rainfall, and glacial sediment is loose, ground-up rock. In less humid climates and on solid bedrock, soils form more slowly. In the soils of the middle portion of the United States, B horizons that developed during the past 10,000 years are still immature, contain little clay, and lack a defined structure. Older soils in the region that date back 100,000 years have B horizons that are rich in clay, have a good structure, and support fertile A layers. Clearly, soil that is lost as a result of farming or other land-clearing activities will not be replaced any time soon.

Soil erosion is normal, resulting from natural changes in topography, climate, or vegetation cover. However, erosion caused by human activities happens so much faster than normal erosion that it overwhelms natural systems in many parts of the world. Farming is becoming ever more intense because each year the world's population grows by an estimated 100 million (births in excess of deaths). Widespread felling of trees has accelerated surface runoff and destabilized soils due to the loss of anchoring roots (Fig. B7.1). Soils in the humid tropics quickly lose their fertility when stripped of their natural vegetation cover for cultivation. In many developing nations, farmers have been forced beyond traditional farmlands and grazing lands onto steep, easily eroded slopes or into semiarid regions.

In the semiarid lands, periodic crop failure is a fact of life, and plowed land is prone to severe wind erosion. At the same time, economic pressures in the more developed countries have increasingly led farmers to shift from ecologically sound land use practices to the planting of profitable row crops that often leave the land vulnerable to increased rates of erosion. So widespread are the effects of soil erosion and degradation that the problem has been described as "epidemic."

In the United States, the amount of farmland soil lost to erosion each year exceeds the amount of newly formed soil by about a billion tons. Soil loss in the former Soviet Union is at least as rapid, whereas in India the soil erosion rate is estimated to be more than that in the United States. For the world as a whole, the estimated loss exceeds 25 billion tons a year. Expressed another way, at the current rate of loss, the world's most productive soils are being depleted at a rate of 7 percent each decade.

**Figure B7.1**
**LOSING GROUND**
Hillslopes in Madagascar, deforested for grazing animals, are now deeply gullied by soil erosion. Streams draining this region turn bright red as they carry away irreplaceable topsoil that is destabilized when anchoring tree roots disappear.

Figure B7.2
**FIGHTING SOIL EROSION**
Contour plowing is an effective method of reducing soil erosion. On this farm, crops have been planted in belts that follow the natural contours of the land, thereby inhibiting the development of rills and gullies that otherwise might be formed perpendicular to the contours.

Soil erosion is a massive worldwide problem affecting you and me. A person eats about three quarters of a ton (750 kg) of food a year, and there now are more than 5.5 billion people on the Earth. Food production leads to an annual loss of 25 billion tons of topsoil, or 4.5 tons (4500 kg) per person. Thus, for every kilogram of food we eat, the land loses 6 kg of soil.

The most serious erosion problems occur on steep hillslopes. In the African country of Nigeria, for example, land with a gentle 1 percent slope[1] that was planted with cassava (a staple food source) lost an average of 3 tons of soil per hectare (2.47 acres) each year. On a moderate 5-percent slope, however, the annual rate of soil loss increased to 87 tons per hectare. At this rate, 15 cm (6 in) of topsoil would disappear in a single generation (about 20 years). On a steep 15-percent slope, the annual erosion rate increased to 221 tons per hectare, a rate that would remove all topsoil within a decade.

Although soil erosion and degradation are having a severe impact on many countries, effective control measures can substantially reduce these adverse effects. Among these measures are contour plowing (Fig. B7.2) and crop rotation. A study in Missouri showed that land that lost 49.25 tons of soil per hectare when planted continually in corn lost only 6.75 tons per hectare when corn, wheat, and clover crops were rotated. In this case, the bare land exposed between rows of corn was far more susceptible to erosion than land that was planted with a more continuous cover of wheat or clover.

---

[1] A 1-percent slope means that over a linear distance of 100 m the land rises 1 m. With a 5-percent slope the land rises 5 m for every 100 m of linear distance (or 5 ft for every 100 ft of linear distance).

# EROSION

As mentioned earlier, *erosion* is a general term that encompasses the transportation of regolith from one place to another and abrasion of the Earth's surface by materials being transported. We draw a distinction between transportation by flowing air, water, or ice—that is, erosion—and **mass wasting**, which is the en masse downslope movement of regolith or bedrock masses due to the pull of gravity. We will discuss these processes separately, starting with erosion.

The ability of flowing air or water to pick up and transport particles depends on its velocity and turbulence. When a fluid flows very slowly, all the fluid particles travel in parallel layers (Fig. 7.18A). This is called *laminar flow*. With increasing velocity, the movement becomes erratic and complex, giving rise to the swirls and eddies that characterize *turbulent flow* (Fig. 7.18B). Laminar flow can transport particles once they are suspended, but turbulent flow is needed to pick particles up or move them along the ground. The velocity at which a flowing fluid becomes turbulent depends partly on its *viscosity,* or resistance to flow. The lower the viscosity, the runnier the fluid and the lower the velocity at which turbulent flow occurs. Air has a very low viscosity, and almost all air flow—that is, wind—is turbulent. Water has a higher viscosity than air, but even so, turbulent flow predominates in stream channels. Only in a thin zone along the bed and channel of a smooth-walled stream, where frictional drag is high, is the velocity low enough for laminar flow to occur.

As the velocity and turbulence of flowing air or water rise, grains of sand start to roll along the ground or stream bottom. The moving grains strike other grains, causing them to move forward in a series of short jumps along arc-shaped paths. This process is called **saltation** (Fig. 7.19).

## Erosion by Water

Erosion by water begins even before a distinct stream has formed. This happens in two ways: by impact, which occurs when raindrops hit the ground and dis-

---

**mass wasting** The downslope movement of regolith and/or bedrock masses due to the pull of gravity.

> **?** Is the definition given here for *viscosity* the same as the one given in chapter 6 for magmas?

**saltation** A mechanism of sediment transport in which particles move forward in a series of short jumps along arc-shaped paths.

---

**Figure 7.18**
**LAMINAR AND TURBULENT FLOW**
*A.* When a fluid flows very slowly, its motion is called *laminar flow;* all the fluid particles travel in parallel layers. *B.* With increasing velocity, the movement becomes erratic and complex, giving rise to the swirls and eddies that characterize *turbulent flow.*

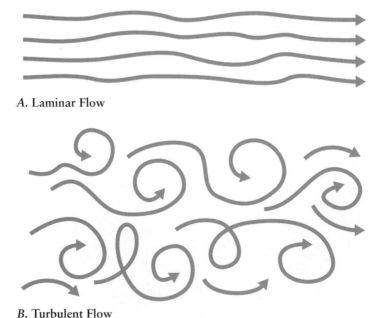

*A.* Laminar Flow

*B.* Turbulent Flow

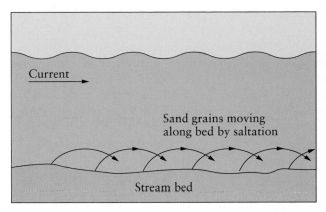

**Figure 7.19**
**SALTATION**
Movement of particles can occur by saltation. Sand on the bottom of a stream moves in places where turbulence reaches the bottom or where suspended grains strike other grains on the bottom. Once lifted, the grains follow arc-shaped paths as gravity pulls them back toward the bottom, where they strike other grains.

lodge small particles of soil, and by overland flow, which occurs during heavy rains. When the water starts flowing in a channel, particles are moved in two ways: Coarse particles move along the stream bed by saltation (the **bed load**), whereas fine particles are suspended and transported as a **suspended load**. Of course, streams also carry a *dissolved load* of soluble materials released by chemical weathering.

The muddy character of many streams is due to the presence of fine particles of silt and clay moving in suspension (Fig. 7.20). Most of the suspended load is derived from fine-grained regolith that has been washed from areas without vegetation and from sediment that has eroded from the stream's own banks. The Yellow River (or Huang He) in China is yellow because of the huge load of yellowish silt it transports to the sea. Before the construction of several major dams along its course, the Colorado River was also extremely muddy; people who lived near it sometimes remarked that the river was "too thin to plow but too thick to drink."

The velocity of upward-moving currents in a turbulent stream exceeds that at which particles of silt and clay can settle under the pull of gravity. Therefore, such particles tend to remain in suspension longer than they would in nonturbulent waters. They settle and are deposited only where velocity decreases and turbulence ceases, as in a lake, in the sea, or in a reservoir.

**bed load** Sediment that is moved along the bottom of a stream.

**suspended load** The part of the total sediment load of a stream that is carried along in suspension.

**?** How would a dam affect the sediment content of a river?

**Figure 7.20**
**THE MUDDY "YELLOW" RIVER**
A large suspended load, eroded from extensive deposits of wind-transported silt, gives the Huang He a very muddy appearance and its English name, Yellow River.

## Erosion by Wind

Because the density of air is about 800 times less than that of water, air cannot move as large a particle as water flowing at the same velocity. In exceptional cases such as hurricanes and tornadoes, winds can reach speeds of 300 km/h (186 m/h) and sweep up coarse rock particles several centimeters in diameter. In most regions, however, wind speeds rarely exceed 50 km/h (31 m/h), a velocity that is described as a strong wind. As a result, most (at least 75%) of the sediment transported by wind occurs through saltation of sand grains (Fig. 7.21). Only the finest particles, the dust, remain aloft long enough to be moved by suspension.

## Erosion by Ice

Ice is a solid. However, it does flow under the influence of gravity—usually quite slowly. A body of ice, consisting largely of recrystallized snow, that is large enough to persist from year to year and within which there is evidence of movement due to the pull of gravity is termed a **glacier.** Compared with water and air, ice is extremely viscous. Glacial ice therefore moves by laminar flow.

Glaciers play a three-part role in erosion: they act like a plow, a file, and a sled. As a plow, a glacier scrapes up weathered rock and soil and plucks out blocks of bedrock (Fig. 7.22A); as a file, it rasps away and polishes firm rock (Fig. 7.22B); as a sled, it carries away the load of sediment acquired by plowing and filing, along with additional debris that falls onto it from adjacent slopes (Fig. 7.22C).

**glacier** A permanent body of ice, consisting largely of recrystallized snow, that shows evidence of movement due to the pull of gravity.

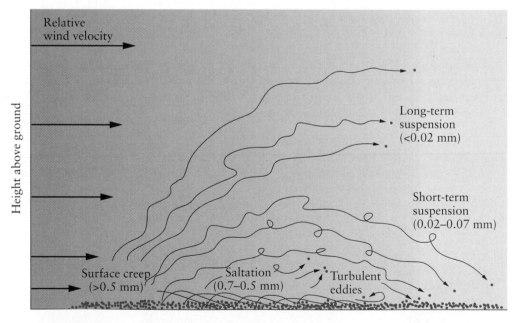

**Figure 7.21**
**WIND MOVING PARTICLES**
Under moderate wind conditions, sand grains larger than 0.5 mm (0.02 in) in diameter move by rolling, while those 0.07 to 0.5 mm (0.003 to 0.02 in) saltate. Smaller grains (0.02 to 0.07 mm (0.0008 to 0.003 in) in diameter) are carried aloft in turbulent eddies and encounter faster-moving air that transports them a long way before they finally settle. The finest dust particles, less than 0.02 mm (0.0008 in) in diameter, remain in suspension as long as the wind blows.

**Figure 7.22**
**EROSION BY GLACIAL ICE**
*A.* This unsorted rocky debris was deposited at the edge of Matunuska Glacier, Alaska. Notice the extreme range in the size of the rock fragments, from 3 m (10 ft) across down to the size of a pinhead. *B.* This polished and grooved surface was produced by the Findelen Glacier in the Swiss Alps. The glacier has melted back in modern times. In the background is the Matterhorn. *C.* Dark bands of rock debris that has fallen from adjacent mountain slopes mark the boundaries between adjacent ice streams. The ice streams merge to form the Kaskawulsh Glacier, Yukon Territory, Canada. The smooth, parallel streams indicate that the flow is laminar.

# MASS WASTING

Landscapes may seem fixed and unchanging, but if you made a time-lapse movie of almost any hillside for a few years you would see that the slope changes constantly as a result of mass wasting. All slopes are subject to mass wasting. Exactly how movement happens and how fast it happens is controlled by the composition and texture of the regolith and bedrock, the amount of air and water in the regolith, and the steepness of the slope. For convenience we divide mass wasting into two categories, as shown in Table 7.2:

TABLE 7.2 Common Mass-Wasting Processes on Land

## *Slope Failures*

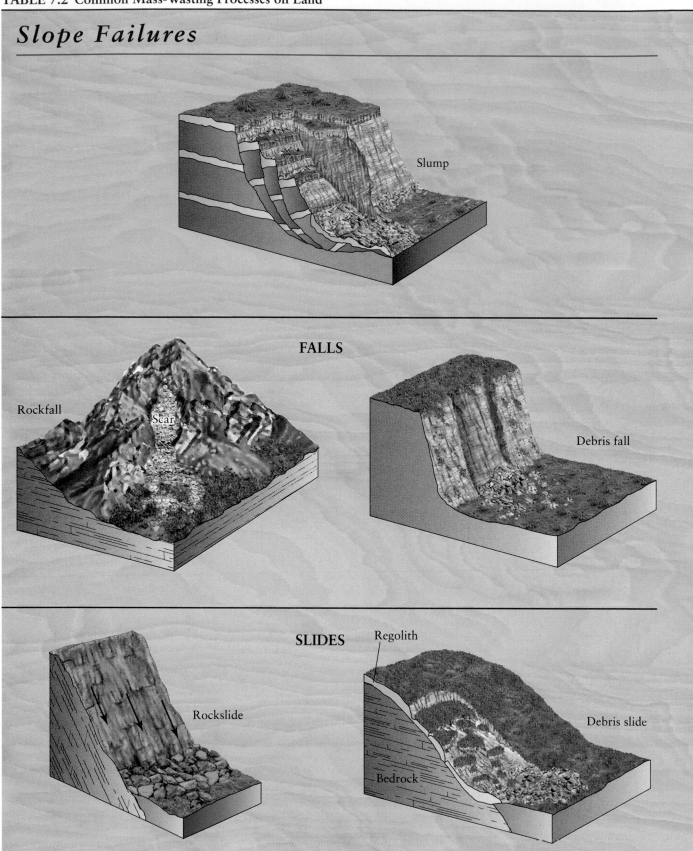

Slump

**FALLS**

Rockfall

Scar

Debris fall

**SLIDES**

Regolith

Rockslide

Debris slide

Bedrock

# *Sediment Flows*

## SLURRY (WET) FLOWS

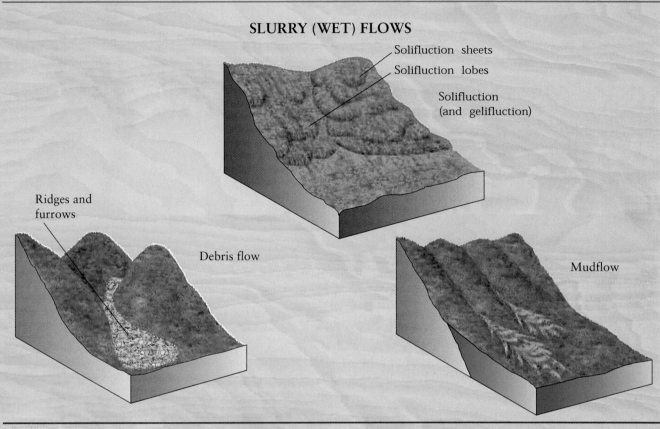

Solifluction sheets

Solifluction lobes

Solifluction
(and gelifluction)

Ridges and
furrows

Debris flow

Mudflow

## GRANULAR (DRY) FLOWS

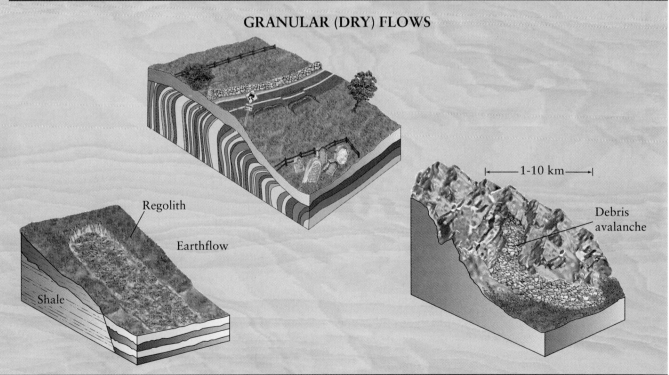

1-10 km

Regolith

Earthflow

Shale

Debris
avalanche

1.  *Slope Failures.* Falling, sliding, or slumping of relatively coherent masses of rock as the result of the sudden failure of a steep slope or cliff.
2.  *Flows.* Downslope flows of regolith.

## Slope Failures

**fall** A type of slope failure that involves a sudden, vertical, or nearly vertical drop of rock fragments or debris.

**slide** A type of slope failure that involves rapid displacement of a mass of rock or regolith down a planar slope.

**slump** A type of slope failure that involves rotational movement of rock or regolith.

If you ask mountain climbers about the greatest danger of their sport, they will likely place falling rock near the top of the list. A **fall** is a sudden, vertical or nearly vertical drop of rock fragments or debris. Rockfalls and debris falls are sudden and usually very dangerous. **Slides** involve rapid displacement of a mass of rock or sediment down a steep or slippery slope. A **slump** is a type of slope failure that involves *rotational* movement of rock and regolith—that is, downward and outward movement along a curved surface. A slide differs from a slump in that it occurs on an essentially planar surface. The amounts of material moved by slumps range from small masses of a few square meters in surface area to huge masses covering tens of thousands of square meters. Slumps sometimes occur accidentally after slopes have been oversteepened for construction of buildings. All of these types of slope failures are illustrated in Table 7.2.

## Flows

**flow** Any mass-wasting process that involves a flowing motion of regolith in which the pores are filled with water and/or air.

Any mass-wasting process that involves a flowing motion of regolith in which the pores are filled with water or air is called a **flow**. The different types of flows are illustrated in Table 7.2. Water-saturated mixtures give rise to *slurry flows.* Slow slurry flow, a process known as *solifluction*, is common in areas with high rainfall. Rapid slurry flows with speeds of up to 160 km/h (100 m/h) sometimes occur and can be very dangerous. When a rapid slurry flow is transporting mostly particles larger than sand, it is a *debris flow*; when the material consists mostly of fine particles, it is a *mudflow.*

**creep** The imperceptibly slow downslope granular flow of regolith.

Flowing regolith that is not water-saturated is called a *granular flow.* Like slurry flows, granular flows can be either slow or fast. The most common, widespread, and important kind of mass wasting is **creep**, the imperceptibly slow downslope granular flow of regolith. It may be possible to see evidence of creep in the region in which you live in the form of curved treetrunks, or old fences, telephone poles, or gravestones leaning at an angle on hillslopes. Steeply inclined rock or layers of regolith may also be bent over in the downslope direction just below the surface of the ground, another sign of creep. Some granular flows are more rapid than creep. They include *earthflows*—which may last days, months, or years but are not life-threatening—and *debris avalanches,* rare, spectacular, and extremely dangerous events in which mixtures of rock and regolith break loose and start as falls or slides. In the case of falls, the material pulverizes on impact, and then continues to travel downslope at velocities of up to several hundred miles per hour (Fig. 7.23).

## Mass Wasting in Cold Climates

**frost heaving** The uplift of surface rock and regolith as a result of the freezing of subsurface water to form ice.

Mass wasting is especially prevalent where average temperatures are very low and much of the landscape is underlain by perennially frozen ground. Ice that is forming in saturated regolith pushes up the ground surface in a process called **frost heaving**. The ground surface is lifted at right angles to the slope. As the ground thaws, each particle tends to drop vertically, pulled downward by gravity. The net result of repeated episodes of freezing and thawing, during which a particle experiences a succession of upward and downward movements, is slow but progressive downslope creep.

Figure 7.23
**DEADLY DEBRIS AVALANCHE**
This view shows Mount Huascarán, Peru, and the debris avalanche that destroyed the villages of Yungay (remains at lower right) and Ranrahirca in May 1970. The debris avalanche was triggered by an earthquake.

In regions that are underlain by frozen ground, a thin surface layer thaws in summer and refreezes in winter. During the summer the thawed layer becomes saturated with meltwater and is very unstable, especially on hillsides. Gravity pulls the thawed sediment slowly downslope in a form of solifluction known as *gelifluction*. Although the rate of movement is generally less than 10 cm/year (about 4 in/y), gelifluction is so widespread on high-latitude landscapes that it is an important agent of mass transport. Hillslopes in arctic Alaska and Canada, for example, are often mantled by sheets or lobes of geliflucted regolith.

**?** What's the next step in the rock cycle, after weathering and erosion?

# WHAT'S AHEAD

Let's return, briefly, to the rock cycle (Fig. 7.1). When rock is uplifted and exposed at the Earth's surface, it is immediately subjected to weathering. Through the processes of chemical and mechanical weathering, regolith is formed. Like a loose-fitting blanket, the regolith slowly moves downslope under the influence of gravity. It is transported by water, ice, and wind and deposited as sediment in some low-lying place—perhaps a lake or the sea—where it eventually becomes sedimentary rock. In this chapter we have learned some things about weathering and erosion and their importance in the rock cycle. The transformation from regolith to sediment and then from sediment to sedimentary rock, the next two legs of the cycle, are discussed in the next chapter.

## Chapter Highlights

**1.** The rock cycle is the continuous cycle of rock formation and transformation that connects the Earth's internal and external processes. It commences when rocks uplifted by tectonic or volcanic processes are exposed at the Earth's surface.

**2. Weathering** extends as far down as air, water, and living organisms readily penetrate the Earth's crust. Water from rain or melted snow enters bedrock along **joints** and other openings. Near the surface, water may freeze and disrupt the rock mechanically, while the slightly acidic rainwater attacks the rock chemically. Weathering has covered most of the Earth's land surface with a blanket of weathered rock called the **regolith.**

**3. Mechanical weathering** is caused by the freezing of ice in rock openings, the growth of salt crystals in confined spaces, spalling and fracturing by fire, and the action of roots.

**4. Chemical weathering** involves the removal of minerals in solution and the transformation of minerals into new minerals that are stable at the Earth's surface. The principal processes of chemical weathering are hydrolysis, solution, and oxidation. Mechanical and chemical weathering involve different processes, but they usually work together.

**5.** The effectiveness of weathering depends on the type and structure of rock, the steepness of the slope, the local climate, the length of time weathering processes operate, and the churning of regolith by burrowing animals. Because high temperature and abundance of water speed up chemical reactions, chemical weathering is far more active in moist, warm climates than in cold, dry ones.

**6. Soil** is regolith, supplemented by organic matter and small organisms, that can support rooted plants. Soils develop distinctive **horizons** whose characters are a function of climate, vegetation cover, parent rock, topography, and time. The A horizon of a soil is rich in humus and has lost soluble material, especially iron and aluminum, through leaching. Clay and the hydroxides of iron and aluminum accumulate in the B horizon. The B horizon overlies the C horizon, which consists of slightly weathered parent rock.

**7.** Soils can be grouped into three large families: **pedalfers,** found in regions of temperate climate and moderate rainfall; **pedocals,** found in dry regions; and **laterites,** found in hot, wet regions.

**8. Erosion** involves the removal and transportation of regolith through the combined action of ice, water, and wind, together with the abrasion of bedrock as a result of the transportation process.

**9.** En masse downslope movement of rock and regolith under the pull of gravity is termed **mass wasting.** Slope failures such as falls, slides, and slumps involve downslope movement of relatively coherent masses of rock or regolith. Flows are mixtures of regolith and water or air. If they are water-saturated, they are called slurry flows; if they are not, they are called granular flows. The most important mass-wasting process is creep, the slow movement of an unsaturated granular flow.

## ▶ The Language of Geology

## ► *Questions for Review*

1. What is regolith, and how is it formed?
2. Explain the difference between the two kinds of weathering.
3. What are joints, how do they form, and what role do they play in weathering?
4. How does frost wedging happen, and why is it most effective in places where temperature frequently passes from above freezing to below freezing?
5. Explain how any four processes of mechanical weathering occur.
6. What effect does acid rain have on chemical weathering?
7. What factors influence the effectiveness of weathering?
8. What is soil, and how does it differ from the rest of the regolith?
9. What are the three large soil families? Under what climatic conditions is each one formed?
10. Why are laterites red?
11. What are the three main agents of erosion?
12. What is mass wasting, and why is it important?

## ► *Questions for Thought and Discussion*

1. Examine the area where you live for evidence of mechanical and chemical weathering. How might you determine the relative importance of the two kinds of weathering in your area?
2. Mars is often called the Red Planet because its regolith is reddish colored. What does the color suggest about regolith-forming processes on Mars?
3. The Moon has neither an atmosphere nor a hydrosphere nor a biosphere, but when the first astronauts landed they discovered that the Moon has a deep regolith. How might the regolith have been formed, and how might it differ from the Earth's regolith?
4. What kinds of mass-wasting processes occur where you live? Can you identify any evidence that would suggest how fast or how slowly mass wasting is moving regolith downslope?

 For an interactive case study erosion of topsoil in the Great Plains, visit our Web site.

# · 8 ·

# FROM SEDIMENT TO ROCK: ROCKS THAT FORM NEAR THE EARTH'S SURFACE

Oil fuels the world's economy. By geologic chance, half the known oil reserves are in the Middle East, a region divided by centuries of intertribal and religious disputes, but a region otherwise blessed with very few natural resources.

When sedimentary rocks rich in organic matter are squeezed and heated under the right conditions, the organic matter is converted to oil. Most of the oil eventually leaks away and is lost. However, if plate tectonic movements buckle sedimentary strata at the very moment the oil is forming, some of the oil can be trapped under the buckles. Around the Persian Gulf, a combination of the right heating and squeezing of sedimentary rocks and the right timing of tectonic buckling has produced the richest entrapments of oil ever discovered.

Unfortunately, geopolitical boundaries and geologic resource boundaries don't coincide. A few countries have as much oil as they need, but most, including the United States, must import part or all that they use. This need has thrust geology and politics into conflict—in 1991 such violent conflict that the United Nations went to war with Iraq to protect oil supplies from Kuwait.

What does the future hold? The Middle East will probably continue to be politically unstable and other wars may be fought, but the importance of the region will eventually decline as continued pumping of the oil diminishes the reserves and the world turns to alternative sources of power.

◆

## In this chapter we discuss

- The many different kinds of sediments
- How sediments become sedimentary rocks
- How sediments provide clues to past environments
- How and why sediments in a single layer vary from place to place

*Oil in the Middle East*

Oil and gas from one of the great Saudi Arabian oilfields are processed in this petrochemical plant.

# Have you ever wondered...

*How do geologists figure out what ancient climates and environments were like?*

...You will know about some of the evidence they use after you have read this chapter.

Like a restless housekeeper, nature is ceaselessly sweeping regolith off the solid bedrock beneath it. The sweepings are carried and deposited as sediment in river valleys, lakes, the sea, and other places. We can see sediment being transported by trickles of water after a rainfall and by every wind that carries dust. The mud on a lake bottom, the sand on a beach, even the dust on a windowsill is sediment—we find sediment nearly everywhere.

We learned in chapter 7 that regolith differs in different places. It differs because its formation, by mechanical and chemical weathering, is determined in part by climate. Climate also influences how regolith is transported by wind, water, or ice. Climate, along with other characteristics of the environment, also determines how and where sediment is deposited. As a result, sediments—and the rocks that are formed from sediments—preserve a geologic record, or archive, of past climates.

**? What kinds of plant or animal fossils indicate that there was once a warm, wet climate in an area that is now dry and hot?**

## ARCHIVES OF THE EARTH'S HISTORY

The Omo River basin of Ethiopia in Africa provides an example of how geology preserves a record of past climates. Today the region is a hot, dry, inhospitable place. Yet it is here, in ancient stream sediments, that archaeologists have discovered fossil remains of some of our earliest human ancestors (see Fig. 8.1). Using radiometric dating techniques (chapter 3) to determine the ages of volcanic rocks interlayered with the fossil-bearing sediments, geologists have discovered that those ancestors lived about 3 million years ago. The stream sediments reveal another interesting story: From the color, size, and rounding of the sediment particles, as well as evidence from bits of fossil plants and animals, it is clear that the Omo region had plenty of water when our early ancestors lived there. There were

Figure 8.1
**SEARCHING FOR HUMAN ORIGINS**
These archeologists are searching ancient sediments in Ethiopia for fossil remains of human ancestors.

Omo region

Figure 8.2
**LAYERS OF ROCK: BEDDING**
When you look at an outcrop of sedimentary rock, one of the first things you notice is layering or *bedding*. Here, layered sedimentary rocks in Capital Reef National Park, Utah, have been exposed by erosion.

lakes and flowing streams, and the region supported a lot of vegetation and small animals. The region clearly had a much more hospitable climate 3 million years ago.

Sediments and sedimentary rocks are the best archives we have of the Earth's past environments and how the Earth system has changed with time. Reading these sedimentary archives is not always easy, but it is a fascinating challenge.

# SEDIMENTS AND SEDIMENTATION

When you look at an outcrop of sedimentary rock, one of the first things you notice is the bedding (Fig. 8.2). **Bedding** is the layered arrangement of strata in a body of sediment or sedimentary rock. Each stratum, or **bed**, within a succession of strata can be distinguished from adjacent beds by differences in thickness or character. The top or bottom surface of a bed is a **bedding surface**. Recall from chapter 3 that the layering of rock—*stratification*—occurs because sedimentary particles are deposited in distinct layers (*strata*). Strata differ because of differences in the characteristics of the particles or in the way they are arranged.

It is the presence of bedding and bedding surfaces that indicates that the rock was once sediment. If you look at individual beds, you will see that there are differences between them. The differences are a result of differences in the formation, transportation, and deposition of the materials that make up the sediment and differences in the processes by which the sediment was converted into rock.

As a first step in deciphering the sedimentary archive, we need to sort out the evidence. We do this by separating sediments and sedimentary rocks into three groups based on the kinds of particles they contain.

1.  **Clastic sediment** (from the Greek word *klastos*, meaning "broken fragment"), which is formed from the loose rock and mineral debris produced by weathering and erosion. Sand is a common clast. So are dust and gravel. This debris is also called *detritus* (from the Latin word for "worn down"), so clastic sediment is also known as *detrital sediment*.

2.  **Chemical sediment,** which is formed by the precipitation of minerals dissolved in lake water or sea water. The salt that coats the surface of dry lakes in Death Valley, California, is chemical sediment.

3.  **Biogenic sediment,** which is composed mainly of the remains of plants and animals.

**bedding** The layered arrangement of strata in a body of sediment or sedimentary rock.

**bed** A *stratum,* or layer, within a succession of sedimentary rock strata.

**bedding surface** The top or bottom surface of a rock layer or bed.

?
**Are all sediments formed in the same way?**

**clastic sediment** Sediment formed from the loose rock and mineral debris produced by weathering and erosion.

**chemical sediment** Sediment formed by the precipitation of minerals dissolved in lake water or seawater.

**biogenic sediment** Sediment that is composed mainly of the remains of plants and animals.

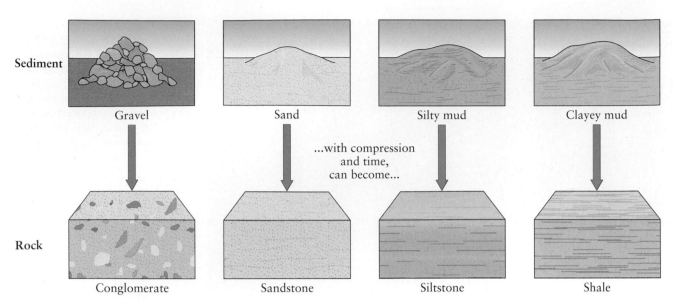

**Figure 8.3**
**CLASTIC SEDIMENTS AND ROCKS**
These are the main kinds of clastic sediment and the sedimentary rocks that form from them. Individual rock fragments in a sediment are called *clasts,* and these clastic sediments are named by clast size. Each sediment size forms a distinctive kind of sedimentary rock.

Let's look briefly at these three types of sediment—what they are like, how they are formed, and how they are deposited.

## Clastic Sediment

**clast** Fragment; the individual particles in a clastic sediment.

Each individual particle in a clastic sediment is a **clast,** or fragment. Clasts can be mineral grains or rock fragments. They range in size from the largest boulders down to clay particles as fine or finer than flour. Clast size (Table 8.1) is the primary basis for classifying clastic sediments and clastic sedimentary rocks (Fig. 8.3).

A poorly sorted sediment contains clasts of many different sizes (Fig. 8.4A); in a well-sorted sediment, all the clasts are similar in size (Fig. 8.4B and C). The range of clast sizes in a clastic sediment reflects a characteristic called *sorting*

**TABLE 8.1    Sediments and Rocks Formed by Clastic Particles**

| Name of Clast | Average Size | Sediment Name | Rock Name |
|---|---|---|---|
| Boulder | Baseball or larger | | |
| Cobble | Chicken's egg | Gravel | Conglomerate[b] |
| Pebble | Dried pea | | |
| Sand | Pin's head | Sand | Sandstone |
| Silt | Grain of table salt | Silt | Siltstone |
| Clay[a] | Particle of flour | Clay | Shale |

[a]Clay refers only to particle size, not to what the particles are made of.

[b]If clasts are angular, the rock is a *breccia* rather than a conglomerate.

# Sorting and Roundness

Figure 8.4

A. Till, a nonsorted clastic sediment deposited by the Matanuska Glacier, Alaska. Clasts range in size from fine dust to boulders 3 m (10 ft) across.

B. Well-sorted and well-rounded grains of quartz sand from the St. Peter Sandstone of Wisconsin. Each sand grain is about the size of a pin's head.

## SORTING

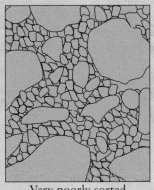
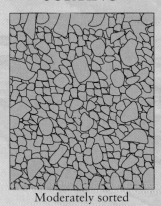
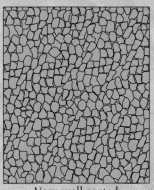

Very poorly sorted          Moderately sorted          Very well sorted

C. Clastic sediments range from very poorly sorted to very well sorted as measured by the range in sizes of clasts.

## ROUNDNESS

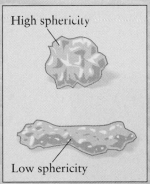
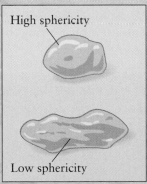
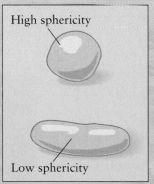

Angular          Intermediate          Rounded

D. Clasts range from angular to rounded, depending on the degree to which sharp edges have been worn off. Poorly sorted grains tend to be angular, and well-sorted grains tend to be rounded.

(Fig. 8.4C). The shapes of clasts also vary, ranging from spherical to non-spherical and from angular to rounded (Fig. 8.4D). Clast sorting and clast shape reflect the mechanisms by which clasts were transported and deposited. Mass-wasted and ice-transported clasts tend to be poorly sorted, with angular or subangular shapes. Poorly sorted—or even nonsorted—clastic sediment that has been transported by ice is called *till* (Fig. 8.4A). On the other hand, clasts that have been transported by water and wind tend to be well sorted, with rounded shapes. As the clasts are transported, they are subject to continuous chemical and physical breakdown. As a result of prolonged weathering, erosion, and deposition, a sediment may eventually consist almost entirely of clasts of quartz, the hardest and most resistant of the rock-forming minerals (Fig. 8.4B).

Various depositional processes also influence the characteristics of clastic sediments. For example, *rhythmic layering* consists of alternating layers of coarse and fine clasts. Such an alternation suggests that some naturally occurring rhythm has influenced the transport and deposition of the sediment. A pair of such rhythmic sedimentary layers deposited during a single year is called a **varve** (a Swedish word meaning "cycle") (Fig. 8.5). Varves are formed by seasonal variations in glacial lakes. In spring and summer, the inflow of glacial meltwater carries coarse sediment. In autumn and winter, the flow of meltwater ceases, the lake freezes over, and any fine sediment that remains in suspension slowly settles to form the fine clast bed of the varve pair.

**Graded beds** are individual beds in which the coarse clasts are concentrated at the bottom of the bed, grading up to the very finest clasts at the top (Fig. 8.6). Graded beds are formed from mixtures of coarse and fine clasts such as might be carried by a stream in flood. When the flowing slurry of water and sediment containing clasts of mixed sizes slows down, the coarse clasts settle quickly, followed by successively finer clasts.

Turbulent flow in streams, wind, or ocean waves produces a type of bedding called **cross bedding**. The resulting beds are inclined with respect to a thicker stratum within which they occur (Fig. 8.7). As they are moved along by wind or water, the particles tend to collect in ridges, mounds, or heaps in the form of ripples, waves, or dunes. These migrate slowly forward in the direction of the current. Particles are carried up and over the top of the pile by the current, forming beds that are inclined at angles as great as 30° to 35°. The direction in which the cross bedding is inclined can tell us the direction in which the water or air currents were forming at the time of deposition.

**Volcanic sediments** are a special kind of clastic sediment. What makes them special is that all of the clasts are volcanic in origin. As discussed in chapter 6, explosive volcanic eruptions blast out large quantities of fragments during an eruption. There is an old saying that explains what is unique about volcanic sed-

**varve** A pair of rhythmic layers of sediment deposited in still water over the course of one year.

**graded bed** Individual beds of sediment or sedimentary rock in which the coarsest clasts are at the bottom of the bed, grading up to the finest at the top.

**cross bedding** Sedimentary strata (beds) that are inclined with respect to a thicker stratum within which they occur.

**volcanic sediment** A special kind of clastic sediment in which all of the clasts are volcanic in origin.

Figure 8.5
**SEDIMENT CYCLES CALLED VARVES**
Some sediment layers are deposited in annual cycles, called varves. Each pair of layers in a sequence of varves represents an annual deposit. The light-colored silty layers were deposited in the warm months, and the dark-colored clayey layers accumulated in the cool months. These varves were deposited in a glacial lake in Connecticut.

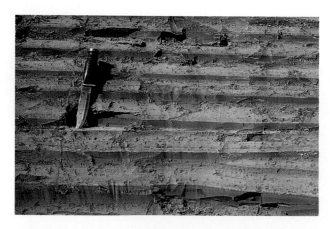

# Graded Bedding

Rain

Turbidity current carries slurry of water and clasts.

As velocity of flow drops, largest clasts settle first.

**Figure 8.6**
Graded beds form when a rapidly flowing turbid slurry of water and mixed clasts slows down and the largest clasts settle first. In this graded bed from California, the largest clasts are the size of a penny; the smallest, at the top of the bed, are clay sized.

iments—"they are igneous on the way up but sedimentary on the way down". Because the fragments were hot when formed, they are called *pyroclasts* (from the Latin *pyro*, meaning "fire").[1] The names used for different sizes of pyroclasts are shown in Figure 6.19.

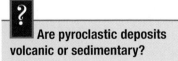

**Are pyroclastic deposits volcanic or sedimentary?**

## Chemical Sediment

Recall from chapter 7 that all surface and groundwater contains dissolved salts, which eventually find their way to a lake or the sea. No water on or in the Earth is completely free from dissolved matter. When dissolved matter is precipitated

---

[1] Recall from chapter 6 that pyroclasts are also called *tephra*.

Figure 8.7
ANCIENT CROSS BEDDING
IN SAND DUNES
These dunes of loose sand became cemented together into
sandstone, preserving the
windswept angles of the sand layers. This is in Zion National
Park, Utah.

from sea or lake waters, *chemical sediment* is the result. This precipitation can happen in two ways:

1. Through the activities of plants and animals in the water. For example, tiny plants living in seawater can change the amount of carbon dioxide dissolved in the water and thereby cause calcium carbonate to precipitate. Many limestones are formed in this way.

2. Through precipitation of a solid as a result of evaporation. For example, if an inland sea is subject to an increasingly warm and dry climate, or if the inflow of fresh water is restricted for some reason, evaporation may exceed the input of fresh water. The sea may then become so shallow and saline that salts that were dissolved in the water will begin to precipitate as solids. A modern example of this process is discussed later in the book in Box 12.1, on the death of the Aral Sea.

## Biogenic Sediment

*Biogenic sediment* is composed of the remains of plants and animals. It may seem similar to chemical sediment formed by the activity of plants and animals in water, but there is an important difference. Chemical sediment was never part of an organism, whereas biogenic sediment consists largely or entirely of material that was once part of living organisms. Solid materials, such as shells and bones, often end up as broken fragments, or clasts. If a biogenic sediment is composed largely of biogenic clasts, it is called a *bioclastic sediment* (Fig. 8.8).

Figure 8.8
**BIOCLASTIC ROCK:**
**AMALGAMATED FOSSILS**
A rock composed largely of
shells and shell fragments is a
bioclastic rock called a co-
quina. This specimen is from
Virginia.

# SEDIMENTARY ROCKS

The term **lithification** (from the verb *lithify*, meaning "turn to stone") refers to a group of processes by which newly deposited, loose sediment is slowly transformed into sedimentary rock. In order for this transformation to take place, the clasts must somehow be bound together, and there are a number of ways this can happen. Lithification is just one result of **diagenesis**, a term that covers all of the chemical, physical, and biological changes that a sediment undergoes from the moment it is deposited up to, during, and immediately after lithification.

## Lithification

The first and simplest change leading to lithification is **compaction.** This change occurs as the weight of accumulating sediment forces the grains together, thereby reducing the *pore space* (the space between grains) and forcing water out of the sediment. As pore water from deep in a pile of sediment rises up, it becomes cooler and substances dissolved in the water precipitate out and cement the grains together. This is called **cementation.** Calcium carbonate is one of the most common cements (Fig. 8.9). Silica, a particularly hard cement, is also common, as is iron hydroxide.

As sediment accumulates, less stable minerals may recrystallize and become more stable. This process—**recrystallization**—is especially common in porous limestones formed from coral reefs. The mineral aragonite, which has the same

**lithification** The group of processes by which newly deposited, loose sediment is transformed into sedimentary rock.

**diagenesis** The chemical, physical, and biological changes undergone by a sediment from the moment it is deposited up to, during, and immediately after lithification.

**compaction** Reduction of pore space in a sediment in response to the weight of overlying sediment accumulation.

**cementation** The process whereby substances dissolved in pore water in a sediment precipitate out and cement the sediment grains together.

**recrystallization** The formation of new crystalline mineral grains in a rock.

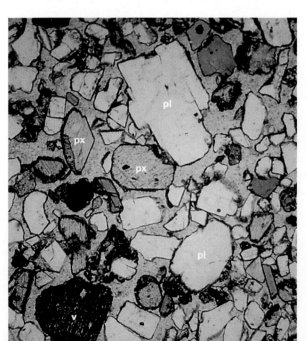

Figure 8.9
**SANDSTONE UNDER**
**A MICROSCOPE**
These are sand grains viewed
in a microscopic section of
sandstone from central Washington. The grains are cemented together by the mineral calcite. Note that the
sandstone is poorly sorted and
that most of the grains are
subangular.

composition as calcite (calcium carbonate), is present in the skeletal structures of living corals and other marine creatures. Over time, aragonite recrystallizes and becomes calcite, a more stable form of calcium carbonate. Like cementation, recrystallization acts to hold together the grains in a sedimentary rock.

## Clastic Sedimentary Rocks

If a sedimentary rock is made up of particles derived from the weathering and erosion of igneous rock, how can we tell that it is sedimentary rather than igneous? In addition to such obvious clues as bedding, the texture of the rock provides evidence. Mineral grains in an igneous rock are irregular in shape, but clasts in a sedimentary rock are usually rounded. The clasts may show signs of the abrasion they received during transport. In addition, clastic sedimentary rock usually contains particles held together by cement, whereas igneous rock consists of interlocking mineral grains without any cement. The presence of fossils also helps distinguish between the two classes of rock. Since no organism can survive the high temperatures at which igneous rocks are formed, the presence of ancient shells or evidence of plant life is an important clue to sedimentary origin.

Clastic sedimentary rocks are classified on the basis of particle size, just as sediments are. The four basic classes are *conglomerate, sandstone, siltstone,* and *shale*. These are the rock equivalents of gravel, sand, silt, and clay (Fig. 8.3). A rock that contains large clasts (larger than 2 mm, or 0.08 in in diameter) is a **conglomerate**. Typically, the large clasts of a conglomerate are surrounded by much finer-grained material called the *matrix*. If the clasts are angular rather than rounded, the conglomerate is called a *breccia*. The presence of angular clasts means that the sediment has only been transported a short distance and was not subjected to a long abrasion process.[2] **Sandstone** is a medium-grained clastic sedimentary rock. Most of the clasts in sandstones are quartz because quartz is such a resistant mineral. However, if the sediment has not been transported very far and the rock still contains a lot of feldspar, we use the term *arkose*. If plentiful rock fragments are also present, the sandstone is called a *graywacke*. **Siltstones** consist primarily of silt-sized particles (rock or mineral fragments with diameters in the range of 0.002 to 0.05 mm, or 0.00008 to 0.002 in.). Rocks with the smallest clast sizes, **shales**, contain abundant clay (mineral fragments with diameters less than 0.002 mm), although a significant amount of silt may also be present. Among these very fine-grained rocks, a distinction is drawn between shales, which break into sheet-like fragments, and *mudstones*, which break into blocky fragments.

## Chemical Sedimentary Rocks

Chemical sedimentary rocks result from the lithification of chemical sediments. As we have seen, such sediments are formed in two ways: by evaporation or by precipitation from solution due to the activities of plants and animals. Most chemical sedimentary rocks contain only one important mineral. That, together with the mode of precipitation, forms the basis for classification of these rocks.

Chemical sedimentary rocks formed by the evaporation of sea or lake water are called **evaporites**. Examples of seawater evaporites are *rock salt* (the mineral halite) and *gypsum*. *Calcite* can also precipitate as a seawater evaporite. Examples of lake water evaporites are sodium carbonate and borax. Many evaporite

---

**?** **What would happen if you left a bucket of seawater in the Sun for a few days?**

**conglomerate** A clastic sedimentary rock consisting of large fragments in a finer-grained matrix.

**sandstone** A medium-grained clastic sedimentary rock, in which the clasts are typically dominated by quartz grains.

**siltstone** A rock that consists primarily of silt-sized particles.

**shale** A very fine-grained clastic sedimentary rock, consisting primarily of clay and silt particles.

**evaporite** A chemical sediment that forms when lake water or seawater evaporates, leaving behind salt deposits.

---

[2] The rate at which clasts become rounded is partly a function of size; large, gravel-sized clasts become rounded more rapidly than sand or silt-sized clasts. Substantial rounding of sand requires considerably more abrasion and a greater distance of transport than is required for gravel.

minerals are mined because they have industrial uses. For example, the gypsum used to make plasterboard is mined from evaporites. In addition, most of the salt we eat comes from evaporites.

Chemical sedimentary rocks formed by the activities of plants and animals include certain kinds of *limestones* and *phosporites*. Phosphorites, which consist largely of the phosphate mineral apatite (calcium phosphate, the same mineral that is present in your bones and teeth) are the principal source of the phosphorous fertilizers. They are formed when the bones of dead marine animals dissolve in deep ocean water. The phosphorous-rich water is brought to the surface by the upwelling of deep ocean currents close to shore. At the surface conditions, apatite precipitates from the water and sinks to the bottom in shallow near-shore waters, where it accumulates.

An unusual but economically important kind of chemical sedimentary rock is **banded iron formation** (Fig. 8.10). Banded iron formations were formed from iron-rich siliceous sediments that were entirely chemical in origin. They are the source of most of the iron mined today. All known banded iron formations are at least 1.8 billion years old.

Banded iron formations are marine in origin; that is, the iron was once dissolved in seawater. Today seawater contains only slight traces of iron because oxygen in the atmosphere oxidizes the iron to an insoluble form. At the time when the banded iron formations were formed, there must have been a very large amount of iron dissolved in seawater. We can conclude, therefore, that when these rocks were formed there was very little oxygen in the atmosphere. Why did the iron precipitate? One possibility is that microscopic organisms floating in the sea released oxygen as a result of photosynthesis, causing the dissolved iron to precipitate.

**banded iron formation** A type of chemical sedimentary rock formed from iron-rich siliceous sediments.

**Figure 8.10**
**BANDED IRON FORMATION**
Iron-rich chemical sediment (dark) is interbedded with siliceous chert layers (white) in the Brockman Iron formation, a 2-billion-year-old banded iron formation in the Hamersley Range of Western Australia. The woman taking the photograph is Dr. Janet Watson, a distinguished English geologist.

*Well, Watson, I will not offend your intelligence by explaining what is most obvious. The gravel upon the window sill was, of course, the starting-point of my research. It was unlike anything in the vicarage garden. Only when my attention had been drawn to Dr. Sterndale and his cottage did I find its counterpart.*

—Sherlock Holmes in *The Adventure of the Devil's Foot* by Sir Arthur Conan Doyle

**The most frightening crimes** *have no witness except the ground on which they were committed. From that alone, forensic geologists (Fig. B8.1) illuminate cases in a way that would impress Sherlock Holmes, the science's first practitioner. The following piece is an excerpt from* The Gravel Page, *by John McPhee, which originally appeared in* The New Yorker, *January 29, 1996.*

Mineral grains and microfossils can narrate a story. A police officer fails to report for work in Harrisburg, Pennsylvania. His private automobile is found in Virginia, its trunk full of blood. The officer remains missing. Harrisburg police search the region for several days and come up with nothing. They turn for help to the F.B.I., which collects the officer's car and turns it over to Special Agent Ronald Rawalt, a forensic geologist. With his petrographic microscope, not to mention his common sense, Rawalt studies the car. He sees a heavy buildup of soil in one wheel well and inside a bumper, and notes that the soil was wet-deposited. He sees also that the car was driven over pavement with water on it between the deposition time and the time the car was recovered. The soil is of one consistency, not the mottled layering from different locations that would usually be found under a fender. Someone stepped on the accelerator and spun

## Biogenic Sedimentary Rocks

**limestone** A biogenic or bioclastic rock that consists primarily of the mineral calcite.

**dolostone** A biogenic or bioclastic rock composed predominantly of the mineral dolomite.

**chert** A biogenic sediment composed of extremely tiny particles of quartz.

**peat** A biogenic sediment formed from the accumulation and compaction of plant remains.

**coal** A combustible rock formed from the lithification of peat.

**Limestone** is the most important biogenic rock. Some limestones are chemical in origin, but by far the majority are biogenic or bioclastic in origin. They consist of lithified shells, or shell fragments, of marine organisms. Because these organisms build their shells of calcium carbonate, limestones are formed chiefly of the carbonate mineral calcite. Calcite is sometimes replaced by the mineral dolomite, and the resulting rock is called a **dolostone.** Another kind of biogenic rock that consists of extremely tiny particles of quartz is **chert.** Bedded chert commonly results from the accumulation of innumerable tiny aquatic organisms that contain siliceous particles.

An important class of biogenic sediment consists of the accumulated remains of trees, bushes, and grasses. As millions of plants in tropical swamps die over thousands of years, their remains accumulate as thick, organic sediments. Over time, and with pressure, this mass gradually becomes **peat,** a sediment made up of unconsolidated plant remains. Eventually, over more time and pressure, peat lithifies and becomes **coal.** Lithification (or *coalification*) involves compaction, release of water, and slow chemical changes that make the coal relatively richer in carbon than the original peat was.

the tire in mud. Like Rawalt, the Harrisburg police have assumed from the beginning that the soil in the wheel well is from a place where the body might be found. They just wonder where to look.

Going through the washed minerals, Rawalt finds microscopic fragments of glass beads and of yellow reflective paint and white reflective paint. The beads could be from any stretch of road, but the presence of both white and yellow paint suggests a hill or a curve. He finds microscopic asphalt. He finds black slag. He knows that Pennsylvania historically has bought slag from its iron smelters and coal-fired furnaces, crushing it for use as an anti-skid on highway curves. He also finds an assemblage of microfossils. He looks them up in a textbook of micropaleontology. The book includes maps. So unusual are these fossils that the pyritic limestone they come from outcrops only in two highly confined localities, one of them in Appalachian Pennsylvania. The limestone in Pennsylvania is just a narrow stringer that comes down off a mountain and crosses a country road south of Harrisburg. Rawalt calls the Harrisburg police. He mentions the road and tells them to stop at a rising curve where both yellow and white paint are present and there's not enough room for a whole car to get off the pavement. Since the missing man is heavy, look a short distance downslope. Next day, the police call Rawalt: "We got him. He was there, under a pile of brush.

**Figure B8.1**
**GEOLOGICAL SLEUTHS AT WORK**
These forensic geologists are making measurements and collecting soil samples from the scene of a hit and run accident. The soil fell from underneath the vehicle fender as a result of the impact and will be used for comparison with samples taken from suspect vehicles.

# DEPOSITIONAL ENVIRONMENTS

Just as history books record the changing patterns of civilization, sedimentary rocks record the environmental history of our planet. Layers of sediment, like pages in a book, show how environmental conditions have changed over more than 3 billion years. We have already seen that the size, shape, and arrangement of particles in sediments, as well as the geometry of sedimentary strata, provide evidence about the geologic environment in which sediments accumulate. These and other clues enable us to demonstrate the existence of ancient oceans, coasts, lakes, streams, deserts, glaciers, and swamps, as well as ancient examples of other places where sediments have been deposited.

**How can geologists tell that an ocean once existed where there is now dry land?**

## Environmental Clues in Sedimentary Rocks

Irregularities formed by currents of water or air moving across sediment can be preserved and later exposed on bedding surfaces. For example, bodies of sand that are being moved by wind, streams, or coastal waves are often rippled, and the ripples may be preserved in sandstones as *ripple marks* (compare Figs. 8.11 A

Figure 8.11
ANCIENT AND MODERN FEATURES IN SEDIMENTARY ROCKS
A. Ripples are forming in shallow water near the shore of Ocracoke Island, North Carolina.
B. These ancient ripple marks are exposed on a bedding surface of a sandstone at Artist's Point,
Colorado National Park, Colorado. C. Mud cracks formed on the surface of a dry lake floor.
D. Fossil mud cracks are preserved on the surface of a shale exposed at Ausable Chasm, New York.

and B). In another example, some shales and siltstones are broken in a random pattern. By comparing them with similar features in modern sediments, we infer that these are *mud cracks* caused by the shrinking and cracking of wet mud as its surface dries (compare Figs. 8.11 C and D). Mud cracks imply the former presence of tidal flats, exposed stream beds, desert lake floors, or some other intermittently wet environments. Footprints and animal trails are often found near ripple marks and mud cracks (Fig. 8.12), suggesting that the animals were hunting for prey stranded by the retreating tide. Even impressions made by raindrops during brief, intense showers may be preserved in strata. Raindrops and mud cracks provide evidence that moist surface conditions were present at the time they were formed.

Fossils provide significant clues about former environments. Some animals and plants are restricted to warm, moist climates, whereas others can live only in cold, dry climates. Using the climatic ranges of modern plants and animals as guides, we can infer the general character of the climate in which similar ancestral forms lived. For example, plant fossils can provide estimates of past rainfall and temperature for sites on land. Fossils of tiny floating organisms can tell us about former surface temperatures and salinity conditions in the oceans. Fossils are also the basis for telling the relative ages of strata; they have played an important role in efforts to reconstruct the past 600 million years of the Earth's history (chapter 3).

*A.* ▲

**Figure 8.12**
**FOOTPRINTS IN THE SAND**
*A.* Seagull footprints are imprinted in sandbars along the Alsek River, Glacier Bay National Park, Alaska. In the background is the Brabazon Range of the Saint Elias Mountains. *B.* Fossil footprints are preserved in sandstone in the Painted Desert, near Cameron, Arizona.

*B.* ▶

**?** **Why are some rocks dark and others bright-colored?**

The color of fresh (that is, unweathered) sedimentary rock can also provide clues to environmental conditions. The color of a rock is determined by the colors of the minerals, rock fragments, and organic matter of which it is composed. Iron sulfides and organic detritus buried with sediment are responsible for most of the dark colors in sedimentary rocks; the presence of these materials implies that the sediment was deposited in a reducing (oxygen-poor) environment. Reddish and brownish colors result mainly from the presence of iron oxides, occurring either as coatings on mineral grains or as very fine particles. These minerals point to oxidizing conditions in the environment.

## Sedimentary Facies

If you examine a vertical sequence of exposed sedimentary rocks, you may notice differences as you move upward from one bed to the next. You may note changes in rock type, color, texture, particle size, and thickness. The differences indicate that the *environmental conditions* in that location changed during the deposition of the sediments. If you trace a single bed for a few miles, you may also notice changes, indicating that at any given time during the deposition of the sediment, conditions differed from one place to another.

The changes in the character of sediment from one environment to another are referred to as changes of *facies* (pronounced *fay-seez*). A **sedimentary facies** is any sediment that can be distinguished from another sediment that accumulated at the same time but in a different depositional environment. One facies may be distinguished from another by differences in grain size, grain shape, stratification, color, chemical composition, depositional structures, or fossils. Adjacent facies can merge into each other either gradually or abruptly (Fig. 8.13). For example, coarse gravel and sand on a beach may gradually pass into finer sand, silt,

**sedimentary facies** Any sediment (or sedimentary rock) that can be distinguished from another sediment that accumulated at the same time, but in a different depositional environment.

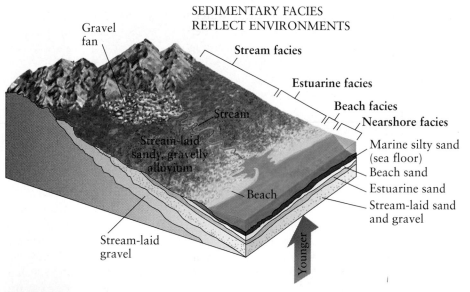

Figure 8.13
**SEDIMENTARY FACIES**
A section extending from a mountain valley across a gravelly plain and estuary to the adjacent ocean shows a variety of depositional environments in which distinctive facies are deposited. Facies adjoin or merge into each other on the land surface. In a vertical section, offshore, they lie one above another. The boundaries between adjacent facies dip seaward. This indicates that over time the boundaries have migrated in the landward direction. This implies that sea level has risen relative to the land.

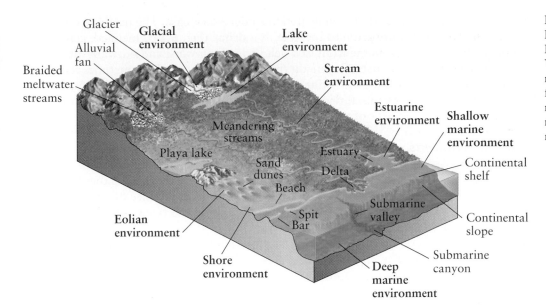

Glacier | Glacial environment | Lake environment
Alluvial fan | Stream environment
Braided meltwater streams
Estuarine environment | Shallow marine environment
Meandering streams
Playa lake | Estuary
Sand dunes | Delta | Continental shelf
Beach | Submarine valley
Eolian environment | Spit Bar | Continental slope
Shore environment | Submarine canyon
Deep marine environment

**Figure 8.14**
**DEPOSITIONAL ENVIRONMENTS**
Various depositional environments are seen while traveling from the crest of a mountain range across the edge of a continent to the adjacent margin of a nearby ocean basin.

and clay on the floor of the sea or a lake. Coarse, bouldery glacial sediment, on the other hand, may end abruptly at the margin of a glacier.

By studying the relationships among different sedimentary facies, and using these characteristics to identify original depositional settings, we can reconstruct a picture of the environmental conditions that prevailed in a region during past geologic times.

## Common Sediments on Land

Sediments on land can be transported by water, ice, and wind, or moved downslope under the influence of gravity. They are deposited wherever the carrying capacity of the transporting agent decreases enough for the sediment to settle out. Thus, common environments for the deposition of sediment on land include stream channels and floodplains, lake shores and bottoms, the margins of glaciers, and areas where the wind is frequently strong, such as beaches or deserts.

*Stream Sediment.* Streams are the main transporters of sediment on land. Sediment deposited by streams differs from place to place depending on type of stream, strength of the water flow, and nature of the sediment load (Fig. 8.14). A typical large, smoothly flowing stream may deposit well-sorted layers of coarse and fine particles as it slowly migrates back and forth across its valley. During spring floods, silt and clay are deposited on the floodplains. By contrast, a meltwater stream flowing from the front of a mountain glacier may dump an abundance of boulders, gravels, sands, silt, and clay. A large mountain stream flowing down a steep valley can transport an abundant load of sediment. When the stream reaches the front of the mountain and is no longer constrained by valley walls, it is free to shift back and forth across the more gentle terrain as the coarser part of its load is dropped. The result is a fan-shaped deposit in which the sediments may range from coarse, poorly sorted gravels to well-sorted, crossbedded sands.

*Lake Sediments.* Sediments that are deposited in a lake (Fig. 8.14) accumulate on the lake shore and on the lake floor. Lakeshore deposits are generally well-sorted sands. The sediment load of a stream entering a lake will be dropped as

**?** **What makes water, ice, and wind drop their sediment loads?**

**delta** A sedimentary deposit, commonly triangular or fan-shaped, that forms where a stream enters a standing body of water.

the stream's velocity and transporting ability suddenly decrease. The resulting deposit, which extends outward into the lake, is a **delta**[3] (Fig. 8.14). Inclined, generally well-sorted layers on the front of a delta pass downward and outward into thinner, finer, evenly laminated layers on the lake floor (Fig. 8.5). Sometimes the water supply to an inland lake is diminished or cut off. If the climate is warm enough, the lake may dry up, leaving behind an evaporite deposit. An example of a modern-day environment of this type is the Great Salt Lake of Utah.

*Glacial Sediments.* Sediment transported by a glacier is either deposited along the glacier's base or released at the margin of the glacier as melting occurs. The sediment is then subjected to further reworking by running water. Debris that has been deposited directly from ice commonly forms a random mixture of particles that range in size from clay to boulders and consist of all the types of rock over which the ice has passed (Fig. 8.4A).

**eolian sediment** Sediment that is carried and deposited by the wind.

*Eolian Sediments.* The activity of wind is referred to as **eolian** (pronounced *ee-OHL-ee-un*), after Aeolus, the Greek god of wind. Sediment carried by the wind tends to be finer than that moved by other erosional agents. Grains of sand are easily moved in places where strong winds are blowing and vegetation is too discontinuous to stabilize the land surface—for example, seacoasts and deserts (Fig. 8.14). In such places, the sand may pile up to form dunes composed of well-sorted sand grains, with bedding inclined in the downwind direction (the direction toward which the air is flowing). Using these characteristics, geologists can easily identify ancient dune sands in the rock record (Fig. 8.7; see also Fig. 1.20). Powdery dust that has been picked up and moved by the wind may travel great distances and be deposited thousands of miles away. Windblown sediment from China has been found on Hawaii, for example, and oceanographers have discovered windblown sediment from the Sahara Desert of North Africa at the center of the Atlantic Ocean. A special kind of yellow-brown windblown sediment is called **loess** (the German word meaning loose and pronounced *luhss*). Consisting predominantly of silt, loess is windblown dust of the Pleistocene age, transported from desert surfaces, glacial sediments, and glacial stream deposits at times of ice-sheet retreat (Fig. 8.15). Loess is thickest and coarsest near its source and becomes progressively thinner and finer with increasing distance downwind.

**loess** A fine, yellow-brown, windblown (eolian) sediment.

## Common Marine Sediments

Rivers transport sediment to the edge of the sea. There it can accumulate near the mouths of streams, be moved along the coast by currents, or be carried seaward to accumulate on the continental shelves, sometimes in thick layers. Spurred in part by the search for large undersea reservoirs of oil and gas, geologists have learned a great deal about the sediments accumulating on the shelves.

**estuary** A semi-enclosed body of coastal water in which seawater is diluted with fresh water.

*Estuarine and Deltaic Sediments.* Much of the load transported by a large river may be trapped in an **estuary,** a semi-enclosed body of coastal water in which seawater is diluted with fresh water (Figs. 8.13 and 8.14). Coarse sediment tends to settle close to land, while fine sediment is carried seaward. Tiny individual particles of clay carried in suspension settle very slowly to the sea floor. When the sediment load is large, deltas may extend outward into the sea (Fig. 8.14). Large deltas are complex deposits consisting of coarse stream-channel sed-

---

[3] Deltas also form along coastlines. A delta can form wherever a stream or river enters a standing body of water.

Figure 8.15
LOESS CLIFFS FOR RENT
Loess deposited during the most recent ice age (18,000 years ago) forms the vertical wall of a valley near Xian in central China. A cave excavated by hand provides a roomy and comfortable home for a Chinese family.

iments, fine sediments deposited between channels, and still finer sediments deposited on the sea floor.

*Beach Sediments.* Quartz, the most durable of the common minerals in continental rocks, is a typical component of beach sands. However, not all ocean beaches are sandy. Any beach consists of the coarsest rock particles contributed by the erosion of adjacent sea cliffs, together with materials carried to it by rivers or by currents moving along the shore. Beach sediments tend to be better sorted than stream sediments of comparable coarseness, and they typically display cross stratification. Particles of beach sediment, dragged back and forth by the surf and turned over and over, become rounded by abrasion.

*Offshore Clastic Sediments.* Most of the world's sedimentary rocks originated as strata on continental shelves. Fresh water flowing through an estuary or past a river mouth will continue seaward across the continental shelf. Most of the coarse sediment whether river-derived or formed by erosion of the shore, is deposited within about 5 km (about 3 mi) of the land. However, some coarse sediment can be found 100 km or more offshore. Such a patchy distribution of coarse shelf sediments reflects changing sea levels in the past. At times when the sea fell below its present level, the shoreline migrated seaward across the shelves, exposing new land. Then, as now, bodies of coarse sediment were deposited in shallow water within a few kilometers of the shore. These were submerged as the sea level again rose across the shelves. As much as 70 percent of the sediment cover on the continental shelves may have been deposited under such conditions of fluctuating sea level.

**?** **Why do beach sands often consist primarily of quartz?**

On the continental shelf of eastern North America, up to a 14-km (almost 9-mi) thickness of fine sediment has accumulated over the last 150 million years. The great bulk of the Earth's sediments are continental shelf strata containing sediment that originated on the adjacent continent. The shelves, in effect, catch weathered continental crust in such a way that it is continually recycled by the processes of plate tectonics and the rock cycle.

*Carbonate Shelves.* Carbonate sediments of biogenic origin accumulate on the continental shelves wherever the climate and surface temperature are warm enough to nurture abundant carbonate-secreting organisms. Carbonate sediments accumulate mainly on broad, flat carbonate shelves that border a continent or rise from the sea floor as platforms (Fig. 8.16).

*Marine Evaporite Basins.* In coastal areas with a sufficiently warm and dry climate, ocean water in a shallow basin may evaporate at such a rate that substances dissolved in the water precipitate and accumulate as marine evaporite deposits. Such deposits are widespread. In North America, for example, marine evaporite deposits of various ages underlie as much as 30 percent of the entire land area.

*Turbidity Currents and Turbidites.* Thick sediments are found at the foot of the continental slope at depths as great as 5 km (3 mi) beneath the surface of the ocean. The origin of these sediments was difficult to explain until marine geologists demonstrated that they could be deposited by **turbidity currents** that originate on the continental shelf. Turbidity currents are turbulent, gravity-driven flows consisting of dilute mixtures of sediment and water, essentially underwater landslides. The density of the mixtures is greater than that of the surrounding water, so they rush swiftly down the continental slope at velocities of up to 90 km/h (60 mi/h). As a turbidity current reaches the ocean floor, it slows down and deposits a graded layer of sediment called *turbidite* (Fig. 8.17). At any site on the continental rise or an adjacent abyssal plain, a turbidite is deposited very infrequently, perhaps only once every few thousand years. In these places, far from the source of the sediment, the deposits are mainly layers no more than 30 cm (about 1 ft) thick. Although deposition is infrequent, over millions of years turbidites can slowly accumulate and form very thick deposits.

*Deep Sea Oozes.* Biogenic calcareous ooze consisting of the remains of tiny sea creatures occurs over wide areas of the ocean floor at low to middle latitudes. In these areas, warm temperatures favor the growth of carbonate-secreting organisms in surface waters. When the organisms die, they fall to the sea floor and accumulate as a muddy ooze. Other parts of the deep ocean floor are mantled with

**turbidity current** A turbulent, gravity-driven flow consisting of a dilute mixture of sediment and water; essentially, an underwater landslide.

Figure 8.16
**CARBONATE SHELF**
Carbonate sediments accumulate in the warm, shallow marine waters of a broad, flat carbonate shelf surrounding the islands of the Bahamas. The sediments show up as white sand bars and consist of fine skeletal debris of tiny sea creatures.

Figure 8.17
**DEEP SEA TURBIDITES**
These deep sea turbidite beds have been tilted, uplifted, and exposed in a wave-eroded bench along the coast of the Olympic Peninsula, Washington.

siliceous ooze from silica-secreting organisms. This material is most common in the equatorial Pacific and Indian oceans and in a belt encircling the Antarctic continent. These are areas where the biological productivity of surface waters is high, owing partly to the upwelling of deep ocean water that is rich in nutrients.

# WHAT'S AHEAD

Refer back to Figure 7.1, which depicts the rock cycle. After sediment has been deposited and lithified into new sedimentary rock, it may be elevated by tectonic processes and again subjected to erosion. Tectonic forces bend, buckle, and break strata, so in the next chapter (chapter 9) we discuss how rock deforms, how deformation is measured, and how it is recorded on geologic maps. Then, in chapter 10, we look at the mineralogic and textural changes that happen to rocks that have been heated and subjected to pressure.

Sediments and sedimentary rock are produced by the Earth's external processes, all of which are driven by solar energy. Rock deformation, by contrast, happens deep in the Earth's crust and is caused by the Earth's internal processes, all of which are powered by the Earth's internal heat. This still leaves open the question of where, how, and why the basins in which sediment has accumulated were formed and why the strata became deformed. We will address that issue in chapter 11.

**?** **What happens *after* sediment has been changed into rock?**

## Chapter Highlights

**1.** Sediments and sedimentary rocks are our best record of how climates and environments have changed throughout geologic history.

**2. Clastic sediment** consists of fragmental rock and mineral debris produced by weathering, together with broken remains of organisms. The fragments, called **clasts**, are classified on the basis of size (from coarsest to finest: boulder, gravel, sand, silt, clay). Clasts may become rounded and sorted during transport by water and wind but not during transport by glaciers. Volcanic sediments are clastic sediments in which all the fragments are volcanic in origin.

**3. Chemical sediment** is formed when substances carried in solution in lake water or seawater are precipitated. The principal means of precipitation are evaporation and biochemically mediated changes to the solution brought about by the activities of aquatic plants and animals.

**4. Biogenic sediment** is composed of the accumulated remains of animals and plants. Sediments that consist of fragments (clasts) of biological materials are referred to as bioclastic sediments.

**5.** Clastic sedimentary rocks, like sediments, are classified mainly on the basis of clast size. **Conglomerate, sandstone, siltstone,** and **shale** are the common rock equivalents of gravel, sand, silt, and clay, respectively. Various arrangements of clasts in sedimentary **beds** lead to sorted beds, **graded beds, varves, cross beds,** and nonsorted layers.

**6.** The most important kinds of chemical sedimentary rocks are **limestone, dolostone,** rock salt, gypsum, phosphorite, and **banded iron formations.** Limestone, dolostone, **chert,** and **coal** are important kinds of biogenic sedimentary rocks. Limestone is formed primarily in warm marine environments.

**7.** When sediment is turned into rock—that is, when the sediment undergoes **lithification**—a number of changes occur, including **compaction, recrystallization,** and **cementation.** Together these changes are referred to as **diagenesis.**

**8.** An extensive body of strata may possess several **sedimentary facies,** each produced in a different depositional environment.

**9.** Most sediments consist of continental detritus that has been transported to the submerged continental margins. Some detritus is trapped in basins on land, but most eventually reaches the sea.

**10.** Common sediments on land are stream, lake, glacial, and **eolian** (wind-deposited) **sediments.**

**11.** Coarse land-derived sediment that reaches the coast is deposited close to the shore. Finer sediment is deposited on the continental shelves and slopes and in the deep sea. Extensive areas of the shelves are covered by sediments that were deposited at times when the sea level was lower.

**12.** Evaporite deposits are widespread. They accumulate in restricted marine basins where the rate of evaporation is high. A third of North America is underlain by marine evaporites. Carbonate shelves and deep sea oozes are other common marine sediments.

**13. Turbidity currents** transport sediment from the edge of the continental shelf to the deeper waters at the foot of the continental slope. Sediments deposited by turbidity currents commonly show graded bedding.

## ▶ The Language of Geology

# ▶ *Questions for Review*

1. What are the main groups of sediment, and on what basis are the kinds of sediment within each group classified?
2. How are clastic sediments classified?
3. What obvious clues can be used to distinguish a clastic sedimentary rock from an igneous rock?
4. Why is quartz the most common mineral found in sandstones?
5. How do chemical sediments form?
6. How is sediment converted into sedimentary rock?
7. What features of sedimentary bedding surfaces provide clues about the environments in which deposition occurred?
8. Describe two kinds of biogenic sediment.
9. How are plants transformed to peat, and peat transformed to coal?
10. Name five different marine environments and the common marine sediments that are deposited in each.

# ▶ *Questions for Thought and Discussion*

1. Estuaries are generally shallow bodies of water, yet there are thick accumulations of estuarine sediments in the geologic record. What hypothesis can you suggest to explain this?
2. Do any sedimentary rocks outcrop in the area where you live? If so, examine them closely, perhaps in a road cut or an old quarry. See if you can recognize the kinds of rock present, and consider the environment or environments in which the sediments were deposited. From a geologic map of the area, find out the age of the rocks. Then check a plate tectonic reconstruction and see what the latitude of your home area was at the time when the sediments were deposited. Does the latitude support your conclusions about the environment of deposition?
3. It is estimated that as much as 25 billion tons of soil are lost through erosion as a result of farming every year. What happens to that soil? Where does the material that was in the soil end up?

4. Exploration for oil has led to the discovery of up to 14 km (almost 9 mi) of sedimentary rock in the Atlantic Ocean on the continental shelf of North America. The oldest of the strata were deposited 150 million years ago during the Jurassic Period. What is the average rate of deposition needed to yield 14 km of sedimentary rock over that period? Do you consider this to be a fast or a slow rate? If you think the number you calculated might be misleading, explain why.
5. Investigate the formation of graded bedding through the following experiment. Fill a large beaker or graduated cylinder with a half-and-half mixture of water and sediment. The sediment should have a variety of grain sizes, from fine clays to sand and even gravel. Shake up the mixture and let it settle quietly until the water is completely clear. Which grains settle out first? Which grains settle out last? What does the final sediment deposit look like?

For an interactive case study on sedimentary geology and energy sources, visit our Web site.

# · 9 ·

# FOLDS, FAULTS, AND GEOLOGIC MAPS

It's late June, and you've landed a great summer job. You're working for the Geological Survey, as part of a mapping team in a wilderness area. The area has never been mapped in detail; it is remote and tree-covered. There is a lot of interest in the geology of this area; mining companies would like to investigate it for possible ore deposits, but environmental groups would like to see it preserved as a national park. It is important to document the natural features of the area.

The team is flown by helicopter to the base camp. Dropping down among the treetops, you get a glimpse of what you're in for—it's pretty rugged terrain! You and your partner are assigned to work around the shoreline of a nearby lake, documenting the rock types and measuring their orientations. Your new rock hammer and geologic compass are swinging from your belt. Your backpack holds your lunch, water bottle, some pens and pencils, rock sample bags, and a clipboard with a base map covered in clear plastic. Your new boots aren't quite broken in, but it feels great to be out hiking in the cool, clear air.

When geologists carry out mapping exercises like this one, they most often find that the rock layers are not perfectly flat and horizontal. Through tectonic movement, mountain-building, and other geologic processes, rocks are deformed. They may be twisted or tilted, squeezed or stretched; they may break or bend, changing their shape, their volume, or both. The presence of such rock structures reveals many important aspects of the geologic history of an area, and geologists record this information on geologic maps.

◆

### In this chapter you will learn about

- How rocks deform in response to stress
- How different geologic conditions lead to different types of rock deformation
- How faults and folds form, and how geologists measure and describe them
- How geologic maps are used to portray information about rock formations and structures
- How rock deformation and erosion combine to create the landscapes we see around us

*A Spectacular Playground*

This small lake seems to perch precariously on top of spectacularly folded and tilted rocks in Dorset, England.

The theory of plate tectonics tells us that lithospheric plates are constantly moving, colliding with one another, and interacting along their margins. These interactions result in deformation of the Earth's crust. **Deformation** refers to all the different ways in which rocks respond to squeezing, stretching, or any other kind of tectonic force. When rocks deform, they may buckle and bend, crack and break, or flatten and change shape. It is easy to find evidence of this kind of activity. If you look at a photograph of the Alps, the Rockies, the Appalachians, or any other great mountain range, you will see rock strata that were once horizontal and are now tilted and bent (Fig. 9.1). Enormous forces are needed to deform such huge masses of rock. These forces come from the protracted movement and interactions of lithospheric plates.

**deformation** The change in shape or volume of a rock in response to squeezing, stretching, shearing, or any other kind of tectonic force.

**?** **How do rocks respond to stress?**

## ROCK DEFORMATION

Usually, we can't see rocks being twisted and bent by tectonic forces. Most rock deformation happens slowly, within the crust or deep in the mantle. We see the deformed rock only when it is uplifted and exposed at the surface by *erosion*, the wearing away of bedrock and transport of loosened particles. Therefore, we must infer how the deformation occurred long ago. Some of the evidence on which we base our inferences comes from laboratory studies, and some comes from direct studies of deformed rocks.

Figure 9.1
**THE CORE OF A GREAT MOUNTAIN**
These tightly folded rocks on South Georgia Island (an island near Antarctica, south and east of the southern tip of South America) show the intense deformation typical of rocks that once formed part of the core of a great mountain chain.

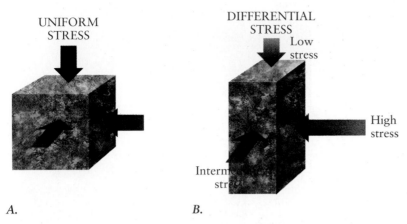

UNIFORM
STRESS

DIFFERENTIAL
STRESS
Low
stress

High
stress

Interm
str

*A.* *B.*

**Figure 9.2**
**UNIFORM AND DIFFERENTIAL STRESS**
*A.* The stress on this block is *uniform,* the same in all directions. *B.* The stress on this block is *differential,* stronger in one direction than in others.

## Stress and Strain

In discussing rock deformation, we often use the word *stress* instead of the related term *pressure.* Stress is a more useful term for describing the causes of rock deformation, because it implies that the pressure on the rock is not necessarily the same in all directions. **Stress** is defined as the force acting on a surface, per unit area. The definition of **pressure** is exactly the same, but the term *pressure* is normally used when the forces acting on the surface of a body are the same in all directions. (Thus, pressure is a specific type of stress.) To be more precise, we can distinguish between *differential stress,* in which the force is greater from one direction than from another, and *uniform stress,* in which the force is equal in all directions. For example, the stress on a small body floating within a liquid is uniform stress—the same from all directions. Uniform stress in rocks can also be called *confining stress* or *confining pressure,* because a rock in the lithosphere is confined by the rocks all around it and is uniformly stressed by those surrounding rocks.[1] The difference between uniform stress and differential stress is shown in Figure 9.2.

In response to stress, a rock will change its shape or its volume, usually both. A change in shape or volume of a rock in response to stress is called **strain.** Differential stress causes rocks to change their shape; it may also cause a change in volume. Uniform stress causes rocks to change their volume. For example, if a rock is subjected to uniform stress by being buried deep in the Earth, its volume will decrease; that is, it will be compressed. If the spaces (*pores*) between the grains become smaller as water is expelled from them, or if the minerals in the rock are transformed into more compact crystal structures, the volume change may be relatively large.

Three different types of stress are shown in Figure 9.3. **Tension** refers to stress that acts in a direction *perpendicular to* and *away from* a surface; this kind of stress pulls or stretches rocks, sometimes causing the volume to increase.

**stress** The force acting on a surface, per unit area.

**pressure** A type of stress, in which the forces acting on the surfaces of a body are the same in all directions.

**strain** A change in shape or volume of a rock in response to stress.

**tension** Stress that acts in a direction perpendicular to and away from a surface.

---

[1] The related terms *lithostatic pressure* and *hydrostatic pressure* also describe uniform stress on a rock, but they convey additional information about how the pressure is transmitted to the rock: by overlying rocks (lithostatic, from *lithos,* the Greek root that means "rock") or by water (hydrostatic, from *hydro,* the Greek root that means "water").

**Figure 9.3**
**THE THREE TYPES OF STRESS**
The shape of a cube of rock changes, depending on the type of stress applied to it. The arrows indicate tensional, compressional, and shear stress. Rocks that are subjected to differential stress—stress that is stronger in one direction than in another—typically respond by changing their shape, as shown by these blocks.

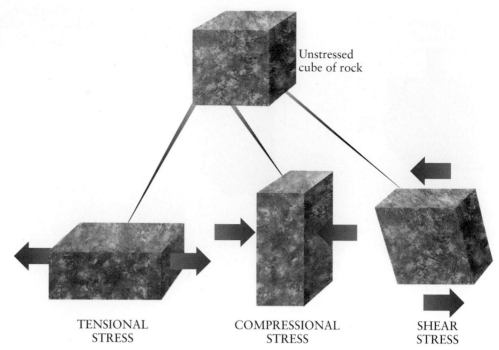

Unstressed cube of rock

TENSIONAL STRESS

COMPRESSIONAL STRESS

SHEAR STRESS

**compression** Stress that acts in a direction perpendicular to and toward a surface.

**shear** Stress that acts parallel to a surface.

**Compression** refers to stress that acts in a direction *perpendicular to* and *toward* a surface; compressional stress squeezes rocks, shortening them or decreasing their volume. **Shear** stress is stress that acts *parallel to* a surface. It causes the rock to change shape by bending, flowing, or breaking. In response to shear stress, different parts of the rock body may slide past each other like cards in a deck. Rocks subjected to *differential stress*—that is, stress that is stronger in one direction than in another—tend to change their shape. Let's learn something about each of these types of deformation and the conditions under which they occur.

## Factors That Influence Deformation

**elastic deformation** Nonpermanent deformation; the rock returns to its original shape and volume after the stress is removed.

When a rock is subjected to stress, it can respond in different ways. **Elastic deformation** is a nonpermanent change in the volume or shape of any solid, including rocks. When the stress is removed, the solid returns to its original shape and size. For example, if you stretch a metal spring and then let go, it will return to its original shape and size; the stretching is elastic deformation. There is a degree of stress—called the *elastic limit*—beyond which the material is permanently deformed; that is, it does not return to its original size and shape when the stress is removed. In the example of the metal spring, if you stretch the spring too far, it won't return to its original shape; its elastic limit has been exceeded.

Rocks, too, will deform elastically up to a point. But if the stress continues and the elastic limit of the rock is exceeded, it will be permanently deformed. Under most circumstances, rocks can withstand only a small amount of elastic deformation before they deform permanently. **Ductile deformation,** also called *plastic* deformation, is one type of permanent deformation in a rock (or other solid) that has been stressed beyond its elastic limit. Rocks that deform in a ductile manner usually change their shape by flowing or bending (Fig. 9.4A).

**ductile deformation** Permanent (nonelastic) deformation, in which a rock may change its shape by flowing or bending.

**brittle deformation** Deformation in which the rock fractures or cracks, instead of flowing or bending.

Another type of permanent deformation is **brittle deformation**, in which the rock fractures, or cracks, instead of flowing or bending (Fig. 9.4B). A brittle material deforms by fracturing, whereas a ductile material deforms by changing its shape. Drop a piece of chalk on the floor and it will break. Drop a piece of cheddar cheese and it will bend or squash instead of breaking. Under the conditions

*A.*                                    *B.*

**Figure 9.4**
**DUCTILE AND BRITTLE DEFORMATION**
These deformed rocks show *A* ductile deformation, in which the rock has responded to
stress by folding and flowing, and *B* brittle deformation, in which the rock has fractured.

of room temperature and atmospheric pressure, chalk is brittle and cheese is ductile. Similarly, some rocks behave in a brittle manner and others in a ductile manner. A rock that is brittle in one set of conditions will be ductile under different conditions. The main factors that affect how a rock deforms are temperature, confining pressure, rate of deformation, and composition. Let's look briefly at each of these.

## Temperature

The higher the temperature, the more ductile and less brittle a solid becomes. At room temperature, for example, it is difficult to bend glass; if we try too hard, it will break because it is brittle (Fig. 9.5A). However, when it is heated slowly over a flame, glass becomes ductile and can easily be bent (Fig. 9.5B). Rocks are like

*A.*

**Figure 9.5**
**GLASS DEFORMING**
At room temperature, glass deforms in a brittle manner, as shown by this lightbulb *A* shattering as it hits a brick floor. However, if glass is heated slowly over a flame *B*, it can bend and flow in a ductile manner.

*B.*

*A.*  *B.*  *C.*

**Figure 9.6**
**ROCK DEFORMATION EXPERIMENTS**
These cylinders show the results of a series of experiments on the effects of confining pressure on rocks. *A* shows an undeformed cylinder of rock. *B* shows a cylinder of rock that was subjected to compression from above, but supported by high confining pressure. The cylinder in *B* deformed in a ductile manner, becoming shorter and fatter in response to compression. In *C*, the rock cylinder was subjected to the same amount of compression, but with a low confining pressure. This cylinder deformed in a brittle manner, that is, it fractured in response to compression.

glass in this respect. They are brittle at the Earth's surface, but they become ductile at deeper levels where temperatures are high.

## Confining Pressure

The effect of confining pressure on deformation is not familiar from our everyday experience. High confining pressure reduces the brittleness of rocks because it hinders the formation of fractures. Figure 9.6 shows the results of a series of experiments in which this is demonstrated. Figure 9.6A shows an undeformed cylinder of rock. In Figure 9.6B, the rock was supported by a high confining pressure and subjected to compression from above. When confining pressure is high, it is easier for a solid to bend and flow than to fracture. The rock in Figure 9.6B deformed in a ductile manner, becoming shorter and fatter. In contrast, the rock in Figure 9.6C had a low confining pressure; when it was subjected to compressing, it fractured. Rocks near the Earth's surface, where confining pressure is low, exhibit brittle behavior and develop many fractures. Deep in the Earth, where confining pressure is high, rocks tend to be ductile and deform by flowing or bending.

## Rate of Deformation

The rate at which a stress is applied to a solid is another important factor in determining how the material will deform. If you take a hammer and whack a piece of ice suddenly, it will fracture. But if stress is applied to the ice little by little over a long period, it will sag, bend, and behave in a ductile manner. The same is true of rocks; if stress is applied quickly, the rock may behave in a brittle manner, but if small stresses are applied over a very long period, the same rock may behave in a ductile manner. The term *strain rate* refers to the rate at which a rock is forced to change its shape or volume. The lower the strain rate, the greater the tendency for ductile deformation to occur.

You can demonstrate the effects of different strain rates using Silly Putty modeling clay, an interesting substance that can undergo both brittle and ductile deformation at room temperature. If you pull the Silly Putty very slowly, it stretches into a long, rubbery string; this is ductile deformation. If you place a blob of Silly Putty on an inclined surface and check it a half-hour later, you will find that it has flowed partway down the surface; this, too, is ductile deformation. But if you pull Silly Putty apart very suddenly, it will break; this is brittle deformation caused by a rapid strain rate.

To summarize, low temperature, low confining pressure, and high strain rates tend to enhance the brittle behavior of rocks. Low temperature and pressure conditions are characteristic of the crust, especially the upper crust. As a result, fracturing is common in upper-crustal rocks. High temperature, high confining pressure, and low strain rates, which are characteristic of the deeper crust and the mantle, reduce the brittle properties of rocks and enhance their ductile properties. The depth below which ductile properties predominate is referred to as the *brittle-ductile transition*. Fractures are uncommon deep in the crust and in the mantle, because rocks at great depths (below about 10–15 km; 6.7–10 mi) behave in a ductile manner.

## Composition

The composition of a material determines the exact point at which its brittle-ductile transition will occur. For example, both chalk and cheese behave brittly at 50 degrees below zero. When warmed to room temperature, however, the cheese behaves in a ductile manner, while the chalk is still brittle. They have different compositions and, therefore, different properties. The same is true of rocks and their mineral constituents. Some minerals—notably quartz, garnet, and olivine—are very brittle. Rocks that contain these minerals, such as sandstone, granite, and granodiorite, tend to behave brittly. The brittle-ductile transition that occurs within the crust is controlled principally by the deformational behavior of quartz-bearing rocks. Other minerals—notably mica, clay, calcite, and gypsum—are more often ductile under natural conditions. Rocks such as limestone, marble, shale, and slate, which contain large quantities of such minerals, tend to deform in a ductile manner. The presence of water in a rock also enhances its ductile properties. It does so by reducing the friction between mineral grains and by dissolving material at points of high stress, permitting it to move to places where the stress is lower.

**? Why is chalk brittle and cheese ductile?**

# STRUCTURAL GEOLOGY

The study of stress and strain, the processes that cause them, and the types of rock structures that result from them are the subject of **structural geology**. Structural geologists study evidence of past rock deformation in order to find out what kinds of deformation have occurred in a particular area. This helps them understand the stresses that prevailed in the area at different times in the past, which in turn allows them to decipher the geologic history of the area. Rock structures are also important from a practical perspective. For example, structures such as faults control the locations of many types of ore deposits. Rock de-

**structural geology** The study of stress and strain, the processes that cause them, and the structures that result from them.

formation and the resulting structures can affect slope stability, influence the flow of groundwater, or trap oil and natural gas deep underground.

To understand and interpret rock structures, structural geologists make observations and measurements to determine what kinds of structures are present, and to specify the orientations and dimensions of those structures. They use these measurements and other sources of information to make inferences—educated guesses—about structures that may lie underground, hidden from view. The most basic structural feature of a rock layer is its orientation—that is, its direction of tilt. To describe this, structural geologists use the concepts of *strike* and *dip*.

## Strike and Dip

The *principle of original horizontality* (see chapter 3) tells us that sedimentary strata are horizontal when they are first deposited. Where such rocks are tilted, we can assume that deformation has occurred. To study and describe this deformation, the geologist starts by measuring the orientation of the rock layer. This information is given by the **strike**, which is the compass direction (north, south, east, west) of the line of intersection between the rock layer and a horizontal

**strike** The orientation of the line of intersection between a rock layer and a horizontal plane.

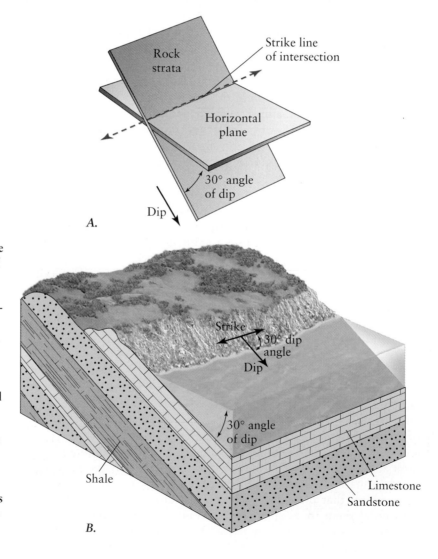

**Figure 9.7**
**STRIKE AND DIP**
Geologists use strike and dip to describe the orientation of tilted layers of rock. *A.* The *strike* is the line formed by the intersection of a rock layer and a horizontal ("level") plane. *B.* In this drawing, the water provides a horizontal plane. The shoreline (that is, the intersection of the water with the rock surface) indicates the strike. The *dip* is the angle at which the rock layers tilt. In this drawing, the rock layers are tilted 30° from the horizontal. The T-shaped symbol tells the direction of strike, as well as the angle and direction of dip. Patterns are used to distinguish among types of rock: the brick-like pattern represents limestone, the tiny dots are sandstone, and the dashes are shale.

plane. For example, Figure 9.7 shows a set of tilted sedimentary rock strata next to a body of water. The surface of the water provides a convenient horizontal plane. The shoreline is the intersection of the rock layer with the horizontal plane of the water surface, so the compass direction of the shoreline is the strike of the rock layer. Commonly, the strike measurement must be corrected for small variations between *magnetic north* (as measured by a compass) and *geographic north* (as shown on a map). This correction, which varies from one location to another, is called *magnetic declination.*

We need another measurement to fully describe the orientation of a tilted rock layer. That is the **dip**, the angle between the tilted surface and a horizontal plane. Dip is measured as an angle downward from the horizontal plane in degrees, using an instrument something like a protractor. In Figure 9.7, the water surface again provides a convenient horizontal plane. In this example the rock strata are tilted at 30°. If they were dipping more shallowly, the angle would be smaller. If they were dipping more steeply, the angle would be greater. A vertical rock layer—one that has been tilted right up on its side—is dipping at 90°. (It makes a right-angle with the horizontal plane.) In Figure 9.7, notice the "T" shaped symbol. The top bar of the "T" is the strike, the stem of the "T" gives the direction of dip, and the number tells the dip angle. Together, these three measurements—strike, direction of dip, and angle of dip—describe the orientation of a tilted rock layer.

As discussed earlier, rocks deform either in a brittle manner, by breaking and fracturing, or in a ductile manner, by flowing and bending. In the following sections we look at *faults* and *joints,* which form by brittle deformation, and *folds,* which form by ductile deformation.

**dip** The angle between a tilted surface and a horizontal plane.

## Faults

**Fractures,** or cracks in rocks, are characteristic of brittle deformation. They occur in all sizes. Some are very tiny—so tiny that you would need a microscope to see where an individual mineral grain has cracked. Cracks this small are sometimes called *microfractures.* A **fault**, already defined in chapter 4, is a fracture in rock along which movement has occurred. Some faults are microscopic, but others are very large. For example, the East African Rift Valley is a huge system of roughly parallel faults, which extends for more than 6000 km (more than 3700 mi) through the countries of East Africa.

There are many types of faults, caused by different kinds of stress. Faults are categorized on the basis of how steeply they dip and the relative direction of movement of the rocks on either side of the fault. Figure 9.8 shows an undeformed block; the most common types of faults and the changes in topography they may produce are shown in Figures 9.9 through 9.13.

**fracture** Any kind of crack or break in a rock.

**fault** A fracture in the crust along which movement has occurred.

**Figure 9.8**
**UNDEFORMED BLOCK**
Compare this undeformed block of rock to the blocks in Figures 9.9 through 9.13, which show the most common types of faults, the stresses that cause them, and the changes in topography that commonly result.

### Normal Faults

Tensional (or *extensional*) stress, that is, stress that stretches or pulls apart the crust, causes **normal faults.** In a normal fault (Fig. 9.9), the block of rock on top of the tilted fault surface, called the *hanging-wall* block, moves down relative to the block on the bottom, the *footwall* block. Normal faulting often produces a cliff-like landform called a *scarp.* Normal faulting causes the crust to lengthen and thin.

Normal faults sometimes occur in pairs. When this happens, the block between the two faults will either drop down or pop up, depending on the direction of dip of the faults. In a *graben* (Fig. 9.10A), the two normal faults are dipping toward each other and the block between them drops down. In a *horst* (Fig. 9.10B), the two normal faults are dipping away from each other, and the block between them is elevated. These structures are common along divergent plate boundaries, where the crust is being stretched. The East African Rift Valley, for example, is a huge system of grabens marking a plate boundary along which the African continent is splitting apart.

### Reverse Faults and Thrust Faults

Compressional stress—stress that pushes blocks together—causes **reverse faults.** In reverse faults, the hanging wall block is pushed over the footwall block (Fig. 9.11), shortening and thickening the crust. The reverse fault in Figure 9.11 dips steeply. When reverse faults dip less than 30°, they are called **thrust faults** (Fig. 9.12). Thrust faults are common in mountain chains along convergent plate boundaries. In large thrust faults, rock in the hanging wall block may move thousands of meters, coming to rest on top of much younger rock in the footwall block.

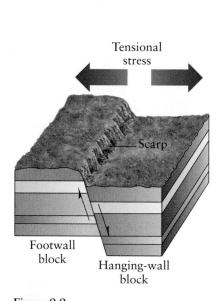

**Figure 9.9**
**NORMAL FAULT**
When the crust is stretched (tension), normal faults occur. The hanging-wall block (the block on top of the fault) moves down relative to the footwall block (the block underneath the fault), creating a fault scarp.

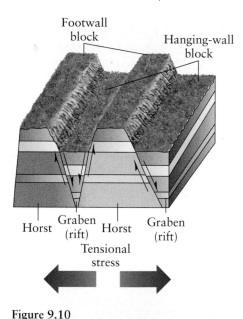

**Figure 9.10**
**HORST AND GRABEN**
When the crust is stretched, it is common for normal faults to occur in pairs. *A.* In a graben, the two faults are dipping toward each other and the block between them drops down. *B.* In a horst, the two faults are dipping away from each other, and the block between them is elevated.

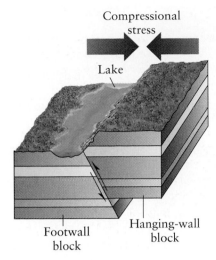

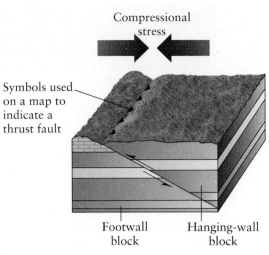

**Figure 9.11**
**REVERSE FAULT**
When the crust is compressed, reverse faults occur. The hanging-wall block (the block on top of the fault) moves up relative to the footwall block (the block on the bottom), causing the crust to shorten and thicken. This particular reverse fault has a steep dip.

**Figure 9.12**
**THRUST FAULT**
When a reverse fault has a very shallow dip, it is called a thrust fault. Thrusting causes older rocks in the hanging wall block to move up and over the younger rocks of the footwall block, sometimes by many kilometers.

## Strike-slip Faults

The movement in **strike-slip faults** is mainly horizontal and parallel to the strike of the fault (Fig. 9.13). These faults are caused by shear stress. One strike-slip fault is so famous that almost everyone has heard of it: it is the San Andreas fault in California. Along this fault the Pacific Plate is moving toward the northwest relative to the North American Plate (Fig. 9.14). (The word "relative" is very important here. In fact, both the Pacific Plate and the North American Plate are moving in a roughly northwesterly direction but the Pacific Plate is moving more quickly, like a fast runner overtaking a slower runner. The Pacific Plate thus appears to be moving toward the northwest relative to the North American Plate. This is called *relative displacement*.) The total extent of relative displacement along the San Andreas fault is not known, but it may amount to more than 600 km (almost 400 mi), having occurred over a period of at least 15 million years.

**strike-slip fault** A fault along which the movement of fault blocks is mainly horizontal and parallel to the strike of the fault.

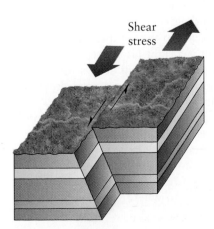

**Figure 9.13**
**STRIKE-SLIP FAULT**
In a strike-slip fault, the movement is mostly horizontal and parallel to the strike of the fault (see Fig. 9.7 to review the concept of strike).

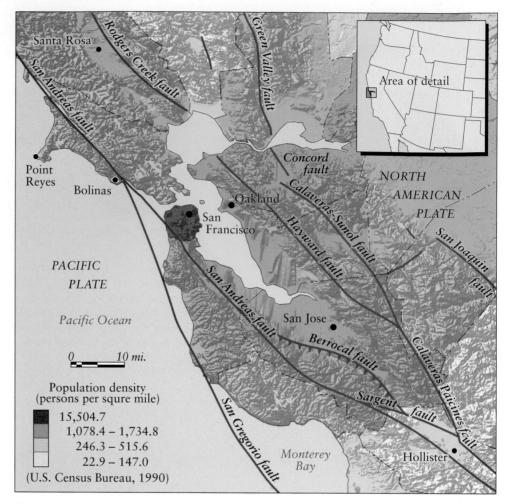

**Figure 9.14**
**THE SAN ANDREAS FAULT SYSTEM**
The San Andreas fault is a well-known example of a large strike-slip fault. It is part of a system of faults along which the Pacific Plate is moving in a northwesterly direction relative to the North American Plate. This map shows some of the faults in the system, along which movement has occurred within the past 10,000 years.

Strike-slip faults can be described according to the direction of relative horizontal motion, as follows: To an observer standing on either block, the movement of the other block is *left-lateral* if it has moved to the left and *right-lateral* if it has moved to the right. The San Andreas fault is a right-lateral strike-slip fault. If you are standing on the east side of the fault and looking across toward the ocean, the block on the other side (the Pacific Plate) is moving very slowly toward the northwest relative to where you are standing (that is, toward your right), carrying part of the California coast along with it. The relative motion in a right- or left-lateral fault is the same regardless of which block the observer is standing on. In other words, if you were to stand on the Pacific Ocean side of the San Andreas fault and look across toward the continental United States, you would have to turn your head to the right (that is, toward the southeast) to track the movement of the North American Plate relative to where you are standing.

# Joints

Because of the weight of overlying rock, rock masses buried deep beneath the Earth's surface are under enormous stress. As erosion wears down the surface, the weight of the overlying rock, and hence the confining stress, are reduced. The rock responds by expanding. As it does so, fractures typically develop (Fig. 9.15). As we discussed in chapter 7, these are called **joints**, fractures in a rock along which no appreciable movement has occurred. Joints are a type of brittle deformation. They are different from faults because the blocks on either side have not moved very far relative to each other along the fracture. Joints can also form by the cooling and subsequent shrinking of thick lava flows. The existence of joints is an important step in rock weathering and the development of sediments and soils, because joints act as pathways for the agents of weathering to penetrate the rock.

**joint** A fracture in a rock, along which no appreciable movement has occurred.

# Folds

When rocks deform in a ductile manner, they bend and flow. The bending may be a broad, gentle warping over many hundreds of kilometers, or a tight flexing of microscopic size, or anything in between. Regardless of the size or shape of the structure, any bending of rocks is referred to as *folding*, and a single bend or warp in a layered rock is called a **fold**.

**fold** A bend or warp in a layered rock.

## Types of Folds

The simplest type of fold is a *monocline*, a local steepening in otherwise uniformly dipping strata (Fig. 9.16A and B). An easy way to visualize a monocline is

Figure 9.15
PRECARIOUS PERCH
Three sets of joints, one horizontal and two vertical, intersect at nearly right angles to form a spectacular rocky vantage point, called the Pulpit, overlooking Lysefjord in southwestern Norway. A widening crack along a vertical joint suggests that the Pulpit will eventually break off and plunge into the icy waters of the fjord far below.

*A.*

**Figure 9.16**
**MONOCLINES**
*A.* This monocline in southern Utah interrupts the generally flat-lying sedimentary strata of the Colorado Plateau. *B.* A monocline is a local steepening in otherwise uniformly dipping rock strata. In the area of maximum bending, the strata are nearly vertical (right-hand side of drawing and photo).

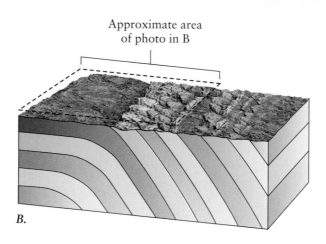

*B.*

**anticline** A fold in the form of an arch, with the rock strata convex upward.

**syncline** A fold in the form of a trough, with the rock strata concave upward.

to lay a book on a table and drape a handkerchief over one side of the book. So draped, the handkerchief forms a monocline. Most folds are more complex than monoclines. Folds are often combinations or variations of two basic types. An **anticline** is a fold in the form of an arch, with the rock strata convex upward. A **syncline** is a fold in the form of a trough, with the rock strata concave upward. Anticlines and synclines often occur in sets, as shown in Figure 9.17. (If you push the edge of a carpet with your foot, it will form a series of anticlinal and synclinal folds.) Sometimes a large area of crust undergoes upwarping or downwarping, which forms broad, gentle folds. Upwarping of this type forms *domes*, while downwarping forms large, bowl-like *basins* (Fig. 9.18A and B).

## Fold Geometry

We need to determine several things in order to describe the geometry and orientation of a fold. First, imagine a plane that divides the fold in half, as symmetrically as possible. This is the fold's *axial plane* or *axial surface,* as shown in Figure 9.17B. The line along which the axial plane intersects a single rock layer in the

A.

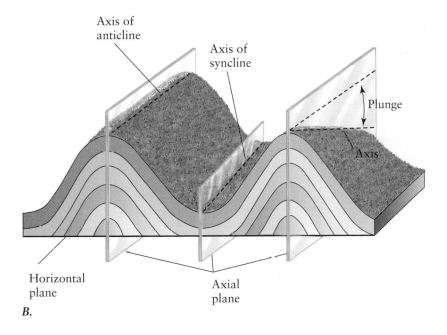

B.

fold is called the *fold axis*. A fold axis is basically a line that separates the two
halves or *limbs* of the fold. Sometimes the fold axis is horizontal. Sometimes the
fold axis is not horizontal, in which case the fold is said to be *plunging*, as in the
righthand fold shown in Figure 9.17B. To fully represent the geometry of a fold,
we show the following elements on a map: (1) the fold axis (a line); (2) the type
of fold (for an anticline, two small arrows pointing away from the axis; for a
syncline, two small arrows pointing toward the axis); (3) the direction in which
the axis is plunging (an arrow); and (4) the measure of the plunge angle. Examples of these symbols are shown in Figure 9.19 and explained in greater detail in
Appendix E.

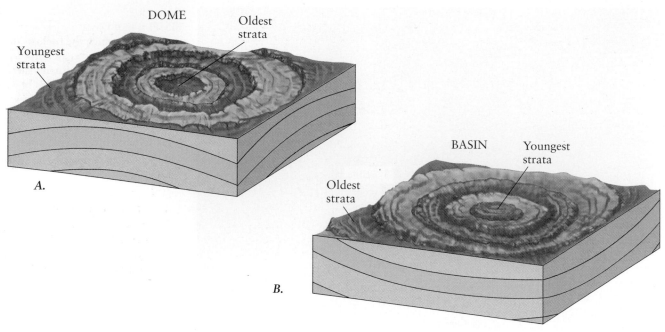

**Figure 9.18**
**DOMES AND BASINS**
*A.* Upwarping of the crust causes the formation of domes. When domes erode, they expose younger rocks in their centers. *B.* Downwarping of the crust causes the formation of basins. In a basin, the youngest rocks will be at the center and the oldest rocks around the outer edge.

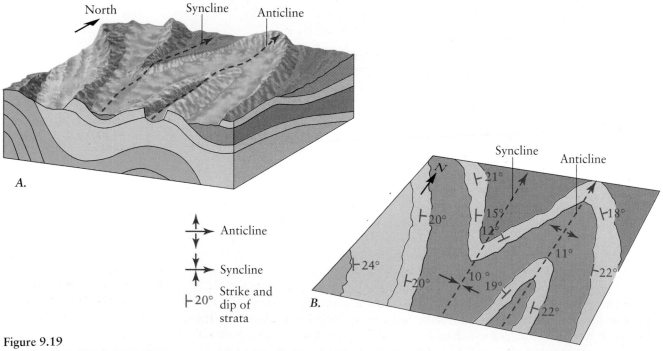

**Figure 9.19**
**SHOWING FOLDS ON MAPS**
*A.* This block diagram shows several different folds (anticlines and synclines) along with the patterns of rock strata on the surface that might result from erosion of these structures. *B.* This is a map of the area in *A*, showing the symbols used for anticlines, synclines, fold axes, and strike-and-dip. Note that the anticlines and synclines shown here are not "level" but plunge below the surface.

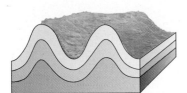

*A.* Open (symmetrical)

*B.* Asymmetrical

*C.* Overturned

*D.* Recumbent

*E.* Isoclinal

**Figure 9.20**
**THE GEOMETRY OF FOLDS**
Five types of fold geometries are
shown: *A.* Symmetrical.
*B.* Asymmetrical. *C.* Overturned.
*D.* Recumbent. *E.* Isoclinal.

Fold geometrics can be quite complicated. Figure 9.20A shows a relatively simple *symmetrical* fold. This is actually a series of anticlines and synclines, in each of which the limbs are symmetrical relative to the fold axes on either side. Figure 9.20B shows an *asymmetrical* fold, in which one limb is dipping more steeply than the other. In an *overturned* fold (Fig. 9.20C), the bottom limb of the fold has been tilted beyond the vertical plane so that it is upside down. Folds that are so strongly overturned that they are almost lying flat are called *recumbent* folds (Fig. 9.20D). Very tight folds with limbs that are nearly parallel are called *isoclinal* (Fig. 9.20E).

# GEOLOGIC MAPS

It is not possible for a geologist to see all the structural details of deformed rocks in a given area; soil, water, vegetation, and buildings cover much of the evidence. The geologist must gather geologic information from *outcrops*—places where bedrock is exposed at the surface (Fig. 9.21A). The type of rock present and the orientation of the layers or presence of structural features such as fractures are examples of important information to be gained from outcrops. These are plotted on a map of the area (Fig. 9.21B). The geologist may also make use of information from other sources, such as drilling, to draw conclusions about what lies beneath the soil, vegetation, water, and buildings in the areas between the outcrops. The result is a **geologic map** that shows the locations and orientations of rock units and structural features. Geologic maps help geologists interpret the geologic history of an area. They are also used in a wide variety of practical applications by geologists from mining companies, oil companies, engineering firms, environmental agencies and consulting firms, and many others. Let's examine the techniques used to portray geologic structures in the map format.

## Making and Interpreting Geologic Maps

There are many things we need to know if we want to understand how information is portrayed on a geologic map. Specifically, we need to know the size or

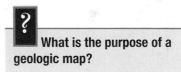

**What is the purpose of a geologic map?**

**geologic map** A map that shows the locations and orientations of rock units and structural features.

# Making a Geologic Map

Figure 9.21

A. This block diagram shows a landscape with tilted rock strata. Much of the rock is covered at the surface by grass and soil, but there are some outcrops where bedrock is exposed. ▼

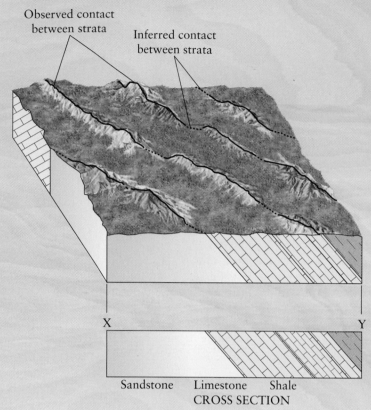

Observed contact between strata

Inferred contact between strata

X     Y

Sandstone    Limestone    Shale
CROSS SECTION

B. The geologist has transferred the information about rock types and contacts between rocks onto a map of the area. In covered areas between outcrops, the geologist has made an educated guess about the types of rock and the locations of contacts. ▼

GEOLOGIC MAP

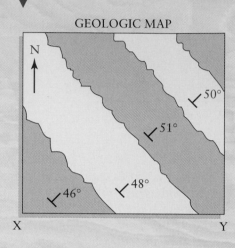

N

50°
51°
48°
46°

X     Y

◄ C. GEOLOGIC CROSS SECTION
The geologic cross section shows what lies beneath the surface.

---

scale of the objects portrayed on the map; how *topographic contour lines* are used to show the shape of the land surface; and what the different colors and symbols on the map indicate.

## Maps Are Scale Models

**scale** The amount by which the size of objects or distances shown on a map has been reduced.

When we draw a map, we are making a scale model, like a model train or car. It would be impossible, for example, to create a full-size map of the Rocky Mountains. The first thing we have to determine is the map's **scale**—the amount by which the size of objects shown on the map has been reduced. A scale of 1:1000 (read "1 to 1000") means that one unit on the map is equal to 1000 of the same unit on the Earth's surface. In other words, 1 km (1000 m) would appear one meter long on such a map. This is still a very large scale; to portray North America on a map of this scale would require a map more than 4 km (2.5 mi) wide!

Therefore, geologic maps commonly use scales that reduce the sizes of objects even further. For example, 1:62,500 is a common scale for geologic maps in the United States. This scale is convenient because one inch on the map is equal to approximately one mile on the ground. Another commonly used scale is 1:125,000. (Note that this is a *smaller* scale than 1:62,500, because just one inch on the map is equal to *two* miles on the ground, instead of *one* mile.) Where the

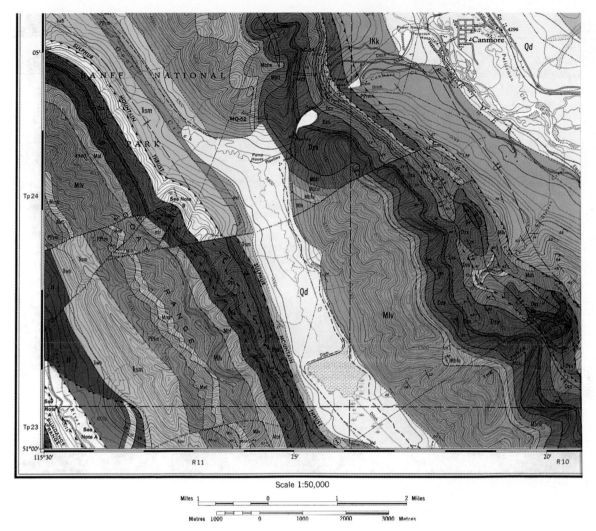

Figure 9.22

INTERPRETING GEOLOGIC MAPS

This is part of a real geologic map, showing the Canmore Quadrangle, in Alberta, Canada. The different rock units are portrayed by different colors. The scale is shown as a bar scale and in numerical form (1:50,000). Topographic contours are shown as light lines, visible beneath the colors of the rock units. Longitude and latitude lines are visible as light gray squares. Shown in black are various symbols for structural features.

metric system is used, as in Canada, Europe, and some U.S. government agencies, scales of 1:100,000 or 1:50,000 are common. Most maps display this type of information as a numerical scale or as a graphical scale, usually in the form of a bar. For example, the numerical scale on the map in Figure 9.22 indicates that the scale is 1:50,000. The bar scale shows that a length of 1 cm on the map is equal to a distance of 50 km on the ground.

## *Topographic Contour Lines*

Commonly, a topographic map is used as the base for a geologic map. **Topographic maps** show the shape of the ground surface, as well as the location and elevation of features like valleys, hills, and cliffs. An important aspect of topography is *relief,* the difference between the lowest and highest elevations in the area. A mountainous region has *high relief,* whereas a flat plain has *low relief.* Topo-

**topographic map** A map that uses contour lines to show the shape and relief of the ground surface, and the location and elevation of surface features.

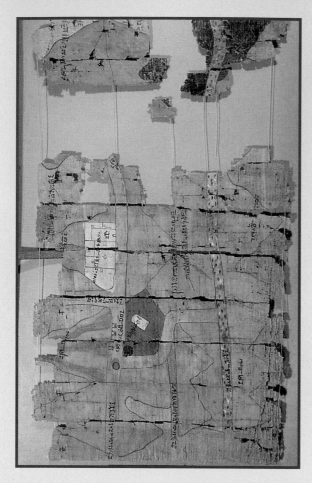

**An ancient Egyptian papyrus scroll** in the Museum of Egypt in Turin, Italy, is the oldest surviving geologic map (Fig. 9B.1). References to Rameses IV on the back of the scroll indicate that the map was made during the six-year reign of this Twentieth Dynasty pharaoh, or about 1150 B.C. It's not the oldest known map, however; that distinction belongs to a Sumerian clay tablet dating from about 2300 B.C. The next oldest geologic map after the Turin Papyrus was produced in Europe in the mid-1700s, 2900 years later.

The Turin map was probably discovered in Egypt in the early 1820s as part of a collection of scrolls buried in the ruins of Deir el-Medina near Thebes and the Valley of the Kings. It was first put on public display in Turin in 1824. It is almost two meters (more than two yards) long, although it must originally have been longer because some of the interior segments and one end of the map appear to be missing.

The map shows topographic and geologic features along a 14-km (8.7 mi) stretch of the dry stream valley

**Figure B9.1**
**THE OLDEST GEOLOGIC MAP**
This photograph shows part of the oldest known geologic map, the Turin Papyrus, discovered in Egypt in the 1920s. It may have been drawn around 1150 B.C. by Amennakht, chief scribe of the village of Deir el-Medina during the reign of Rameses IV.

---

**contour line** A line of equal elevations on a topographic map.

graphic maps use **contour lines,** lines of equal elevation, to portray the shape and relief of the land surface, that is, its *topography*.

To understand how contour lines work, imagine a rectangular tank partly filled with water, with a clay model of a hillside sitting in the tank (Fig. 9.23). The level of the water makes a horizontal line—a line of equal elevation—around the hillside. If you were to place a piece of glass over the tank, you could trace that line onto the glass by looking down through the glass from above. Then you could fill the tank with a bit more water (that is, move the waterline to a higher elevation) and trace the next contour line. If this process were repeated, raising the water level by the same amount each time, you would end up with a set of contour lines drawn on the glass. The contour lines represent the different elevations of the hillslope, giving you a flat, two-dimensional representation of the topography.

The vertical difference—that is, the difference in elevation—between the successive water levels in the tank is called the *contour interval*. If you raised the water level by 1 cm between each step in the process, then the contour interval

(*wadi*) known as Wadi Hammamat in the mountains of Egypt's Eastern Desert. The mountains are shown as stylized conical forms laid out flat on both sides of the *wadi* (a valley or seasonally dry stream bed). The mountains around Bir el-Hammamat are colored black; those to the northeast around Bir Umm Fawakhir and Wadi Atalla are colored pink or, in one case, pink with broad brown streaks. These colors correspond in a general way with the actual appearance of the rocks of which the mountains are composed: purple to dark gray and dark green metamorphic rocks to the southwest and pink granites and rhyolites to the northeast. The location of the granitic rocks corresponds exactly with the pink brown-streaked mountain on the Turin Papyrus. The streaks may represent iron-stained gold-bearing quartz veins mined by the ancient Egyptians.

Other features shown on the map include the locations of a gold mining settlement and the famous bekhen stone quarries, which is the source of many of the monumental stones used for buildings and statues in ancient Egypt. Annotations in hieratic script comment on the gold and silver content of the surrounding mountains, the destinations of several roads, travel distances, and the dimensions of stone from the quarries. All indications are that the map was drawn primarily as an aid to the exploitation of bekhen stone, rather than gold, as was previously thought.

The map may have been drawn by Amennakht, who was the chief scribe during the reign of Rameses IV in the Village of the Craftsmen. This 3100-year-old document is not only the oldest surviving geologic map but also the earliest example of geologic thought. Amennakht, then, was the first geologist and *cartographer* (map-maker) of whom we have any record.

---

*Source:* From "Oldest Geologic Map Is Turin Papyrus," by J. A. Harrell and V. Max Brown, *Geotimes,* March 1989, pp. 10–11. Used with permission by the American Geological Institute.

would be 1 cm. If you raised the water level by 1 inch between each step in the process, then the contour interval would be 1 inch. On a real topographic map, the contour interval might be 10 ft, 10 m, 50 m, or even 100 m. In Figure 9.22 the contour interval is 100 ft. Between one contour line and the next on the map, the elevation of the land surface changes by one contour interval. Figure 9.24 shows an example of a *topographic profile*—a side view of the topography. Note that where the land surface is steep, the contour lines tend to be close together. This is because the elevation changes a lot over a short distance. In contrast, where the land surface is flatter, the contour lines will be farther apart. In Appendix E you will find some rules that will help you interpret topographic maps.

## Legend, Units, and Symbols

We also need to understand the meaning of the colors, patterns, and symbols on a geologic map. Most maps provide a legend or key, as does the map in Figure 9.22. Note that the legend refers to rock units as *formations*. A **formation** is a unit of rock that can be mapped on the basis of rock type and recognizable

**formation** A unit of rock that can be mapped on the basis of rock type and recognizable boundaries with other rock units.

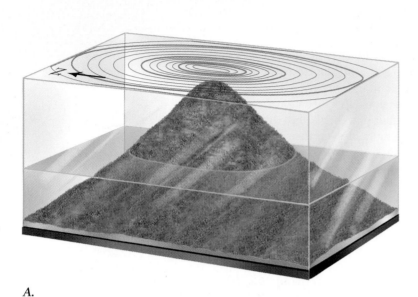

**A.**

**B.**    TOPOGRAPHIC MAP

**Figure 9.23**
**TOPOGRAPHIC CONTOURS**
This model of a hillside shows how topographic contour lines work. *A.* The surface of the water in the tank is level and horizontal. The edge of the water traces a line of equal elevation around the hillside. *B.* If you could lay a piece of glass on top of the tank and trace the waterline at different levels, you would end up with a flat, two-dimensional representation of the topography of the hillside.

boundaries, or *geologic contacts,* with other rock units. The colors used to portray rock formations on maps are standardized to a certain extent. Sedimentary rocks are often portrayed in green, blue, brown, or gray; recent (Quaternary) sediments in yellow; and igneous and metamorphic rocks in red, purple, or pink. Sometimes patterns are used instead of colors. For example, small dots usually indicate sandstone; a brick-like pattern indicates limestone; small Vs indicate granite; and wavy lines indicate metamorphic rocks such as schist or slate. In Appendix E you will find a legend of patterns that are commonly used to indicate different types of rock on geologic maps.

The colors of the rock units in Figures 9.21 and 9.22 do not reveal whether the underlying rock strata are vertical or are tilted in one direction or another. More information is needed. This information can be indicated by symbols showing the locations and orientations of folds, faults, and other geologic features; you can see some examples on the maps in Figures 9.19, 9.21, and 9.22. The meanings of these symbols are also given in Appendix E. In some cases, it is also helpful for the geologists to make inferences about geologic structures under the ground surface. This is done with *geologic cross sections.*

## Geologic Cross Sections

**? How do geologists use maps to figure out what's going on beneath the Earth's surface?**

A geologic map shows the locations of all rock outcrops on the ground surface and the orientations of layering, structures, and other geologic features within them, as well as the geologist's educated guess as to what lies under the soil, vegetation, and buildings between the outcrops. A geologic map can also be used to make inferences about what happens to the rock layers just under the ground. Do they bend completely around and come back to the surface? Do they level out and become flat? Do they grade into a different type of rock? We need to know these things to figure out the geologic history of the area; to determine whether there are structures beneath the surface that could contain oil or mineral deposits; to assess the geologic stability of the area for construction purposes; and

Figure 9.24
## HILLS AND VALLEYS FROM A TOPOGRAPHIC MAP
This diagram shows how to visualize the ups and downs of the land from the contour lines on a topographic map. Here's how geologists construct a *topographic profile* from a map: We draw a line (A-A') across the map. Wherever our line crosses a contour line on the map, it tells us the elevation of that point of land. From that point, we extend a dashed line down to the same elevation on the profile. Then we simply connected those elevations, drawing the profile or side view of the land. Note that the land isn't really this rugged. To make the differences in elevation stand out, we exaggerated the height.

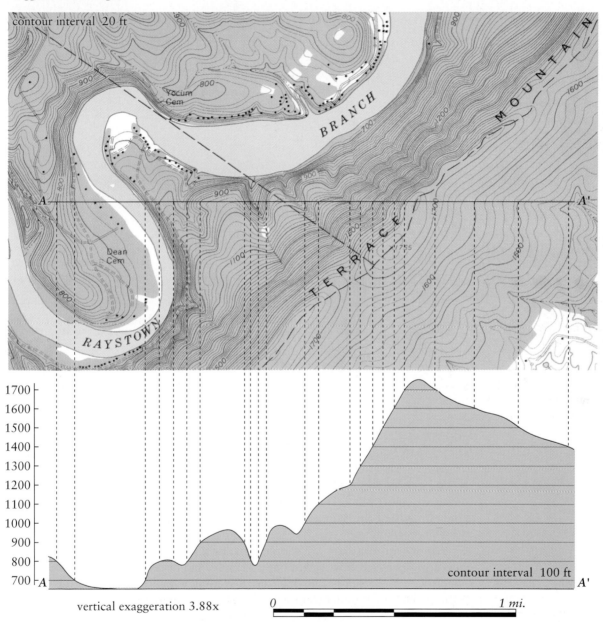

# The Alps

Figure 9.25

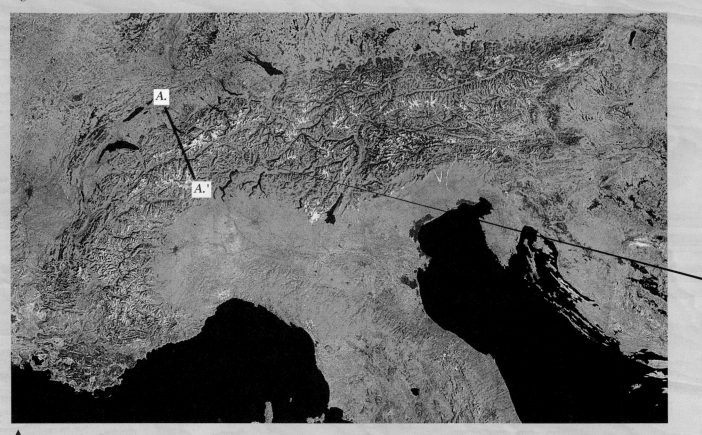

▲
A. The photo shows part of the Alps, a great mountain chain formed by crustal deformation and compressive forces.

for many other reasons. In other words, we must try to visualize the area in 3-D, as shown in Figure 9.21, even though we may only have information concerning the rocks at the surface.

We can do this 3-D visualization by constructing a cross section like the one shown in Figure 9.21. The line marked X-Y will be the location of our cross section. Along that line we mark each contact between rock units, noting the rock types and orientations (the direction and degree of tilt) based on the strike-and-dip symbols on the map. We supplement the information from the geologic map with any information we might have from drilling or remote investigations of the subsurface. The result is a **geologic cross section**, a diagram showing the geologic features that occur underground on the plane containing line X-Y.

**geologic cross section** A diagram showing geologic features that occur underground.

## Rock Deformation and Landscapes

Many of the landscapes we see around us are the results of rock deformation and the action of erosion on rock strata of differing resistivity. Nowhere is this more evident than in the great mountain ranges. The Alps, for example, were formed by intense compression and resulting crustal deformation—both folding and faulting, as shown in Figure 9.25. This type of deformation is present in all the

> **?**
> **How does rock deformation create landforms?**

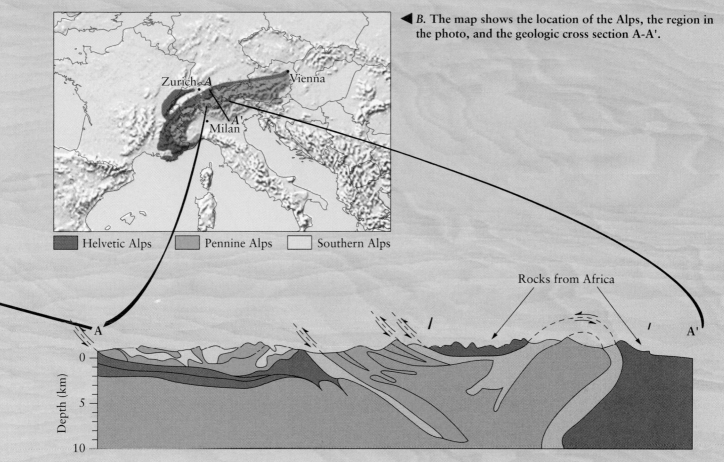

B. The map shows the location of the Alps, the region in the photo, and the geologic cross section A-A'.

Zurich A
Vienna
A'
Milan

| Helvetic Alps | Pennine Alps | Southern Alps |

Rocks from Africa

A
A'
Depth (km)
0
5
10

▲ C. The cross section reveals the intense folding and faulting that have resulted in the mountainous landscape characteristic of the Alps.

great mountain chains of the world, including the Himalayas (in Nepal and Tibet), the Urals (in Russia), and the North American Cordillera (of which the Rockies are a part).

Even in ancient mountain chains like the Appalachians (Fig. 9.26A), which have been worn down by erosion, we can still see clear evidence of compression and crustal deformation. Differences in the way adjacent strata erode have created distinctive topographic patterns that reveal the presence of folds. It is important to understand, however, that synclines do not always form valleys and anticlines do not always form ridges. In the Appalachians, soft, easily eroded rocks (mostly limestones and shales) underlie the valleys, and rocks that are resistant to erosion (mostly sandstones and conglomerates) form the ridges. Figure 9.26B is a geologic map, and Figure 9.26C is a geologic cross section of part of the Valley and Ridge Province of the Appalachians; this area is close to that shown in the satellite image in Figure 9.26A. The cross section reveals how folding, faulting, and erosion have combined to create the distinctive landscape of this region.

Distinctive landforms can also form as a result of normal faulting. The East African Rift Valley, the valley of the Rio Grande in New Mexico, and the valley of the Rhine River in western Europe are examples of valleys that were formed by deformation in response to tensional stresses. The Basin and Range Province

265

# Rock Deformation in the Appalachians

Figure 9.26
The mountainous terrain of the Appalachians, now worn down by the action of erosion, resulted from extensive and repeated periods of crustal deformation, volcanism, and uplift along a convergent plate margin.

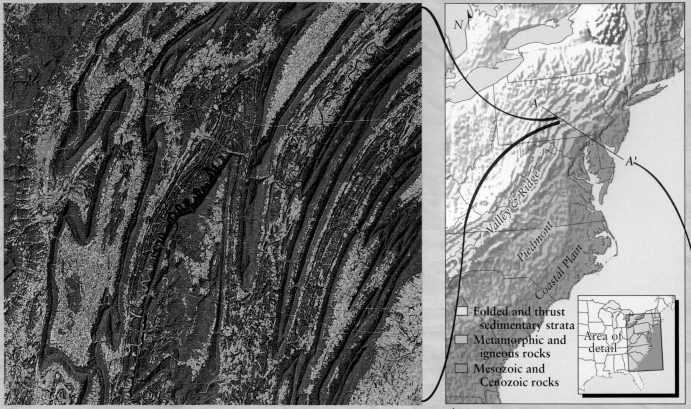

Folded and thrust
sedimentary strata

Metamorphic and
igneous rocks

Mesozoic and
Cenozoic rocks

Area of detail

▲ A. The remnants of the folding and faulting that built this mountain chain can be clearly seen in this LANDSAT image.

▲ B. The map includes the area in the satellite image.

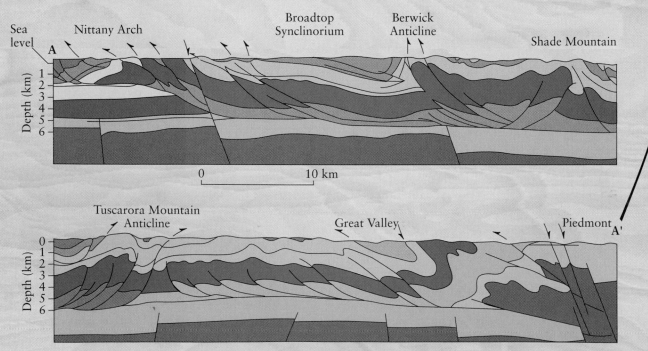

▲ C. The cross section shows the folding, faulting, and erosion of different types of rock that have created the distinctive landforms of this region.

Figure 9.27
**BASIN AND RANGE**
This photograph looks east toward the Lost River Range, Idaho, from Pioneer Mountain, part of the Basin and Range Province. The landforms in the photo are caused by underlying extensional fault structures. The valley in the foreground is a graben, and the Lost River Range is a horst.

in Utah, Nevada, and Wyoming is a good example of rugged topography created by normal faulting. There, movement on a series of north-south striking normal faults has formed alternating horsts and grabens. The horsts are now mountain ranges, and the grabens are sedimentary basins. The Basin and Range Province, which is bounded in the east by the Wasatch Range and continues westward to the Sierra Nevada, exhibits some spectacular scenery, as seen in Figure 9.27.

# WHAT'S AHEAD

In this chapter we have learned how rocks deform as a result of compression, tension, and shearing caused by plate movement. But some important questions remain: Why do mountain ranges form where they do? What kinds of rocks make up these enormous structures? What else, besides structural deformation, happens to rocks during the process of mountain-building? What are the effects of high pressures and temperatures on rocks and minerals? Where do volcanoes fit in this story? How do all of these processes work together to create the familiar landforms we see every day? We will address these questions in the next two chapters.

## Chapter Highlights

**1.** Rock **deformation** results from **stress** placed on rocks by the movements and interactions of lithospheric plates. Most deformation happens deep within the Earth, and we see the deformed rock only when it is uplifted and exposed at the surface by erosion. **Struc**tural geology is the field of geology that is concerned with the study of rock deformation.

**2.** Stress can be uniform (the same in all directions) or differential (stronger in one direction than in another). **Compression** results from forces that push rocks together. **Tension** results from forces that stretch rocks or pull them apart. **Shear** stress causes different parts of a body of rock to slide past each other.

3. In response to stress, a rock may undergo **strain;** that is, it may change its shape or its volume, or both. A nonpermanent change is **elastic deformation.** A permanent change that involves folding or flowing is **ductile (or plastic) deformation.** A permanent change that involves fracturing is **brittle deformation.**

4. Low temperature, low confining pressure, and high rate of strain enhance the brittle properties of rocks. Failure by fracture is common in upper-crustal rocks, where temperature and pressure are low. High temperature, high confining pressure, and low rate of strain, which are characteristic of the deeper crust and the mantle, enhance the ductile behavior of rocks. At depths greater than about 10 to 15 km, ductile deformation becomes increasingly common. The composition of a rock also affects whether it will deform in a brittle or ductile manner.

5. The **strike** of a rock layer gives the orientation of the intersection of the layer with a horizontal plane. The **dip** is the angle between the tilted surface and a horizontal plane, measured down from the horizontal in degrees. Together, the strike and the angle and direction of dip completely describe the orientation of a tilted rock layer.

6. A **fault** is a fracture in a rock along which movement occurs; faults are characteristic of brittle deformation. **Normal faults,** in which the hanging wall block moves down relative to the footwall block, are caused by tensional stress. In **reverse faults,** caused by compression, the hanging wall block moves up and over the footwall block. Shallowly dipping reverse faults are called **thrust faults.** In **strike-slip faults** the movement is mainly horizontal and parallel to the strike of the fault; these are caused by shear stress.

7. **Joints** are fractures along which no appreciable movement has occurred. They form when the confining stress on a rock is reduced as a result of uplift and erosion of overlying rocks. They can also form by cooling and shrinkage of thick lava flows.

8. A **fold** is a bend in a rock; it is characteristic of ductile deformation. An **anticline** is a fold in the form of an arch; a **syncline** is a fold in the form of a trough. Many folds are combinations or variations of these two basic types. To fully portray the geometry of a fold, we must describe the fold axis and axial surface, the type of fold, the direction in which the axis is plunging, and the dip of the plunging axis.

9. Geologists use **geologic maps** to portray information about the type and orientation of rock **formations** and structures found at the Earth's surface. It is common to use a **topographic map** as the base for a geologic map. Topographic maps use **contour lines** to portray the shape and relief of the land surface. Maps also provide a legend or key to explain the symbols, colors, and patterns used to portray different rock units and structures.

10. A geologic map provides information about rock formations and structures at the surface, but it can also be useful to visualize the area in three dimensions. This is done by constructing a **geologic cross section,** a diagram showing the geologic features that occur under the ground on a designated plane.

11. Rock deformation and erosion combine to create the landscapes we see around us. The great mountain chains contain abundant evidence of compressional deformation, both folding and faulting. Extensional deformation can result in landforms such as great rift valleys.

## ▶ *The Language of Geology*

- anticline 254
- brittle deformation 244
- compression 244
- contour line 260
- deformation 242
- dip 249
- ductile deformation 244
- elastic deformation 244
- fault 249
- fold 253
- formation 261
- fracture 249
- geologic cross section 264
- geologic map 257
- joint 253
- normal fault 250
- pressure 243
- reverse fault 250
- scale 258
- shear 244
- strain 243
- stress 243
- strike 248
- strike-slip fault 251
- structural geology 247
- syncline 254
- tension 243
- thrust fault 250
- topographic map 259

## ► *Questions for Review*

1. What is the difference between pressure and stress?
2. What is the difference between stress and strain?
3. What is the difference between elastic, ductile, and brittle deformation?
4. What conditions enhance the brittle properties of rocks? What conditions enhance the ductile properties of rocks?
5. Explain how strike and dip are measured.
6. What is the difference between a fault and a joint?
7. In the type of fold called a(n) _____, the rock strata are arch-shaped or convex upward. In the type of fold called a(n) _____, the rock strata are trough-shaped or concave upward.
8. Draw and label a simple diagram illustrating the difference between normal and reverse faults.
9. What do the contour lines represent on a topographic map? What does it mean when the contours are close together? What does it mean when they are far apart?
10. Use Appendix E to find the symbols that are used to portray the following structures on geologic maps: (a) vertical strata; (b) strike-slip fault; (c) axis of a plunging syncline.

## ► *Questions for Thought or Discussion*

1. Find real examples of plate boundaries along which each of the following types of stress predominates: (a) compression; (b) tension; and (c) shearing. Try to find different examples from those used in the text.
2. Think of some examples of everyday objects that are brittle or ductile (like the chalk, cheese, and Silly Putty examples used in the chapter). Would these objects deform differently under different temperatures or rates of strain?
3. Look closely at Figure 9.26C, the geologic cross section of part of the Appalachian Mountains. Are the faults mostly normal faults or reverse faults? What type of stress do you think was responsible for most of the deformation shown in the cross section—compression, tension, or shearing?
4. Geologic cross sections show the orientations of rocks underground. To construct them, geologists use the information on geologic maps, often supplemented with information about the subsurface provided by remote study techniques. Drilling is mentioned in the chapter; what are some of the other techniques used to study rocks beneath the Earth's surface? Refer to chapter 5 if you need to refresh your memory.
5. In this chapter we have provided some examples of practical reasons for studying structural geology and producing geologic maps and cross sections. One example was the role of structural geology in determining the locations of ore deposits. What were some of the other examples? Go to the map library, if there is one at your school or in your community, or speak with someone at your school who uses geologic maps. Try to find out about some other practical uses and applications of structural geology, geologic maps, and cross sections.

For an interactive case study on the structural geology of Wasatch Fault, visit our Web site.

# · 10 ·

# METAMORPHISM: MAKING NEW ROCK FROM OLD

The Pietà—Mary holding the lifeless body of Jesus—is one of the world's great works of art. To carve the statue, Michelangelo Buonarroti needed a block of perfect marble 2 m across. The year was 1498, the place was Rome, and such blocks were not quarried on the hope of a sale. Michelangelo had resigned himself to a visit to the quarries in distant Carrara when word came that a perfect block had been cut on order, shipped to Rome, but not paid for. Michelangelo's response is described by Irving Stone in his book *The Agony and the Ecstasy*:

> It tested out perfect against the hammer, against water, its crystals soft and compacted with fine graining. He . . . .watched the rays of the rising sun strike the block and make it transparent as pink alabaster, with not a hole or hollow or crack or knot to be seen in all its massive white weight. His Pietà had come home.

A white marble starts out as a sediment composed of shelly fragments, most of them calcite. When subjected to heat and pressure, the calcite recrystallizes, all traces of the shells disappear, and the rock is metamorphosed to a crystalline white marble.

Many believe the Pietà to be the most perfect, most deeply spiritual work of Michelangelo's career. It stands to this day in St. Peter's Cathedral in the Vatican.

◆

## In this chapter we discuss

- Effects of pressure and temperature on rocks
- How to recognize rock that has been changed as a result of increased pressure and temperature
- Different kinds of metamorphic rocks
- Where and why metamorphism happens

*Mother and Child*

The Pietà, Michelangelo's most famous statue, was carved from marble, a metamorphic rock from the Carrara quarries in Italy. The statue is in St. Peter's Cathedral in the Vatican.

When Michelangelo sculpted, he required only the purest, whitest marble, but...

Where does the sculptor's marble come from?

...You will find out in this chapter.

?

Sediments are deposited at the very top of the crust. If they are buried, they will be converted to sedimentary rock within the upper 5 km (about 3 mi) of the continental crust. If the rock is buried as deeply as 35 to 40 km (22 to 25 mi), melting and igneous processes will begin. In between—that is, between a depth of about 5 km and about 40 km, which is most of the thickness of the crust—rock being buried deeper and deeper will remain solid, but its character will slowly change as a result of rising temperature and increasing pressure. As the temperature rises, minerals begin to recrystallize and form larger grains; they also react chemically to form new minerals. Recrystallization and the growth of new minerals changes the texture of the rock. This and the influence of increasing pressure to squeeze and perhaps distort and bend the rocks can cause completely new textures to develop. Rocks that have grown new mineral assemblages or have developed new textures—in most cases both have happened together—are said to have been *metamorphosed*. The rock Michelangelo used for the Pietà is an example of a sedimentary rock (limestone) that was metamorphosed to marble.

**metamorphism** The mineralogical, chemical, and structural adjustments of solid rocks to physical and chemical conditions at depths below the region of sedimentation and diagenesis.

?

**How do metamorphic rocks "remember" the changes they have undergone?**

# WHAT IS METAMORPHISM?

Metamorphism comes from two Greek words: *meta*, meaning "change," and *morphe*, meaning "form," hence, "change of form." We define **metamorphism** as all the mineralogical, chemical, and structural adjustments of *solid rocks* to physical and chemical conditions at depths below the region of sedimentation and diagenesis (refer back to chapter 8 for a discussion of diagenesis).

We emphasize that metamorphism involves solid rocks because this is what makes metamorphic rocks so interesting. As tectonic plates move and collide, rocks are squeezed, stretched, bent, heated, and changed in complex ways (Fig. 10.1). But even if a rock has been altered two or more times, some hints of its earlier forms are usually preserved because the rock remains solid as the changes occur. Solids, unlike liquids and gases, preserve clues to the events that changed them. For example, if you throw a stone in a pond, the splash and ripple are soon smoothed out. Throw a stone at a window, however, and the result is permanently cracked glass. Metamorphic rocks retain a record of all the heatings, stretchings, bumpings, and grindings that have happened to them throughout geologic history. Deciphering that record is an exceptional challenge for geologists, but the rewards in understanding are high. For example, when tectonic plates collide, distinctive kinds of metamorphic rocks are formed along the margins of the plates. Geologists determine where the boundaries of ancient continents once were by studying those rocks. They also use evidence from metamorphic rocks to determine how long plate movements have been occurring. So far the evidence indicates that plate tectonics has been operating for at least 2 billion years!

## The Limits of Metamorphism

The heat that causes metamorphism is the Earth's internal heat. We know from drilling deep gas and oil wells, and from deep gold mines, that temperature in the continental crust increases with depth at a rate of about 30°C/km (about

Figure 10.1
**SEDIMENTARY ROCK CHANGED BY METAMORPHISM**
Original clay-rich (dark) and quartz-rich (light) layers in a sediment have been recrystallized and the once horizontal sediment layers have been bent and controlled into a complex pattern during metamorphism deep in the crust. This specimen is 12 cm (about 5 in) wide.

85°F/mile). At a depth of about 5 km (about 3 mi), the temperature is about 150°C (about 300°F). This is the dividing line between the processes of diagenesis that change sediments into sedimentary rocks (below 150°C) and processes we call metamorphic (above 150°C).

The pressure at sea level on the Earth's surface is 1 atmosphere (1.03 kg/cm² = 14.7 lb/in²). Pressure in the crust increases with depth at a rate of about 300 atm/km (about 480 atm/mi). At a depth of 5 km, pressure is 1500 times greater than atmospheric pressure. Five kilometers is the depth in the crust where metamorphism starts because at that depth both temperature and pressure are high enough for recrystallization and the growth of new minerals to start.

The upper temperature limit of metamorphism is about 800°C (about 1470°F). We say "about" because the onset of partial melting, which is what separates metamorphic processes from magmatic processes, involves more than just temperature (as you learned in chapter 6); it depends on the pressure, the composition of the rock, and the amount of fluid present, especially water and carbon dioxide.

Figure 10.2 shows the range of temperature and pressure over which metamorphism occurs. In the plot on the righthand side of Figure 10.2, pressure is

**?** **Where does diagenesis end and metamorphism begin? Where does metamorphism end and melting begin?**

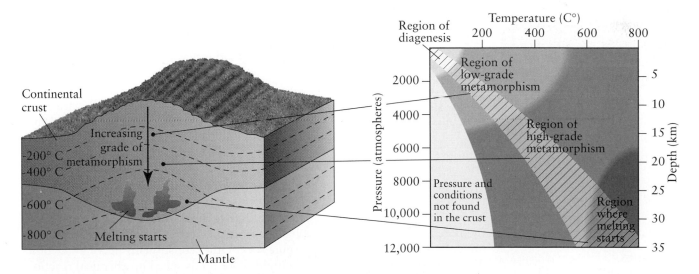

Figure 10.2
**TEMPERATURE AND PRESSURE CONDITIONS OF METAMORPHISM**
This diagram shows the conditions of pressure and temperature in which diagenesis, metamorphism, and melting occur. On the left is a sketch showing where in the crust metamorphism and melting occur. On the right is a plot of pressure versus temperature, with the pressure–temperature regimes of metamorphism indicated. The broad, curved band shows how pressure and temperature increase with depth through much of the continental crust. Pressure is shown along the lefthand side of the diagram, and depth, equivalent to pressure in the crust, is shown on the righthand side.

**low-grade** Rocks that are metamorphosed under temperature and pressure conditions up to about 400°C and 4000 atm.

**high-grade** Rocks that are metamorphosed under temperature and pressure conditions higher than about 400°C and 4000 atm.

represented on the vertical axis; it increases downward, just as pressure increases with depth in the Earth. Temperature is represented on the horizontal axis, increasing from left to right. The pressure and temperature conditions under which sediments are formed and changed into sedimentary rocks during diagenesis occupy the upper lefthand corner of the diagram. From the region of diagenesis up to a temperature of about 400°C (750°F) and a pressure of about 4000 atms—that is, to a depth of about 15 km (about 9 mi)—metamorphism occurs but is not very intense. We refer to rocks formed under such conditions as **low-grade** metamorphic rocks. At higher temperatures and pressures, extending to the onset of melting, **high-grade** metamorphic rocks are formed.

# FACTORS THAT INFLUENCE METAMORPHISM

**How is metamorphism like cooking?**

One way to think about metamorphism is to compare it with cooking. When you cook, the results depend on the ingredients and how you cooked them. So too with rocks; the end product is controlled by the composition of the rock and the metamorphic (or cooking) conditions. Rock composition plays an important role in the mineral assemblage that is formed, but so do changes in temperature and pressure. The effects of temperature and pressure are not straightforward, however. They are influenced by the presence or absence of fluids; by how long a rock is subjected to high pressure or temperature; and by whether the pressure is uniform or is different in different directions so as to twist or break the rock. Let's look briefly at each of these factors and how they influence metamorphism.

## The Influence of Pore Fluids

As we discussed in chapter 8, the open spaces between clasts in a sediment are called pores. The same term is used for all the tiny open spaces in a rock, such as those between the grains in a sedimentary rock and the tiny fractures in igneous and metamorphic rocks. Most pores are filled either by a gas, such as carbon dioxide, or by a watery fluid. The watery fluid is never just pure water. It always has small amounts of gases and salts dissolved in it, together with traces of all the mineral constituents that are present in the enclosing rock. Moreover, at high temperature the pore fluid is more likely to be a gas than a liquid. Regardless of their specific character, however, pore fluids play a vital role in metamorphism.

Pore fluids enhance metamorphism in two ways. First, the presence of a pore fluid permits material to dissolve from one place, move through the pore fluid, and be precipitated in another place. In this way, pore fluids speed up recrystallization. The second way pore fluids enhance metamorphism is by acting as a reservoir during the growth of new minerals. When the temperature and pressure of a rock that is undergoing metamorphism change, so does the composition of the pore fluid. Some of the dissolved constituents move from the fluid to the new minerals growing in the metamorphic rock. Other constituents move in the other direction, from the minerals to the fluid. Pore fluids speed up chemical reactions in much the same way that water in a stew pot speeds up the cooking of a tough piece of meat. Metamorphism proceeds rapidly when pore fluids are present, but when pore fluids are absent, or are present in tiny amounts, metamorphic reactions occur very slowly.

As pressure increases and metamorphism proceeds, the amount of pore space decreases and the pore fluid is slowly driven out of the rock. As the temperature

Figure 10.3
**QUARTZ VEIN IN META-MORPHIC ROCK**
This is a vein of quartz in a metamorphic rock. The quartz was dissolved in pore water that was squeezed out of the rock during metamorphism.

Factors That Influence

Chap

276

Figure 10.
FLATT
CON
Diff
m

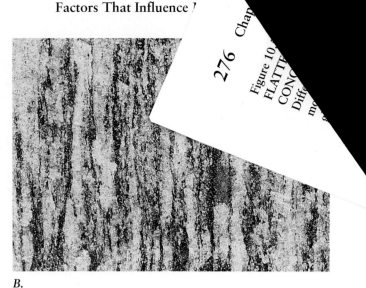

*A.*

*B.*

**Figure 10.4**
**UNIFORM AND DIFFERENTIAL STRESS**
These photos show the effects of stress on the textures of rocks with the same mineral assemblages.
*A.* This granite, consisting of quartz (glassy looking), feldspar (white), and biotite (dark), crystallized from magma under conditions of uniform stress. Note that the biotite grains are randomly oriented. *B.* This high-grade metamorphic rock developed a mineral assemblage of quartz, feldspar, and biotite under differential stress. The biotite grains are parallel, giving the rock a pronounced layered texture.

of the rock increases, *hydrous* minerals (minerals that contain water) begin to expel water. Water released in this way joins the pore fluid and is slowly driven out of the metamorphic rock. Escaping pore fluid carries with it small amounts of dissolved mineral matter. If such a fluid flows through fractures in the rock, some of the dissolved minerals might be precipitated; the result is a *vein* (Fig. 10.3).

## Effects of Pressure

The effects of pressure on a rock are a bit more complex than they may appear at first. They are complex because solids have strength and can be squeezed more strongly in one direction than in another. Because liquids and gases lack strength, pressure within a liquid or gas at rest is always uniform—that is, equal in all directions. In considering the effects on solids, it is necessary to specify the direction in which pressure is applied. For this reason geologists prefer to use the term *stress*[1] because it implies direction. The textures in most metamorphic rocks record *differential stress* (meaning stress that is not equal in all directions). Most igneous rocks, by contrast, have textures that were formed under conditions of *uniform stress* (meaning stress that is equal in all directions), because igneous rocks crystallize from liquids.

The most striking effect of metamorphism under differential stress is seen in minerals belonging to the *mica* family. Micas grow as flat sheets, which line up perpendicular to the direction of maximum stress. Compare Figures 10.4A and B. Figure 10.4A shows a granite with a typical texture of randomly oriented mineral grains that grew in conditions of uniform stress as the granite magma (a liquid) crystallized. Figure 10.4B, on the other hand, shows a high-grade metamor-

> **?** **Is pressure as important as temperature in metamorphism?**

---

[1] Recall that *stress* was also defined and discussed in chapter 9 in the context of rock deformation.

**NED
LOMERATE**
rential stress during meta-
rphism has deformed this con-
lomerate. Sandstone pebbles,
originally round, have been
squeezed and flattened.

phic rock; it contains the same minerals as the granite, but in this case they grew as new minerals in a solid rock and they did so under conditions of differential stress. In the metamorphic rock (Fig. 10.4B), all the mica grains are parallel, giving the rock a distinctive texture. Another example of metamorphism under differential stress is shown in Figure 10.5. This shows a conglomerate that originally consisted of uniformly rounded pebbles; squeezing of the pebbles during metamorphism distorted them, creating flattened discs. Note that it is the texture of a metamorphic rock—not its mineral assemblage—that is influenced by the kind of stress to which it is subjected.

## Foliation

**foliation** A planar metamorphic rock texture.

The changes that occur during metamorphism produce a texture called **foliation** (Fig. 10.6), from the Latin word *folium,* meaning "leaf." Foliated metamorphic rocks tend to split into thin, leaf-like flakes. Foliation may be either pronounced or subtle, but if present in a rock it is strong evidence that the rock has been metamorphosed. During the earliest stages of low-grade metamorphism, as layer after layer of sediment is deposited and the pressure at the base of a pile of sedimentary rock increases, maximum stress tends to be vertical (Fig. 10.7A). The new micas, and therefore the foliation, tend to be parallel to the bedding planes of the sedimentary rock that is undergoing metamorphism. But when compression due to a plate collision deforms the flat sedimentary layers into folds, the direction of maximum stress is no longer perpendicular to bedding so that new micas and the foliation are no longer parallel to the bedding (Fig. 10.7B).

Low-grade metamorphic rocks tend to be so fine-grained that the new mineral grains can be seen only under a microscope. Foliation in a low-grade meta-

**Figure 10.6
FOLIATION UNDER
THE MICROSCOPE**
By gluing a rock chip to a glass microscope slide and then grinding the chip until it is thin enough for light to pass through, the presence or absence of foliation can be seen under a microscope. This sample, which is 1 cm across, shows a nonfoliated rock consisting mainly of quartz.

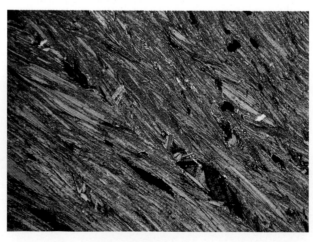

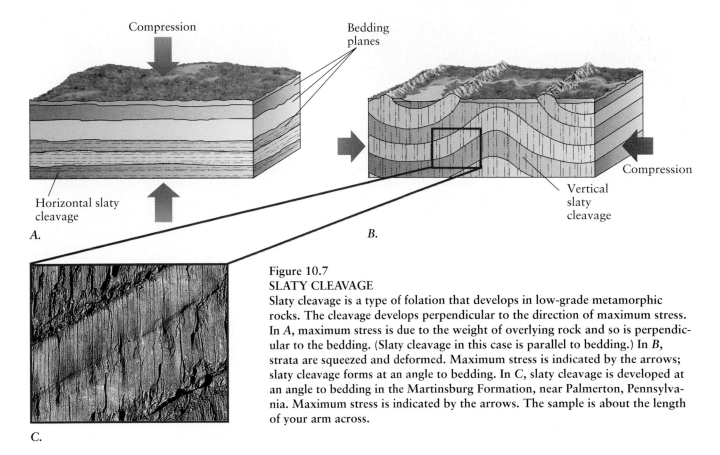

Figure 10.7
**SLATY CLEAVAGE**
Slaty cleavage is a type of foliation that develops in low-grade metamorphic rocks. The cleavage develops perpendicular to the direction of maximum stress. In *A*, maximum stress is due to the weight of overlying rock and so is perpendicular to the bedding. (Slaty cleavage in this case is parallel to bedding.) In *B*, strata are squeezed and deformed. Maximum stress is indicated by the arrows; slaty cleavage forms at an angle to bedding. In *C*, slaty cleavage is developed at an angle to bedding in the Martinsburg Formation, near Palmerton, Pennsylvania. Maximum stress is indicated by the arrows. The sample is about the length of your arm across.

morphic rock produces a distinctive style of fracture; the rock tends to break into flat, plate-like fragments, a property called **slaty cleavage** (Fig. 10.7C). The fracture planes are defined by the parallel alignment of microscopic-sized platy minerals, especially clays and micas. Cleavage in a rock is similar to cleavage in a mineral (see chapter 2), but there is an important difference. Cleavage in a mineral grain involves a single crystal lattice, and it happens because the bonds between atoms are weaker in some directions than in others. Cleavage in a rock involves a great many crystal lattices, and it happens because the crystal lattices, and therefore the cleavage directions, in many minerals are approximately parallel.

The size of the mineral grains that form under conditions of high-grade metamorphism is large enough that individual grains can be seen with the naked eye. Foliation in coarse-grained metamorphic rocks is called **schistosity,** from the Greek word *schistos,* meaning "cleaves easily." Schistosity differs from slaty cleavage mainly in the size of the mineral grains. In addition, schistosity is not necessarily planar, as slaty cleavage generally is; high-grade metamorphic rocks tend to break along wavy or distorted surfaces.

**slaty cleavage** In low-grade metamorphic rocks, the tendency to break into flat, plate-like fragments.

**schistosity** Foliation in coarse-grained metamorphic rocks.

## Effects of Temperature

When a mixture of flour, salt, sugar, yeast, and water is baked, the high temperature causes a number of chemical reactions. The original ingredients break down, new compounds are formed, and a loaf of bread is created. When rocks are heated, some of the original minerals recrystallize, others are involved in chemical reactions that form new minerals, and a metamorphic rock is created. Both recrystallization and the growth of a new mineral assemblage are controlled mainly by temperature. Of course, stress and the presence or absence of pore flu-

# GEOLOGY AROUND US

## Metamorphism and the Game of Billiards

**The modern game of billiards, and its close relative, pool,** are played on a table made from metamorphic rock. Billiards is a very old game; historical records reveal that it was played in France during the reign of King Louis XI (1461–1483). The first billiard table known to be in North America was brought to Florida in 1565 by the Spaniards.

Early billiard tables resembled modern ones in that they were rectangular in shape, surrounded by edges of elastic cushions, and were fitted with six pockets. But there the similarity ends. The tables were made entirely of wood, and the balls were struck with little mallets, shaped something like a modern golf club. In about 1735, a straight wooden stick resembling the modern cue began to replace the mallet. Replacement was complete by about 1800, by which time a leather tip had been added to the end of the stick, thus creating the modern cue.

Using a cue, the players direct balls around a billiard table with great accuracy, provided that the table is smooth and free from vibrations. Wood can be smoothed, but it vibrates. By about 1825, manufacturers of billiard tables had discovered that if slate were used for the table (and provided the stone was laid so that the slaty cleavage was parallel to the table top), it was both smooth enough and free of vibrations (Fig. B10.1).

Slate has been widely quarried in the United States, and it has many uses besides billiard tables; for example, slate is used for roofing and for floor tiles, for good quality chalk boards, and for insulation of panels on which to mount electrical equipment. Much of the slate produced in North America has been mined in a belt of rocks that runs along the eastern seaboard from Georgia to Maine. Slate in this region is found in low-grade metamorphic rocks in which slaty cleavage was developed as a result of continental collisions during the plate tectonic assembly of Pangaea during the Paleozoic Era.

**Figure B10.1**
**METAMORPHISM PRODUCES SLATE FOR BILLIARD TABLES**
Slate, resulting from low-grade metamorphism of shale (*A*, *B*, *C*), has properties of strength and freedom from vibration which are ideal for billiard tables. After slate is quarried (*D*), it is split along the slaty cleavage, and it is the cleavage surface that becomes the table top (*E*).

ids also play roles by helping to determine the extent of recrystallization and by influencing the size of the individual mineral grains in the new mineral assemblage that is formed during metamorphism. For example, plentiful pore fluids generally lead to large mineral grains.

As temperature and stress increase, different mineral assemblages are formed. For any given rock composition, each mineral assemblage is produced under a specific set of metamorphic temperatures and pressures. Figure 10.8 illustrates the way mineral assemblages change during the metamorphism of shale, a sedimentary rock. Let's consider the meaning of the information in Figure 10.8. We read the diagram from left to right, starting with conditions of diagenesis; then as temperature and pressure increase, diagenesis and sedimentary rock are

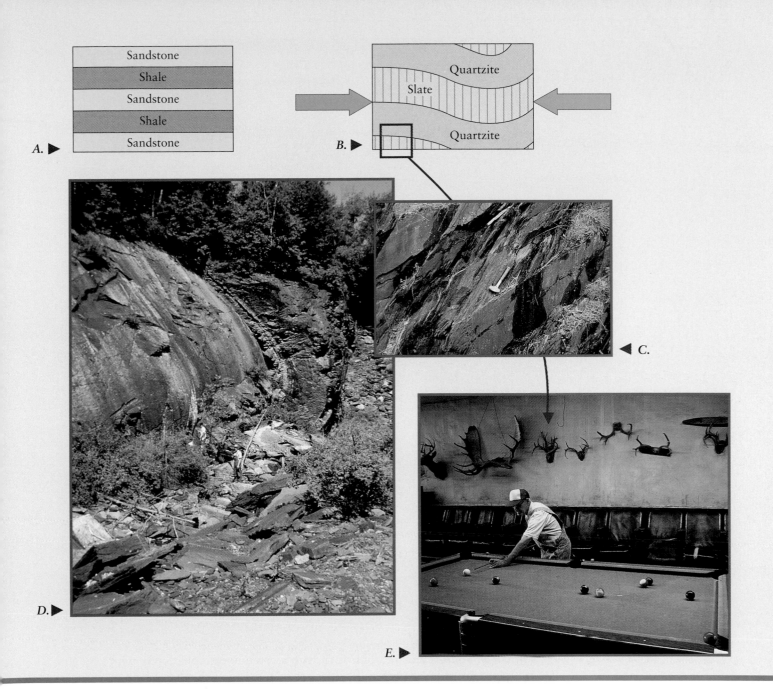

A.

| Sandstone |
|---|
| Shale |
| Sandstone |
| Shale |
| Sandstone |

B.

Slate    Quartzite

Quartzite

C.

D.

E.

succeeded first by low-grade and then by high-grade metamorphic rocks. As conditions change, so does the mineral assemblage.

Shales are composed principally of clay minerals and quartz. All of the mineral grains are so tiny that they can be seen only through a microscope. When a shale is metamorphosed, quartz recrystallizes and new minerals such as plagioclase feldspar, muscovite, and chlorite grow as a result of chemical reactions involving the clay minerals and pore fluids. Low-grade metamorphism of a shale produces *slate*. The new minerals in a slate are too small to be seen with the naked eye.

Continued metamorphism of a slate as a result of increasing temperature and pressure leads to the disappearance of chlorite and the growth of biotite. The re-

279

# Shale to Schist

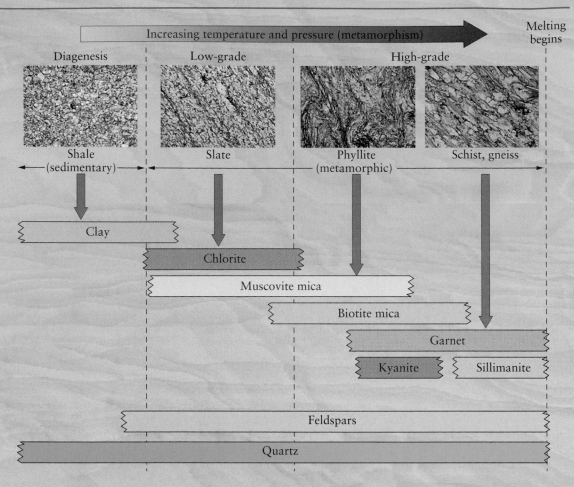

**Figure 10.8**
The bars in this diagram show the changes in mineral assemblage that occur as a shale is metamorphosed from low to high grade. Metamorphism progresses from left to right on the diagram. Before metamorphism occurs, the shale is a sedimentary rock made of clay particles and quartz grains. The first metamorphic rock to develop is a low-grade slate, then a phyllite, and finally high-grade schist and gneiss. The photographs show the rocks under a microscope; each photo shows a field of view approximately 3 mm (0.12 in). The minerals kyanite and sillimanite have the same composition ($Al_2SiO_5$) but different crystal structures—they are found only in metamorphic rocks.

sult is a *phyllite*, a metamorphic rock with a pronounced foliation in which individual grains of muscovite and biotite are just visible with a hand-held magnifying glass. At still higher grades of metamorphism, muscovite disappears and minerals such as potassium feldspar, garnet, kyanite, and sillimanite are formed. High-grade metamorphic rocks often have strongly developed schistosity; they are *schists* if mica-rich, *gneisses* if mica-poor.

One important temperature-controlled effect—the beginning of rock melting—is greatly influenced by the presence of pore fluid. The effect of fluid on rock is the same as the effect of salt on ice—it lowers the melting temperature. The upper temperature limit of metamorphism therefore depends on the amount of pore fluid present. When a tiny amount of fluid is present, only a small amount of melting occurs and the magma remains trapped in small pockets in the metamorphic rock. When the rock cools, so do the pockets of magma. The result is a composite rock—metamorphic but with a small igneous component (Fig. 10.9). Such rocks are called **migmatites**. When abundant pore fluid is pre-

**migmatite** A high-grade metamorphic rock in which a small amount of melting has occurred, and pockets of trapped melt have cooled and solidified within the rock.

Figure 10.9
MIGMATITE
Migmatite is a composite rock, partly metamorphic, partly igneous. Such rocks form when a metamorphic rock reaches a temperature and pressure at which melting commences—part of the rock melts, part remains solid. The fraction melted is igneous (pink); the fraction that remained unmelted is metamorphic (dark grey). The photograph is a migmatite in northern Wisconsin.

sent and large volumes of magma develop, the magma rises and intrudes into overlying metamorphic rock. As a result, we observe that batholiths of granitic rock (discussed in chapter 6) and large volumes of metamorphic rock tend to be closely associated. This igneous-metamorphic rock association occurs along subduction and collision margins of tectonic plates.

## The Influence of Time

Chemical reactions can occur rapidly or slowly. Some reactions, such as the burning of natural gas to produce carbon dioxide and water, happen so fast that they can cause an explosion if not handled carefully. At the other end of the scale are reactions that take millions of years to complete. Most of the reactions that happen during metamorphism are of the latter kind. Despite the slowness of these reactions, scientists have been able to demonstrate that high temperature, high pressure, abundant pore fluids, and long reaction times produce large mineral grains. Thus, we can conclude that coarse-grained rocks—those with mineral grains the size of a thumbnail or larger—are products of long-sustained metamorphic conditions (possibly over millions of years) at high temperatures and high pressure. On the other hand, fine-grained rocks—those with mineral grains the size of a pin's head or less—either were produced under conditions involving lower temperature and lower pressure, or formed under higher temperatures and pressures but under conditions such that pore fluids were scarce or reaction times were short.

**?** How long does metamorphism take?

# THREE KINDS OF METAMORPHISM

The processes that cause changes in texture and in mineral assemblages in metamorphic rocks are *mechanical deformation* and *chemical recrystallization*. Mechanical deformation includes grinding, crushing, and fracturing. Figure 10.5, which shows flattened pebbles in a conglomerate, is an example of metamorphism due largely to mechanical deformation. Chemical recrystallization includes changes in mineral composition, growth of new minerals, recrystallization of old minerals, and changes in the amount of pore fluid due to chemical reactions that occur when a rock is heated and squeezed. Metamorphism always involves both mechanical deformation and chemical recrystallization, but their relative importance varies greatly.

**?** How does metamorphism work?

Figure 10.10
CONTACT METAMORPHISM
This diagram shows a limestone, which consists of calcite, being changed by hot fluids released from a cooling magma. An areole of contact metamorphic rock surrounds the granite intrusion. Grading in from the unaltered limestone, the limestone has been metamorphosed to a marble. Inside the marble is a zone in which, by the addition of material from the magma, chlorite and serpentine develop. Inside the chlorite and serpentine zone, adjacent to the magma, garnet and a green pyroxene called diopside grow.

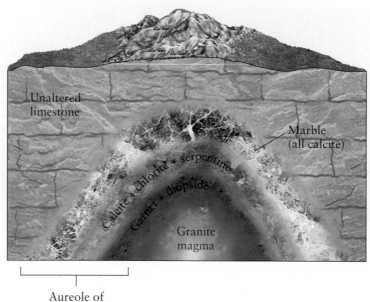

Aureole of
metamorphic
rock

We can distinguish among several kinds of metamorphism on the basis of the dominant processes (that is, the relative importance of mechanical deformation and chemical recrystallization) and the tectonic environments in which they occur. The three most important kinds of metamorphism are *contact metamorphism*, *burial metamorphism*, and *regional metamorphism*.

## Contact Metamorphism

**contact metamorphism** Metamorphism that occurs when rocks are heated and chemically changed by the intrusion of a body of hot magma.

**Contact metamorphism** occurs adjacent to bodies of hot magma that have intruded into cooler rocks. Such metamorphism involves mainly chemical recrystallization in the intruded rocks and occurs in response to a pronounced increase in temperature and to the involvement of chemically active pore fluids released by the magma. Mechanical deformation (and therefore foliation) is minor because the stress around a mass of magma tends to be uniform or nearly so.

The extent of contact metamorphism depends on the size of the intrusive body, the amount of pore fluid involved, and the kind of rock being metamorphosed. The zone or *aureole* of contact metamorphism surrounding a small intrusion such as a dike or sill may extend only a few centimeters, especially when the magma intrudes into a relatively impervious rock such as a shale, or when little pore fluid is released by the magma. However, a large intrusion contains more heat energy than a small one and may also give off a lot of pore fluid as it cools. When intrusions are very large—perhaps many kilometers across (the size of stocks and batholiths)—and when they intrude into highly reactive and relatively pervious rock, such as limestone, contact metamorphic effects may extend for hundreds of meters. Several concentric zones may be formed with different mineral assemblages, each one characteristic of a certain temperature range (Fig. 10.10).

## Burial Metamorphism

**burial metamorphism** Metamorphism that occurs after diagenesis, as a result of the burial of sediments in deep sedimentary basins.

Sedimentary rocks, sometimes with interbedded pyroclastic rocks, may reach depths of 10 km (6.2 mi) and temperatures of 300°C (572°F) or more when buried in a sedimentary basin. **Burial metamorphism** is the first stage of metamorphism after diagenesis. The metamorphic processes caused by burial begin at about 150°C (300°F) and are speeded up by the abundant pore water in most sedimentary rocks. Rock that is water-saturated is weak, and differential stress is not pronounced; the maximum stress exerted during burial metamorphism tends to be vertical, so that foliation, if present, is parallel to bedding. Burial metamorphism involves mainly chemical recrystallization. Burial metamorphism is usually observed in deep sedimentary basins, such as trenches along the margins of tectonic plates. As temperature and pressure increase, burial metamorphism grades into regional metamorphism.

## Regional Metamorphism

The most common kinds of metamorphic rocks of the continental crust—slates, phyllites, schists, and gneisses—are found in areas extending over tens of thousands of square kilometers. The process by which they are formed is called **regional metamorphism.** Regionally metamorphosed rocks are found in mountain ranges and in the eroded remnants of former mountain ranges (Fig. 10.11). They are formed as a result of subduction and collisions between masses of continental crust. During a collision, sedimentary rock along the margin of a continent is subjected to intense differential stresses. The foliation that is characteristic of regionally metamorphosed rocks is a consequence of such stresses.

In order to understand what happens during regional metamorphism, consider a segment of the crust that is subjected to a maximum stress in the direction of plate convergence as a result of a collision between two plates. Rock in the crust folds and buckles. This causes the crust to be regionally thickened, as shown in Figure 10.12. The bottom of the thickened mass is pushed downward, displacing hot mantle rock. Eventually, the bottom part of the thickened crustal mass will become as hot as the mantle rock it displaces. As a result, rock near the bottom of the pile is subjected to greater stress and higher temperature, new mineral assemblages form and new textures develop. However, because rock is a poor conductor of heat, the heating-up process can be slow. If the folding and

**regional metamorphism** Metamorphism of an extensive area of the crust, as a result of the stresses and high temperatures associated with plate convergence, collision, and subduction.

**Figure 10.11**
**A MOUNTAIN RANGE OF METAMORPHIC ROCK**
This is an aerial view of the Himalaya Mountains in Tibet. The rocks, once marine sediments, have been subjected to low-grade regional metamorphism.

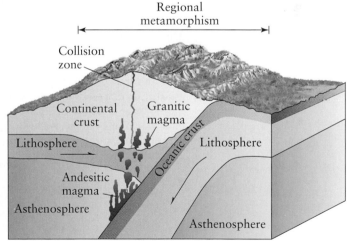

**Figure 10.12**
**REGIONAL METAMORPHISM: THE TECTONIC SETTINGS**
Regional metamorphism occurs when compressive stress deforms and thickens the crust where rocks are squeezed between two colliding plates. The base of the thickened crust is pushed so deep into the Earth that melting commences, forming granite magma.

thickening are very slow, the rate at which the pile heats up may keep pace with the temperature of adjacent parts of the crust and mantle. In this case, the maximum temperature will coincide with the maximum pressure. However, if burial occurs very fast, as when sediment is dragged down in a subduction zone, there is not enough time for the pile to heat up and the pressure will be very high but the temperature will not be so high. Depending on the rate of burial, therefore, the same original rock can yield different kinds of metamorphic rocks because maximum temperature and maximum pressure are reached at different times.

## Isograds and Metamorphic Facies

The first geologists to make a systematic study of regionally metamorphosed rocks did so in the Scottish Highlands. They studied a large area of metamorphic rocks with the chemical composition of shale. Even though the rocks had the same overall composition, the mineral assemblages differed from one place to another because of differences in the grade of metamorphism they had undergone. The geologists found that the rocks could be subdivided on the basis of the mineral assemblages they contained. They selected characteristic *index minerals* that marked the appearance of each new mineral assemblage in a progression from low-grade metamorphic rock to higher-grade rock (as in Fig. 10.8). Thus, the first occurrence of biotite in the rocks marked a transition from low-grade rocks containing chlorite to a mineral assemblage caused by a slightly higher-grade metamorphism. The geologists noted these localities on a map. The first occurrence of garnet marked a transition to even higher-grade rock, and so on for kyanite and sillimanite.

The geologists soon discovered that it was possible to draw a line on the map connecting all points of first occurrence of each index mineral. Such a line is called an *isograd* (from the Greek *isos*, meaning "equal," and *grade*, referring to grade of metamorphism). This method of mapping index minerals, first used in Scotland a century ago, has been successfully tested on different types of rock and is now used around the world in studies of metamorphic rocks (Fig. 10.13).

**Do rocks change their chemical composition when they are metamorphised? (A tricky question...)**

Figure 10.13
**REGIONAL METAMORPHISM IN SCOTLAND (ISOGRADS)**
This map shows isograds in a body of regionally metamorphosed rocks in Scotland. The isograds mark the first appearance of the index minerals. Rocks between the isograds are said to be in a particular metamorphic zone. For example, rocks lying between the biotite and garnet isograds are in the biotite zone.

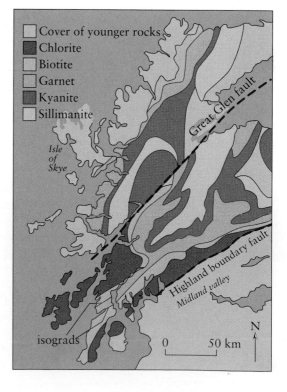

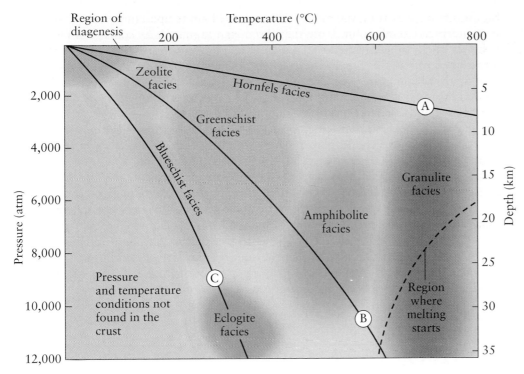

**Figure 10.14**
**METAMORPHIC FACIES**
Regions of pressure and temperature for different metamorphic facies. The zeolite facies is characteristic of burial metamorphism, and the hornfels facies is characteristic of contact metamorphism. Blueschist and eclogite facies are characteristic of subduction zones. Greenschist, amphibolite, and granulite facies occur in continental crust thickened by plate collisions. The lines marked *A*, *B*, and *C* are the way temperature and pressure change under conditions of contact metamorphism, in continental crust thickening by plate collisions, and in subduction zones.

It should be noted that the first appearance of a mineral does not necessarily mean the last appearance of some other mineral. For example, the first appearance of biotite in low-grade metamorphic rocks does not coincide with the last appearance of chlorite. Geologists refer to the metamorphic rocks lying between one isograd and the next higher isograd as being in a *metamorphic zone*. For example, in Figure 10.13, all rocks lying between the biotite isograd and the garnet isograd are in the *biotite zone*. The concepts of index minerals, isograds, and metamorphic zones are equally applicable to burial, contact, and regional metamorphism.

Careful study of metamorphic rocks around the world has demonstrated that the chemical compositions of most rocks are little changed by metamorphism. The main changes that occur are the addition or loss of volatile constituents such as water and carbon dioxide; the solid, nonvolatile constituents may be locally redistributed, as in the formation of veins, but except for water and carbon dioxide, overall the chemical compositions of rocks remain approximately constant. The principal chemical changes brought about during metamorphism, then, are changes in the mineral assemblage, not changes in the overall chemical composition of the rock. We can conclude, therefore, that for a given rock composition the mineral assemblage must be determined principally by the temperature and stress to which the rock is subjected during metamorphism.

On the basis of this conclusion, geologists have developed the concept of **metamorphic facies,** which is that the different mineral assemblages that are formed in rocks of different composition under a given set of temperature and stress conditions belong to the same metamorphic facies. Each of the different metamorphic facies has been given a name, as you can see in Figure 10.14. The names reflect some obvious feature, such as the color of a common rock (blueschist and greenschist); the presence of a distinctive mineral (zeolite and amphibole); an unusual type of rock (eclogite); or a distinctive texture (granulite). You can also see from Figure 10.14 that each metamorphic facies represents a characteristic range of temperature and pressure. For example, the blueschist facies is representative of relatively high-pressure, low-temperature metamorphism;

**metamorphic facies** The set of metamorphic mineral assemblages that are formed in rocks of different composition under given temperature and stress conditions.

the hornfels facies is representative of relatively high-temperature, low-pressure metamorphism; and so on. Note that the metamorphic facies concept is just as valid for rocks subjected to burial and contact metamorphism as it is for regionally metamorphosed rocks. In Figure 10.14, the region of the zeolite facies corresponds to the conditions of burial metamorphism, and the region of the hornfels facies corresponds to the conditions of contact metamorphism; all the other facies correspond to conditions of regional metamorphism.

Having discovered that temperature, stress, and the original composition of the rock determine metamorphic mineral assemblages, geologists discovered that they could prepare a detailed grid, like the one shown in Figure 10.14, on which all the different metamorphic facies could be plotted in terms of temperature and stress. For any given composition of rock, it is possible to prepare a grid showing the specific mineral assemblages characteristic of each of the different metamorphic facies. The three pressure–temperature gradients (A, B, and C) shown in Figure 10.14 are approximately the gradients to be expected during metamorphism under conditions of contact metamorphism (A), burial metamorphism that is followed by regional metamorphism during a continental collision (B), and subduction-related regional metamorphism (C).

# KINDS OF METAMORPHIC ROCK

**How are metamorphic rocks named?**

The names of metamorphic rocks are based partly on their texture and partly on their mineral assemblage. The most widely used names are those applied to metamorphic rocks derived from the sedimentary rocks, shale, sandstone, and limestone, and the igneous rock, basalt. This is because shales, sandstones, and limestones are the most abundant of the sedimentary rocks, while basalts are by far the most abundant of the igneous rocks. We have already encountered some of these names in Figure 10.8, which portrays the metamorphism of a shale. You may also notice that some of the names given to metamorphosed basalts are similar to the names of the metamorphic facies portrayed in Figure 10.14. That's because some of the first work on index minerals, isograds, metamorphic zones, and metamorphic facies was carried out by geologists working in an area of metamorphosed basalts.

In the following section, we discuss first the naming of rocks that always have some degree of foliation and then the naming of certain kinds of rock that often occur without foliation.

## Rocks with Foliation

*Slate.* As discussed previously, the low-grade metamorphic product of shale is **slate**. Slate forms under conditions of regional metamorphism or, less commonly, burial metamorphism. The major minerals present in a shale are clay minerals and quartz. Under conditions of low-grade metamorphism (zeolite facies and the beginning of greenschist facies), muscovite or chlorite forms from the clay minerals and pore fluids. Although the rock may still look like a shale, the tiny new mineral grains produce slaty cleavage. The presence of slaty cleavage is clear proof that a rock has been transformed from a sedimentary rock to a metamorphic rock (Fig. 10.15).

*Phyllite.* Continued low-grade metamorphism of a slate produces both larger grains of mica and a changing mineral assemblage (greenschist facies). The rock develops pronounced foliation and is called **phyllite** (from the Greek word *phyl-*

**slate** The product of low-grade metamorphism of shale.

**phyllite** A metamorphic rock with pronounced foliation, produced by continuing metamorphism of slate.

# Foliated Metamorphic Rocks

SHALE
A sedimentary rock

SLATE
A low-grade metamorphic rock

PHYLLITE
A medium-grade
metamorphic rock

**Figure 10.15**
This series of photographs shows the sequence
of foliated metamorphic rocks developed by
regional metamorphism of a shale. Each of the
photographs shows an area 2.5 x 1.5 cm (1 x
0.6 in). Note how the size of individual mineral
grains increase as the grade of metamorphism increases.

SCHIST
A mica-rich high-grade
metamorphic rock

GNEISS
A mica-poor high-grade
metamorphic rock

*lon*, meaning "leaf"). In a slate the new grains of mica cannot be seen without a
microscope, but in a phyllite they are just large enough to be visible (Fig. 10.15).

***Schist and Gneiss.*** Still further metamorphism to conditions of high-grade
metamorphism (amphibolite and granulite facies) produces **schist**, a coarse-
grained rock with pronounced schistosity (Fig. 10.15). The most obvious difference among slate, phyllite, and schist is in the size of the mineral grains, but increasing grain size is only one of a number of changes—the mineral assemblages
also change (refer to Fig. 10.8). At the high grades of metamorphism that are
characteristic of schists, minerals may start to segregate into separate bands. A
high-grade rock with coarse grains and pronounced foliation, but with layers of
micaceous minerals segregated from layers of minerals such as quartz and
feldspar, is a **gneiss** (pronounced "nice," from the German *gneisto*, meaning "to
sparkle"). The rock shown in Figure 10.4B is a gneiss.

The names *slate* and *phyllite* describe textures. They are commonly used
without adding mineral names as adjectives. The names of the coarse-grained
rocks, schists and gneiss, are also derived from textures, but in these cases mineral names are commonly added as adjectives; for example, we might refer to a
quartz-plagioclase-biotite-garnet gneiss. The difference arises because minerals in
coarse-grained rocks are large enough to be seen and readily identified.

***Greenschist.*** The main minerals in basalt are olivine, pyroxene, and plagioclase, each of which is *anhydrous* (that is, contains no water). When a basalt is
subjected to metamorphism under conditions in which pore water can enter the

**schist** A coarse-grained metamorphic rock with pronounced schistosity.

**gneiss** A coarse-grained, high-
grade, strongly foliated metamorphic rock with micaceous bands.

287

*A.*

*B.*

Figure 10.16
METAMORPHISM OF BASALT
*A.* Low-grade metamorphism of basalt produces a distinctively colored rock called a greenschist, so-called because of the presence of the green, mica-like mineral, chlorite. *B.* Under high-grade metamorphism, a greenschist becomes an amphibolite, a rock devoid of chlorite but rich in amphibole. The area photographed in each case is 2.5 x 1.5 cm (1 x 0.6 in).

rock and form hydrous minerals, distinctive mineral assemblages develop. Under low grades of metamorphism, the resulting rock is equivalent in metamorphic grade to a slate but has a very different appearance. Like a phyllite, it has pronounced foliation, but it also has a distinctive green color because of the presence of chlorite (a green mica-like mineral). It is therefore called *greenschist* (Fig. 10.16A).

***Amphibolite and Granulite.*** When a greenschist is subjected to high-grade metamorphism, chlorite is replaced by amphibole; the resulting coarse-grained rock is an *amphibolite*. Foliation is present in amphibolites but is not pro-

*A.*

*B.*

Figure 10.17
METAMORPHISM DURING SUBDUCTION
When oceanic crust is subducted, it is metamorphosed under conditions of high pressure and moderate temperature.
*A.* Blueschist, the low-grade metamorphic derivative of basalt, owes its color to a bluish-colored amphibole called glaucophane. Blueschists are weakly foliated. *B.* Eclogite, the high-grade metamorphic equivalent of blueschist, contains the pyroxene mineral jadeite, and the red pyrope garnet. The area photographed in each case is 2.5 x 1.5 cm (1 x 0.6 in).

nounced because micas (the platy minerals) are usually absent (Fig. 10.16B). At the highest grade of metamorphism, amphibole is replaced by pyroxene, and a weakly foliated rock called a *granulite* develops.

*Blueschist and Eclogite.* Metamorphism of a basalt under conditions of high stress and moderate temperature, such as the conditions reached by rapidly subducted oceanic crust, produces two very distinctive rocks. *Blueschist* is the low-grade metamorphic product of basalt. Like its close relative, greenschist, blueschist forms under circumstances in which pore water can enter and form hydrous minerals. One of the hydrous minerals is an amphibole called glaucophane, which has a pronounced bluish color (Fig. 10.17A). At the highest grade of metamorphism (which in this case means very high stress but moderate temperature), blueschist is replaced by *eclogite*, a distinctive rock that contains an unusual kind of pyroxene called jadeite, and a bright red garnet, called pyrope (Fig. 10.17B). Jadeite is the mineral name of the prized gemstone, jade (Fig. 10.18).

## Rocks Without Foliation

Two kinds of sedimentary rock sometimes consist almost entirely of a single mineral species and therefore are said to be *monomineralic*. The first is sandstone, a clastic sedimentary rock that is sometimes made up almost entirely of quartz grains. The second is limestone, a chemical sedimentary rock whose only essential constituent is calcite. Neither a pure quartz sandstone nor a pure limestone contains the ingredients necessary to form micas or other minerals that might impart foliation to a metamorphic rock. As a result, quartzite and marble—the metamorphic rocks derived from quartz sandstone and limestone, respectively—often lack foliation.

*Marble.* **Marble** is composed of a crystalline, interlocking network of calcite grains. During the recrystallization of a limestone, the bedding planes, fossils, and other features of sedimentary rocks are largely obliterated. The end result, as shown in Figure 10.19A, is an even-grained rock with a distinctive, somewhat sugary texture. Pure marble is snow white in color and consists entirely of pure grains of calcite. Such marbles, formed as a result of regional metamorphism, are favored for statuary, marble gravestones, and statues in cemeteries, perhaps because white is thought to symbolize purity. It was this kind of pure white marble that Michelangelo chose for the Pietà. Many marbles contain impurities such as organic matter, pyrite, limonite, and small quantities of silicate minerals, which impart various colors to the rock.

*Quartzite.* Sandstone is transformed into **quartzite** when the spaces between the original grains are filled in with silica and the entire mass is recrystallized (Fig. 10.19B). Sometimes the ghost-like outlines of the original sedimentary grains can be seen, even though recrystallization may have almost completely rearranged the original grain structure.

# METASOMATISM

The metamorphic processes that we have discussed so far involve essentially fixed rock compositions and relatively little fluid. The amount of fluid is small

**Figure 10.18**
**METAMORPHIC GEMSTONE**
Jadeite, a mineral characteristically formed under conditions of high pressure and medium temperature during metamorphism associated with subduction zones. This 10 3/4-inch high Olmec specimen is an axe with a carved jaguar face.

**marble** The product of metamorphism and recrystallization of limestone.

**quartzite** The product of metamorphism, recrystallization, and silica cementation of sandstone.

**?** **If pore fluids are important in metamorphism, what happens when a lot of fluid is involved?**

Figure 10.19
NONFOLIATED
METAMORPHIC ROCKS
*A.* Marble is a metamorphic rock composed entirely of calcite. Left is a hand-sized specimen from Tate, Georgia. Note the uniform size of the grains. Right is a microscopic view of an area about 0.6 cm (0.25 in) across, photographed in polarized light. Note the uniformity of grain sizes and the way the grains pack tightly so that there is little or no pore space. All vestiges of the original sedimentary texture in the original limestone have disappeared. *B.* Quartzite is a metamorphic rock composed entirely of quartz. Left is a hand-sized specimen from Minnesota. Right is a microscopic view of an area about 0.6 cm (0.25 in) across. Arrows point to faint traces of the original rounded grains of quartz.

A.

B.

**metasomatism** The process whereby the chemical composition of a rock is altered by the addition or removal of material by solution in fluids.

because the pore space in rock that is undergoing metamorphism tends to be small and because even the most abundant minerals that contain water have relatively small amounts. Geologists use the expression "a small water–rock ratio" to describe the typical metamorphic environment. What they mean is that the weight ratio of fluid (mainly water) to rock is about 1:10 or less. This is enough fluid to facilitate metamorphism but not enough to dissolve a lot of the rock and thus change the rock composition noticeably.

In a few circumstances, however, large water–rock ratios occur. One example is an open rock fracture through which a lot of fluid flows. Under such conditions the water–rock ratio can be 10:1 or even 100:1. The rocks adjoining the fracture can be drastically altered by the addition of new material, the removal of material to solution, or both. The term **metasomatism** (from the greek *meta*, meaning "change," and *soma* from the Greek, meaning "body") is applied to the process whereby the chemical compositions of rocks are distinctively altered by the addition from, or the removal of, material in solution. Note the difference in meaning between *metamorphism*, which refers to the change of *form* (meaning mineral assemblage and texture), and *metasomatism*, which is the change in *body* (meaning the composition) of a rock that results from the addition or subtraction of material. Note too, that a rock may be simultaneously subjected to both metamorphism and metasomatism.

Metasomatism is commonly associated with contact metamorphism, especially when the rocks being metamorphosed are limestones, as in the example shown in Figure 10.10. Metasomatic fluids released by a cooling magma move outward, passing through the pores of the rock undergoing contact metamorphism. Because the fluids may carry constituents such as iron and magnesium in

Figure 10.20
**METASOMATISM**
Metasomatism of a limestone produced this colorful rock. White is calcite, red is garnet, silvery grey is muscovite. The sample is 2.5 × 1.5 cm (1 × 0.6 in).

solution, the composition of limestone that is close to the cooling magma can be drastically changed. Limestone that is distant from the magma—beyond the reach of invading fluids—remains unchanged.

Figure 10.20 is a photograph of a contact metamorphic rock that was originally a limestone. Without the addition of new material, the limestone would have become a marble, but through metasomatism it was changed into an assemblage of garnet, a green pyroxene called diopside, and calcite. Metasomatic fluids may also carry valuable minerals in solution and form mineral deposits, a topic that is discussed in chapter 16.

# WHAT'S AHEAD

The rock cycle (Fig. 7.1) works because two great energy sources, the Sun and the Earth's internal heat, interact. Sediment and sedimentary rock are formed as a result of processes such as weathering, erosion, and diagenesis that operate on or near the surface of the continental crust, and the energy that drives near surface processes is the Sun's heat energy. When sediment accumulates, it does so in basins created by tectonic processes. The formation of basins and the conversion of sedimentary rock into metamorphic rock happens as a result of the Earth's internal heat. The tectonic forces that elevate metamorphic rocks in mountain ranges, thereby exposing the rocks to weathering and erosion, and starting the material in the rocks through the rock cycle again, are also driven by the Earth's internal heat. Based on the fact that metamorphic rocks derived from sedimentary rocks are present among the most ancient rocks ever found on the Earth (4.0 billion years old), we have to conclude that the processes of the rock cycle have been operating throughout most of the Earth's history.

In chapter 11 we will discuss in greater detail the evidence that links plate tectonics, metamorphism, and the other processes of the crustal rock cycle. We will also discuss why sedimentary rocks are formed in specific places, why different kinds of metamorphic rocks are found at particular locations, and what relationships exist among plate tectonics and igneous, sedimentary, and metamorphic rocks.

# Chapter Highlights

**1.** New rock textures and new mineral assemblages develop when rocks are subjected to elevated temperatures, most often in association with elevated pressure. All such changes that take place while the rock remains solid are said to be the result of **metamorphism.**

**2.** Metamorphic changes, which are caused by the Earth's internal heat, start at about 150°C (300°F) (equivalent to a depth of 5 km or 3.1 mi) and cease at about 800°C (1472°F), where melting and igneous processes start.

**3.** Sedimentary rocks are affected by diagenetic changes down to a depth of about 5 km below the Earth's surface. From 5 km (3.1 mi) to 15 km (9.3 mi), **low-grade** metamorphic rocks are formed. From 15 km (9.1 mi) to the depth at which melting starts, between 30 and 40 km (18.6 and 24.9 mi), is the region of **high-grade** metamorphism.

**4.** The results of metamorphism are influenced by the initial composition of the rock, as well as by temperature, pressure (stress), the presence or absence of pore fluids, and how long the rock is heated and squeezed.

**5.** A distinctive planar texture in metamorphic rocks is known as **foliation**; it is caused by differential stress during metamorphism. Foliation developed in low-grade metamorphic rocks is termed **slaty cleavage**. Foliation developed in high-grade metamorphic rocks is termed **schistosity**.

**6.** The growth of new mineral assemblages during metamorphism is controlled primarily by temperature.

**7.** High temperature, high pressure, abundant pore fluid, and long reaction times produce metamorphic rocks with large mineral grains. Metamorphic rocks containing small mineral grains usually form at lower temperature and pressure, but can also form under more extreme conditions of temperature and pressure if pore fluid is scarce or reaction time is short.

**8.** Metamorphism due to elevated temperature around an igneous intrusion is called **contact metamorphism.** Under such conditions, stress tends to be uniform or nearly so; therefore, contact metamorphic rocks are not strongly foliated.

**9.** Sedimentary rocks that are saturated with water and subjected to sufficiently elevated temperature and pressure due solely to their burial under overlying strata undergo **burial metamorphism.**

**10.** **Regional metamorphism** occurs under differential stress and elevated temperatures as a result of collisions between tectonic plates. Regionally metamorphosed rocks commonly extend over areas of thousands of square kilometers.

**11.** For a given rock composition, the assemblages of minerals that are formed under a given set of temperature and stress conditions are the same, no matter where in the world metamorphism happens. This fact has allowed geologists to develop the concept of **metamorphic facies,** which is that the different mineral assemblages that form in rocks of different chemical composition, under a given set of temperature and stress conditions, belong to the same metamorphic facies.

**12.** The names of metamorphic rocks are based partly on texture and partly on composition. Foliated metamorphic rocks derived from shales are **slate, phyllite, schist,** and **gneiss.** (Each of these names refers to a higher metamorphic grade.) Schists and gneisses are of the same grade, but schists are strongly foliated, whereas gneisses are less so. Foliated rocks derived from basalts are greenschists and amphibolites. Weakly foliated metamorphic rocks derived from basalts under conditions of high pressure and moderate temperatures are blueschists and eclogites. Nonfoliated metamorphic rocks are **marble** and **quartzite.**

**13.** Rocks whose composition is distinctly changed during metamorphism as a result of high water–rock ratios have undergone **metasomatism.**

## ▶ The Language of Geology

# ▶ Questions for Review

1. How do metamorphic rocks differ from sedimentary and igneous rocks?
2. At what approximate temperature and pressure in the Earth does metamorphism commence?
3. Name five factors that influence the rock texture and mineral assemblages produced by rock metamorphism.
4. What roles do pore fluids play in metamorphism?
5. Describe how uniform stress and differential stress differently influence the development of metamorphic textures.
6. How does slaty cleavage differ from schistosity?
7. Some metamorphic rocks consist of large mineral grains (thumbnail size or larger), whereas others are made up of mineral grains so small they can hardly be seen. What controls the size of mineral grains in metamorphic rocks?
8. What are the differences among contact, burial, and regional metamorphism?
9. How are index minerals and isograds employed in the mapping of regionally metamorphosed rocks?
10. Why are some metamorphic rocks foliated while others are nonfoliated?
11. What is the difference between a limestone and a marble? A schist and a gneiss? A shale and a slate?
12. What is metasomatism and what role does it play in metamorphism?

# ▶ Questions for Thought and Discussion

1. Why and how do the pressure and temperature of rocks that are undergoing regional metamorphism change? Sketch how pressure and temperature might change with time for rocks being subjected to contact metamorphism and subduction-related regional metamorphism.
2. Suppose a large meteorite struck a pile of clastic sedimentary rocks. How would pressure and temperature change in the rocks involved? Would the effects be the same for rocks on the Moon as for rocks in a sedimentary basin on the Earth?
3. Geologists often say, "Look at the texture. It will tell you whether or not the rock has been subjected to metamorphism." Why would they make such an assertion?
4. Use Figure 10.8 to find mineral assemblages characteristic of slate, phyllite, and high-grade schist.
5. Compare the concept of metamorphic facies to that of sedimentary facies (chapter 9). In what ways are they similar? In what ways are they different?

# · 11 ·

# THE ROCK CYCLE REVISITED

**Charles Darwin's name is forever entwined with evolution.** His seminal ideas came from observations made during the globe-encircling voyage of the H.M.S. *Beagle* in the 1830s. But Darwin made important geological observations too—in fact, he dedicated the diary of his voyage to Charles Lyell, the most famous geologist of the time.

On February 20, 1835, in Valdivia, Chile, Darwin experienced a great earthquake; he immediately investigated the changed landscape and discovered "putrid mussel shells still adhering to the rocks, ten feet above high-water mark; the inhabitants had formerly dived at low-water spring-tides for the shells." This, he realized, could explain an earlier observation, made near Valparaiso, Chile, where he had found loose shells in marine mud 400m (1300 feet) above sea level. The Andes, he realized, must have been pushed up bit by bit, earthquake by earthquake, and he wondered at the nature of the "force which has upheaved these mountains." He immediately understood that slow uplifting of the Andes could explain the origin of the vast sediment-covered plains he had seen in Patagonia, Argentina. At the time he had wondered "how any mountain-chain could have supplied such masses, and not have been utterly obliterated." He realized that a connection existed between tectonic uplift and the deposition of sediment.

By observation and deduction, Darwin came to understand that over geologic time there is a balance between the tectonic cycle that raises mountains and the erosive forces of the rock cycle that wears them away.

◆

### In this chapter you will learn about

- The processes that change and shape the Earth's surface
- The complex interactions between the rock cycle and the tectonic cycle
- How and where different kinds of igneous rocks form
- How plate tectonics controls where different kinds of rocks form
- Why metamorphism occurs along the margins of continents
- Where sediment accumulates, and why

*Snow Fun*

Skiers enjoy the deep snow and spectacular scenery of the Gremlin's Cap Patagonian range in Chile.

Space exploration has changed forever the way we think about the Earth. Our planet is special in a number of ways, as we learned in chapter 1. But it has much in common with the other rocky planets. The same kinds of processes that slowly change and shape the Earth also change and shape the other rocky planets. There are four kinds of planet-shaping processes, and they differ greatly in importance from one planet to another, but they operate on all the rocky planets and rocky moons in the solar system. We begin this chapter with an overview of the great planet-shaping processes.

## PLANET-SHAPING PROCESSES

These are the four kinds of planet-shaping processes:

1. Impact cratering, which is caused by fast-moving extraterrestrial objects such as meteorites, asteroids, and comets.
2. Tectonic processes, which are driven by a planet's internal heat energy; these include continental drift and the uplift of mountain ranges.
3. Magmatic processes such as the eruption of basaltic lava flows and the formation of new oceanic crust at midocean ridges.
4. The combined effects of weathering, mass wasting, and erosion, through which highlands are gradually worn down.

Together, these planet-shaping processes produce the tectonic cycle and the rock cycle, the two most important cycles in the Earth system. In earlier chapters you have learned about the Earth's internal structure, about rocks and minerals, and about basic rock-forming and rock-modifying processes. In this chapter we bring together all of this information. We will look at each of the major groups of rocks and place them in the context of the rock cycle and the tectonic environments in which they are formed. First, however, let's look briefly at the relative importance of the two most fundamental planet-shaping processes: impact cratering and plate tectonics.

### The Role of Impact Cratering

Recall from chapter 1 that the rocky planets and the rocky moons were formed by the accretion of small mineral particles that condensed from a cosmic cloud of gas. During the later stages of growth, the rocky planets were ceaselessly and violently pelted by objects of many shapes and many sizes. This period of solar system history, which is sometimes called the time of *early intense bombardment*, played a very important role in the rocky planets' early history. The impacts contributed mass to the growing planets; the energy of the collisions transferred vast amounts of heat to them; and they disrupted and fragmented newly solidified crust. On Earth, these early collisions may even have influenced the origin and early evolution of life.[1]

---

[1] Complex organic (carbon-based) molecules—which may have been the precursors of life on Earth—have been identified in some types of meteorites. This suggests that meteorites may have delivered organic material to the Earth's surface. On the other hand, by disrupting the newly forming crust and causing devastating environmental changes, they may also have delayed the development of life-supporting habitats on the Earth.

**Figure 11.1**
**MESSENGER FROM SPACE**
This meteorite fell near Bruderheim, Alberta, Canada, March 4, 1960. The black skin formed as a result of intense heating as the meteorite plunged through the atmosphere. The fresh interior of the meteorite can be seen where a sample was cut off for scientific study.

By about 4.5 billion years ago, when the Earth had nearly reached its present size, the rate of collisions had greatly decreased. But the process of planetary accretion did not stop then. It still continues today, though at a very slow pace. Hundreds of small extraterrestrial bodies moving at speeds of up to 100,000 km/hr (62,000 mi/hr) reach the Earth every day. Most of them are very small particles; these are heated so strongly as they are slowed down by the atmosphere that they vaporize completely in a burst of light—we see them as "shooting stars." The atmosphere has a lesser effect on the much larger objects that collide with the Earth once in a while (not every day). Only the outermost parts of such objects are vaporized during entry, and the inner part of such an object may have been only slightly heated by the time it reaches the ground. (Remember that rock is not a good conductor of heat.) We call these fragments *meteorites* (Fig. 11.1). Once in a long while a very speedy extraterrestrial object is large enough to punch right through the atmosphere, slam into the ground, and create a huge, circular **impact crater** (Fig. 11.2). Some of the largest impact craters on the Earth are more than 200 km (about 125 mi) in diameter, and even larger ones have been found on other planets.

High-speed impacts can shatter huge amounts of crust, causing instantaneous high-pressure metamorphism, sometimes even rock melting. They also eject vast amounts of shattered rock from the crater. One famous crater, the Ries

**impact crater** An excavated depression formed on a planetary surface by the collision of a large meteorite.

**Figure 11.2**
**IMAGINE IF IT HIT A CITY!**
Meteor Crater, near Flagstaff, Arizona, was formed by a meteor impact about 50,000 years ago, more than 30,000 years before humans are thought to have arrived in North America. The crater is 1.2 km in diameter and 200 m deep. Note the raised rim and the blanket of broken rock thrown out of the crater. Many even greater impacts are believed to have occurred during the Earth's early history (see Fig. 1.8).

**?**

**What kind of evidence would you look for if you suspected that a large circular structure might be an impact crater?**

Crater in southern Germany, is home to the old walled town of Nördlingen. The crater is 24 km (about 15 mi) in diameter and was formed about 15 million years ago as a result of an impact so intense that many of the rocks in and around the crater now contain high-pressure minerals, one of which is diamond. The diamonds are microscopic in size and too small to be economically useful, but many of the buildings in Nördlingen are built of diamond-bearing rock. The cathedral at Nördlingen (Fig. 11.3) is estimated to contain as much as 5 kg, or 25,000 carats, of diamond!

The intensely cratered surfaces of other rocky planets and of many moons provide clear evidence that impact cratering was once a major planet-shaping process everywhere in the solar system (Fig. 11.4). Early in the Earth's history, during the Hadean Eon, and possibly even into the earliest part of the Archean Eon, impact cratering was the dominant surface-shaping process on the Earth, too. More than two hundred impact craters have been discovered on the Earth, some as much as 2 billion years old. But evidence from other planets suggests that thousands, even tens of thousands, of impact craters must once have existed on the Earth. Apparently, volcanism, tectonics, weathering, and erosion have been very efficient in removing craters and thus obscuring the evidence of impact.

Geologists are now trying to answer a difficult question: If weathering and erosion remove the craters, or if a crater is later buried beneath lava flows or piles of sedimentary strata, what other evidence of the planet-shaping effects of ancient impacts might remain? One hypothesis is that very large impacts can lead to the extinction of numerous biological species and that the fossil record might provide evidence of ancient impacts. That is a likely explanation for the demise of the dinosaurs at the end of the Cretaceous Period. Of course, evidence of major extinction events in the fossil record can best be found in rocks of the Phanerozoic Eon, the only eon for which there is an abundant fossil record (as

**Figure 11.3**
**HOLY DIAMONDS**
Nördlingen Cathedral, Germany, sits in the center of an ancient impact crater. It is constructed of rocks that were metamorphosed by the heat and high pressure created by the impact. One of the high-pressure minerals formed by the impact event is diamond. The building stones of the Cathedral are estimated to contain 25,000 carats of tiny diamonds.

**Figure 11.4**
**ANCIENT CRATERS**
The face of the Moon, as seen from the Earth, provides a record of the intensity and frequency of impact events in the early history of the solar system. Early telescope observers called the dark-colored lowland areas *maria*, Latin for "seas." The basaltic lavas filled craters produced by exceptionally large impacts. The sites of the six *Apollo* manned landings during 1969–72 are indicated. Note the young crater on the lower left of the photograph. The rays spreading out from the crater are made of debris splashed out by the impact.

you learned in chapter 3, and about which you will learn more in chapter 15). Another hypothesis is that some great outpourings of lava, such as those that formed the Deccan Traps in India, might have happened when deep-seated magma rose through holes punched in the crust by giant impacts. New evidence to support these and other hypotheses about impact cratering awaits further geologic research.

## The Tectonic Planet

Although impact cratering may once have been the Earth's dominant surface-shaping process, that honor now belongs to the movement of tectonic plates. As you learned in chapter 4, tectonic plates move as a result of the steady escape of internal heat through the process of convection. Heat escapes from all the rocky planets, and their surfaces indicate long histories of tectonic deformation through such processes as uplift and faulting. However, no other planet seems to have been shaped by Earth-style plate tectonics.[2]

Exactly when the shift from shaping by impacts to shaping by plate tectonics occurred is not known, but it must have happened a long time ago. Geologic evidence suggests that plate tectonics has been the dominant Earth-shaping process for the last 3 billion years and probably even longer. Evidence of present-day plate tectonics comes mainly from oceanic crust, but as shown in chapter 4, all of today's oceanic crust is young—less than 200 million years old. Therefore, evi-

**?** **Where does the energy needed to move tectonic plates come from?**

___
[2] A planet's style of tectonic deformation depends on factors such as the size and composition of the planet, and the rate of heat loss from the planet's interior.

Figure 11.5
**THE STRUCTURE OF A CONTINENT**
An assemblage of ancient cratons and orogens, all older than 1.6 billion years, forms the core of North America. Surrounding the core are younger orogens. The Grenville is about 1 billion years old, while the Caledonide, Appalachian, Cordilleran, and Innuitian orogens are all less than 600 million years old.

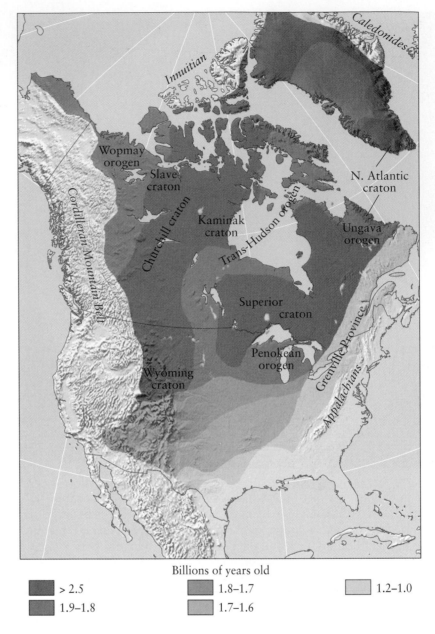

Billions of years old

> 2.5    1.8–1.7    1.2–1.0
1.9–1.8    1.7–1.6

**orogen** An elongate region of the crust that has been intensely deformed and metamorphosed during a continental collision.

dence of plate tectonics occurring more than 200 million years ago must be sought in the older rocks of the continental crust. Part of this evidence is found in the way continents are constructed. They are made up of two kinds of structural units, called *cratons* and *orogens* (Fig. 11.5).

**Orogens** are elongate regions of the crust that have been intensely deformed and metamorphosed during a continental collision. (The terms *orogen* and *orogenesis*—the process of mountain-building—come from the Greek words *oros*, meaning "mountain" and *genesis* meaning "to come into being.") The Alps and the Himalayas are orogens that are still active and still being deformed and uplifted. Through radiometric dating, some orogens have been found to be very old—perhaps as old as 4.0 billion years. This provides evidence confirming the antiquity of lithospheric plates, plate collisions, and the processes of mountain-building. Ancient orogens, now deeply eroded, betray their history through a combination of deformational and metamorphic features (Fig. 11.6).

Figure 11.6
ANCIENT OROGEN
This is the eroded remains of a 300 million year old orogen. The Appalachian Mountains of central Pennsylvania are part of an orogen formed in the Paleozoic Era. The false color satellite image shows dense forest growth in red. Forested ridges are underlain by rocks such as sandstone and quartzite that resist erosion. Pale blue denotes mixed agricultural land in the valleys. Soft, easily eroded rocks such as shale and limestone underlie the valleys.

Orogens are subject to vertical motions because of **isostasy**, the flotational balance of the lithosphere on the asthenosphere. Remember that the asthenosphere is so hot and weak that it is easily deformed and acts like a liquid in comparison to the rigid lithosphere overlying it. Refer back to Figure 1.9 for a reminder that continental lithosphere consists of low-density continental crust sitting on higher density mantle rock, and that the boundary between the lithosphere and the asthenosphere is in the upper mantle. The density of continental lithosphere—that is, lithosphere capped by continental crust—is less than the density of the underlying asthenosphere, so the lithosphere floats on the asthenosphere.

Continental collisions cause the continental crust to thicken. Beneath every great mountain range is a root of thickened continental crust that gives that portion of the continental lithosphere enough extra buoyancy to support the weight of the mountain range. Thus, a mountain range is a thickened capping of continental crust on a block of lithosphere. Like an iceberg floating in the sea, thickened continental lithosphere floats high.

As erosion weathers away the top of the mountain range, like sunshine melting the top of an iceberg, the entire block of continental lithosphere, including the root at the base of the range, rises so as to maintain flotational—that is, isostatic—balance. This is why mountain ranges can still show some topographic relief after millions of years of erosion. Seawater lacks strength, so a melting iceberg immediately bobs up. The hard, rocky lithosphere has a lot of strength, however, and a root pushing up from below an eroded mountain range has to overcome the strength of the rock on either side of the root—a very slow process. Therefore, an orogen—even though deeply eroded—may still retain a root and still be subject to isostatic adjustments hundreds of millions of years after the collision that formed it (Fig 11.7).

**isostasy** The flotational balance of the lithosphere on the asthenosphere.

**Why are the Appalachians still mountainous hundreds of millions of years after they were formed?**

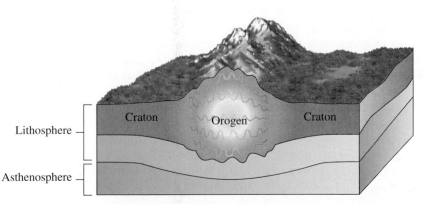

Figure 11.7
ADJUSTING THE HEIGHT OF CONTINENTS BY ISOSTASY
The rigid lithosphere floats on the plastic asthenosphere. In a collision of two cratons, the lithosphere is thickened and the lithosphere-asthenosphere boundary is pushed down. As erosion slowly reduces the height of the mountains in the orogen, the root slowly rises to compensate for the removal of mass. This continuing adjustment to maintain buoyancy, like a cork finding its level in water, is called isostasy.

**craton** A portion of continental crust that has been stable and free from deformation for a very long time.

**Cratons** are portions of the continental crust that have been stable and free from deformation for a very long time (at least a billion years). By stable we mean that they have reached a condition of isostatic balance. Rocks in cratons may or may not have once been parts of mountain ranges. Once weathering and erosion have reduced the surface almost to sea level, and any root that may once have existed has disappeared, there is no tendency for any further isostatic adjustment to occur. A craton is like a log of wood floating in a still pond. Cratons form the cores of all continents. As the cratons were assembled into larger masses of continental crust, rocks between the cratons formed orogens.

Let's summarize. Long ago, plate tectonics succeeded impact cratering as the principal planet-shaping process on the Earth. Ancient orogens establish the antiquity of plates, plate collisions, and mountain-building processes. The isostatic balance of ancient cratons proves that isostatic adjustment, too, has been going on for an exceedingly long time. If the lithosphere did not float on the asthenosphere, and if the lithosphere did not break into plates, plate tectonics could not happen. The formation of most kinds of igneous, metamorphic, and sedimentary rocks is controlled either directly or indirectly by plate tectonic processes. Therefore, the distribution of most kinds of rocks on the Earth can best be understood in terms of plate tectonics.

# IGNEOUS ROCK AND PLATE TECTONICS

Igneous rock is crystallized magma. Therefore, the distribution of igneous rock reflects how and where different magmas were formed. We will first review some of the key aspects of magma formation (refer back to chapter 6 for more details). These basic concepts will come in handy as we examine different types of igneous rocks in the context of the tectonic environments in which they occur.

Recall that magma is formed by the melting of rock, which happens in two ways (Figs. 6.4 and 11.8):

1. *When the temperature rises.* At a high enough temperature, all rocks will melt.

2. *When the pressure drops.* The temperature at which any dry sample of rock will begin to melt increases with pressure. When a hot mass of rock is under pressure and the pressure suddenly decreases, *decompression melting* can occur.

Two special factors require comment. The first concerns the role of water, which lowers the melting temperature. Just as salt lowers the melting temperature of ice, water lowers the melting temperature of rock. This is called *wet melting* (as opposed to *dry melting*, which occurs in the absence of water). Pressure enhances the effect of water on melting.

The second factor, partial melting, also plays an important role in magma formation (see Fig. 6.6). Because rocks are assemblages of minerals, they do not melt at a single temperature, the way ice (a single mineral) does. Rather, rocks melt over a temperature range, called a *melting interval*, that may span several hundred degrees centigrade. The ratio of melted rock (magma) to unmelted rock slowly changes across the melting interval. At the lowest temperature there will be only a tiny amount of magma. (The melting curves shown in Fig. 11.8 are, for simplicity, the "beginning of melting" curves.) At the highest temperature, when melting is almost complete, only a small amount of solid material will remain. As

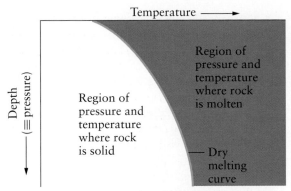

A. Pressure—The *dry melting temperature* of a rock is pressure sensitive. The higher the pressure, the higher the melting temperature. The *dry melting curve* is the transition line that divides two temperature-pressure regions: where the rock is solid (left) and where it is molten (right).

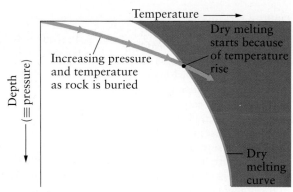

B. Temperature—When rock is buried, both temperature and pressure rise. At a sufficiently high temperature, dry melting will start.

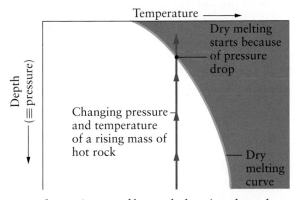

C. Pressure drop—A mass of hot rock that rises through the mantle remains at high temperature, but the pressure continually decreases. At some depth, the pressure becomes low enough for dry melting to start. This is called *decompression melting*.

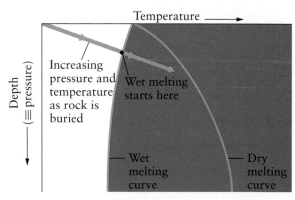

D. Add water—The presence of water changes the melting curve of a rock dramatically. Water makes it possible for rocks to melt at much lower temperatures, or nearer the surface. Under some conditions, wet melting can start in the crust at depths as low as 25 km (about 16 mi).

Figure 11.8
CONDITIONS FOR ROCK MELTING INSIDE THE EARTH
Inside the Earth, rock melts under different conditions of temperature, pressure, and the presence of water. Four situations are shown, each drawn on a diagram in which temperature is on the horizontal axis and pressure, which is proportional to depth, is on the vertical axis. Rock melts across a temperature interval; for simplicity, the melting curves in the diagrams show the beginning of melting.

the ratio of magma to unmelted rock changes across the melting interval, the composition of the magma changes too. The composition of magma therefore is determined by the proportion or fraction of a rock that melts. Geologists loosely refer to a "10-percent *partial melt*" or a "10-percent *fractional melt*" when they mean that 10 percent of a rock has melted to form a magma.

## Igneous Rocks of the Ocean Basins

Underlying the ocean basins is a crust of basaltic composition. New crust is formed at spreading centers which coincide with the midocean ridges; these

Figure 11.9
**MAKING MORB**
The magma that makes MORB (midocean ridge basalt) is formed by decompression melting of the asthenosphere. When the lithosphere splits to create a new spreading center, (1) the asthenosphere rises into the fracture; (2) pressure is reduced and melting begins; and (3) the emerging lava chills and solidifies to form new oceanic lithosphere.

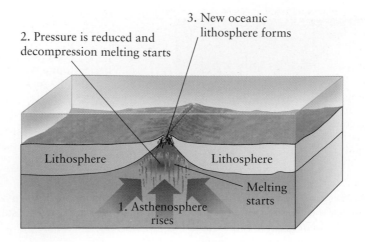

2. Pressure is reduced and decompression melting starts

3. New oceanic lithosphere forms

Lithosphere

Lithosphere

Melting starts

1. Asthenosphere rises

**MORB** An acronym for midocean ridge basalt, the magma that rises at seafloor spreading centers and solidifies to form oceanic crust.

**?** **What is the most common type of igneous rock on the Earth's surface?**

**ophiolite** Fragments of oceanic crust found on continents.

spreading centers are places where the lithosphere is splitting apart (see chapter 4). Beneath a midocean ridge the boundary between the lithosphere and the asthenosphere rises close to the surface, as you can see in Fig. 11.9. Hot rock of the asthenosphere moves upward, squeezing into the fracture zone. As it does so, decompression melting produces magma (Fig. 11.9). The magma that solidifies to form the basalt of the oceanic crust varies little in composition around the world and is simply referred to as **MORB,** an acronym for *MidOcean Ridge Basalt.*

## MORB

Even though its composition is nearly constant around the world, MORB can reveal much to geologists about the processes by which oceanic basins and oceanic crust are formed. The decompression melting that produces MORB magmas starts in the mantle when the upward-moving rocks of the asthenosphere reach a depth of about 65 km (40 mi). A 5-percent partial melt of mantle rock produces a magma of MORB composition. (Remember that the degree of partial melting influences the composition of magma; if a larger or smaller fraction of the mantle rock melts, the resulting magma will not be MORB.) The reason MORB is similar everywhere is the similarity of the tectonic environment in which these magmas are formed—the depth of decompression melting and the degree of partial melting vary little from one midocean ridge to another.

The ocean basins cover 71 percent of the Earth's surface. Consequently, MORB is the most common igneous rock at the surface. Common though it is, MORB is not easy to observe; the ridge and sea floor are everywhere covered by water except in a few places such as Iceland, where the midocean ridge stands above sea level. There is, however, one circumstance in which it is possible to see MORB above sea level: in places where a plate collision has caught up and crushed a fragment of oceanic crust between two colliding continental masses, like a nut between two sides of a nutcracker.

## Ophiolites

When MORB is found on land, it usually consists of the fragmented remnants of a portion of oceanic crust that had been caught up in the collision process. In this process, the minerals that are characteristic of basalt are transformed into an assemblage dominated by a distinctive green, fibrous mineral called *serpentine* (Fig. 11.10). These serpentine-dominated fragments of oceanic crust found on continents are called **ophiolites** from the Greek word for serpent, *ophis.*

Figure 11.10
**SERPENTINE REVEALS AN OLD
COLLISION ZONE**
Serpentine is formed by alteration of olivine and pyroxene present in the parent rock. This specimen is from an
ophiolite in Marin County, California. The green color
is typical of the mineral serpentine, and the shattered
nature of the rock reflects the grinding and crushing of
rocks caught in a subduction zone. The photograph is
about 30 cm (1 ft) across.

Ophiolites are studied intensively because they provide evidence of ancient
plate collisions. They also supply convenient samples of oceanic crust, which
otherwise is rather difficult to study. Ophiolites tend to be quite similar wherever
they are found. Figure 11.11 is an idealized cross section of an ophiolite—and,
therefore, an idealized cross-section through a typical segment of oceanic crust.
At the top is a thin veneer of sediments that were deposited on the ocean floor.
Beneath the sediments are layers of *pillowed* basaltic lavas—pillows indicate that
the basaltic lava was extruded under water. Still deeper are sills of gabbro (the
plutonic equivalent of basalt). Cutting through the sills and pillow lavas are innumerable vertical dikes of gabbro. The basalts and gabbros are similar in com-

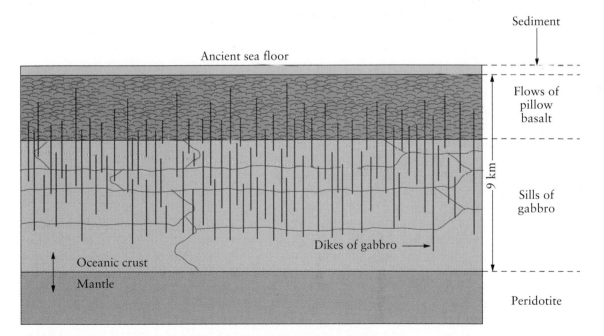

Figure 11.11
**OPHIOLITE: FRAGMENTS OF OCEANIC CRUST ON LAND**
This is an idealized section of oceanic crust reconstructed from observations of an ophiolite. A thin layer of
sediment overlies pillow basalts and a thick pile of gabbro sills. Both the lavas and the sills are intruded by
gabbro dikes. The composition of the basaltic lava flows, sills, and dikes is the same: MORB.

# The Fascinating World of the Midocean Ridge

**How curious it is that the most widespread kind of volcanism** on the Earth is also the least spectacular and the most difficult to see. The location of all this volcanism is the midocean ridge, a vast underwater volcanic mountain chain that snakes for 65,000 km (about 40,000 mi) through the oceans of the world and marks the spreading edges of the tectonic plates. If we could somehow remove all the water from the ocean, we would see that the midocean ridge stands, on average, 0.6 km (0.37 mi) above the deep sea floor and is the world's dominant topographic feature. We don't know how many active volcanoes there are beneath the sea, but given what has been discovered about active volcanic centers in the few well-studied parts of the ridge, there must be many thousands, possibly even tens of thousands. By contrast, there are only about 800 active volcanoes on land.

Down the center of the midocean ridge runs a narrow rift, generally less than 3 km (1.8 mi) wide and about 400 m (440 yds) deep. The rift is not a smooth valley; rather, it is rough, broken by innumerable fractures in the brittle oceanic crust, covered by tumbled piles of pillow lava, and every so often capped by a large volcanic mound (Fig. B11.1). With so much volcanism at the bottom of the sea, it is hardly surprising that there are also lots of submarine hot springs along the midocean ridge. The hot springs happen when seawater enters the oceanic crust through open fractures, becomes heated by the magma that gives rise to the intense volcanism, and then rises convectively to the seafloor where it vents as jets of superhot water at temperatures of up to 400°C (750°F) (Fig. B11.2). The combination of submarine volcanism and hot springs releases so much heat to the oceans that the midocean ridges are where about 10 percent of the Earth's internal heat escapes.

**Figure B11.1**
**VOLCANIC SEAFLOOR**
The midocean ridge is split by a long, narrow valley, and in the valley volcanism is very active. This photo of the seafloor taken from *Alvin*, a deep-diving submarine, shows a pile of basaltic pillows formed as a result of a submarine eruption.

Submarine hot springs bring more than just hot water to the sea floor. As the water in the fractures becomes heated, it dissolves chemicals, especially silica and metals such as iron, copper, cadmium, and zinc, from the rocks with which it is in contact. When the hot water rises to the sea floor, it carries a veritable cocktail of dissolved matter, some of which precipitates when the hot water meets the cold water of the sea. The precipitate makes the rising plume of hot water a murky black color, giving rise to the popular name for the hot springs, "black smokers." The dissolved matter that does not precipitate mixes into seawater, and in this way midocean ridge volcanism plays a major role in balancing the chemical composition of seawater—another example of a critical interaction between different parts of the Earth system.

Figure B11.2
**SUBMARINE HOT SPRING**
This is a submarine hot spring at a depth of 3080 m (10,160 ft) below sea level on the Mid-Atlantic Ridge. Viewed from the deep-diving submarine *Alvin*, the hot spring is known as a "black smoker" because the hot water precipitates a cloud of tiny mineral particles as it is cooled by the seawater. The temperature of the spring water is 185°C (365°F).

More startling and more profound than the hot springs is the presence of a huge assemblage of exotic animals that live around the springs—shrimp, curious crabs, giant worms, and clams the size of dinner plates, but also myriad tiny primitive bacteria (Fig. B11.3). Their world is one of Stygian darkness, and many of them are blind. They rely on the heat and the minerals dissolved in the hot waters for energy and food, and they live a life that is so different from their near surface relatives that biologists have suggested that among the microscopic forms may be some of the most primitive living creatures. Perhaps here, on the mid-ocean ridge, may be the evidence needed to answer the question, "Where and how did life arise?" The argument goes as follows: If, as many scientists now suspect, life arose as long ago as 4 billion years, it must have done so in an environment where it was at least partly sheltered from the early, intense bombardment of the Earth by meteorites; where better than on the sea floor, sheltered by at least 2000 m (1.2 mi) of seawater, and adjacent to a nurturing hot spring!

Figure B11.3
**LIVING IN A HELLISH ENVIRONMENT**
Unusual life forms abound in the seafloor rift environment. Here, a colony of giant tube worms clusters around a hot spring.

**?** **Can you remember the name of the boundary between the crust and the mantle? (See chapter 5.)**

position to MORB and clearly originated from the same magma. In fact, some of the gabbro dikes and sills were probably conduits through which the basaltic magma ascended from the upper mantle to the ocean floor.

Beneath the gabbro sills in an ophiolite there is often rock with a very different composition from that of MORB. This is *peridotite,* an olivine- and pyroxene-bearing rock that is characteristic of the mantle. Ophiolites, therefore, provide samples of the oceanic crust and, in a few cases, rock of the upper mantle. The contact between the gabbro at the base of the oceanic crust and the peridotite of the upper mantle is, of course, the Moho (chapter 5).

### Mantle Plumes

In the oceans there are many shield volcanoes, some of which rise above sea level to form volcanic islands. The Hawaiian islands are probably the best known examples, but there are many others. These volcanoes have many distinctive features, but two are especially important. First, the magma of these volcanoes is similar in composition to the magma from which MORB forms and therefore is probably of similar origin. The second distinctive feature is that each of today's active oceanic shield volcanoes is simply the youngest in a chain of volcanoes that were formed as the plate carrying the volcanoes moved over a stationary magma source. (For a more detailed explanation, refer to the discussion of the Hawaiian Island chain in Box 4.1.)

Why are these features significant? The magmas that form shield volcanoes like those of Hawaii are similar to MORB magmas but not exactly the same; ordinary MORB magmas contain less magnesium and more aluminum than the basaltic magmas of the oceanic shield volcanoes. The differences in composition cannot be fully explained by differing degrees of partial melting. Geochemical experiments have shown that the basaltic shield volcano magmas must have been formed at higher pressures, and therefore deeper in the mantle, than MORB magmas. A long chain of volcanoes all originating from the same kind of magma implies a long-lived, stationary, deep-seated magma source. The most plausible explanation is that a *plume* of hot rock rises in a long, thin stream from deep in the mantle, perhaps as deep as the core–mantle boundary. Reaching a depth of 100 to 200 km (about 60 to 120 mi) below the surface, the rock in these plumes begins to undergo decompression melting. The rising plumes, rooted in their sources deep within the mantle, are thought to be part of the great mantle convection system discussed in chapter 4.

**?** **What is the most common type of rock in the solar system?**

Ocean-floor basalts can also tell us some interesting things about rocks on other planets. Space exploration has revealed that basalt is the most abundant type of igneous rock on each of the terrestrial planets. This suggests that basalt is the most common igneous rock in the solar system and that decompression melting is the most important magma-forming process in the solar system. However, there is no evidence that Earth-like plate tectonics operates on the other planets, and the distinction between oceanic and continental crust seems to be uniquely terrestrial. This means that the origin of extraterrestrial basalts is more likely to be plume-related than spreading center-related. There is a good deal of evidence to support this hypothesis. For example, the Martian volcano, Olympus Mons, is the largest volcano discovered in the solar system so far (Fig. 11.12). It is a basaltic shield volcano, similar to the Hawaiian volcanoes but much larger. Olympus Mons could have reached its giant size only if the Martian lithosphere was stationary—not mobile, like the Earth's lithosphere. The volcanic edifice of Olympus Mons was formed by magma fed to the surface by a very long-lived plume deep within the planet.

**?** **What does Hawaii have in common with Mars?**

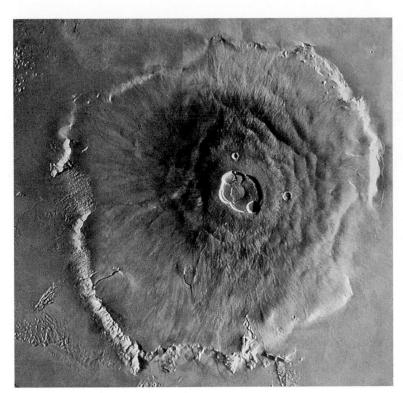

**Figure 11.12**
**GIANT VOLCANO ON MARS: LARGEST IN THE SOLAR SYSTEM**
Olympus Mons, a giant basaltic shield volcano on Mars, is the largest volcanic edifice discovered so far in the solar system. The volcano is 600 km (372 mi) across and stands 27 km (16.7 mi) high. The largest volcano on the Earth, Mauna Loa in Hawaii, is only one-third the size—225 km (about 140 mi) in diameter—and it stands 9 km (5.6 mi) above the sea floor.

# Igneous Rocks along Convergent Plate Margins

From the tectonic cycle we know that new oceanic crust is formed at spreading centers and that the seafloor spreads laterally, forming an ocean basin. Eventually, it will meet another fragment of crust—either oceanic or continental—along a convergent plate margin, and subduction will occur.[3] The special tectonic environment of active convergent plate margins gives rise to magmas that are much more variable in composition than MORB magmas.

## Subduction Zone Volcanism

Along a convergent plate margin, oceanic lithosphere that is subducted into the mantle carries with it a capping of cool basaltic crust that has been in contact with seawater for millions of years. The sinking process is rapid; although the slab heats up as it sinks, it still remains cooler than the mantle rocks around it. As it sinks, hydrous (water-bearing) minerals in the oceanic crust begin to release water. This water induces wet partial melting in the part of the mantle that is immediately adjacent to (and overlying) the subducting slab. Melting begins when the slab reaches a depth of about 100 km (62 mi). The subducting slab of oceanic crust itself may also undergo a small amount of wet fractional melting.

Magma that is formed by the subduction of oceanic crust and wet partial melting of adjacent mantle rocks is andesitic in composition. It rises through the overlying mantle and erupts to form stratovolcanoes. These volcanoes occur in chains parallel to the deep ocean trench that marks the boundary between two

---

[3] This cycle of production of new oceanic crust, opening of the ocean basin, lateral movement of oceanic crust, consumption of the crust through subduction, and eventual closure of the ocean basin is called the *Wilson cycle* after Canadian geophysicist J. Tuzo Wilson, who first proposed it in the 1960s as an explanation for the geologic history of the Appalachian Mountains.

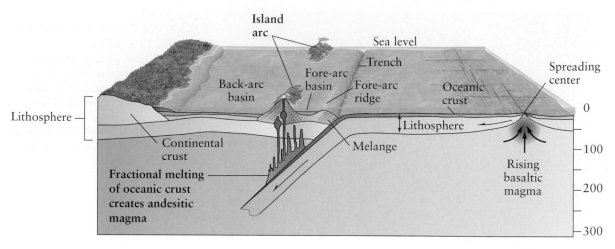

**Figure 11.13**
**VOLCANIC ISLAND ARC**
Wet partial melting of oceanic crust at a subduction zone produces andesitic magma, which rises to form a volcanic island arc of stratovolcanoes parallel to the oceanic trench.

**volcanic island arc** A chain of islands, each of which is an andesitic stratovolcano formed along an ocean–ocean plate subduction boundary.

plates, one of which (the subducting plate) is sinking (Fig. 11.13). Such a chain of islands, each of which is an andesitic stratovolcano formed along a plate subduction edge, is a **volcanic island arc**. The Aleutians islands and the Kuril islands are examples. As volcanic island arcs mature, the volcanoes grow larger and coalesce to form a larger land mass, or simply a *volcanic arc*. Japan and the Kamchatka Peninsula of Russia are striking examples of volcanic arcs (Fig. 11.14). Volcanoes in the Aleutians, the Kurils, Japan and Kamchatka have been formed where oceanic crust is subducting beneath another oceanic plate (an ocean–ocean plate boundary). Andesitic stratovolcanoes can also be formed on continental margins

**Figure 11.14**
**THREE RUSSIAN STRATOVOLCANOES**
Three andesitic stratovolcanoes on Russia's Kamchatka Peninsula—Klyuchevskoy (foreground), Kamen, and Bezymianny—are part of a volcanic arc. In this photo Klyuchevskoy is erupting volcanic ash.

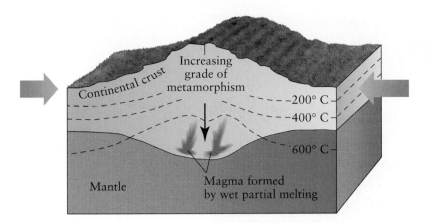

**Figure 11.15**
**COMPRESSION GENERATES GRANITIC MAGMA**
Compression causes the crust to deform and thicken. In the lowest part of the thickened crust, wet partial melting produces granitic magma. It is so viscous and flows so reluctantly that most of it ends up as granite plutons; little erupts at the surface.

if the oceanic crust is subducting under continental crust (an ocean-continent plate boundary). The chains of andesitic stratovolcanoes that make up the Andes and the Cascades are examples of **continental volcanic arcs.**

If there is continental crust on part of the plate that is being subducted, continued subduction will consume the ocean basin that lies between such continental crust and the volcanic arc. Eventually, the continents on either shore of the ocean basin may collide to form a great orogenic belt, as discussed earlier. Commonly, small fragments of oceanic lithosphere—ophiolites—have been squeezed up and preserved in such orogenic belts. Granitic magmas are formed at the base of continental crust that has been thickened as a result of a plate tectonic collision—that is, in an orogen (Fig. 11.15). Unlike MORB, granites vary widely in composition, and that fact alone makes it unwise to generalize too much about the formation of granite. However, we know that granitic magma is formed as a result of wet partial melting of the continental crust. The resulting magma is very rich in silica and extremely viscous. In fact, granitic magma is so viscous, and moves upward so slowly, that most of it cools and crystallizes as great batholiths of intrusive **orogenic granite**, like the granites we see today in the Sierra Nevada Mountains in California. The small fraction of granitic magma that does reach the surface is erupted as silica-rich rhyolitic lavas and tuffs.

**continental volcanic arc** A chain of andesitic volcanoes formed along an ocean–continent subduction zone plate boundary.

**orogenic granite** Huge batholiths of granite that form in orogens as a result of wet partial melting at the base of the continental crust.

## Igneous Rocks of Continental Interiors

Some igneous rocks are formed in the interior parts of continents, far removed from the intense activities occurring at plate margins. We refer to them as intraplate magmas; some are well understood, whereas others remain a geologic puzzle.

### Continental Hot Spots

If a plate capped by continental crust passes over a plume-related mantle hot spot like the one that feeds the Hawaiian volcanoes, the rising basaltic magma will force its way through the continental crust. The result is a vast outpouring of very fluid basaltic lava onto the continent. This lava forms a flat sheet of volcanic rock called **plateau basalt** (also known as *flood basalt*). Thus, plateau basalts, like shield volcanoes in the ocean, are products of plume-related decompression melting. The Columbia River basalts of Oregon and Washington (Fig. 11.16) were formed in this fashion, as were the plateau basalts of the Snake River region farther to the east, in Idaho. In fact, the Columbia and Snake River basalts were

**?** **What happens when a continent passes over a hot spot in the mantle?**

**plateau basalt** A vast, flat sheet of basalt that forms when continental crust passes over a plume-related mantle hot spot.

Figure 11.16
**DISSECTION OF A BASALT PLATEAU**
Flow upon flow of basaltic lava underlie the Columbia Plateau, one of the world's great basaltic plateaus. Here the plateau is being dissected by the Grande Ronde River in southeastern Washington. In the distance are the snow-capped peaks of the Wallowa Range.

probably formed by the same plume-generated magma source. They are geographically separated because the North American Plate is moving westward over the hot spot.

The hot spot that formed the Columbia and Snake River plateau basalts is still active and now lies beneath Yellowstone National Park. Although there is a lot of basaltic lava at Yellowstone, much of the region is covered by a thick blanket of rhyolitic volcanic ash. The rhyolitic magma is thought to have been formed as a result of wet partial melting of the continental crust when it was heated by the rising hot basaltic magma (Fig. 11.17).

Some rhyolitic magma of the kind that has erupted at Yellowstone doesn't make it all the way to the surface. It remains in the crust and crystallizes, forming intrusive bodies of granite. When such granites are exposed by erosion, it is clear that unlike the orogenic granites discussed earlier, they were not formed in an orogen as a result of subduction or collision. They are therefore called **anorogenic granites,** meaning "not orogenic."

**anorogenic granite** Granitic batholiths that form in areas of the crust not related to an orogen.

### Continental Rifts and Other Oddities

Besides plume-related magmas, other examples of magmatic activity in plate interiors are associated with rifting. Some continental rifts that are active today are spreading centers that have so far failed to develop into new ocean basins and may never do so, in which case they will be failed rifts. Examples include the East African Rift Valley and the Rio Grande Valley of New Mexico. There are many older failed rifts in which all spreading motion has ceased. Examples are the Newark basin of New Jersey and the Hartford basin of Connecticut. Rifts form

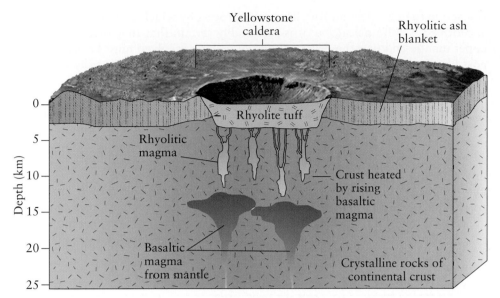

**Figure 11.17**
**VOLCANIC YELLOWSTONE PARK**
This is a section through the crust beneath the caldera in which Yellowstone National Park sits. Basaltic magma formed by decompression melting in a rising plume locally melts the crust and creates rhyolitic magma. The magma erupted explosively in the past, forming great blankets of volcanic ash.

when lithosphere is stretched and thinned. In response to the thinning of the lithosphere, the asthenosphere moves upward. Basaltic magma formed by decompression melting then rises, forming basaltic lava flows and gabbro sills and dikes. An interesting suite of other igneous rocks may also be formed.

Among the most puzzling of the rift-related igneous rocks are those that are composed largely or entirely of calcite or dolomite; these are known as *carbonatites*. For many years it was not even certain that carbonatites were igneous rocks. (The rock textures were recognized as igneous, but it's hard to imagine where a magma with such an extreme composition, totally lacking in silica, could come from.) Then, in 1962, carbonatite magma rich in sodium carbonate, with a composition similar to that of baking soda, was observed erupting from a

**Figure 11.18**
**WHERE DIAMONDS ARE FOUND**
The Finsch diamond pipe in South Africa was mined from this steep walled pit down to a depth of 430 m (1420 ft). Subsequent mining has been underground from a shaft in the complex of buildings to the upper left. This photograph was taken in September 1989.

volcano in the East African Rift. The exact process by which carbonatite magmas are formed is still unclear, but it is reasonably certain that they originate in the upper mantle and that the degree of partial melting involved must be very small, possibly less than 1 percent. Other rift-related magmas have equally extreme compositions, including unusually potassium- and sodium-rich magmas.

*Kimberlite* is another rare igneous rock (Fig. 11.18). It is named for the town of Kimberley in South Africa. Kimberlites are very poor in silica but rich in potassium, an unusual combination. They have the distinction of being the primary source of large diamonds. Kimberlite magma apparently is formed very deep in the mantle and only beneath cratons. The magma contains a lot of dissolved carbon dioxide, and it is intruded so forcefully that, like a cork coming out of a bottle of champagne, the magma can punch a pipe-like hole through the crust. Kimberlite pipes, often called *diamond pipes*, range in diameter from a few meters to a few hundred meters. Kimberlites and carbonatites are often found together. Kimberlites, like carbonatites, are thought to be formed by very small degrees of partial melting of mantle rocks beneath old cratons.

# METAMORPHISM ALONG PLATE MARGINS

As we discussed in chapter 10, *metamorphism* is the term used to describe changes in mineral assemblage and rock texture that are caused by increased temperature and pressure. Rocks that are subjected to such changes may be sedimentary, igneous, or even previously formed metamorphic rocks. The changes all take place while the rock is in the solid state—that is, at temperatures and pressures below those at which melting begins.

Recall from chapter 10 that there are three basic kinds of metamorphism:

1. *Contact metamorphism,* which embraces all the changes induced in cooler rocks that are intruded by a mass of hot magma. Contact metamorphism can be observed in rocks adjacent to most intrusive igneous bodies.

2. *Burial metamorphism,* which includes all the changes that result from increasing temperature and pressure in a deepening pile of sedimentary and volcanic rocks.

3. *Regional metamorphism,* which includes all the changes that occur when rock masses are squeezed and heated as a result of subduction or a plate collision.

Contact metamorphism can occur in any situation in which hot rocks come into contact with cool ones, but burial and regional metamorphism occur in specific tectonic environments. Figure 10.14 shows the relationship between the three kinds of metamorphism, the associated metamorphic facies, and the geothermal gradient that is characteristic of each kind of metamorphism. You may find it useful to refer back to Figure 10.14 as you read the following paragraphs. You may also wish to compare Figure 10.14 with Figure 11.19, which shows the geologic and tectonic environments in which the different kinds of metamorphism occur.

## Passive Continental Margins and Burial Metamorphism

A *passive* continental margin is the trailing edge of a continent. The Atlantic coast and the Gulf coast of North America are examples of passive margins. The

**? Which edge of North America is "active" and which edge is "passive"?**

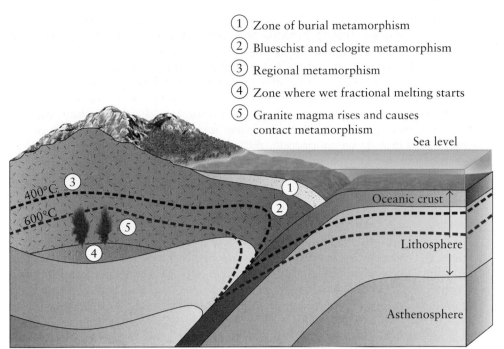

① Zone of burial metamorphism

② Blueschist and eclogite metamorphism

③ Regional metamorphism

④ Zone where wet fractional melting starts

⑤ Granite magma rises and causes contact metamorphism

**Figure 11.19**
**TYPES OF METAMORPHISM**
This cross-sectional view of a convergent plate boundary reveals various types of metamorphism at work. The dashed lines are isotherms—lines of equal temperature.

North American Plate is moving westward, carrying the continent along as it moves away from the spreading margin of the North American Plate, represented by the Mid-Atlantic Ridge. By contrast, the western coast of North America—where the North American Plate encounters the Pacific Plate and two smaller plates—is an *active* continental margin along which there is a great deal of seismic and volcanic activity. When the North Atlantic Ocean formed more than 160 million years ago, it did so as the result of rifting within a great continent (Pangaea). The two margins of the rift are today's eastern coastal region of North America and the coastal region of northwestern Africa. Other rift valleys were also formed as a result of the Atlantic rifting. They are failed rifts that did not become oceans; these rifts are found on both sides of the Atlantic inboard from the continental margins. One of the rift valleys is represented by the Newark basin in New Jersey, as mentioned earlier. Now the spreading activity is occurring mainly at the Mid-Atlantic Ridge, and the present Atlantic and Gulf coasts of the continent itself are far removed from the geologic activity that characterizes an active plate margin.

Along passive continental margins like those on both sides of the north Atlantic, thick piles of sediment accumulate just offshore on the continental shelf and slope. Burial metamorphism occurs in the lower portions of the sediment pile (Fig. 11.19). Wells drilled to depths of 5 km (3 mi) or more—for example, to seek oil in the great pile of sediment that has accumulated off the mouth of the Mississippi River—have encountered large volumes of sedimentary rock altered by burial metamorphism.

## Subduction Margins and Blueschist Metamorphism

When rock undergoes subduction, it is subjected to high levels of heat and pressure. But changing the actual temperature and pressure of a rock is more complex than it might seem. If you squeeze a cube of rock—any rock—in a vise, all parts of the rock are immediately subjected to increased pressure. Now heat one side of the rock with a flame (Fig. 11.20). It will take a long time for a ther-

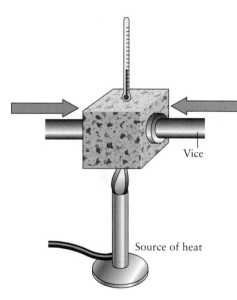

**Figure 11.20**
**QUICK PRESSURE, SLOW HEAT**
When a block of rock is squeezed, pressure is transmitted throughout the rock immediately. But when the rock is heated on one side, it takes a long time for the entire block to get hot. A thermometer on the side away from the heat source will still register a low temperature after many minutes of heating.

**?** What kind of metamorphism is caused by low temperatures and high pressures? What kind is caused by high temperatures and low pressures?

mometer on the other side of the rock to register a rise in temperature. This is because heat moves through a solid, rigid rock by conduction, and conduction through rock is a slow process.

From the simple experiment just described we can conclude that when rocks are buried, the pressure will rise more rapidly than the temperature. This is exactly what happens during subduction. When solid, cold lithosphere sinks through the hot, weak asthenosphere, the pressure on the sinking lithosphere is equal to the pressure of the enclosing asthenosphere. However, because heat can enter the slab only by conduction, the sinking lithosphere remains cooler than the adjacent asthenosphere. Note in Figure 11.19 how the contours of equal temperature (called *isotherms*) plunge steeply downward with the sinking lithosphere.

As we discussed in chapter 10, the combination of low temperature and high pressure in a subduction zone produces distinctive conditions of metamorphism and, hence, distinctive kinds of metamorphic rocks. At intermediate depths, where pressure and temperature conditions of the blueschist facies are reached, the bluish-colored amphibole mineral called glaucophane forms and the metamorphic rock in which it occurs is termed a *blueschist*. Blueschists can be seen in many parts of the Coast Ranges of California, which are exhumed parts of an old subduction zone. At greater depths, and therefore at higher pressure, blueschists are replaced by *eclogites*. Eclogites contain a beautiful ruby-colored garnet and a pyroxene called *jadeite*, one of the most prized of all gems. Eclogites, and therefore jadeite are rare; they are found only in places where some tectonic event such as a plate collision has caused a fragment of subducting lithosphere to be brought back to the surface.

## Continental Collisions and Regional Metamorphism

When two continental masses collide, the result is a long, arcuate mountain range like the Himalayas or the Alps. The mountains stand high because the light continental crust is thickened by the collision process. Because of isostasy there is a deep root beneath every mountain chain as discussed earlier in this chapter. Just as it does during subduction, the pressure rises more rapidly than the temperature in a thickening mass of crust. However, there is a limit to the thickening process, and after some millions of years the temperature finally catches up with the pressure. The result is the formation of regionally metamorphosed rocks—commonly slates, schists, and gneisses—all of which have distinctive foliations as a result of the squeezing associated with the collision process.

It should be apparent from the preceding discussion that regionally metamorphosed rocks are formed in orogens. If the collision process that forms the orogen makes the crust so thick that wet partial melting starts at the base, granitic magma is formed. The magma then rises and intrudes into the overlying rocks forming a batholith of orogenic granite. The existence of ancient orogens of regionally metamorphosed rocks with intrusions of granite batholiths provides some of the most convincing evidence that plate tectonics has been operating for a very long time.

# WHEN CONTINENTS ERODE

Recall from chapter 8 that there are three kinds of sediment: clastic, chemical, and biochemical. Chemical and biochemical sediments are most commonly formed in shallow lakes and seas, wherever water evaporates or organic material accumulates. For example, an ocean basin trapped between converging plates may grow shallower and shallower. Eventually, it may dry up altogether, leaving behind evaporite deposits, a type of chemical sediment.

Clastic sediments are far more abundant than chemical or biochemical sediments, and the locations where they are formed are largely controlled by plate tectonics. Clastic sediments are formed from the rock and mineral debris produced by weathering and erosion of continental masses. Once the sediment is formed it is removed by erosion, eventually accumulating in low-lying areas—troughs, trenches, and basins of various types. Rates of weathering and erosion are lowest in regions of low relief, such as the centers of cratons, and highest in regions of active tectonics where topographic relief is greatest—that is, in present-day orogenic belts. Let's look briefly at the principal tectonic environments in which clastic sediments accumulate.

## Rift Valleys

In the Newark basin and the East African Rift, we observe thick wedges of clastic sediments brought in by stream transport. In the Red Sea, marine sediments now cover the clastic sediments that were deposited before the Red Sea was wide enough for the sea to enter.

Elongate belts of low-lying **rift valleys** are formed when a continent is split apart by tensional forces. The East African Rift is an example of such an environment. A rift valley may eventually become a passive continental margin, or the rifting may cease and the valley becomes a failed rift. The Newark basin of New Jersey is an example of a failed rift, as previously mentioned. The East African Rift is an example of a rift that is still active, the Red Sea is a young rift where a new ocean is forming, and the Atlantic Ocean is a mature rift. Thick wedges of clastic sediment accumulate as streams transport sediment to the growing ocean basin. On the Atlantic Ocean margin of North America, sediment that has eroded from the adjacent continent has accumulated over many millions of years to a thickness of over 10 km (about 6 mi) (Fig. 11.21A). Most of the strata deposited on the shelf of a passive continental margin, like the Atlantic Ocean margin of North America, are shallow-water marine sediments. The continental shelf slowly subsides as accumulation takes place, and the pile of sediment grows thicker and thicker. *Diagenesis* (the processes that collectively turn sediment into sedimentary rock) occurs because of increasing temperature and pressure as the clastic sediments are buried progressively deeper in the pile, and at still higher temperatures and pressures deep within the pile burial metamorphism takes place.

## Intermontane Valleys and Structural Basins

A variety of low-lying regions are found within and along the edges of high mountain ranges. Many of these are structural basins or troughs caused by faulting, folding, and downwarping associated with the process of mountain-building. Coarse stream sediments eroded from a rising mountain range will accumulate in these low areas as vast thicknesses of conglomerates, gravels, and sands (Fig. 11.21B). The Himalayas provide an example. Here, thick sequences of conglomerates and coarse sandstones flank the southern edge of the range, while the finest-grained sediments have been transported all the way to the sea by the many streams that flow to the Indian Ocean.

**?** **What do Africa and New Jersey have in common?**

**rift valley** A valley that has formed along a rift, that is, along a divergent plate boundary or spreading center.

**Figure 11.21**
**SITES OF SEDIMENTATION**
Clastic sediment accumulates in low-lying areas in specific plate tectonic settings. *A.* Thick sedimentary wedges accumulate in rift valleys and along passive continental margins that are formed when continental crust rifts and a new ocean basin opens. *B.* Sediments accumulate in structural basins along the edges of mountain ranges formed by continental collisions. *C.* Sediments are shed from continents into deep-sea trenches above subduction zones. The sediment forms a wedge that is compressed and crushed as the oceanic plate is subducted.

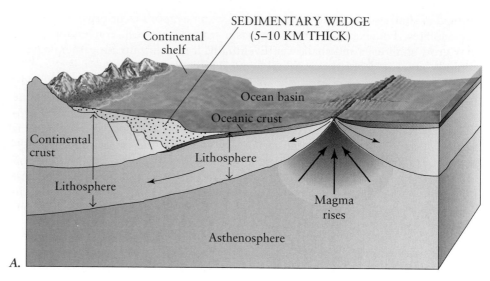

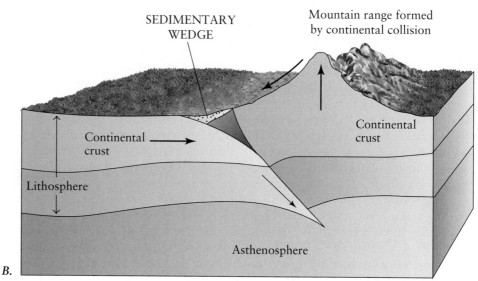

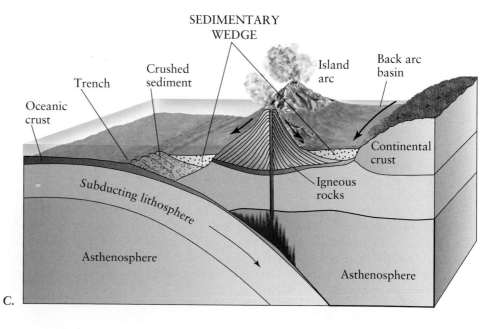

## Deep Ocean Trenches

The lowest-lying points on the Earth's surface are the long, deep trenches on the ocean floor that are the topographic expression of subduction zones. In places where subduction zones are formed near continental margins, such as along the western margin of South America, the greatest topographic relief on the Earth is observed—from the bottom of the trench to the top of the volcanoes on the edge of the continent. Sediment is transported from the continent to the adjacent trench by streams and turbidity currents, and it accumulates in the trench as turbidites (Fig. 11.21C). The presence of volcanoes on the land ensures that a lot of volcanic debris is present in the sediment. These wedge-shaped accumulations of clastic sediment are called *accretionary wedges*.

Some of the water-soaked sediment in deep ocean trenches may be subducted into the mantle along with the sinking slab of oceanic lithosphere, sometimes releasing water that facilitates wet melting of the surrounding mantle. The rest of the sediment and volcanic debris is scraped off the subducting slab, ending up in piles of chaotically folded, jumbled, and broken-up material on the ocean floor. These scraped-off collections of sediment and volcanic debris are called *mélanges* from the French word for "mixture."

Two other low-lying topographic features are usually associated with subduction-zone trenches. A linear, trough-like depression is often formed between the trench and the volcanic arc; this is called a *fore-arc basin*. Tensional forces created by the pull of the sinking slab can also create a depression behind the volcanic arc; this is a *back-arc basin*. Fore-arc and back-arc basins (Fig. 11.13), though not as dramatic topographically as the deep ocean trenches, are also low-lying areas where ocean-floor sediments accumulate.

# WHAT'S AHEAD

As we have seen throughout this book, plate tectonics explains many things. It explains why continents are assemblages of cratons and orogens, why today's continents and ocean basins are located and shaped the way they are, and why the many different kinds of rocks are found where they are. As discussed in earlier chapters, plate tectonics and the continual rearrangement of the lithosphere are caused by the Earth's internal heat energy.

When rocks are uplifted and exposed as a consequence of plate tectonics, they immediately start to weather and break down to create regolith. The energy that produces weathering comes from the Sun. Like two giant sculptors working on the same statue—one from the inside, the other from the outside—plate tectonics and weathering constantly change and reshape the Earth's surface. One consequence of the constant reshaping is the rock cycle. We observe that the kinds of rocks being formed today are the same as those rocks we can see in the ancient cratons and orogens. From this we conclude that plate tectonics and the rock cycle have been operating more or less as they are today for billions of years.

As you have learned in this chapter, many of the processes in the rock cycle are directly influenced by plate tectonic activity. But weathering, erosion, and deposition, and therefore the rock cycle, involve the atmosphere and the hydrosphere. To fully appreciate the intricate interactions between the various parts of the Earth, we must investigate these spheres too. In Part IV, therefore, we discuss the effects of air, wind, water, and ice on the Earth's surface, and the vital roles they play in making the Earth the dynamic place it is.

## Chapter Highlights

**1.** Four kinds of processes have shaped all the rocky planets in the solar system: impact cratering, tectonics, magmatic activities, and weathering and erosion.

**2. Impact cratering** by fast-moving bodies such as meteorites, asteroids, and comets played a highly significant role in the early shaping of the Earth. Although it continues today at a slow rate, cratering is no longer very important. Plate motion is the most important cause of the tectonic process shaping the Earth today.

**3.** Continents are assemblages of cratons and orogens. **Cratons** are ancient bodies of continental crust that have not been subjected to deforming forces for a very long time. **Orogens** are elongated regions of the continental crust that have been intensely folded, faulted, and metamorphosed during continental collisions.

**4. Isostasy** is the flotational balance of the lithosphere on the asthenosphere; different parts of the lithosphere have different degrees of buoyancy. Isostasy is one of the factors that makes plate tectonics possible.

**5.** Magma is formed in two ways: by melting of rock at a sufficiently high temperature and by decompression melting. Partial melting and the presence of water (wet melting) influence both ways of melting. The distribution of igneous rock reflects how and where different magmas are formed, as well as the roles played by partial melting and wet melting.

**6.** Midocean ridge basalt (**MORB**) is formed at oceanic spreading centers as a result of decompression melting in the mantle. MORB is the most common igneous rock on the Earth. **Ophiolites** are ancient fragments of oceanic crust that have been thrust onto continental crust during plate collisions. Oceanic shield volcanoes are also built from magma formed by decompression melting. The source of this magma is a plume of hot rock that rises from deep in the mantle.

**7.** Andesitic volcanism is a result of wet partial melting associated with the subduction of oceanic crust. Andesitic volcanoes occur in **volcanic island arcs** and **continental island arcs** as a result of subduction beneath oceanic and continental crust, respectively.

**8. Orogenic granites** are formed by wet partial melting of continental crust near the base of tectonically thickened orogens. **Anorogenic granites** are also formed by wet partial melting, but they are associated with centers of rhyolitic volcanism generated by plume-related melting in continental crust.

**9.** Burial metamorphism occurs in thick piles of sediment that accumulate along passive continental margins. Blueschist facies metamorphism occurs in the special high-pressure, low-temperature conditions that are characteristic of subduction zones. Regional metamorphism occurs along plate collision boundaries. Today regional metamorphism is occurring in orogens such as the Himalayas and the Alps.

**10.** Clastic sediments are formed from the loose rock and mineral debris produced by weathering of continental masses. The locations where these sediments are formed are largely controlled by plate tectonics. Low-lying areas in which clastic sediments commonly accumulate are rift valleys and passive continental margin shelves; intermontane valleys and structural basins; and deep ocean subduction trenches and associated back-arc and fore-arc basins.

## ▶ The Language of Geology

- anorogenic granite 312
- continental volcanic arc 311
- craton 302
- impact crater 297
- isostasy 301
- MORB 304
- ophiolite 304
- orogen 300
- orogenic granite 311
- plateau basalt 311
- rift valley 317
- volcanic island arc 310

## ▶ Questions for Review

1. What are the four major groups of planet-shaping processes?
2. Why was impact cratering so important in the Earth's early history, and why is plate tectonics so much more important today?
3. How and why do cratons differ from orogens?
4. Where would you go to find examples of modern orogens?
5. What is the most common type of igneous rock on the Earth, and how is its parent magma formed?
6. Describe the origin of an ophiolite. How are ophiolites related to MORB?
7. Why does MORB differ from plume-generated basalt?
8. Discuss the origin of arcuate belts of andesitic volcanoes and their relationship to plate tectonics.
9. How does an orogenic granite differ from an anorogenic granite?
10. Where and how does regional metamorphism occur?
11. What are the three main tectonic settings in which clastic sediments accumulate?

## ▶ Questions for Thought and Discussion

1. Space exploration of the rocky planets and moons has revealed that basalt is by far the most abundant kind of igneous rock on each of them. What hypothesis can you offer to explain this observation?
2. Suppose you are a member of a research team studying images of a planet that orbits a nearby star. The unmanned spacecraft sends back images on which features as small as 1 km (0.6 mi) can be seen. What kind of evidence would you look for to determine whether plate tectonics is active on this planet?
3. At times when sea level is high, a cover of marine sedimentary strata may have been deposited over parts of old, deeply eroded cratons. When we observe such strata today they are always horizontal or nearly so. Can you explain why this is so?
4. Study a plate tectonic map of the world and find four places where you would expect blueschist metamorphism to be occurring.
5. Why is isostasy such an important property, and how does it play a role in the surface topography of the Earth?
6. Volcanic island arcs typically have formed parallel to the deep ocean trenches that mark where one plate is subducted during the ocean–ocean plate convergence, but the volcanoes are 100 km (60 mi) or more behind the trench. Why aren't they formed right at the trench?

For an interactive case study on the role of geology and plate tectonics in shaping the landscape, visit our Web site.

# The Art of Geology

*Thomas Moran,* "Grand Canyon of the Yellowstone," 1872

The rhythm of the rocks beats very slowly, that is all. The minute hand of its clock moves by the millions of years. But it moves. And its second hand moves by the ceaseless eroding drip of a seep spring, by the stinging flight of sand particles on a grey and windy evening, by the particle-on-particle accretion of white travertine in warm blue-green waters—by the same ticking seconds that our watches record. And if you listen carefully—when you have immersed yourself long enough, physically and mentally, in enough space and enough silence and enough solitude—you begin to detect, even though you are not looking for it, something faintly familiar about the rhythm. You remember hearing that beat before, point and counter-point, pulsing through the inevitable forward movement of river and journey, of species and isolated Indian community, of lizard and of flowering plant…And you grasp at last, in a fuller and more certain way than you ever have before, that all these worlds move forward, each at its own tempo, in harmony with some unique basic rhythm of the universe.

—*Colin Fletcher*
**The Man Who Walked Through Time**

He who with pocket-hammer smites the edge

Of luckless rock or prominent stone, disguised

In weather-stains or crusted o'er by Nature

With her first growths, detaching by the stroke

A chip or splinter—to resolve his doubts;

And, with that ready answer satisfied,

The substance classes by some barbarous name,

And hurries on; or from the fragments picks

His specimen, if but haply interveined

With sparkling mineral, or should crystal cube

Lurk in its cells—and thinks himself enriched,

Wealthier, and doubtless wiser, than before!

—*William Wordsworth*
**The Excursion: Book Third**

# PART FOUR

# WATER WORLD

One characteristic that makes the Earth unique is the presence at its surface of water in all three states—solid (ice), liquid (water), and gas (vapor or steam). In particular, liquid water dominates the Earth system: more than 70 percent of the surface of the Earth is covered by water. Water shapes the surface of the land and sustains life. The ocean is the largest reservoir for water in the hydrologic cycle, but the role of the ocean extends much beyond this. Interactions between the ocean and atmosphere drive many of the processes that determine global climate. The permanently frozen parts of the hydrosphere—the glaciers—are also important in the global climate system. Geology contributes significantly to our understanding of global climatic change, past and present. This is especially important because human actions may be contributing to climatic change, and we need to understand the consequences of these actions.

> ▶ All of these topics and more are addressed in the chapters of Part IV, which are:

Chapter 12:    Water On and Under the Ground

Chapter 13:    Oceans, Winds, Waves, and Coastlines

Chapter 14:    Deserts, Glaciers, and Climatic Change

*Sport on the Hydrosphere*
These people are sea kayaking in Nootka Sound, a glacially carved embayment in British Columbia.

# · 12 ·

# WATER ON AND
# UNDER THE GROUND

California's Mono Lake, located in the Sierra Nevada Mountains near Yosemite Park, is a unique and strikingly beautiful natural habitat. As a result of geologic uplift, volcanism, and glacial activity over the past million years, the lake is surrounded by hills and has no natural outlets. The lake is the perfect habitat for brine shrimp, which provide food for many kinds of migratory birds. Tall calcium carbonate (*tufa*) spires that used to be submerged now rise hauntingly from the lake. They were deposited thousands of years ago, when the water level in the lake was much higher than it is today.

In 1941 the city of Los Angeles began to divert large quantities of water from the tributaries of Mono Lake, removing the water before it reached the lake. Since that time, the lake has lost about half its volume and has doubled in salinity. By the late 1980s, it became clear that the lake was close to ecological collapse. Environmental groups obtained a court order to stop the diversions, hoping to increase the water level and allow the ecosystem to restore itself. However, Los Angeles, which was drawing 17 percent of its water supply from the lake at that time, fought the injunction. In 1993, an agreement was reached in which the California state legislature agreed to provide funds for the preservation of the lake and to compensate Los Angeles for the loss of this source of water. The fragile and unique ecosystem of Mono Lake may have been saved just in time.

### In this chapter you will learn about

- The hydrologic cycle, that is, the pathways and processes whereby water moves among different reservoirs in the Earth system
- The factors that control the flow of water in streams
- Why rivers flood, and the effects of flooding on human interests
- How water moves and is stored under the ground
- The importance of water as a resource
- How surface and groundwater bodies can become contaminated

*Mono Lake*

Tall pillars of calcium carbonate form part of the strange, almost other-worldly landscape of Mono Lake, California, an island lake with a unique ecosystem.

?

The presence of liquid water makes the Earth unique among planetary bodies, as far as we know. Seen from space, the Earth appears mostly blue and white because large portions of it are covered by water, snow, ice, and clouds. $H_2O$ is not known to be present as a liquid anywhere except on the Earth. Elsewhere on many other bodies in the solar system $H_2O$ has been detected, but only in the form of ice or vapor. There may once have been liquid water on Mars, but because of its very low temperatures and pressures, it can exist there now only in the form of vapor or ice. The surface of Venus is so hot that water (what little there is) can exist there only as a vapor. Ganymede, the largest of Jupiter's

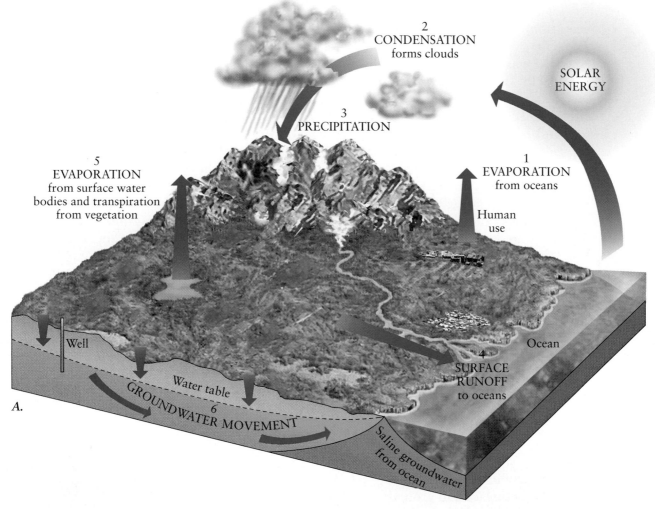

**Figure 12.1**
**THE WATER CYCLE**
*A.* The water cycle, or *hydrologic cycle,* is the movement of water from one reservoir to another in the Earth system.
Solar energy causes water to evaporate (1) from the ocean, transforming it into water vapor. The vapor condenses (2) to
form clouds and then precipitates (3) as rain or snow, onto both sea and land. Precipitation on the land flows back
to the sea as surface runoff (4) or as glacial ice, or it evaporates (5) back to the atmosphere from the land surface and from
plant leaves, or it infiltrates (6) into the ground, joining the groundwater reservoir.

moons, is so frigid that it is covered by a thick "lithosphere" of ice. Many of the particles that make up Saturn's rings are fragments of water ice. Even the Earth's Moon is now known to contain $H_2O$—but only in the form of ice.

In chapter 1 we named the four great reservoirs that make up the Earth system: the lithosphere, hydrosphere, biosphere, and atmosphere. The **hydrosphere** is the part containing the ocean, lakes, and streams; underground water; and snow and ice, including glaciers. This definition is straightforward, but it fails to mention that water is present in all parts of the Earth system, not just in the hydrosphere. Water vapor is an important part of the atmosphere. Water is a constituent of many common minerals (such as micas and clays) in the lithosphere, where it is tightly bonded in their crystal structures. And, of course, water is a fundamental component of living things in the biosphere. Strictly speaking, these occurrences of water are not part of the hydrosphere, but they are part of the *hydrologic cycle* (Fig. 12.1).[1]

**hydrosphere** The part of the Earth system that contains the ocean, lakes, and streams; underground water; and snow and ice, including glaciers.

---

[1] *Hydro* comes from the Greek word for water. Scientists who study the movement and distribution of water in the natural environment are called *hydrologists*.

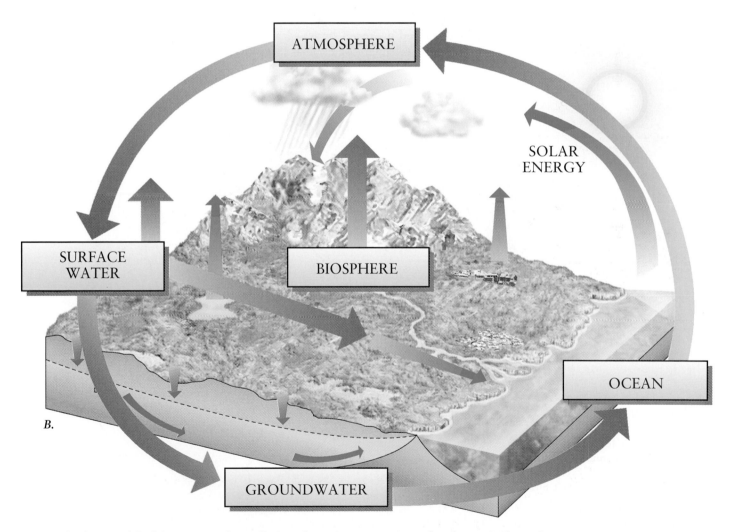

*B.* This box model of the water cycle emphasizes the various reservoirs and pathways in the cycle.

# THE WATER CYCLE

**hydrologic cycle** The set of interconnected reservoirs and processes whereby water moves around in the Earth system.

The *water cycle,* also known as the **hydrologic cycle,** is the movement of water from one reservoir to another in the Earth system (Fig. 12.1). The cycle consists of a set of interconnected reservoirs and pathways. The *reservoirs* are like storage tanks—places in the Earth system where water resides for varying lengths of time. The *pathways* are the processes by which water moves back and forth from one reservoir to another, both within the hydrosphere and among the other reservoirs of the Earth system. The hydrologic cycle maintains a *mass balance,* meaning that the total amount of water in the cycle is fixed. There are fluctuations on a local scale—sometimes quite large fluctuations, such as those that cause floods in one area and droughts in another—but on a global scale the fluctuations balance each other out. The hydrologic cycle is also linked to the rock cycle and tectonic cycle, as shown in Figure 12.2. All of the chapters in Part Three focus on the processes and reservoirs in the highlighted area of this figure. Let's begin our investigation of water on and under the ground by looking more closely at the reservoirs and pathways in the hydrologic cycle.

## Pathways and Processes

The pathways and processes by which water moves in the Earth system are the dynamic part of the hydrologic cycle. The movement of water is powered mainly

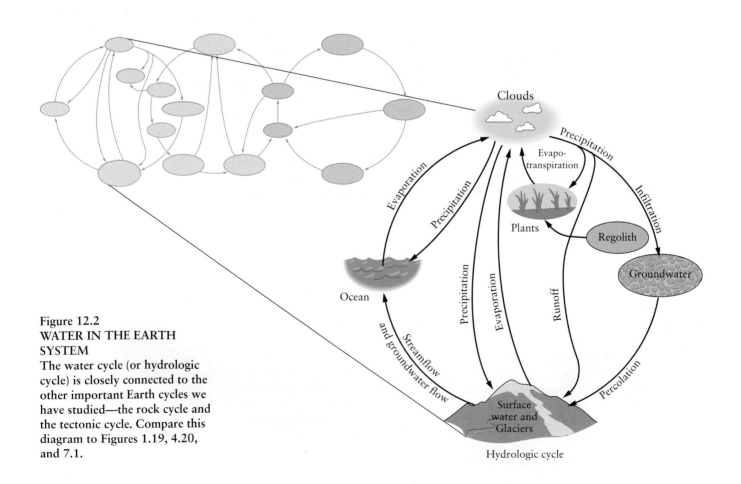

Figure 12.2
**WATER IN THE EARTH SYSTEM**
The water cycle (or hydrologic cycle) is closely connected to the other important Earth cycles we have studied—the rock cycle and the tectonic cycle. Compare this diagram to Figures 1.19, 4.20, and 7.1.

by heat from the Sun, as shown in Figure 12.2. The Sun causes **evaporation,** the process by which water changes from a liquid to a vapor. Water evaporates from the ocean, from surface water bodies, from vegetation, and from land. Water taken up by the roots of plants also can pass directly into the atmosphere; this is called **transpiration.** Water vapor produced by evaporation and transpiration enters the atmosphere and moves with the flowing air. Depending on local conditions of temperature, pressure, and humidity in the atmosphere, some of the water vapor will undergo **condensation**—that is, change from a vapor into a liquid or solid. In a liquid or solid form it can fall as **precipitation** (rain, snow, or hail) on the land or ocean.

Rain that falls on land may be evaporated directly, or it may be intercepted by vegetation, eventually returning to the atmosphere through transpiration. Or it may drain off over the land or in streams, becoming **surface runoff.** It may be stored temporarily in surface water bodies such as lakes or swamps. Or it may work its way into the ground, passing through small openings and channels in the soil and other surface materials in a process known as **infiltration.** Some of the water that infiltrates will eventually become part of the vast reservoir of underground water. Snow may remain on the ground for one or more seasons until it melts and the meltwater flows away. Water in the snow and ice of glaciers remains locked up much longer, sometimes for thousands of years, but eventually it too melts or evaporates and returns to the ocean.

## Reservoirs

The largest reservoir of water in the hydrologic cycle is the ocean, which holds 97.5 percent of all the Earth's water (Fig. 12.3). This means that most of the water in the hydrologic cycle is saline, not fresh. This fact has important implications for humans because we depend on fresh water as a resource for drinking, agriculture, and industrial use. The largest reservoir of fresh water is the permanently frozen polar ice sheets, which contain almost 74 percent of all fresh water. The ice sheets are a long-term holding facility; water may be stored there for thousands of years before it is recycled. Of the remaining unfrozen fresh water, almost 98.5 percent resides in the next largest reservoir, groundwater. Only a small fraction of the water in the hydrologic cycle resides in the atmosphere or in surface freshwater bodies such as streams and lakes. An even smaller amount is stored in the biosphere at any given time.

There is a correlation between the size of a reservoir and the *residence time,* that is, the average length of time spent by a water molecule in that reservoir. Residence times in the large-volume reservoirs, the ocean and ice sheets, are many thousands of years. The time spent by water in the groundwater system may amount to tens or hundreds of years or more. In the small-volume reservoirs the residence time of water is shorter—a few weeks in streams and rivers, a few days in the atmosphere, and a few hours in living organisms.

Although water is continuously being cycled from one reservoir to another, the total volume of water in each reservoir is approximately constant over short time intervals. Over long intervals, however, the volume of water in the different reservoirs can change dramatically. During glacial ages, for example, vast quantities of water evaporate from the ocean and are precipitated on land as snow. The snow slowly accumulates to build ice sheets that are thousands of meters thick and cover vast areas. At such times, the amount of water removed from the ocean is so large that the world sea level falls by 100 m (330 ft) or more and the expanded glaciers greatly increase the ice-covered area of the Earth.

**evaporation** The process by which water changes from a liquid into a vapor.

**transpiration** The process by which water taken up by the roots of plants passes directly into the atmosphere.

**condensation** The process by which water changes from a vapor into a liquid or solid.

**precipitation** The process by which water that has condensed in the atmosphere falls as rain, snow, or hail on the land or ocean.

**surface runoff** Precipitation that drains off over the land or in stream channels.

**infiltration** The process by which water works its way into the ground, passing through small openings and channels in the soil and other surface materials.

**?  How much of the water on the Earth is salt water?**

# Reservoirs in the Water Cycle

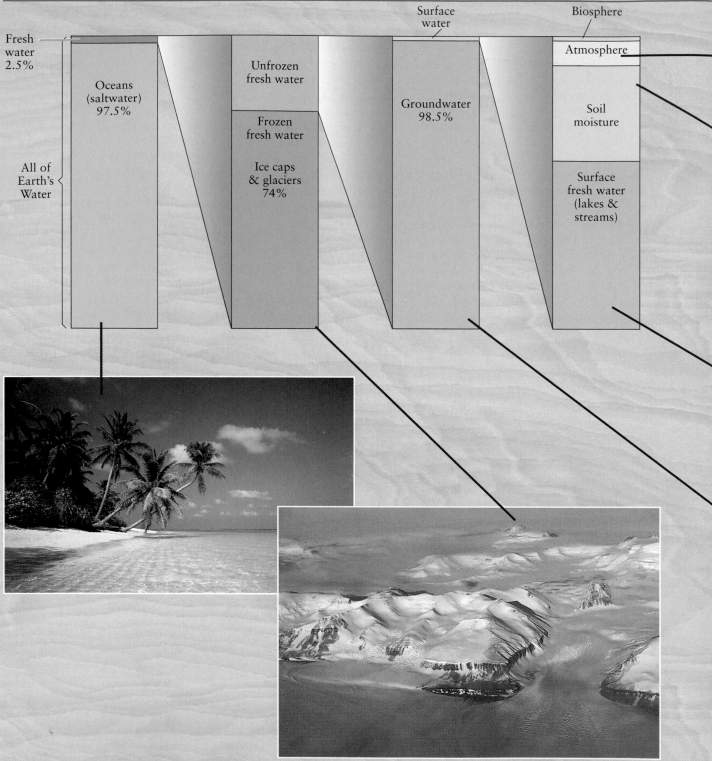

**Figure 12.3**
The largest reservoir for water in the hydrologic cycle is the ocean, which contains more than 97.5 percent of all the water in the system—and it is all salty. Drinkable fresh water is only a tiny percentage of our planet's water. Three-fourths (74 percent) of fresh water is frozen, most of it located in polar ice sheets. Of the unfrozen fresh water, about 98.5 percent is found in the next largest reservoir, groundwater. Only a very small fraction of the water passing through the hydrologic cycle resides in the atmosphere or in streams, rivers, and lakes. An even smaller amount cycles through the biosphere.

## Why Is the Water Cycle Important?

Water is a crucial resource; without it, life on Earth would not be possible. Most early cities and towns were founded close to streams that would serve as a source of water (Fig. 12.4). The processes that control the movement and distribution of water are also important in everyday life. Where infiltration rates are low and precipitation is high, floods may occur. During extended periods with below-average precipitation, there are droughts. Water also performs services, such as diluting or carrying away wastes. This can be troublesome; for example, rain falling on a landfill may infiltrate and dissolve noxious chemicals, carrying them away to contaminate a stream or groundwater supply. Moving water also carries sediment, which sometimes creates problems: we lose valuable topsoil, or a stream channel that is needed for transportation becomes clogged with sediment. But sometimes the movement of sediment by water is beneficial. For example, before the Nile River was subjected to heavy engineering for flood-control and irrigation purposes, its annual flooding deposited mineral-rich sediment. This created one of the most fertile agricultural regions on the Earth.

Bodies of water—especially the ocean—control the weather and influence the distribution of climatic zones. The varied landscapes we see around us are another important consequence of the hydrologic cycle. The erosional and depositional work of streams, waves, and glaciers, combined with tectonic movement, volcanism, and deformation of crustal rocks, have produced a diversity of landscapes that make the Earth's surface unlike that of any other planet in the solar system (Fig. 12.5). In all these ways and more, the hydrologic cycle influences our everyday lives.

# FRESH WATER ON THE SURFACE

Although surface freshwater bodies are small compared to other reservoirs in the hydrologic cycle, they are important simply because of their accessibility. People in many parts of the world depend on rivers, streams, or lakes for their water

**Figure 12.4**
**LIFE ON THE WATER'S EDGE**
Many of the world's great cities are built on rivers or coastlines. This is the River Seine, running through Paris, France.

Figure 12.5
WATER SHAPES
THE LANDSCAPE
Over millions of years, flowing water in the Colorado River has sculpted this magnificent landscape, viewed here from Dead Horse Point in Utah.

supply. Many of the hydrologic processes that affect people on a daily basis—such as flooding and the transport of sediment—involve the movement of surface water. Therefore, it is important to learn about the factors that govern the flow of surface water.

## Streams and Streamflow

A **stream** is a body of water that flows downslope along a clearly defined natural passageway, transporting particles and dissolved substances. The stream's passageway is its **channel.** A **river** is a stream with a considerable volume and a well-defined channel. Rivers and streams are vital geologic agents. They carry most of the water that goes from the land to the sea and transport billions of tons of sediment to the ocean each year. They carry soluble salts released by the weathering of rocks; these play an essential role in maintaining the salinity of seawater.

If you stand outside during a heavy rain, you can see that water initially tends to move downhill in a process called *overland flow* (or *sheet flow*, since the flowing water often takes the form of a thin, broad sheet). After traveling a short distance, overland flow becomes concentrated into well-defined channels, thereby becoming **streamflow.** Overland flow and streamflow together constitute surface runoff, the portion of precipitation that flows away over the surface of the land.

Several factors control the way a stream behaves. The most important are (1) *gradient*, the steepness of the channel;[2] (2) the channel's cross-sectional area

**stream** A body of water that flows downslope along a clearly defined natural passageway, transporting particles and dissolved substances.

**channel** The passageway of a stream.

**river** A stream with a considerable volume and a well-defined channel.

**streamflow** That part of surface runoff that travels in stream channels.

---

[2] Technically, gradient is the vertical distance that a channel falls per unit of horizontal distance. A steep mountain stream may have a gradient of 60 m/km (about 300 ft/mi) or even more. Near the mouth of a large river, such as the Missouri River, the gradient may be less than 0.1 m/km (0.5 ft/mi).

**discharge** The amount of water passing by a point on a channel's bank during a unit of time.

**load** The material carried along by a stream.

(stream width × stream depth); (3) the velocity of water flow; (4) **discharge,** the amount of water (stream width × stream depth × velocity of flow) passing by a point on the channel's bank during a unit of time; (5) the roughness of the stream bottom (called the *bed*); and (6) **load,** the amount of material carried along by the stream. The load has three parts: *bed load,* particles that move along the bottom; *suspended load,* particles and organic debris that are suspended in the water; and *dissolved load,* dissolved substances that are a product of rock weathering.

These factors are interrelated. For example, if the gradient of a stream becomes steeper along a particular stretch of channel, the velocity of flow is likely to increase as well. If the velocity is high, a greater load can be carried. If the discharge increases, the channel must handle more water in a given period; as a result, both the velocity of flow and the depth of the water in the stream will increase. In some cases, the width and depth of the channel may also increase. A flowing stream may increase the cross-sectional area of its channel by scouring the banks and the bottom. This scouring not only deepens and widens the channel but also adds sediment to the load. When velocity eventually decreases, the sediment settles out, filling in the channel and allowing it to return to its original size.

## Stream Landforms and Drainage Patterns

**erosion** The wearing away of bedrock and transport of loosened particles.

**deposition** The laying down of sediment.

Streams create landforms through two processes: **erosion,** the breakdown, removal, and transport of materials; and **deposition,** the laying down of sediment. The ability of streams to erode and carry particles of sediment is related to the way water moves through the channel. If velocity is high, the water will move in the complex swirls and eddies that characterize *turbulent* flow. The higher the ve-

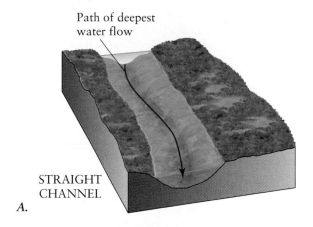

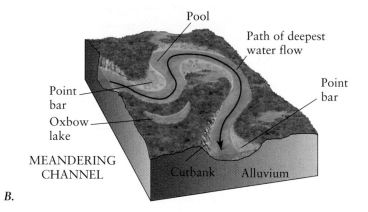

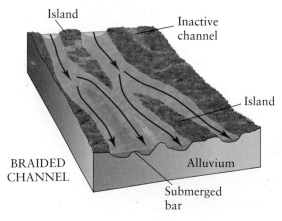

**Figure 12.6**
**FEW STREAM CHANNELS ARE STRAIGHT**
This drawing shows common types of stream channels: *A* straight, *B* meandering, and *C* braided. The arrows show the direction of flow and trace the path of the deepest water. Even in the "straight" channel, the flow of the deepest water curves from one bank to the other.

locity and the more turbulent the flow, the greater the ability of the water to pick up particles and carry them away. When a stream loses velocity because of a change in gradient or a decrease in discharge, its transporting power drops and it deposits part of its load. We looked at the processes of erosion and deposition of sediment in chapters 7 and 8; now let's focus on the landforms created by streams through these processes.

## Types of Channels

Stream channels vary widely in width, depth, and shape. The shape of a channel reflects variations in both erosion and deposition. The factors that affect stream-flow—gradient, discharge, and load—are all important determinants of a channel's shape. The underlying rocks are important, too. For example, a stream may take a sudden bend, or its gradient may increase when it passes from rock that is resistant to erosion into rock that is easily eroded. Because all of these factors interact in different ways, no two channels are exactly alike.

If a channel has many curves, we refer to it as *sinuous*. *Straight* channels are rare, and they usually occur only along short segments of a channel. Close examination of a straight segment of natural channel will show that it is like a sinuous channel in some respects. A line connecting the deepest parts of the channel typically does not follow a straight path but wanders back and forth across the channel (Fig. 12.6A). The water velocity tends to be highest in the deepest parts of the stream. In places where the deepest water lies at one side of a channel, a deposit of sediment (a point *bar*) tends to accumulate on the opposite side because velocity is lower.

In many streams the channel forms a series of smooth bends called *meanders*. They are named after the Menderes River (from the Latin word *Meander*) in southwestern Turkey, which is noted for its winding course. If you try to wade or swim across a meandering stream, you will find that the velocity of the flowing water is not uniform. When the water rounds a bend, the zone of highest velocity swings toward the outside of the channel (Fig. 12.6B). As water sweeps around the bend, turbulent flow causes undercutting and erosion of material where the fast-moving water meets the steep outer bank. Meanwhile, along the inner side of each meander, where the water is shallow and velocity is low, sediment accumulates to form a *point bar* (Figs. 12.6B and 12.7). As a result, mean-

**?**   **Why are there different types of stream channels?**

**Figure 12.7**
**A MEANDERING STREAM**
This meandering river near Phnom Penh, Cambodia, flows past agricultural fields that thrive on the rich soil of the floodplain. Light-colored sandy bars lie on the inner banks of the meander bends. Two oxbow lakes, the products of past meander cutoffs, lie adjacent to the present channel.

Oxbow Lake

Point Bar

Oxbow Lake

ders slowly change shape and shift position along a valley as sediment is subtracted from and added to the banks. Sometimes the water finds a shorter route downstream, bypassing a meander by cutting it off. As sediment is deposited along the banks of the new channel route, the cutoff meander may be blocked and converted into a curved *oxbow lake* (Fig. 12.7).

If a stream is unable to move all the available load, it deposits the excess sediment as bars. The bars divide the flow and concentrate it into deeper channels on either side. The water repeatedly divides and reunites as it flows through interconnected channels separated by bars or islands (Fig. 12.6C). This is referred to as a *braided channel*. Large braided rivers typically have many short-lived islands and constantly shifting channels (Fig. 12.8). Braided patterns tend to form in streams with a highly variable discharge and easily eroded banks that can supply abundant sediment to the system.

## Stream Deposits

Stream deposits form along channel margins, valley floors, mountain fronts, and the margins of a lake or the ocean. These are all places where changes in stream energy—and, therefore, changes in the stream's ability to carry its load—take place. Sediment that has been deposited by a stream in fairly recent times is called **alluvium**. A point bar is one example of a stream deposit. Others include *floodplains, alluvial fans,* and *deltas.*

When a stream rises during a flood, water overflows the banks and inundates the adjacent valley floor, called the **floodplain** (Fig. 12.9). As sediment-laden water flows out of the channel, its depth, velocity, and turbulence decrease abruptly at the margins of the channel. This results in sudden, rapid deposition of the coarser part of the suspended load along the margins, building up a broad, low ridge of alluvium atop each bank called a *natural levee*. Farther away, the finer particles settle out in the quiet water covering the valley. This creates the

**alluvium** Sediment that has been deposited by a stream in fairly recent times.

**floodplain** The relatively flat valley floor adjacent to a stream channel, which is inundated when the stream overflows its banks.

**Figure 12.8**
**THE BRAHMAPUTRA, A LARGE BRAIDED STREAM**
This satellite image shows the intricate braided pattern of the Brahmaputra River where it flows out of the Himalayas en route to the Ganges Delta. Noted for its huge sediment load and constantly shifting channels, the river may be 8 km (5 mi) wide during the rainy monsoon season.

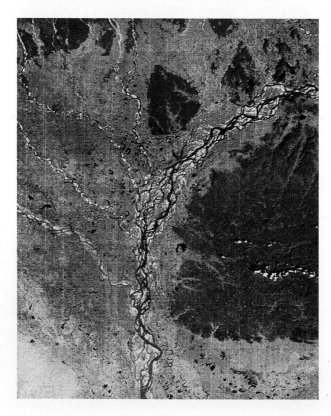

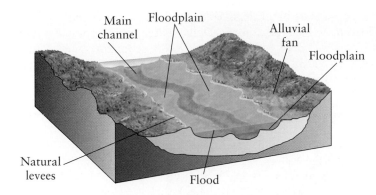

Main channel · Floodplain · Alluvial fan · Floodplain · Natural levees · Flood

**Figure 12.9**
**STREAM VALLEY DEPOSITS**
This drawing shows the main types of landforms that are created when streams deposit their loads; these landforms include floodplains, natural levees, and alluvial fans.

broad, flat, fertile land that is typical of floodplains. If a stream flowing through a steep upland valley suddenly emerges onto the floor of a much broader valley, it will experience a decrease in slope, a drop in velocity, and a decrease in its ability to carry sediment. It will deposit the part of its load that can no longer be transported. This deposit is called an *alluvial fan* (Fig. 12.10).

**Figure 12.10**
**ALLUVIAL FAN**
Where this stream emerges from the mountains, it abruptly slows and deposits its sediment load. This has formed a symmetrical alluvial fan out over the margin of Death Valley, California. A braided system of channels covers the fan surface.

Figure 12.11
THE NILE RIVER DELTA
This LANDSAT image shows Egypt and the Nile Delta as viewed from space. The Nile River is the thin, wavy, dark line. The direction of flow is north-ward, from the bottom of the page to the top. The reddish, tri-angular region at the top is the Nile Delta, where the river emp-ties into the Mediterranean Sea. (The red color indicates the pres-ence of vegetation, which appears red in satellite images that record infrared radiation.)

A deposit that forms where a stream flows into standing water is called a *delta,* so named because of its shape, which is most commonly triangular (resembling the Greek letter delta "Δ"). Each delta has its own peculiarities, which are determined by the stream's discharge, the character and volume of its load, the shape of the bedrock coastline near the delta, the offshore topography, and the intensity and direction of currents and waves. In places where strong currents and wave action redistribute sediment as quickly as it reaches the coast, deltas cannot form. However, if the rate of sediment supply exceeds the rate of coastal erosion, a delta will form. When the stream first enters the standing water, it quickly loses velocity and the heaviest particles drop out, forming a coarse, thick, steeply sloping layer. Most of the fine suspended load is carried farther seaward, eventually settling out to form a gently sloping delta front. This depositional se-quence may be repeated many times as the delta is constructed. Most of the world's great rivers, including the Nile, Ganges-Brahmaputra, Huang He, Ama-zon, and Mississippi, have built massive deltas (Fig. 12.11).

## Drainage Basins and Divides

**drainage basin** The total area from which water flows into a stream.

Every stream is surrounded by its **drainage basin,** the total area from which water flows into the stream. Drainage basins range in size from less than a square kilometer to vast areas the size of subcontinents. In general, the greater a stream's annual discharge, the larger its drainage basin. The vast drainage basin of the Mississippi River encompasses more than 40 percent of the total area of the con-tiguous United States (Fig. 12.12). As a drainage system develops, a stream may gain or lose *tributaries*—smaller streams that drain into the larger stream. Thus, just as the stream channel and its discharge and load are constantly changing, so too is the drainage system changing.

**divide** The topographic high that separates adjacent drainage basins.

The line that separates adjacent drainage basins is a **divide.** Divides are topo-graphically higher than the basins they separate (Fig. 12.13). On continents, great mountain chains separate streams that drain toward one side of the conti-

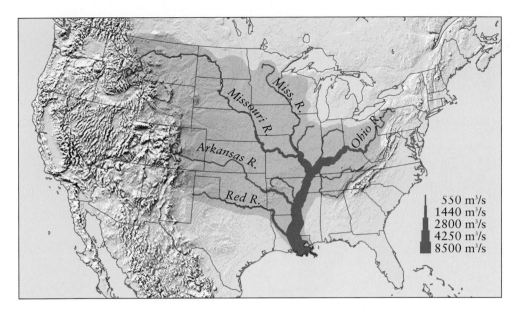

**Figure 12.12**
**THE VAST MISSISSIPPI RIVER DRAINAGE BASIN**
The drainage basin of the Mississippi River encompasses a major portion of the central
United States and extends into southern Canada. In this diagram the widths of the river
and its major tributaries are exaggerated to represent the discharge in m³/s, as shown in
the legend (lower righthand corner).

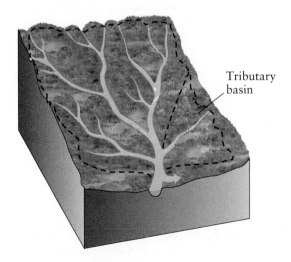

**Figure 12.13**
**DRAINAGE BASINS AND DIVIDES**
A drainage divide is topographically "high" ground that separates adjacent drainage
basins. In this drawing, the divide is highlighted with a dashed line. Each of the
smaller tributaries shown here has its own, smaller drainage basin within the larger
drainage basin.

**Figure 12.14**
**CONTINENTAL DIVIDES IN THE AMERICAS**
This map of North and South America shows the location of continental drainage divides.

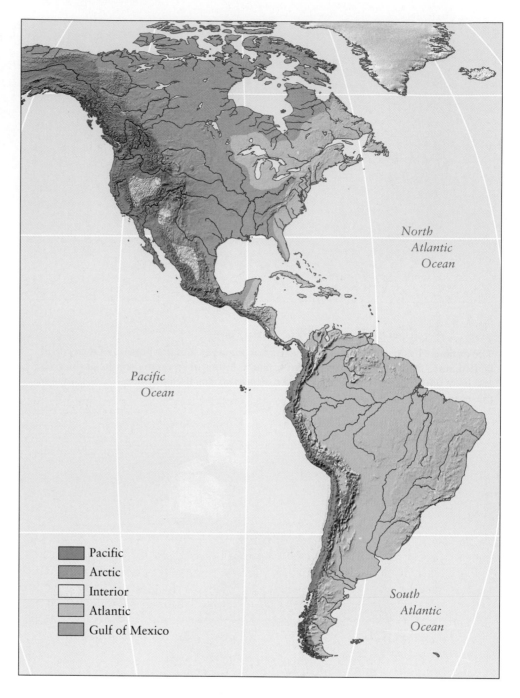

Pacific
Arctic
Interior
Atlantic
Gulf of Mexico

North Atlantic Ocean

Pacific Ocean

South Atlantic Ocean

nent from streams that drain toward the other side. These are *continental divides*. The continental divide of western North America lies along the length of the Rocky Mountains (Fig. 12.14). Streams to the east of this divide ultimately drain into the Atlantic Ocean, while those to the west drain into the Pacific.

Not surprisingly, there is a close relationship between drainage patterns and the nature of the underlying rocks. The ease with which a particular rock is eroded depends on its structure and composition. The course a stream takes across the land is strongly influenced by these factors. Structures such as faults and fractures in the underlying rocks can influence the direction of flow and the type of drainage pattern that develops. Figure 12.15 shows some of the most common drainage patterns and the geologic factors that control them.

DENDRITIC   RADIAL

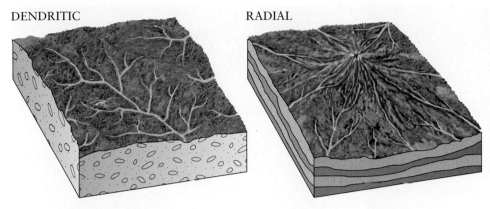

RECTANGULAR   DERANGED

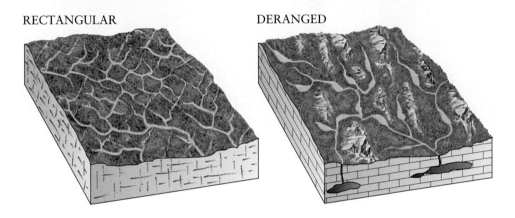

Figure 12.15
NOT ALL DRAINAGE
PATTERNS ARE ALIKE
These drawings show some common drainage patterns. The sides of the blocks reveal the relationship of the drainage patterns with underlying rock types and structures.

## Lakes

A **lake** is a water body that has an upper surface exposed to the atmosphere and has no appreciable gradient (that is, the surface is flat). Other surface water bodies are *ponds* (small, shallow lakes) and *wetlands* (areas of poor surface drainage, such as marshes and swamps, which may contain standing water). These can be included under the general definition of a lake. Lakes receive water input from streams, overland flow, and groundwater, and they lose water by evaporation. Many lakes also lose water at an outlet, where water drains over a dam (natural or constructed) to become an outflowing stream.

Lakes are quite important from the human viewpoint. They are frequently used as sources of fresh water (as in the opening story about Mono Lake, California), and they also support ecosystems that provide food for humans. Where dammed to a high level above the outlet stream, they can provide hydroelectric power. Lakes, ponds, and wetlands are also important recreation sites and sources of natural beauty. The scientific study of lakes and other inland bodies of water is called *limnology,* from the Greek word *limne* meaning "lake" or "pond."

Natural basins occupied by lakes are created by a number of geologic processes. For example, crustal faulting creates many large, deep lakes. Lava flows often form a dam in a river valley, causing water to back up as a lake. Landslides suddenly create lakes by blocking valleys. Throughout formerly glaciated regions of North America and Europe, plains of glacial sand and gravel contain natural pits and hollows left by the melting of stagnant ice masses that were buried in the sand and gravel deposits. The pits later fill with water to form

**lake** A water body that has an upper surface exposed to the atmosphere and has no appreciable gradient.

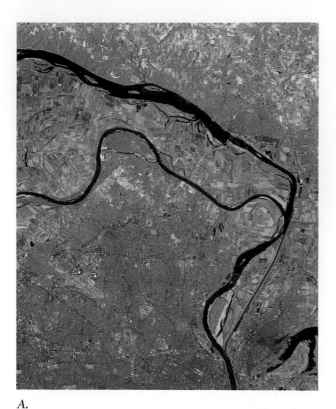

*A.*

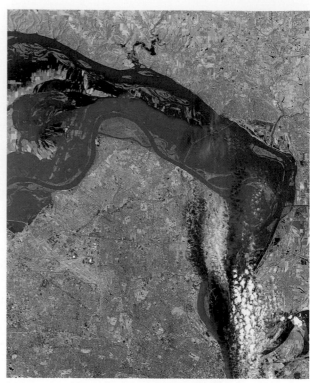

*B.*

**Figure 12.16**
**FLOOD!!**
A pair of satellite images shows the region where the Missouri River joins the Mississippi River at St. Louis, Missouri. *A.* This photo shows a dry summer with low flows (July 1988). *B.* This photo shows the river during the disastrous flood of July 1993. Weeks of rain hundreds of kilometers away caused the rivers to overflow protective levees. Numerous towns and vast areas of farmland were inundated.

**flood** An event in which a water body overflows its banks.

kettle lakes, common in New England. (Thoreau's *Walden* is an·example of a small kettle lake.)

An important characteristic of lakes in general is that they are short-lived features on the geologic time scale. They disappear by one of two processes, or a combination of both. First, lakes that have stream outlets will be gradually drained as the outlets are eroded to lower levels. Second, lakes accumulate inorganic sediment carried by streams entering the lake and organic matter produced by plants within the lake. Eventually, they fill up, forming a boggy wetland with little or no free water surface. This process is called *eutrophication.*

Lakes can also appear and disappear as a result of differing climatic conditions. In moist climates, the water level of lakes and ponds coincides with the water table in the surrounding region. Seepage of groundwater into the lake, as well as direct runoff of precipitation, maintains these free water surfaces permanently throughout the year. If temperature increases or precipitation decreases, evaporation may exceed input, and the lake will shrink. In regions where climatic conditions consistently favor evaporation over precipitation, many lake beds are dry or only intermittently filled with a shallow layer of water. Streams bring dissolved solids—salts—to these ephermeral lakes. Since evaporation removes only pure water, the salts remain behind and salinity levels may build up. Eventually, the salts may be precipitated as solid *evaporites,* a number of which are of economic value.

## Flooding

Uneven distribution of rainfall through the year causes many streams to rise in flood from time to time. A **flood** occurs when a stream's discharge becomes so great that it exceeds the capacity of the channel, causing the stream to overflow its banks (Fig. 12.16). Lakes can also flood, as can oceanic coastal zones.

Lakeshore and oceanic coastal flooding are most often related to the occurrence of an intense storm, with strong winds and sometimes high tides leading to high-water conditions and storm waves battering the coastline. Both coastal flooding and stream flooding often take people by surprise, but geologists view them as normal and inevitable events. The geologic record shows that floods have been occurring throughout most of Earth history.

The unusually high discharge of a flood appears as a peak on a *hydrograph*, a graph in which stream discharge has been plotted against time. In the example in Figure 12.17, a passing storm generated a brief interval of intense rainfall. As the runoff moved into the stream channel, the discharge quickly rose. The *crest* of the resulting flood—the time when the peak flow passed the hydrologic station at which the measurements were made—occurred about two hours after the storm. It took another eight hours for the flood runoff to pass through the channel and for the discharge to return to its normal level.

**?**   **What happens when there is too much water in one part of the hydrologic cycle?**

## Hazards Associated with Flooding

Major floods can be disastrous events, causing both loss of life and extensive property damage (Table 12.1). The Huang He in China, sometimes called the Yellow River because of the yellowish-brown color produced by its heavy load of silt, has a long history of catastrophic floods. In 1887, the river inundated 130,000 km$^2$ (more than 50,000 mi$^2$) and swept away many villages in the heav-

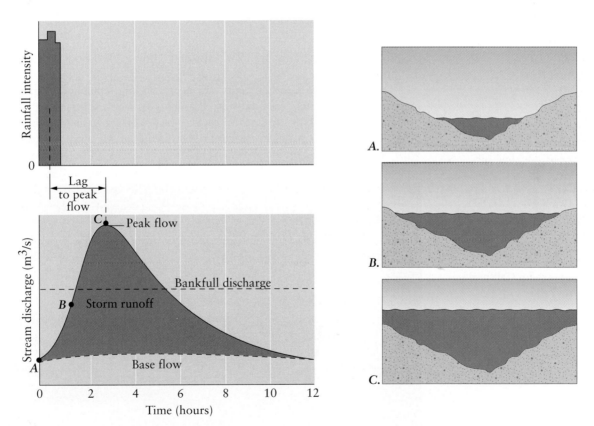

**Figure 12.17**
**A HYDROGRAPH OF STREAM DISCHARGE AFTER A STORM**
This is the hydrograph of a stream following a brief, intense storm. A rainfall event (top) causes a rise in stream discharge (bottom) as the runoff collects and runs down the stream channel. Peak discharge lags about 2.5 hours behind the peak rainfall in this example.

**TABLE 12.1** Fatalities from Some Disastrous Floods

| River | Type of Flood | Date | Fatalities | Remarks |
|---|---|---|---|---|
| Huang He, China | River flooding | 1887 | ~900,000 | The flood inundated 130,000 km² (50,000 m²) and swept many villages away. |
| Johnstown, Pennsylvania, USA | Dam failure | 1889 | 2200 | After the dam failed, a wave 10–12 m (33–40 ft) high rushed down the valley. |
| Chang Jiang, China | River flooding | 1911 | ~100,000 | The flood formed a lake 130 km (80 mi) long and 50 km (30 mi) wide. |
| Huang He, China | River flooding | 1931 | ~3,700,000 | The flood extended over 800 km (almost 500 mi), leaving millions homeless. |
| Vaiont, Italy | Dam failure | 1963 | 2000 | A landslide into the reservoir caused a wave that overtopped the dam and inundated villages below. |
| Bangladesh | Cyclone | 1970 | ~300,000 | This low-lying country is very susceptible to flooding from cyclones that form in the Bay of Bengal. |
| Bangladesh | Cyclone | 1991 | ~200,000 | A repeat of the disaster of 1970. |
| Mississippi River, USA | River flooding | 1993 | 50 | Fatalities were low but damages were very high, topping $10 billion. The flood inundated more than 55,000 km² (21,200 mi²) of land, including towns and farmland. |
| Red River, USA and Canada | River flooding | 1997 | — | The city of Winnipeg, Manitoba was saved by flood protection engineering works surrounding the city, but many outlying towns and villages were destroyed. |

ily populated floodplain. In 1931, another Huang He flood killed a staggering 3.7 million people.

The strength of flood currents can be surprising, tearing out trees or knocking buildings off their foundations (Fig. 12.18). The increased velocity of flow during a flood enables the stream to carry not only a greater load but also larger particles. Damage caused by moving debris is one of the main hazards associated with flooding. One of the most frightening episodes associated with the great

Figure 12.18
**THE POWER OF FLOOD WATERS**
Trees were pulled out by their roots and this pickup truck was overturned by raging flood waters in northern California, December, 1996.

Figure 12.19
AFTER THE FLOOD:
CLEANING UP
When flood waters finally receded after the great Mississippi River flood of 1993, they left behind a gooey blanket of muddy sediment, polluted with sewage, agricultural chemicals, and any fluid that leaked from ruptured tanks.

Mississippi River flood of 1993 involved a group of large propane tanks that were ripped from their moorings by the force of the rising water. If the propane vapor leaking from the tanks had ignited, much of south St. Louis could have been obliterated. Fortunately, after the flood waters began to recede, emergency crews managed to tether the tanks and repair the leaks.

The massive load of finer particles carried by a stream during flooding represents a different type of hazard. After flood waters recede, people may find themselves almost knee-deep in mud (Fig. 12.19). Flood waters can also concentrate garbage and pollutants, as described in this eyewitness account of the aftermath of a destructive flood:

> *Well, there was mud all over the house, no place to sleep. There was mud in the beds, about a foot and a half of it on the floors. And I had all that garbage. It was up over the windows. In the yard I had poles and trees and those big railroad ties from where they had washed out . . . furniture, garbage cans, anything that would float in the water.*[3]

Disruption of services such as electricity, water, and gas is another common hazard associated with flooding. Road, bridge, and rail closures can contribute to the shortage of basic supplies in flooded communities. Floods can also have long-term effects, such as loss of wildlife habitat or farmland, or permanent changes in the river's channel. During the 1993 Mississippi River flood, a major concern was that the river might abandon its main channel north of New Orleans and seek a shortcut to the Gulf of Mexico via the Atchafalaya River. So far, this has been prevented by engineering structures at the juncture of the two channels.

## Flood Prediction

Because floods can be so damaging, much harm can be avoided by predicting and preparing for them. To do this, the frequency of occurrence of past floods of different sizes is plotted on a graph, producing a *flood-frequency curve* (Fig.

---

[3] from Erikson, K. T. (1976). *Everything in its Path—Destruction of a Community in the Buffalo Creek Flood.* Simon & Schuster, N.Y.

12.20). The average time interval between two floods of the same magnitude is called the *recurrence interval*. In the case of the Skykomish River at Gold Bar, Washington, a flood with a discharge of 1750 m³/s (about 62,000 ft³/s) has a recurrence interval of 10 years. This means that there is a 1-in–10 (that is, 10 percent) chance that a flood of this magnitude will occur in any given year. This is referred to as a "10-year flood." A flood with a discharge of 2500 m³/s (or 88,000 ft³/s) is a "50-year flood" for this particular stream, having a recurrence interval of 50 years and a 1-in–50 (2%) chance of occurring in any given year.

**Figure 12.20**
**PREDICTING FLOODS**
*A.* The graph shows the frequency of floods of different sizes on the Skykomish River at Gold Bar, Washington. A flood with a discharge of 1750 m³/s has a recurrence interval of 10 years and, thus, a 1-in–10 chance of occurring in any given year (a 10-year flood). *B.* The landscape diagram shows the normal, nonflood discharge for a hypothetical stream in the darkest blue. In lighter blues, the diagram shows what larger floods might be like—the 2-year, 10-year, 50-year, and 100-year floods for this stream.

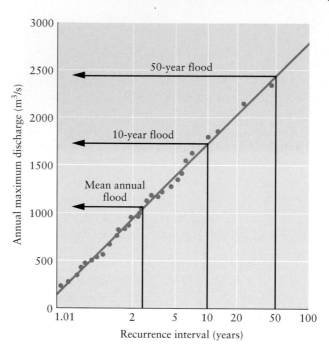

A.

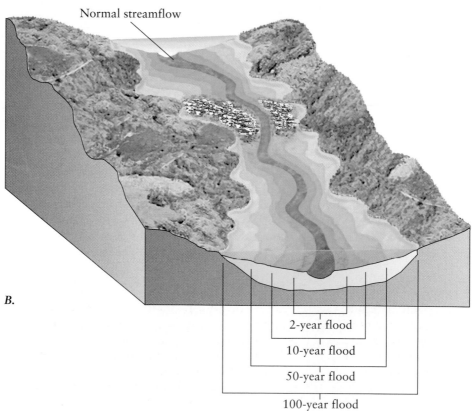

B.

Another important aspect of flood prediction is the real-time monitoring of storms. Hydrologists use the term *forecast* to refer to a short-term prediction that specifies the magnitude of a flood's peak and the time when it will pass a particular location. Accurate forecasting combines information about the progress of the storm with knowledge of the area, including topography, vegetation, and impermeable ground cover. This information is used to predict the amount of surface runoff, its velocity, and its probable course, and to issue an early warning to communities that may be affected.

## Human Intervention

Flooding is highly susceptible to the effects of human intervention. River channels are often modified or "engineered" for the purpose of flood control and protection, as well as to increase access to floodplain lands, facilitate transportation, enhance drainage, and control erosion. The modifications usually consist of some combination of straightening, deepening, widening, clearing, or lining of the natural channel. These are collectively referred to as *channelization*. In the context of flood control, channelization is generally undertaken with the aim of increasing the channel's cross-sectional area. Given the relationships among discharge, velocity of flow, and cross-sectional area, an increase in the width and depth of a channel should enable it to handle a greater discharge at a higher velocity.

**? Why do people modify natural river channels?**

Channelization is a controversial process, however. Many opponents believe that channel modifications interfere with natural habitats and ecosystems. The aesthetic value of the river can be degraded, the groundwater regime disrupted, and water pollution aggravated by channelization. Paradoxically, although channelization may control flooding in the immediate area, it can actually contribute to more intense flooding further downstream. Perhaps the most important thing to understand regarding channelization is that *any* modification of a channel's course or cross-section renders invalid any series of hydrologic data collected there in the past. During the Mississippi River floods of 1973 and 1993, the actual water levels were much higher than had been predicted for the discharges recorded during those events. Most analysts agree that extensive modifications of the river's channel over many years contributed significantly to the failure to predict the extent of damage from those floods.

Urban development can also contribute to flooding in a variety of ways. Urban construction on unconsolidated, compressible sediments, often accompanied by withdrawal of groundwater, can lead to subsidence and urban flooding. The increase in impermeable ground cover associated with urbanization can add substantially to surface runoff in urban areas. Storm sewers can also be important contributors to flooding because they allow the runoff from paved areas to reach the river channel more quickly. Floods in urbanized basins often have higher peak discharges and higher total discharges and reach their peak more quickly than floods in undeveloped basins.

## Surface Fresh Water as a Resource

A reliable water supply is critical not only for survival and health but also because of its role in industry, agriculture, and other economic activities. Globally, crop irrigation accounts for 73 percent of the demand for water, industry for 21 percent, and domestic use for 6 percent, although the proportions vary from one region to another. Demand in each of these sectors has more than tripled since 1950. Population growth is partly responsible for the increasing demand, but improvements in standards of living have also contributed to the large increase in per capita water use over the past few decades. The total amount of water being withdrawn or diverted from rivers, lakes, and groundwater for human use is now

**The Aral Sea,** on the border between Kazakhstan and Uzbekistan, is shrinking so rapidly that once-prosperous fishing villages now lie 50 km (30 mi) from the shore (Fig. B12.1). Human activities caused the change. Thirty years ago, the Aral Sea was the fourth largest lake in the world after the Caspian Sea, Lake Superior, and Lake Victoria. The sea covered 68,000 km$^2$ (26,250 mi$^2$), had an average depth of 16 m (52.5 ft), and yielded 45,000 tons of fish a year. Today the sea is only the sixth largest lake. It now covers 40,000 km$^2$ (15,440 mi$^2$), has an average depth of 9 m (29.5 ft), and is so salty that its fishing industry is dead. It is disappearing so fast that by 2010 it will be a waterless desert.

The Aral Sea is fed by two large rivers, the Amu Dar'ya and the Syr Dar'ya, which carry meltwater across the desert from the snowy mountains of northern Afghanistan, Tadzhikistan, and Kirghizia. The sea is shrinking now because the inflow of water has declined, mainly because the extent of irrigation (which has been practiced in the river valleys for millennia) has increased dramatically in modern times. By 1960, so much water was being taken from the two rivers that inflow to the Aral Sea had declined to a trickle. The sea has been shrinking steadily ever since.

The people who planned the irrigation systems expected the Aral Sea to shrink. What they did not anticipate were the side effects. The sea, it is now realized, exerts a major influence on the local climate. Because it is shrinking, local rainfall is declining, the average temperature is rising, and wind velocities are increasing. Most of the newly exposed sea bottom is covered with salt. The wind blows the salt around and has created withering salt storms. Supplies of potable water have declined, and various diseases, especially intestinal diseases, are afflicting the local population at alarming levels.

The situation could be reversed by greatly reducing the amount of irrigation. The problem is that the irrigated area is now extremely prosperous, so the ultimate solution will probably have to be a compromise between returning the sea to its original size and keeping all the irrigated land.

**Figure B12.1**
**THE SHRINKING SEA**
This used to be part of the Aral Sea. These fishing boats were stranded in the desert created when the Aral Sea shrank as a result of the diversion of water from the rivers that once fed into it.

**Figure 12.21**
**A DYING LAKE**
Why is this lake not blue and clear? It has turned green and mucky-looking because of the growth of algae stimulated by excessive plant nutrients (probably from sewage or farm fertilizers). Eventually, the algae will use up all the oxygen in the water, making it impossible for other life to exist in the lake; fish and other aquatic life will suffocate. This is called *eutrophication.*

about 4340 km$^3$/yr (1040 mi$^3$/yr), eight times the annual streamflow of the Mississippi River!

Often water can be drawn from a nearby lake or stream to serve a local population. But sometimes, because of population growth and development, regions with the greatest demand for water do not have an abundant and readily available supply of surface water. For this reason, surface water is often transferred from one drainage basin to another, sometimes over long distances. Aside from raising political issues related to water rights, such *interbasin transfer* can have negative environmental impacts. The diversion of surface water may affect the flow and salinity of the water, the amount of sediment carried by the stream, and even the local climate. Ecosystems and water users both upstream and downstream may be affected. The unexpected environmental impacts of surface freshwater diversion are illustrated by the case of the Aral Sea, discussed in Box 12.1.

Aside from supply and the environmental impacts of diversion, another source of concern is the quality of surface water resources. Because of their accessibility, surface water bodies are highly susceptible to contamination. Sometimes wastes are intentionally discharged into lakes and streams. Many cities dump sewage directly into nearby water bodies, sometimes with little or no treatment. The accumulation of organic material (the main component of sewage) leads to accelerated *eutrophication,* in which the water becomes clogged with decaying organic material such as mucky green algae.[4] As the algae and other organic materials decay, they use up the supply of dissolved oxygen in the water, eventually making it impossible for other life to exist there (Fig. 12.21). Surface runoff from agricultural areas, which may contain fertilizers and other nutrients and organic materials, can also lead to eutrophication. Industrial wastewaters, poorly engineered landfill sites, and urban runoff are other sources of surface water contamination.

---

[4] Eutrophication can happen in any water body in which the demand for oxygen (called *biological oxygen demand,* or *BOD*) exceeds the supply of dissolved oxygen in the water. This is part of the natural process whereby a lake turns into a swamp. When the increased oxygen demand is caused by anthropogenic inputs, it may be referred to as *cultural eutrophication.*

# FRESH WATER UNDERGROUND

**groundwater** Subsurface water, that is, the water contained in spaces within bedrock and regolith.

> **?** **Does water flow in rivers under the ground?**

Less than 1 percent of all water in the hydrologic cycle is **groundwater,** which may be loosely defined as all subsurface water, that is, the water contained in spaces within bedrock and regolith. Although the percentage of groundwater sounds small, its volume is 40 times greater than that of all the water held in freshwater lakes or flowing in streams. Water is present everywhere beneath the Earth's surface, even in hot deserts. More than half of it is near the surface, occurring within a depth of 750 m (2460 ft) below ground. At greater depths the amount of groundwater decreases gradually. A small amount of water may be present in crustal rocks at depths of many kilometers. However, the pressure exerted by overlying rocks is so great and the openings in rocks are so small that the water cannot move freely through the enclosing rocks. For our purposes, therefore, we will think of groundwater as the water found between the land surface and a depth of about 750 m.

## The Water Table

Much of what we know about groundwater has been learned from the accumulated experience of hundreds of generations of people, who have dug or drilled millions of wells. This experience tells us that a hole penetrating the ground ordinarily passes first through a layer of moist soil and then into a zone in which the spaces between the grains in regolith or bedrock are filled mainly with air. This is the *zone of aeration* (Fig. 12.22). It is also called the *unsaturated zone,* for although water may be present, it does not completely saturate the ground.

After passing through the zone of aeration, the hole enters the *saturated zone,* in which all openings are filled with water. The top of the saturated zone is the **water table,** which tends to imitate the shape of the land surface above (Fig. 12.22). The water table is high beneath hills and low beneath valleys. Water tends to move toward low points in the topography, where the pressure is lowest, but it moves very slowly because it has to move through tiny spaces in the rocks. If all rainfall were to cease, the water table would slowly flatten. Seepage of water into the ground would diminish and then cease, and streams would dry up as the water table fell. During droughts, the depression of the water table is evident from the drying up of wells. When a well becomes dry, we know that the water table has dropped to a level below the bottom of the well. Repeated rainfall, which soaks the ground with fresh supplies of water, maintains the water table at a normal level and keeps surface water bodies replenished.

**water table** The top surface of the saturated zone.

Whether it is deep or shallow, the water table represents the upper limit of readily usable groundwater. For this reason, a major aim of groundwater specialists and well drillers is to determine the depth and shape of the water table. To do this they must first understand how groundwater moves and what forces control its distribution underground.

## How Groundwater Moves

Most groundwater is in motion. Unlike the swift flow of rivers, however, which is measured in kilometers per hour, the movement of groundwater is so slow that it is measured in centimeters per day or meters per year. The reason is simple: Whereas the water of a stream flows through an open channel, groundwater must move through small, constricted passages, often along a tortuous route. Therefore, the rate of groundwater flow is dependent on the nature of the rock or sediment through which the water moves, especially its porosity and permeability.

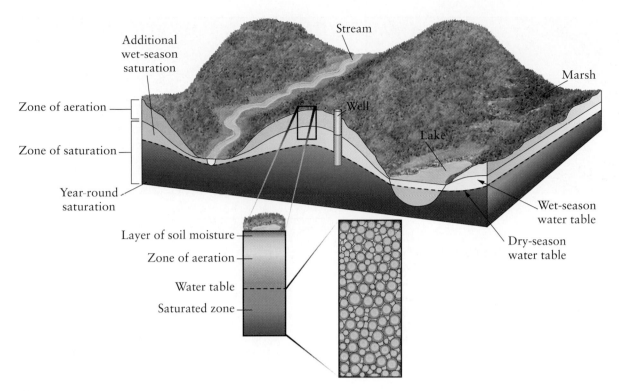

**Figure 12.22**
**WATER UNDER THE GROUND**
If you dug underground, you would first excavate through soils and sediments that have pores filled with both air and water. This is the zone of aeration. If you kept digging, eventually you would reach the water table. Below the water table the pore spaces in rock and soil are completely filled with water—no air.

## Porosity and Permeability

**Porosity** is the percentage of the total volume of a body of rock or regolith that consists of open spaces, or *pores*. Porosity determines the amount of fluid a sediment or rock can contain. The porosity of sediments is affected by the sizes and shapes of the rock particles and the compactness of their arrangement (Fig. 12.23A and B). The porosity of a sedimentary rock is also affected by the extent to which the pores have been filled with cement (Fig. 12.23C). Plutonic igneous rocks and metamorphic rocks, which consist of many closely interlocking crystals, generally have much lower porosities than sediments and sedimentary rocks. However, if crystalline rocks have many joints and fractures, their porosity will be higher.

Permeability is a measure of how easily a solid allows fluids to pass through it. A rock with low porosity is likely also to have low permeability. However, high porosity does not necessarily mean high permeability, because the size and continuity of the pores (that is, the extent to which the pores are interconnected) also influence the ability of fluids to flow through the material.

**porosity** The percentage of the total volume of a body of rock or regolith that consists of open spaces, or *pores*.

**permeability** A measure of how easily a solid allows fluids to pass through it.

## Percolation

Water from a rain shower soaks into the soil (the process of infiltration). Some of the water in the soil evaporates, and some of it is taken up by plants. The remaining water seeps downward under the influence of gravity, a process called

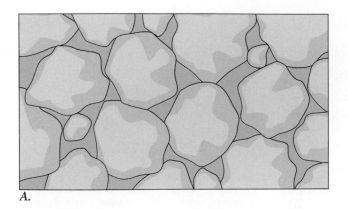

A.

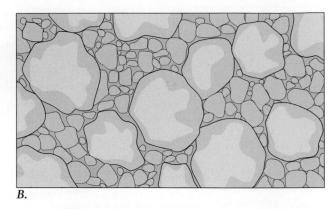

B.

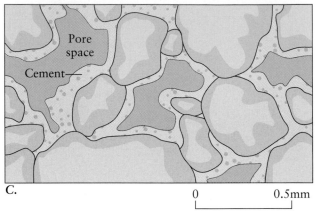

C.

0        0.5mm

**Figure 12.23**
POROSITY IN DIFFERENT KINDS OF SEDIMENTS
In these examples, all of the pore spaces are filled with water, as they would be in the saturated zone. In the zone of aeration at least some of the pore spaces would be filled by air. *A.* The porosity is about 30 percent in this sediment with particles of uniform size. *B.* The porosity is 15 percent in this sediment, in which fine grains fill the spaces between larger grains. *C.* Porosity can be reduced by the presence of cement binding grains together and filling the pores.

**percolation** The process by which groundwater seeps downward under the influence of gravity.

**percolation,** until it reaches the water table. The movement of groundwater in the saturated zone is similar to the flow of water that occurs when a water-soaked sponge is squeezed gently. Water moves slowly through very small pores along parallel, threadlike paths. The water flows from areas where the water table is high toward areas where it is lower. In other words, it generally flows toward surface streams or lakes (Fig. 12.24). Some of the flow paths turn upward and enter the stream or lake from beneath, seemingly defying gravity. This upward flow occurs because groundwater is under greater pressure beneath a hill than beneath a stream or lake. Since water tends to flow toward points where pressure is low, it flows toward bodies of water at the surface, where the pressure is low.

### Recharge and Discharge

**recharge** Replenishment of groundwater.

**discharge** The process by which subsurface water leaves the saturated zone and becomes surface water.

**spring** A flow of groundwater that emerges naturally at the ground surface.

**Recharge** or replenishment of groundwater occurs when rainfall and snowmelt infiltrate the ground and percolate downward to the saturated zone (Fig. 12.24). The water moves slowly along its flow path toward zones where **discharge**[5] occurs, that is, where subsurface water leaves the saturated zone and becomes surface water. In discharge zones, subsurface water either flows out onto the ground surface as a **spring** (a flow of groundwater that emerges naturally at the ground surface) or joins bodies of water such as streams, lakes, ponds, swamps, or the ocean. Pumping groundwater from a well also creates a point of discharge. The

---

[5] Note that *discharge* has two meanings: (1) the amount of water (width × depth × velocity of flow) passing by a point on a channel's bank during a unit of time; and, (2) a zone where subsurface water leaves the saturated zone and becomes surface water.

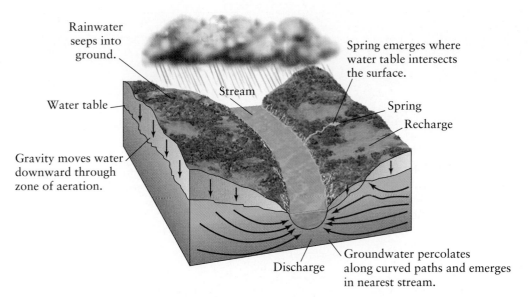

**Figure 12.24**
**HOW GROUNDWATER FLOWS**
Rainwater and snowmelt soak into the soil. The long, curved arrows represent a few
of the possible flow paths for groundwater. Surface bodies of water occur wherever
the water table intersects the land surface. Some of the groundwater paths turn upward
where they flow into surface water bodies, because the pressure is lower there than
under hills.

amount of time water takes to move through the ground to a discharge area depends on distance and rate of flow. It may take as little as a few days or as much
as thousands of years.

# Caves, Caverns, and Karst

In regions underlain by rocks that are highly susceptible to chemical weathering,
groundwater creates extensive systems of underground caverns. In such areas, a
distinctive landscape forms on the surface. *Karst topography*, named for the
Karst region of the former Yugoslavia, is characterized by many small, closed
basins and a disrupted drainage pattern (Fig. 12.25A). Streams disappear into
the ground and eventually reappear elsewhere as large springs. Karst is most typical of regions underlain by limestone and dolostone. Let's take a few minutes to
consider the mechanisms whereby groundwater creates underground cave systems and karst topography.

## *Dissolution*

As soon as rainwater infiltrates the ground, it begins to react with the minerals in
regolith and bedrock, causing chemical weathering (as discussed in chapter 7).
An important part of this process is *dissolution*, in which minerals and rock materials are dissolved into the groundwater and carried away in solution.

Among the rocks of the Earth's crust, the carbonate rocks (those consisting
of minerals based on the $CO_3^{2-}$ anion, such as calcite and dolomite) are most
readily attacked by dissolution. Carbonate rocks, such as limestone, dolostone,
and marble, are almost insoluble in pure water but are readily dissolved by *carbonic acid* ($H_2CO_3$). You may recall from chapter 7 that carbonic acid is a com-

# When Groundwater Dissolves Rocks

**Figure 12.25**
**A.** Karst landscape regions underlain by easily dissolved carbonate rocks often develop karst topography, which is characterized by many small, closed basins and a disrupted drainage pattern. This is the Karst region in former Yugoslavia.

mon constituent of rainwater, formed when the rainwater reacts with carbon dioxide ($CO_2$). The weathering attack occurs mainly along fractures and other openings in the carbonate bedrock, often with impressive results. When limestone weathers, nearly all of its volume may be dissolved in slowly moving groundwater. In some carbonate terrains, the rate of dissolution is faster than the average rate of erosion of surface materials by streams and mass wasting.

## Caves and Sinkholes

**cave** Underground open space.

**Caves** are underground open spaces. A large cave or system of interconnected cave chambers is called a *cavern*. The Carlsbad Caverns in southeastern New Mexico (Fig. 12.25B) include a chamber that is 1200 m long, 190 m wide, and 100 m high (4000 ft long × 620 ft wide × 330 ft high). The recently discovered Good Luck Cave on the island of Borneo includes a chamber so large that it could accommodate not only the world's largest previously known chamber (in Carlsbad Caverns) but also the largest known chamber in Europe (in Gouffre St. Pierre Martin, France) and the largest chamber in Britain (Gaping Ghyll).

Caves are formed when carbonate rock is dissolved by circulating groundwater. Most caves appear to be excavated near the top of a seasonally fluctuating water table. The process begins with dissolution along interconnected fractures and bedding planes. A cave passage then develops along the most favorable flow

**B. CAVE**
For many thousands of years, groundwater has percolated through these carbonate rocks, dissolving and carrying away material to form the great Carlsbad Caverns.

**C. SINKHOLE!**
This sinkhole formed in Winter Park, near Orlando, Florida. The crater appeared at 7:00 P.M., May 8, 1981, and grew to be 100 m wide (330 ft) wide by noon the following day.

route. Carbonate formations are chemically precipitated drop by drop like icicles on the cave walls and ceiling, while a stream occupies the floor. After the stream has stopped flowing, similar formations are deposited on the floor. These carbonate deposits are *stalactites* (hanging from the ceiling) and *stalagmites* (projecting upward from the floor).

The rate of cave formation is related to the rate of dissolution. In areas where the groundwater is acidic, the rate of dissolution increases with increasing velocity of flow. As a passage grows and the flow of groundwater becomes more rapid and turbulent, the rate of dissolution also increases. Still, the development of a continuous passage by slowly moving groundwater may take up to 10,000 years, and the further enlargement of the passage by more rapidly flowing groundwater to create a fully developed cave system may take an additional 10,000 to 1 million years.

Caves are dissolution cavities that are closed to the surface or have only a small opening. In contrast, a *sinkhole* is a dissolution cavity that is open to the sky. Some sinkholes are formed when the roofs of caves collapse, whereas others are formed at the surface where rainwater is freshly charged with carbon dioxide and hence is most effective as a solvent. Some sinkholes form slowly; others form catastrophically. An example of the latter is the Winter Park sinkhole in Florida (Fig. 12.25C). This large sinkhole formed over a period of only a few hours. It began at 7:00 P.M. on May 8, 1981, when a small tree in a vacant lot suddenly disappeared. By the time the sinkhole had stabilized, it had swallowed up part of a house, six commercial buildings, and the municipal swimming pool, as well as several automobiles. The damages totaled over $2 million. Events as dramatic as the Winter Park sinkhole are rare, but sinkhole collapse is a common occurrence in areas underlain by carbonate rocks.

## Groundwater as a Resource

**aquifer** A body of water-saturated, porous and permeable rock or regolith.

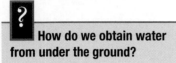

**?** **How do we obtain water from under the ground?**

When we wish to find a reliable supply of groundwater, we search for an **aquifer** (Latin for *"water carrier"*), a body of water-saturated, porous, and permeable rock or regolith. Gravel and sand generally make good aquifers, for they tend to be both porous and permeable. Many sandstones are also good aquifers. An aquifer in which the water is free to rise to its natural level is an *unconfined aquifer* (Fig. 12.26). In a well drilled into an unconfined aquifer, the water will rise to the level of the surrounding water table, the top of the saturated zone. The water in an unconfined aquifer is in contact with the atmosphere, through the porosity of the overlying rocks or sediments.

In contrast, a *confined aquifer* is overlain by impermeable rock units, called *aquicludes* (Fig. 12.26). The water in a confined aquifer is held down by the pressure of the overlying impermeable unit. If a well is drilled into the aquifer, the pressure will cause the water to rise in the well. If the ground surface is lower than the level of the water table in the recharge zone, water may flow out of the well without having to be pumped. This is called an *artesian well* (Fig. 12.27). Although water from artesian wells is sometimes touted as having health-giving qualities, it is no different from any other groundwater.

As mentioned earlier, a spring is a flow of groundwater that emerges naturally at the ground surface. The simplest kind of spring is one that issues from a place where the land surface intersects the water table (Fig. 12.26). A change in permeability often gives rise to a spring or a line of springs. Such a change may be due to the presence of an aquiclude. In Figure 12.28A, for example, a porous limestone overlies an impermeable shale, and a spring discharges at the point of contact between the two types of rock. Springs may also issue from lava flows,

**Figure 12.26**
**AQUIFERS, CONFINED AND UNCONFINED**
In an unconfined aquifer the top of the saturated zone is the water table, and it is open to the atmosphere through pores in the rock and soil above the aquifer. In contrast, the water in a confined aquifer is trapped between impermeable rock layers.

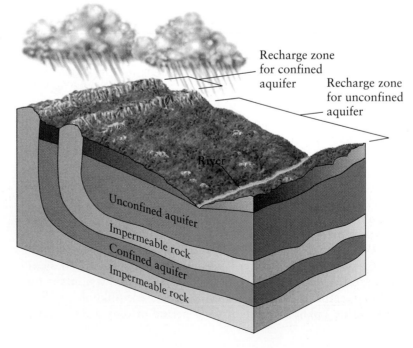

Recharge zone for confined aquifer

Recharge zone for unconfined aquifer

River

Unconfined aquifer

Impermeable rock

Confined aquifer

Impermeable rock

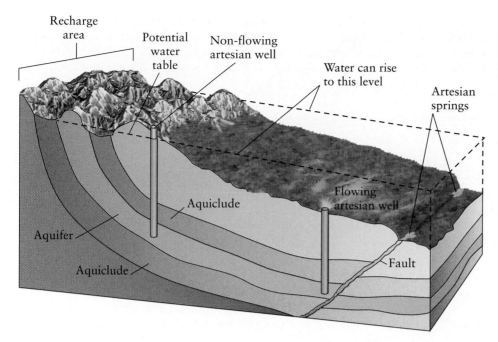

Recharge area

Potential water table

Non-flowing artesian well

Water can rise to this level

Artesian springs

Aquiclude

Flowing artesian well

Aquifer

Aquiclude

Fault

**Figure 12.27**
**ARTESIAN WATER**
Two conditions are necessary for an artesian system to exist: a confined aquifer and sufficient water pressure in the aquifer to make the water in a well rise above the aquifer. The water in a well drilled into the aquifer rises to the height of the water table in the recharge zone, indicated by the dashed line. If this height is above the ground surface, the water will flow out of the well without being pumped.

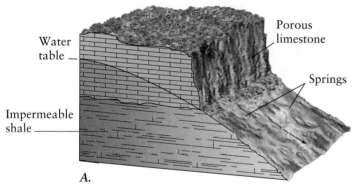

Water table

Porous limestone

Springs

Impermeable shale

*A.*

**Figure 12.28**
**WHAT CAUSES SPRINGS?**
Examples of springs formed under different geologic conditions. Springs often occur at places where there is a change in rock permeability. In *A*, a porous limestone overlies an impermeable shale unit, an aquiclude. Consequently, a line of springs occurs along the hillside where the two rock units meet. In *B*, springs issue from the contact between a highly jointed (and therefore permeable) lava flow and the underlying impermeable mudstone. In *C*, springs flow from the place where a fault intersects the ground surface.

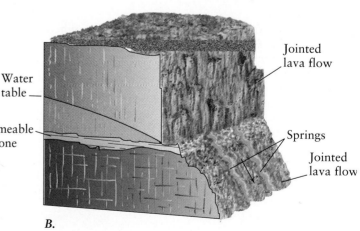

Water table

Impermeable mudstone

Jointed lava flow

Springs

Jointed lava flow

*B.*

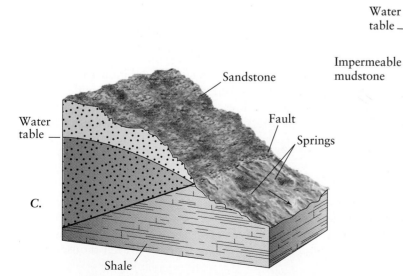

Sandstone

Fault

Springs

Water table

*C.*

Shale

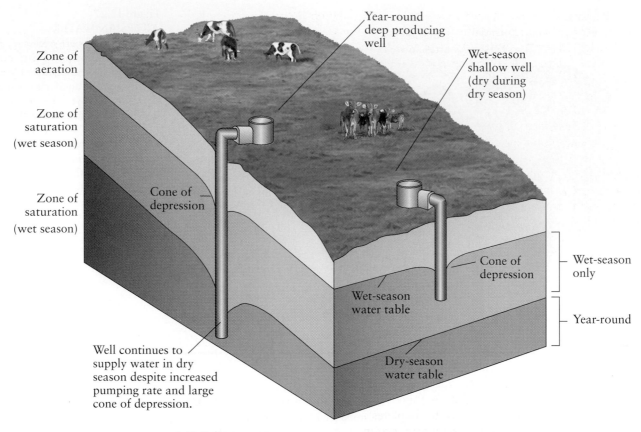

Zone of aeration

Zone of saturation (wet season)

Zone of saturation (wet season)

Cone of depression

Well continues to supply water in dry season despite increased pumping rate and large cone of depression.

Year-round deep producing well

Wet-season shallow well (dry during dry season)

Cone of depression

Wet-season water table

Dry-season water table

Wet-season only

Year-round

**Figure 12.29**
**WELLS: YEAR-ROUND AND SEASONAL**
Seasonal changes affect the height of the water table. During the wet season, recharge is high, so the water table rises high enough to reach even shallow wells. During the dry season, the water table falls. The deeper well still reaches below the water table, so it continues to supply water, but the shallow well runs dry. Increased pumping has created a large cone of depression around the deeper well.

especially where a jointed lava bed overlies an aquiclude (Fig. 12.28B) or along the trace of a fault (Fig. 12.28C).

A well will supply water if it intersects the water table. As shown in Figure 12.29, a shallow well may become dry during periods when the water table is low, whereas a deeper well may yield water throughout the year. When water is pumped from a well, a *cone of depression* (a cone-shaped dip in the water table) will form around the well (Fig. 12.29). In most small domestic wells, the cone of depression is hardly discernible. Wells pumped for irrigation and industrial uses, however, withdraw so much water that the cone may become very wide and steep and lower the water levels in surrounding wells. When large cones of depression from pumped wells overlap, the result is regional depression of the water table.

## Groundwater Mining

Groundwater is a major source of water for human consumption, especially in dry regions where there are few streams. If the rate of withdrawal exceeds the rate of natural recharge, the volume of stored water steadily decreases; this is sometimes called *groundwater mining*. It may take hundreds or even thousands of years for a depleted aquifer to be replenished. The results of excessive withdrawal include lowering of the water table, drying up of springs and streams, compaction of the aquifer, and subsidence (lowering of the ground surface). Sometimes it is possible to recharge an aquifer by pumping water into it. In other cases the effects may be permanent. When an aquifer suffers *compaction*—that is, when its mineral grains collapse on one another because the pore water that held them apart has been removed—it is permanently damaged and may never be able to hold as much water as it originally held.

Urban development can increase the rate of depletion of groundwater, not only by increasing the demand for water but also by increasing the amount of impermeable ground cover in the area. When a recharge area is covered by roads, parking lots, buildings, and sidewalks, the rate of groundwater recharge can be substantially reduced. Subsidence caused by withdrawal of groundwater is a problem in many urban areas, such as Mexico City, Bangkok, and Venice. The weight of buildings also contributes to the compaction of compressible sediments.

Unfortunately, laws and policies relating to groundwater rights are very complicated, partly because water law, in general, has a complicated history. But the application of water law to groundwater is even more complicated than for surface water because groundwater is hidden from view. It is difficult to monitor the flow of groundwater and regulate its use. There are some complex issues associated with groundwater rights. For example, if you drill a well into an aquifer underlying your property, are you entitled to withdraw as much water as you need from that well? Should you withdraw water only for your own purposes, or should you be permitted to withdraw the water and sell it elsewhere? What happens if withdrawing the groundwater depletes the aquifer and your neighbor's well runs dry? In most parts of the world, groundwater use is unlimited and unregulated, which sometimes leads to serious conflicts between adjacent landowners. Similar problems arise when a landowner's activities cause an aquifer to become contaminated; in some areas groundwater contamination is not considered to be a problem *unless* the contaminant migrates underground and crosses a property boundary.

## Groundwater Contamination

Many of the types and sources of contaminants that affect surface water also cause groundwater contamination (Fig. 12.30). Because of its hidden nature,

**Can an aquifer be permanently depleted?**

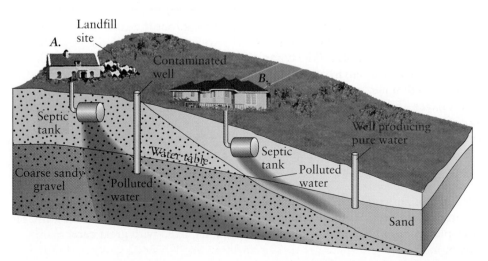

**Figure 12.30
CONTAMINATED
GROUNDWATER**
A groundwater system can become contaminated in many ways. *A.* Rainwater leaches toxic chemicals in an unlined landfill, percolates downward, and contaminates an underlying aquifer. *B.* A leaking septic tank contaminates groundwater, and any wells or streams that lie downslope. Agricultural chemicals and leaking underground storage tanks are common sources of groundwater contamination.

Figure 12.31
REMOVING "LUST"
Removing a leaking underground storage tank at a gas station in Silver
Spring, Maryland.

however, groundwater contamination is more difficult to detect, control, and clean up. The most common source of water pollution in wells and springs is untreated sewage. Agricultural pesticides and fertilizers, significant sources of surface water pollution, are also common contaminants in groundwater. Harmful chemicals leaking from waste disposal facilities can infiltrate into groundwater reservoirs and contaminate them. Leaking underground storage tanks (commonly referred to by the acronym LUST) at gas stations, refineries, and other industrial settings are among the most serious groundwater contamination problems in North America. Gasoline contains thousands of potentially toxic compounds, and it has been estimated that as many as 25 percent of underground storage tanks are leaking. In other words, at any intersection with a gas station on each corner, the chances are high that one of the four stations may have a leaking storage tank (Fig. 12.31).

# WHAT'S AHEAD

As we begin to understand the complex interconnections between various Earth systems and their environmental functions, we also begin to appreciate the broader economic, geologic, and ecologic functions performed by water, both on the surface and underground. Water plays a role in almost every important geologic process in and on the Earth. We rely on natural waters to support wildlife habitats and fish populations, dilute and disperse pollutants, maintain soil moisture, generate electricity, transport goods, and provide recreational sites. Today 26 countries worldwide, with a total population of 232 million people, are designated as *water-scarce*. The lack of water in these countries places serious constraints on agricultural production, economic development, health, and environmental protection. Water—the material that is responsible for sustaining life on this planet—is a resource to be cherished and managed with great care.

In this chapter, we have focused principally on the freshwater reservoirs in the hydrologic cycle, as well as on the surface and subsurface processes that govern the movement and distribution of fresh water. In the next chapter (chapter 13), we turn our attention to the ocean, the largest water reservoir on this planet, and the atmosphere, another of the great Earth systems. The atmosphere and ocean are inextricably linked, two parts of a huge, complex machine that determines the Earth's climate and weather. We will look at how the ocean, atmosphere, and solid Earth evolved together to their present state. And we will examine what happens when the ocean meets the coastline—one of the most dynamic geologic environments on the Earth.

## Chapter Highlights

**1.** The water cycle or **hydrologic cycle** is the movement of water from one reservoir to another in the Earth system. The ocean is the largest reservoir, followed by the polar ice sheets. The largest reservoir of unfrozen fresh water is groundwater. Surface water bodies, the atmosphere, and the biosphere are much smaller water reservoirs.

**2.** The pathways or processes by which water moves from one reservoir to another include **evaporation, condensation, precipitation, transpiration, infiltration, percolation,** and **surface runoff.** The hydrologic cycle maintains a mass balance. The total amount of water doesn't change; the water simply moves from place to place within the system.

**3.** A **stream** is a body of water that flows downslope along a clearly defined natural passageway, the **channel,** transporting particles and dissolved substances. A **river** is a stream of considerable volume with a well-defined channel. Factors that influence the behavior of a stream include the gradient, the cross-sectional area of the channel, velocity of water flow, **discharge,** and **load.** All of these factors are interrelated.

**4.** Streams create landforms through **erosion** and **deposition.** Point bars, alluvial fans, deltas, and **floodplains** are depositional landforms. Common types of channels are sinuous, straight, meandering, and braided.

**5.** Every stream is surrounded by its **drainage basin,** the total area from which water flows into the stream. The line that separates adjacent drainage basins is a **divide.** The type and structure of rock underlying a drainage basin influence the type of drainage pattern that will form.

**6.** A **flood** occurs when a stream's discharge becomes so great that it exceeds the capacity of the channel, causing the stream to overflow its banks. Prediction of flooding is based on analyses of the frequency of occurrence of past events and on real-time monitoring of storms.

**7.** Surface water bodies are susceptible to contamination. A common problem is eutrophication, the accumulation and consequent decay of organic material, resulting in depletion of oxygen. Agricultural runoff and sewage, which contain nutrients and organic material, can lead to eutrophication. Industrial wastewater, poorly engineered landfill sites, and urban runoff are other significant sources of surface water contamination.

**8. Groundwater** is subsurface water contained in spaces within bedrock and regolith. More than half of all groundwater occurs within about 750 m (2460 ft) of the Earth's surface. In the zone of aeration (unsaturated zone), water is present, but it does not completely saturate the ground. In the saturated zone, all openings are filled with water. The top of the saturated zone is the **water table.**

**9.** The rate of groundwater flow is dependent on the characteristics of the rock or sediment through which the water must move. **Porosity** is the percentage of the total volume of a body of rock or regolith that consists of open space. **Permeability** is a measure of how easily a solid allows fluids to pass through it. Groundwater in the saturated zone moves slowly by **percolation** through very small pores from areas where the water table is high to areas where it is lower, generally toward surface streams or lakes.

**10.** Groundwater **recharge** occurs when rainfall and snowmelt infiltrate and percolate downward to the saturated zone. The water moves slowly along its flow path toward zones where **discharge** occurs, that is, where subsurface water leaves the saturated zone and becomes surface water.

**11.** An **aquifer** is a body of highly permeable rock or regolith lying in the zone of saturation. An aquifer in which the upper surface of the saturated zone coincides with the water table is an unconfined aquifer. An aquifer that is bounded by impermeable rock units—aquicludes—is a confined aquifer. If the rate of groundwater withdrawal from an aquifer exceeds the rate of natural recharge, the volume of stored water decreases steadily. The results of excessive withdrawal can include lowering of the water table, drying up of springs and streams, compaction, and subsidence.

**12.** Many of the types and sources of contaminants that affect surface water also cause groundwater contamination. The most common source of water pollution in wells and springs is untreated sewage. Agricultural pesticides and fertilizers, significant sources of surface water pollution, are also common contaminants in groundwater. Harmful chemicals leaking from waste disposal facilities can infiltrate groundwater reservoirs and contaminate them.

# ▶ The Language of Geology

- alluvium 338
- aquifer 358
- cave 356
- channel 335
- condensation 331
- deposition 336
- discharge (two meanings) 336, 354
- divide 340
- drainage basin 340
- erosion 336
- evaporation 331

- flood 344
- floodplain 338
- groundwater 352
- hydrologic cycle 330
- hydrosphere 329
- infiltration 331
- lake 343
- load 336
- percolation 354
- permeability 353
- porosity 353

- precipitation 331
- recharge 354
- river 335
- spring 354
- stream 335
- streamflow 335
- surface runoff 331
- transpiration 331
- water table 352

# ▶ Questions for Review

1. What is the largest reservoir for unfrozen fresh water in the hydrologic cycle?
2. What does it mean when we say that the hydrologic cycle maintains a mass balance?
3. Use Figure 12.6B to explain how meanders form.
4. In what ways do the types and structures of underlying rock influence drainage patterns? Use both the text and Figure 12.15 in arriving at your answer.
5. What kinds of changing climatic and geologic conditions can make lakes appear or disappear?
6. In the discharge hydrograph shown in Figure 12.17, the peak discharge lagged about 2.5 hours behind the peak rainfall. Use the figure to determine how long it took for the storm runoff to pass through the system and for the discharge to fall to its prestorm level.
7. What is the top surface of the saturated zone called?
8. What is the difference between a confined aquifer and an unconfined aquifer?
9. What is the difference between porosity and permeability?
10. What are the environmental side-effects of excessive groundwater withdrawal? What are the common environmental side-effects of surface diversion and interbasin transfer?
11. Describe the common pathways by which groundwater may become contaminated.

## ► *Questions for Thought and Discussion*

1. List as many ways as you can in which we depend on the availability of fresh water in our daily lives.
2. Investigate the causes and effects of the 1997 flooding of the Red River and its tributaries in northern United States and southern Canada.
3. The concept of residence time is very important in geology. It applies not only to water but also to any substance that moves from one reservoir to another in the Earth system. Why is it important to understand the factors that determine how long a substance will remain in a reservoir? Can you think of any other substances, besides water, that move around in the Earth system? Why do you think we might want to monitor those substances and keep track of their residence times in different reservoirs?
4. Visit a stream before and after an intense rainfall. What changes do you notice? Keep in mind the factors that control the behavior of streams.
5. Where does your community obtain its water supply? Is it a groundwater source or a surface freshwater source? Is either the quantity or the quality of the water threatened?

## *Virtual Internship: You Be the Geologist*

In the Groundwater Contamination internship on your CD-ROM, you are an environmental geologist. Your task is to discover the source of contaminated groundwater found in the basement of a small-town school.

For an interactive case study on floods, saline lakes, and water management in dry land regions, visit our Web site.

# · 13 ·

# OCEANS, WINDS, WAVES, AND COASTLINES

The rhythm of the seasons determines how we live, from the food we eat and the clothes we wear to the sports we play and our places of travel. But why are there seasons? The answer, of course, is that the Earth rotates around an axis that is tilted at an angle to the plane of the Earth's orbit around the Sun. If that were not the case, the Sun would always be directly overhead along the Equator. The axis of rotation is tilted 23.5° away from perpendicular to the plane of the Earth's orbit. As a result, the place where the Sun is directly overhead at noon changes a little bit each day.

What an influence that tilt of 23.5° has on the environment! Plants bloom in the spring and go dormant in the fall. Animals and birds migrate great distances to ensure food supplies and warm weather. We sedentary humans have a different response; we burn fossil fuels to heat our houses in the winter and cool them in the summer.

The seasons play a very important role in wind patterns, in ocean currents, in the climate, and in such things as weathering of rocks, formation of soils, and distributions of plants and animals. Look at the photograph of the Earth on the facing page. The great swirling masses of clouds attest to global wind patterns, and the belts of brown and green across Africa are due to belts of rainfall. If the tilt were more or less than 23.5°, the world would be a very different place.

◆

### In this chapter we discuss

- The atmosphere, its composition, and its structure
- Global wind patterns
- The Earth's climate zones
- The composition and structure of the ocean
- The origin of tides, ocean currents, and waves
- Coastal erosion and coastal landforms

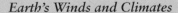
*Earth's Winds and Climates*

Viewed from space, the Earth's winds are revealed by the swirling patterns of clouds. Climate zones are revealed by the brown color of desert areas contrasted to the greens of tropical regions. The equator runs approximately through the center of the green belt that crosses Africa.

Can you imagine the Earth without an ocean? An oceanless Earth would be dry, dusty, and probably lifeless. Fortunately for life as we know it, however, our planet does have an ocean; seawater covers 71 percent of the Earth's surface. Far from being a dry planet, the Earth is a very wet planet. As we saw in chapter 2, one of the unique characteristics of the Earth is that $H_2O$ is present in three states—water, ice, and water vapor. This makes the Earth habitable and distinguishes it from the other rocky planets. Mars is so cold that only ice and water vapor are present. Venus is so hot that only water vapor is present. Scientists call the Earth a "Goldilocks planet" because, as in the story of Goldilocks and the Three Bears, "everything is just right—not too hot and not too cold."

In this chapter we discuss both the oceans and the atmosphere because they are interdependent parts of the Earth system. They are the two great water reservoirs fundamental to the functioning of the global hydrologic cycle and to all the biogeochemical cycles. The ocean plays a critical role in climate and weather systems, particularly in regulating the temperature and humidity of the lower atmosphere, and the climate plays a critical role in weathering and the formation of soils. Atmospheric circulation (wind), in turn, drives ocean waves and currents. Let's begin by learning about the structure of the atmosphere and its processes.

**?** **What is the difference between "the air" and "the atmosphere"?**

**air** The gaseous envelope that surrounds the Earth.

# THE ATMOSPHERE

An *atmosphere* is the gaseous envelope that surrounds a planet or any other celestial body (see chapter 1). **Air** is the gaseous envelope that surrounds one particular planet, the Earth. In other words, air is the Earth's atmosphere (Fig. 13.1).

## What Is Air Made Of?

Air is an invisible, odorless mixture of gases and suspended particles. Its composition varies slightly from place to place and even from time to time in the same place because of the presence of *aerosols* and the presence of *water vapor*.

- *Aerosols* are liquid droplets or solid particles that are so small that they remain suspended in the air. Water droplets in fogs are liquid aerosols. Examples of solid aerosols are tiny ice crystals, smoke particles from fires, sea-salt crystals from ocean spray, dust stirred by winds, and volcanic ash emitted during eruptions.

- *Water vapor* is always present in the air, and the amount is termed the *humidity*. On a hot, humid day in the tropics, as much as 4 percent of the air by volume may be water vapor. On a crisp, cold day, less than 0.3 percent may be water vapor.

Both the water vapor and aerosol contents of air vary widely. For this reason, the relative amounts of the remaining gases are generally reported as if the air were entirely lacking in water vapor and aerosols. When these two components are ignored, the relative proportions of the remaining gases in the air—termed *dry air*—turn out to be essentially constant. As shown in Figure 13.2, three gases—nitrogen (78%), oxygen (21%), and argon—make up 99.96 percent

Figure 13.1
**THIN LAYER OF AIR**
Viewed from the space shuttle Columbia, the atmosphere can be seen as the thin, bluish-colored layer
above the curved horizon. The view is looking north across the eastern end of the Mediterranean Sea.
The curved dark band in the lower left is the Nile Valley; the larger dark strip starting at the lower right
corner is the Gulf of Suez; the smaller dark strip on the right is the Gulf of Aqaba.

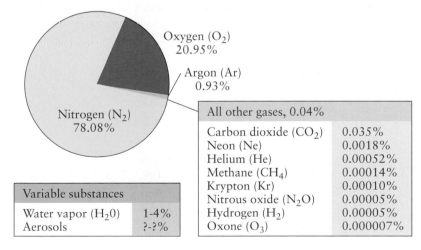

| All other gases, 0.04% | |
| --- | --- |
| Carbon dioxide ($CO_2$) | 0.035% |
| Neon (Ne) | 0.0018% |
| Helium (He) | 0.00052% |
| Methane ($CH_4$) | 0.00014% |
| Krypton (Kr) | 0.00010% |
| Nitrous oxide ($N_2O$) | 0.00005% |
| Hydrogen ($H_2$) | 0.00005% |
| Oxone ($O_3$) | 0.000007% |

| Variable substances | |
| --- | --- |
| Water vapor ($H_2O$) | 1-4% |
| Aerosols | ?-?% |

Figure 13.2
**WHAT AIR IS MADE OF**
This chart shows the composition of air that contains no water vapor or aerosols. Three
gases—nitrogen, oxygen, and argon—make up 99.96 percent of the air by volume. The
volume of water vapor and aerosols varies widely.

of dry air by volume. The remaining gases (carbon dioxide, 0.035%; neon, 0.0018%; and six others) are present in very small quantities. However, some of these minor gases are profoundly important for life on the Earth because they absorb certain wavelengths of sunlight. They act both as a warming blanket and as a shield against deadly ultraviolet radiation.

## Layers in the Atmosphere

As the Sun's radiant energy passes through the atmosphere, some wavelengths are absorbed by water vapor and some of the minor gases. Absorption of the Sun's radiation raises the temperature of the Earth's surface and of the air itself. For this reason, scientists use temperature as the main variable defining the state of the atmosphere. Scientists have discovered that the atmosphere is composed of four layers with distinct temperature profiles, each separated by thermal boundaries called *pauses* (Fig. 13.3). From the bottom up, these layers are (1) the *troposphere*, from the ground surface up to about 15 km (about 9 mi); (2) the *stratosphere*, up to 50 km (31 mi); (3) the *mesosphere*, up to 90 km (49 mi); and (4) the *thermosphere*, up to 700 km (435 mi).

### The Troposphere and the Greenhouse Effect

Humans live at the bottom of the troposphere. Even the Earth's highest mountain, Mount Everest, reaches little more than halfway through the troposphere. The troposphere contains 80 percent of the actual mass of the atmosphere, including virtually all the water vapor and clouds. The atmosphere that astronauts can see from space, as in Figure 13.1, is mainly the troposphere. Almost all weather-related phenomena originate in the troposphere. Although very little

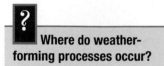

**Where do weather-forming processes occur?**

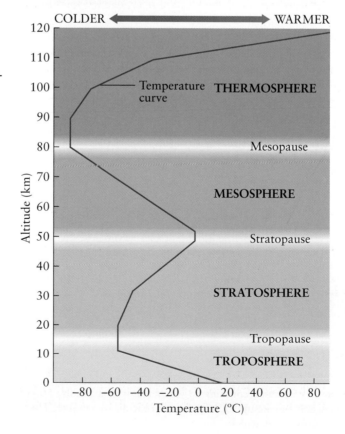

Figure 13.3
**FOUR MAJOR LAYERS**
Temperature varies with altitude in the atmosphere. The atmosphere is divided into four temperature zones, defined by altitude, where temperature changes markedly. These altitudes are called *pauses*.

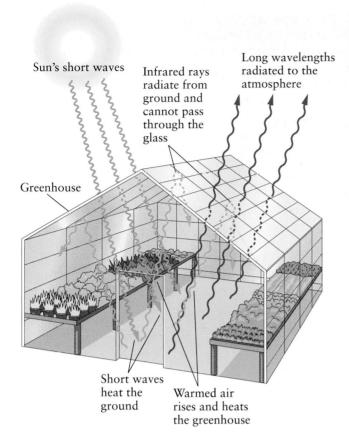

Sun's short waves

Infrared rays radiate from ground and cannot pass through the glass

Long wavelengths radiated to the atmosphere

Greenhouse

Short waves heat the ground

Warmed air rises and heats the greenhouse

**Figure 13.4**
**THE WAY A GREENHOUSE WORKS**
The glass roof of a greenhouse allows short-wavelength radiation from the Sun to pass through and warm the air and the ground inside. The glass then slows down outgoing long-wavelength heat energy. The atmosphere plays the same role for the Earth as the glass roof does for a greenhouse.

mixing occurs between the troposphere and the stratosphere, the troposphere itself is constantly moving and thoroughly mixed by winds.

The troposphere contains most of the heat-absorbing gases (also called *radiatively active gases* or *greenhouse gases*) that play a role in warming the Earth's surface. Incoming solar energy is mostly in the form of short-wavelength radiation. About 30 percent of the incoming solar radiation is simply reflected back into space by clouds and the ocean, but the remaining 70 percent is absorbed by the ocean, atmosphere, land, and biosphere. Virtually all of the absorbed radiation becomes heat (thermal energy) within the materials that absorb it. (An exception is the energy used in photosynthesis.) Eventually, all of this energy is re-radiated in the form of long-wavelength infrared radiation from the Earth back into outer space. Some of the outgoing radiated energy encounters heat-absorbing gases that slow down its escape. As a result, the heat energy is retained a little longer in the lower atmosphere, causing the temperature at the surface to be higher than it would otherwise be. A comparable effect explains why the temperature of air in a glass greenhouse is warmer than that of the air outside: the glass allows short-wavelength radiation to come in but slows down outgoing, long-wavelength, heat energy (Fig. 13.4). We call this natural atmospheric process the **greenhouse effect,** and we discuss the effect further in chapter 14. Without the greenhouse effect, the surface of the Earth would be a cold and inhospitable place.

### The Stratosphere and the Ozone Layer

The stratosphere contains 19 percent of the atmosphere's total mass, so that the troposphere and the stratosphere combined contain 99 percent of the atmosphere. Despite their small masses compared to the troposphere, the thermos-

**greenhouse effect** The process through which long-wavelength (infrared) heat energy is absorbed by gases in the atmosphere, thereby warming the surface of the Earth.

Figure 13.5
**PROTECTION FROM DEADLY RADIATION**
The ultraviolet radiation coming in from the Sun is harmful. Fortunately, the atmosphere protects us from the rays because three kinds of oxygen, O, $O_2$, and $O_3$ (ozone), absorb the lethal radiation.

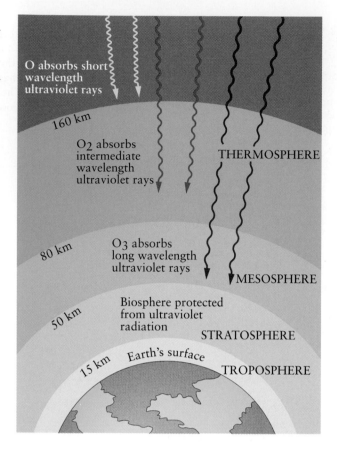

O absorbs short wavelength ultraviolet rays

160 km

$O_2$ absorbs intermediate wavelength ultraviolet rays

THERMOSPHERE

80 km

$O_3$ absorbs long wavelength ultraviolet rays

MESOSPHERE

50 km

Biosphere protected from ultraviolet radiation

STRATOSPHERE

15 km Earth's surface

TROPOSPHERE

**ozone layer** A zone in the stratosphere in which ozone ($O_3$) is concentrated.

phere, mesosphere, and stratosphere play very important roles for life on the Earth. Oxygen in various forms absorbs harmful ultraviolet rays coming from the Sun (Fig. 13.5). Short ultraviolet wavelengths are absorbed by the thermosphere, intermediate ultraviolet wavelengths by the mesosphere, and long ultraviolet wavelengths by the stratosphere. In the stratosphere it is the gas ozone ($O_3$), present in tiny but vital amounts, that absorbs the most dangerous of the ultraviolet rays. This concentration of $O_3$ in the stratosphere is called the **ozone layer.** Recently, there has been much concern about the breakdown of stratospheric ozone caused by chemical pollutants, thereby raising the possibility that the amount of harmful long-wavelength ultraviolet radiation reaching the Earth's surface may increase.

## Movement in the Atmosphere

Two things energize the atmosphere: the *Sun's heat* and the *Earth's rotation*. Because the Earth is a sphere, the Sun's heat does not reach the surface in equal amounts everywhere. In places where the Sun is exactly overhead, the incoming rays are perpendicular to the surface and a maximum amount of heat is received per unit area. But because of the Earth's curvature, at all other locations the surface is at an angle to the incoming rays. Therefore, less heat reaches each unit of surface area (Fig. 13.6). Winds and ocean currents are the natural processes by which the Earth system attempts to smooth out the temperature differences that result from this unequal heating of its surface. Winds and ocean currents move

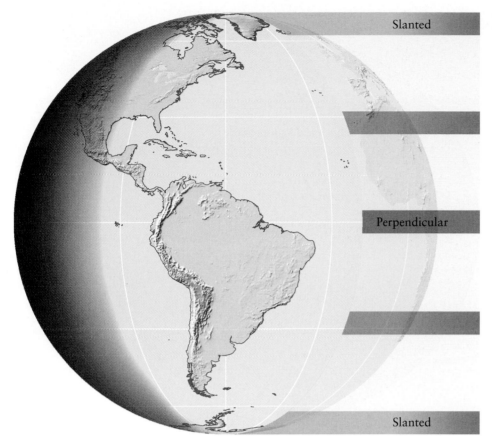

Slanted

Perpendicular

Slanted

**Figure 13.6**
**HOT AT THE EQUATOR,**
**COLD AT THE POLES**
The Sun's rays deliver their maximum heat energy per unit area when they are perpendicular to the Earth's surface, as at the equator. Away from the equator the rays hit the Earth at a slanting angle and *less* energy reaches the surface per unit area.

heat from the equator—where the input of solar heat is greatest—toward the poles, where it is least.

Unequal heating of the Earth's surface causes *convection currents* in the atmosphere.[1] Heated air near the equator expands, becomes lighter, and rises. Near the top of the troposphere it spreads outward toward the poles. As the upper air travels northward and southward toward the poles, it gradually cools, becomes heavier, and sinks. Upon reaching the surface, this cool air flows back toward the equator, warms up, and rises, thereby completing a convective cycle. Convection in the atmosphere is illustrated in Figure 13.7.

## The Coriolis Effect

If the Earth did not rotate, the convection currents would simply flow from the equator to the poles and back again. But of course the Earth does rotate, and its rotation complicates the convection currents in the atmosphere (and in the ocean). The **Coriolis effect** is named for the nineteenth-century French mathematician who first analyzed the effect. It causes anything that moves freely with respect to the rotating Earth (such as a plane, a missile, or the wind) to veer off

**Coriolis effect** The phenomenon whereby anything that moves freely with respect to the rotating Earth is caused to veer off-course.

---

[1] To understand atmospheric circulation, remember that materials expand when heated, becoming less dense. In contrast, when materials cool they become more dense. Thus, warm air rises and cool air sinks. The study of how fluids (including air and water) move in response to such changes is called *fluid dynamics*.

Figure 13.7
**GLOBAL ATMOSPHERIC CIRCULATION**
Huge convection cells transfer heat from equatorial regions, where the input of solar energy is greatest, toward the poles, where the solar input is least. Because the Earth is rotating, the flow of air toward the poles and the return flow toward the equator are influenced by the Coriolis effect. The combination of convection and the Coriolis effect produces large cells or belts of warm, moist air and cool, dry air.

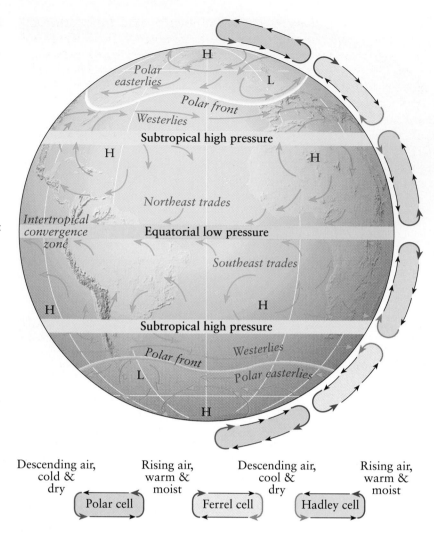

course. It is like trying to throw a ball to a friend when you are both on a spinning merry-go-round. In the Northern Hemisphere the Coriolis effect causes a moving mass to veer to the right of the direction in which it is moving, and in the Southern Hemisphere to the left. Both flowing water and flowing air respond to the Coriolis effect and are diverted by it. The global pattern of wind systems is strongly influenced by the Coriolis effect (Fig. 13.7).

## Wind Systems

The Coriolis effect breaks up the flow of convective air between the equator and the poles into belts (Fig. 13.7). For example, a large belt or cell of circulating air lies between the equator (0°) and about 30° latitude in both the Northern and Southern Hemispheres. Warm air rises at the equator, creating a low-pressure zone called the *intertropical convergence zone*. The air rises to the top of the troposphere and begins to flow toward the poles, but it veers off course as a result of the Coriolis effect. By the time it reaches a latitude of 30°, the high-altitude air mass has cooled and started to sink. The descending air flows back across the Earth's surface toward the equator. As it flows, the land and sea warm the air so that it eventually becomes warm enough to rise again. The low-latitude cells (from 0° to 30° N and S) created by this circulation pattern are called *Hadley cells*. The prevailing surface winds in Hadley cells are northeasterly in the North-

**? Why do the tradewinds blow from east to west?**

ern Hemisphere (i.e., they flow *from* the northeast toward the southwest), whereas in the Southern Hemisphere they are southeasterly. These winds are called *tradewinds* because their consistent direction and flow carried trade ships across the tropical oceans at a time when winds were the chief source of power.

A second set of convecting air cells, called *polar cells*, lies over the polar regions. In a polar cell frigid air flows across the surface away from the pole and toward the equator, slowly being warmed as it moves. When the polar air has reached about latitude 60° N or S, it has warmed sufficiently to rise convectively high into the troposphere and to flow back toward the pole where it cools and descends again, thereby completing the convection cell. Because of the Coriolis effect, the cold air that flows away from the poles is deflected, giving rise to a wind system called *polar easterlies*.

Between the Hadley cells and the polar cells, a third, but less well-defined set of convection cells is found in the mid-latitudes between about latitude 30° to 60° N and S. In these mid-latitude cells, called *Ferrel cells*, westerly (i.e., *from* the west) winds prevail. These circulation patterns will become clearer if you take some time to study Figure 13.7.

# GLOBAL CIRCULATION AND THE EARTH'S CLIMATE SYSTEM

The global patterns of air flow ultimately control the variety and distribution of the Earth's climatic zones (Fig. 13.8). Those patterns, in turn, are influenced by the nonuniform heating of the Earth's surface, the Coriolis effect, the distribution of land masses and oceans, and the topography of the land. If the Earth had no mountains and no oceans to affect the moving atmosphere, the major climate zones would lie parallel to the equator. However, the pattern of climate zones is

**Figure 13.8**
**THE EARTH'S CLIMATE ZONES**
In a simplified version of the climate classification system devised by Köppen, an expert on meteorology and plants, there are six basic types of climates; (1) tropical, (2) dry, (3) temperate-humid, (4) cold-humid, (5) polar, and (6) highland (mountain).

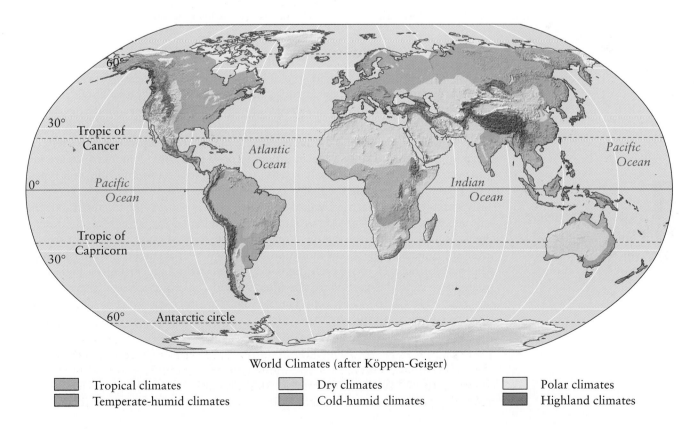

World Climates (after Köppen-Geiger)

| | | |
|---|---|---|
| ▨ Tropical climates | ▨ Dry climates | ▨ Polar climates |
| ▨ Temperate-humid climates | ▨ Cold-humid climates | ▨ Highland climates |

distorted by the distribution of oceans, continents, high mountains, and plateaus. As a result, average temperature, precipitation, cloudiness, and windiness vary greatly from one place to another and give rise to an array of distinct regions or zones.

**What is the difference between weather and climate?**

**weather** Local atmospheric conditions at any given time.

**climate** Weather patterns averaged over a long time.

## The Difference Between Weather and Climate

Before proceeding, we need to distinguish between two commonly confused terms, *weather* and *climate*. **Weather** refers to local atmospheric conditions at a particular time. It is usually described in terms of temperature, pressure, humidity, cloud cover, precipitation, and wind velocity. For example, we say "The weather today is hot, but yesterday it was wet." Weather patterns averaged over a long time are called **climate**. We say, "The climate, if you live in the tropics, is hot and humid," by which we mean there are more hot, humid days than any other kind of days in the tropics.

The role of the atmospheric circulation system is to smooth out differences in surface temperature by transporting heat and moisture from one part of the globe to another. Local changes in weather and the overall restlessness of the atmosphere are the result of this natural—but neverending—process. The interactions among air, land, and water that create weather are not only complex but also highly sensitive to changing conditions. They make weather prediction a very tricky business.

Remember that climate is determined by averaging weather patterns over a significant period. The patterns and processes that we think of as weather—wind, rain, snow, sunshine, storms, and even floods and droughts—are just temporary local variations when viewed against the more stable, longer-term background of global climate. The climate classification for a given area takes into account average or "normal" weather conditions as well as the types of weather extremes that occur there, the likelihood of occurrences and their frequency.

## The Earth's Climate Zones

*Climatologists*—scientists who study climate—use several classification schemes to describe and categorize the Earth's climate zones. One of the most widely used classification schemes is the *Köppen-Geiger climate system* originally devised in 1918 by the Austrian scientist Vladimir Köppen. Köppen's classification (Fig. 13.8) defines climate zones on the basis of temperature and precipitation, two aspects of climate that have a major influence on vegetation. Climate zones are important because they are closely related to such things as weathering, soil types, and boundaries between the major types of vegetation.

**What causes "extreme" weather?**

Some weather that we think of as "extreme" is actually characteristic of the climate zone in which it occurs and can be explained by examining the movement of air masses in the region. Consider the torrential rains and violent storms that are typical of the Asian monsoon. For half a year during the winter months (in the Northern Hemisphere), the wind blows across India from the high, cold plateau of central Asia toward the intertropical convergence zone, where warm air rises (Fig.13.9A). The cold plateau air flowing over India is dry. Consequently, winters in India are dry.

But during the summer the pattern is reversed. With the Sun now overhead on land in the Northern Hemisphere, the intertropical convergence zone shifts northward. The land mass of Asia heats up and is covered by low-pressure systems. Warm, moisture-laden winds now blow in the opposite direction, from the

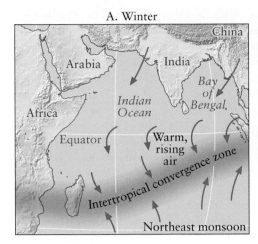
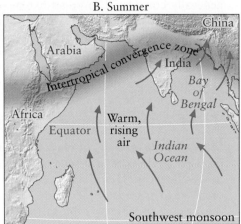

**Figure 13.9**
**THE ASIAN MONSOON**
The Asian monsoon is characterized by reversing winds. *A.* During the winter months when the Sun is overhead in the Southern Hemisphere, winds flow offshore from the northeast toward the intertropical convergence zone. *B.* During the summer months the land heats up, the intertropical convergence zone moves north, and winds reverse, flowing from the southwest across Asia. This brings moisture-laden air from over the Indian Ocean, and copious rainfall.

Indian Ocean onto the land (Fig. 13.9B). Consequently, summer in India is a time of hot, humid weather and heavy rains. This weather starts in southern India in late May, progresses to central India by mid-June, and reaches China by late July. The reversing winds that characterize the *monsoon* take their name from the Arabic word *mausim*, which means "season."

# THE OCEANS

The atmosphere and the oceans are closely interconnected. Water and carbon dioxide are exchanged freely between them via evaporation and precipitation. The oceans play a critical role in climate and weather systems, particularly in regulating the temperature and humidity of the lower part of the atmosphere. Atmospheric circulation, in turn, drives ocean waves and currents. Let's now turn our attention to the second great reservoir in the ocean–atmosphere system.

**? How old is the ocean?**

## The Ocean Basins

The oldest rocks so far discovered on the Earth (about 4.0 billion years) are gneisses that were once sedimentary strata. The strata were deposited in water and are similar to strata being deposited today. We are sure, therefore, that 4.0 billion years ago the Earth had liquid water on its surface, and we can be reasonably certain, as a result, that the ocean formed sometime between 4.56 billion years ago (when the Earth was formed) and 4.0 billion years ago. Where the water in the oceans came from is still an open question, but most likely it condensed from steam produced during primordial volcanic eruptions.

Most of the water on our planet is contained in four huge interconnected basins—the Pacific, Atlantic, Southern, and Indian Oceans. (The Arctic Ocean is generally considered to be an extension of the North Atlantic.) The Pacific, Atlantic, and Indian oceans are connected with the Southern Ocean, the body of water that encircles Antarctica. Collectively, these four vast interconnected bodies of water, together with a number of smaller ones such as the Mediterranean Sea, Hudson Bay, and the Persian Gulf, are often referred to as the *world ocean* and cover 71 percent of the Earth's surface.

## The Composition of Seawater

**salinity** Saltiness.

**Salinity** is the saltiness of seawater. It is expressed in *units per mil* (‰ = parts per thousand) rather than percent (% = parts per hundred). The salinity of seawater ranges between 33 and 37‰, which is quite small, yet very significant—you can easily taste it. Chemical elements in seawater are present mainly as ions; the main ones that contribute to the salinity are sodium and chlorine. Not surprisingly, when seawater is evaporated, more than three quarters of the dissolved matter is precipitated as common salt (NaCl, the mineral halite). Seawater contains most of the other natural elements as well, but many of them in such low concentrations that they can be detected only by extremely sensitive analytical instruments. The salinity of seawater is a nice example of balance between input and outflow of one of the Earth system's major reservoirs.

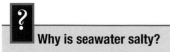

**Why is seawater salty?**

The elements dissolved in seawater come from two principal sources, weathering and volcanic eruptions. Chemical weathering releases soluble materials such as salts of sodium, potassium, and sulfur. The soluble compounds are leached out and become part of the dissolved load in river water flowing to the sea. Volcanic eruptions, both on land and beneath the sea, also contribute soluble compounds via volcanic gases and hot springs.

The salinity of seawater varies within narrow limits. It is balanced by the interplay among several processes, of which the most important are the following:

- *Evaporation.* Evaporation removes water, so the remaining water is saltier. This is important in desert regions.

- *Precipitation.* Rainfall and snowfall add fresh water and thus make the sea less salty. This effect is most pronounced near the equator and other places where rainfall is high.

- *Inflow of fresh water from rivers.* The addition of fresh water makes the sea less salty. This effect is most pronounced off the mouths of the world's great rivers, such as the Amazon, Mississippi, and Congo. The rivers dissolved loads contribute salts to ocean water, as discussed above.

- *Freezing of sea ice.* When seawater freezes, salts are excluded from the ice, leaving the unfrozen seawater saltier. This effect is most pronounced in polar regions.

The quantity of dissolved matter that has been added to the sea over billions of years far exceeds the amount now dissolved in the ocean. Why, then, isn't the sea saltier? The reason is that chemical substances are being removed from seawater at the same rate that they are being added. Some elements, such as silicon, calcium, and phosphorus, are withdrawn from seawater by aquatic plants and animals to build their shells or skeletons. Other elements, such as potassium and sodium, are absorbed and removed by clay particles and other minerals as they settle slowly to the sea floor. Still others, such as copper and lead, are precipitated as sulfide minerals. The processes that add elements to seawater are balanced by the processes that remove them. Therefore, the composition of seawater remains essentially unchanged. This is the problem John Joly ran into when he tried to use the salinity of seawater to estimate the age of the Earth (chapter 3).

## Layers in the Ocean

The salinity and temperature of seawater combine to control its density: high salinity means high density; low temperature also means high density. In other words, cold salty water will sink, while warm, less salty water will rise. Ocean scientists have discovered three major layers or zones in the ocean, in which the density of the water differs. The differences are caused by changes in both temperature and salinity, with temperature being the major factor (Fig. 13.10). The *surface zone,* typically extending to a depth of about 100 m (about 330 ft), consists of relatively warm water and therefore is low in density. Below the surface zone lies another zone in which the temperature decreases rapidly with depth. This zone, called the *thermocline,* reaches the surface at high latitudes and reaches a maximum depth of about 1500 m (3920 ft) near the equator. Below the thermocline lies the *deep zone,* which contains the bulk of the ocean's volume. The temperature in the deep zone is low—about 2°C (36°F)—and the density is therefore high. The importance of these layers in the oceans lies in the roles they play in the ocean currents.

## Ocean Currents

In 1492, when Christopher Columbus set sail across the Atlantic Ocean in search of China, he took an indirect route. Instead of sailing due west from Spain, which would have made his voyage shorter, he took a longer route southwest toward the Canary Islands and then west on a course that carried him to the Caribbean Islands where he first sighted land. In choosing this course, he was following the path of the prevailing winds and surface ocean currents. Instead of fighting the westerly winds and currents at 40° N latitude, he drifted with the Canary Current and the North Equatorial Current as the northeast trade winds

**When he embarked on his voyage of discovery, why did Columbus sail south rather than due west?**

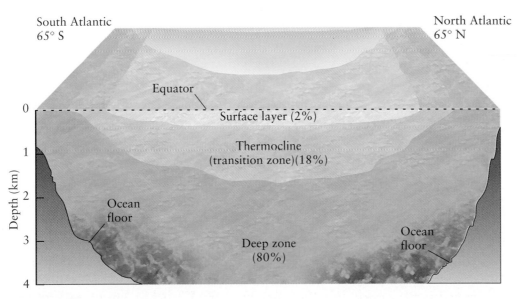

**Figure 13.10**
**STRUCTURE OF THE OCEAN**
There are three major layers in the ocean. The *surface layer,* typically extending to a depth of 100 m (330 ft), consists of relatively warm, low-density water. Although it only contains 2 percent of the water in the ocean, the surface layer is the major life-zone of the sea, home to the majority of marine plants and animals. Just below the surface layer, in the *thermocline,* the temperature of the water increases rapidly with depth. Below the thermocline lies the *deep zone,* in which the temperature is uniformly low. Most of the water in the ocean (80%) is in the deep zone.

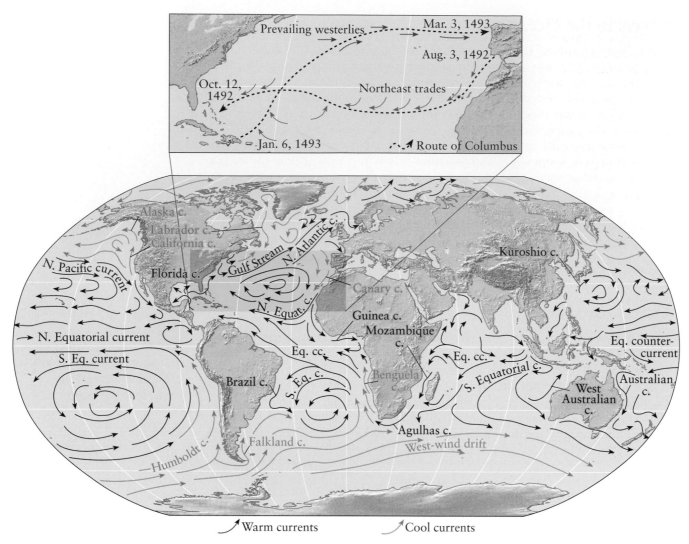

Warm currents          Cool currents

**Figure 13.11**
**SURFACE OCEAN CURRENTS**
Surface ocean currents form a distinctive pattern, curving to the right (clockwise) in the Northern Hemisphere and to the left (counterclockwise) in the Southern Hemisphere. Christopher Columbus used these currents to good advantage in 1492.

filled the sails of his three small ships. When he sailed back to Spain to announce his great discovery, he sailed a more northerly route where he was propelled by strong westerly winds and the North Atlantic Current. His route is shown by the black arrows in Figure 13.11.

Ocean currents, like those Columbus followed, are broad, slow drifts of water that are set in motion by the prevailing winds (Fig. 13.11). Air that blows across the sea drags the water forward, creating a current of water that is as broad as the current of air but only 50 to 100 m (165 to 330 ft) deep (that is, confined to the surface zone). Compare Figure 13.7, showing the global wind pattern, and Figure 13.11, showing the surface ocean currents. You can see that the wind directions and the surface ocean current directions are roughly parallel, providing strong evidence that surface currents are caused by winds. (For a discussion of one striking result of a surface ocean current see the Box on p. 382.)

But the ultimate energy source for the motion of the surface ocean currents is not the wind but the Sun, for it is solar energy that sets in motion the planetary wind system. Thus, ocean circulation results from the interplay of several key elements of the Earth system: (1) radiation from the Sun, which provides heat energy to the atmosphere; (2) nonuniform heating, which generates winds; and (3) the winds, which in turn, drive the movement of water in the surface layer of the ocean.

Deep currents also arise as a result of interactions between the atmosphere and the ocean. Seawater near Antarctica and in the Arctic Ocean is very cold; it is also quite saline due to formation of sea ice. As we saw earlier, low temperatures and high salinity mean high density. In polar regions cold, salty, dense water sinks and propels a deep circulation system (Fig. 13.12). The surface and deep ocean currents are the main mixing forces in the ocean; they keep the temperature and salinity of the ocean in balance.

## The Ocean Helps Regulate Climate

The ocean differs from the land in the amount of heat it can store. When the Sun's rays strike the land, only the surface is warmed. In the ocean, mixing by currents and waves means the Sun's heat affects much more than just the surface layers. In addition, as we saw in chapter 2, the latent heats involved with evaporation and freezing of water are large. As a result, the ocean can absorb and release large amounts of heat with very little change of temperature. When sea ice forms, the ocean releases latent heat; when seawater evaporates and releases water vapor to the atmosphere, the ocean absorbs latent heat. As a result of these processes—mixing in the ocean and the latent heats of freezing and evaporation—both the total range and the seasonal changes in ocean temperatures are much less than those occurring on land. For example, the highest recorded land temperature is 58°C (136°F), measured in the Libyan Desert; the lowest, measured at Vostok Station in central Antarctica, is –88°C (–126°F). The total range, therefore, is 146°C (262°F). By contrast, the highest recorded ocean temperature is 36°C (97°F), measured in the Persian Gulf, and the coldest, measured in the polar seas, is –2°C (28°F), a range of only 38°C (68°F).

The annual change in temperature at the ocean's surface is 0–2°C (0–4°F) in the tropics, 5–8°C (9–15°F) in middle latitudes, and 2–4°C (4–8°F) in the polar regions. But corresponding seasonal temperature ranges on the continents can exceed 50°C (122°F). The ocean thus has a strong moderating influence, and in-

Figure 13.12
**DEEP OCEAN CURRENTS**
Deep ocean currents arise as a result of interactions between the atmosphere and the ocean. Seawater near Antarctica and in the Arctic Ocean is very cold, dense, and saline. In polar regions this cold, salty, dense water sinks and spreads slowly across the ocean floor, propelling a deep circulation system. Eventually, the cold water wells up, becoming shallower and warmer. It takes about 1000 years for water to complete the circuit.

The fishing grounds off the coast of Peru, among the richest in the world, are sustained by upwelling cold waters filled with nutrients. Each year around Christmas time a slightly tepid current appears off the coast of Peru and Ecuador, reducing the fish population a little and giving the fishermen a rest over the holidays. To this phenomenon the fishermen long ago gave the name El Niño, meaning "The Child," in reference to the infant Jesus. At irregular intervals of 2 to 7 years, an exceptionally warm and long-lived current appears and lasts for up to 20 months. The name El Niño has now been transferred to the irregular warm current.

During El Niño years, upwelling is markedly reduced and the fish population declines; this decline is accompanied by a great die-off of the coastal bird population, which depends on the fish for food. Because the Peruvian fishery is among the most important in the world, the occurrence of an El Niño event constitutes a local economic catastrophe. At the same time, very heavy rains fall in normally arid parts of Peru and Ecuador, Australia experiences drought conditions, anomalous cyclones appear in Hawaii and French Polynesia, the seasonal rains of northeast Brazil are disrupted, and the Indian monsoon is greatly reduced. During exceptional El Niño years, weather over much of Africa, eastern Asia, and North America is affected. In North America, unusually cold or unusually mild winters can result in the northeastern United States, while the Southeast becomes wetter; in California abnormally high rainfall can produce major flooding and landslides.

What happens during an El Niño event is reasonably well understood. Normally, the southeast tradewinds (see Fig. 13.7) blow from east to west and fan the surface of the Pacific. The tradewinds push warm surface water away from Peru and Ecuador and allow cold water to well up (Fig. B13.1A). The warm surface water, which piles up on the western side of the Pacific near northern Australia, the Philippines, and Indonesia, creates a large area of heavy rainfall.

An El Niño event begins when the tradewinds slacken, the piled-up warm water slowly sloshes back across the ocean toward South America, and the upwelling off Peru and Ecuador slows down. The eastward movement of warm water causes the zone of high rainfall to shift to the central Pacific near the international date line, simultaneously bringing drought conditions to Indonesia and Australia (Fig. B13.1B). At the peak of an event, equatorial surface water moves from west to east and also toward the poles. This flow gradually reduces the equatorial pool of warm water, leading to intensification of the tradewinds and an eventual return to normal conditions.

When a major El Niño event occurs, as it did in 1997–98, its effect on cli-

habitants of coastal areas benefit from this natural regulation. Along the Pacific coast of Washington and British Columbia, for example, winter air temperatures seldom drop to freezing, while east of the coastal mountain ranges they can plunge to –30°C (–22°F) or lower. In the interior of a continent summer temperatures may exceed 40°C (104°F), whereas along the ocean margin they typically remain below 25°C (77°F). Here, then, is a fine example of the interaction of the hydrosphere, atmosphere, and land surface: ocean temperatures affect the climate, both over the ocean and over the land; climate, in turn, controls weathering of rocks, formation of soil, and distribution of plants and animals.

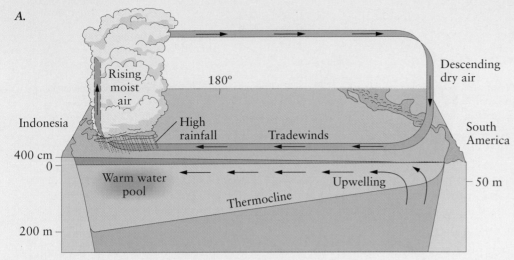

A.

"Normal" conditions in the tropical Pacific

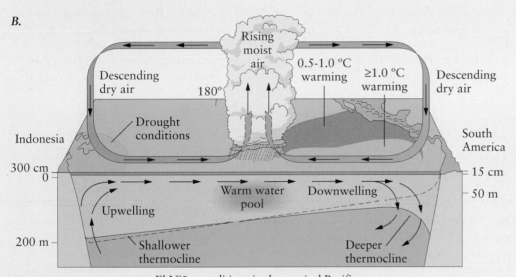

B.

El Niño conditions in the tropical Pacific

Figure B13.1
THE EL NIÑO CYCLE
A. During normal years, persistent tradewinds blow westward across the tropical Pacific from a zone of upwelling water off the coast of Peru. This water warms up as it is transported westward, forming a large pool of warm water above the thermocline in the western Pacific. The warm water causes the moist maritime air to rise and cool, bringing abundant rainfall to Indonesia.

B. During an El Niño event, the tradewinds slacken and the pool of warm water moves eastward to the central Pacific. Descending cool, dry air brings drought conditions to Indonesia, while rising moist air above the warm-water pool greatly increases rainfall in the mid-Pacific. Surface waters in the eastern Pacific become warmer, and downwelling shuts off the supply of deep-water nutrients, adversely affecting the normally productive fishing ground off the coast of Peru.

mate is felt over at least half of the Earth. It provides us with an especially instructive example of the interactions among atmosphere, hydrosphere, and biosphere because it not only involves the tropical oceans and atmosphere but also directly influences precipitation and temperature on major land areas, thereby also affecting plants and animals.

# WHERE OCEAN MEETS LAND

The majority of the world's population lives within 100 km of the ocean. This reflects our dependence on the oceans as well as the richness of resources in coastal zones. However, the concentration of large numbers of people in coastal areas means that the coastal environment must absorb the impacts of a wide range of human activities. It also means that human vulnerability to hazards can be particularly high in coastal zones; infrequent events such as large storms can cause major loss of life and damage to property. It is important to understand the geo-

**Figure 13.13**
**WHAT CAUSES OCEAN TIDES?**
The daily rise and fall of the sea along a shore results from two interacting forces: the Moon's gravitational attraction and the inertia of the rotating Earth–Moon system. On the side of the Earth that is facing the Moon, both forces combine to distort the water level, creating a tidal bulge. On the opposite side, where inertial forces are greater than the Moon's gravitational force, a tidal bulge forms in the opposite direction. As the Earth rotates on its axis, the tidal bulges remain essentially stationary and the Earth rotates through them twice a day, creating two high tides and two low tides for each full rotation of the Earth.

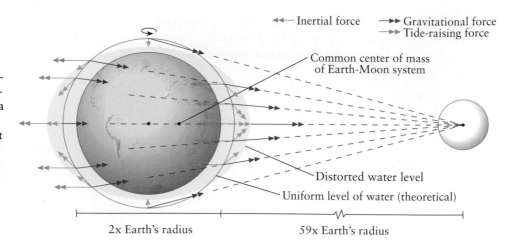

**tides** The cycles of regular rise and fall in the level of water in the ocean and other large bodies of water, which result from the gravitational interaction of the Moon, the Sun, and the Earth.

> **?** Why are tides low in the middle of the ocean but high along the coastline?

logic processes that characterize coastal zones as well as the hazards to which the inhabitants of these areas are vulnerable.

If you visit almost any coastal zone on two occasions a year apart, you will see changes. Sometimes the changes are small, but often they are substantial. Large sand dunes may have shifted. Sand may have built up behind barriers or been eroded away. Steep sections of coastline may have collapsed. Channels may have broken through from the sea to lagoons on the landward side, where there were no channels before. The energy driving these continual changes comes from tides and waves, as well as occasional storms.

## Tides and Changes in Sea Level

**Tides** are the cycles of regular rise and fall in the level of water in oceans and other large bodies of water like the Great Lakes. They result from the gravitational attraction of the Moon and the Sun acting on the Earth. Gravitational attraction causes ocean water to bulge upward on the side of the Earth nearest to the Moon. On the opposite side of the Earth, *inertia* (the force that tends to maintain a body in uniform linear motion) created by the Earth's motion around the common center of mass of the Earth–Moon system[2] also causes ocean water to bulge, but in the opposite direction (Fig. 13.13). The result is two *tidal bulges*, one on either side of the Earth. The Sun's gravitational force also affects the tides, sometimes opposing the Moon by pulling at a right angle to the Moon's pull and sometimes aiding it by pulling in the same direction or in the opposite direction. However, the Sun is so far from the Earth that it is only half as effective as the Moon in producing tides.

To visualize how tides work, consider the tidal bulges oriented with their maximum height lying along a line running through the center of the Earth and the center of the Moon, as shown in Figure 13.13. While the Earth rotates around its axis, the tidal bulges remain stationary beneath the Moon. Thus, any given coastline will move eastward through both tidal bulges each day. Every time a land mass encounters a tidal bulge, the water level along the coast rises. As the Earth continues to rotate, the coast passes through the highest point of the tidal bulge (high tide) and the water level begins to fall. Along most coastlines two high tides and two low tides are observed each day.

---

[2] The motion referred to here is not the Earth's rotation around its axis. It is a slow rotation of the Earth–Moon system; gravity holds them together like two ends of a dumbbell and they slowly rotate around their common center of mass.

In the open sea the effect of tides is small. However, the shape of a coastline can greatly influence tidal *runup* height, the highest elevation reached by the incoming water. Narrow openings into bays, rivers, estuaries, and straits can amplify normal tidal fluctuations. At the Bay of Fundy in Nova Scotia, a tidal range (the difference between high and low tide) of up to 16 m (52 ft) is reported (Fig. 13.14). The bay is very long and narrow. This causes the incoming tide to rush in, forming a steep-fronted, rapidly moving wall of water called a *tidal bore*. The extreme tidal range at the Bay of Fundy makes it one of the few places in the world that may be suited for generating electricity from tidal energy. However, a wall of water moving as fast as 25 km/h (16 m/h) can easily move large quantities of sediment. Minimizing the impacts of water-borne sand is a major engineering challenge in the development of tidal energy technologies.

Tides are a daily cause of fluctuations in water level in the ocean. Over longer time periods global sea level changes in response to a wide variety of factors, including the advance and retreat of polar ice caps and even changes in the size and shape of the ocean basins. Other short-term changes are caused by wind-driven atmospheric forces, earthquakes, and submarine landslides. Of these factors, glaciation is the most important. At the height of the most recent glacial period, about 18,000 years ago, sea levels were about 120 m (400 ft) lower than they are today, reflecting the fact that a great quantity of water was tied up in ice sheets at the time. Changes in sea level that are global in extent produce unconformities in the sediments that accumulate on continental shelves, and these provide a detailed record of sea level fluctuations through geologic time.

## Ocean Waves

Like surface ocean currents, ocean waves receive energy from winds. The size of the largest wave depends on how fast, how far, and how long the wind blows across the water surface. A gentle breeze blowing across a bay may ripple the water or form low waves less than a meter high. By contrast, storm waves produced by intense winds blowing for days across hundreds or thousands of kilometers of open water may become so high that they tower over ships unfortunate

**Figure 13.14**
**EXTREME TIDES**
The tidal range in the Bay of Fundy in eastern Canada is one of the greatest in the world.
*A.* This is the coastal harbor of Alma, New Brunswick, at high tide. *B.* This is the same view at low tide.

*A.*

*B.*

**Figure 13.15**
**STORMY SEAS**
A ship in the open ocean struggles to maintain course through huge storm waves that tower over its deck.

**? What causes waves?**

enough to be caught in them (Fig. 13.15). Large waves may travel across an entire ocean before running into a coastline. If you look carefully at the waves coming into a beach, you can usually see them arriving from two or more directions. Each set of waves originated from a storm-related wind far beyond your line of sight.

In deep water each small parcel of water in a wave moves in a loop, returning very nearly to its former position as the wave passes (Fig. 13.15). The distance between successive wave crests or troughs is the *wavelength*. Downward from the surface the movement of the water decreases, and this is expressed as a decrease in the diameter of the looplike motion of the water parcels. Eventually, at a depth equal to about half of one wavelength, the motion of the water becomes negligible. This effective lower limit of wave movement (which, by extension, is also the lower limit of erosion by the bottoms of waves) is called the **wave base**.

**wave base** The effective lower limit of wave movement.

## Wave Action Along Coastlines

As a wave approaches the shore, it undergoes a rapid transformation. The circular loops that characterize wave motion in deep water become flatter as the water becomes shallower (Fig. 13.16). Where water depth becomes less than half a wavelength, the circular wave motion is influenced by the increasingly shallow sea floor, which restricts vertical movement. Shallow sea floor interferes with wave motion and distorts the wave's shape, the wave height increases, and the wavelength decreases. Now the front of the wave is in shallower water than the rear and is also steeper than the rear. Eventually, the front becomes too steep to support the advancing wave. As the rear part continues to move forward, the wave collapses, or *breaks* (see Fig. 13.17).

When a wave breaks, the motion of the water instantly becomes turbulent, like that of a swift river. Such "broken water," called **surf**, is found between the line of breakers and the shore, forming an area known as the *surf zone*. Each wave finally dashes against the rocks or rushes up a sloping beach until its energy is expended; then the water flows back toward the open sea. Water that has piled up against the shore returns seaward in an irregular and complex way, partly as a

**surf** The "broken," turbulent water found between a line of breakers and the shore.

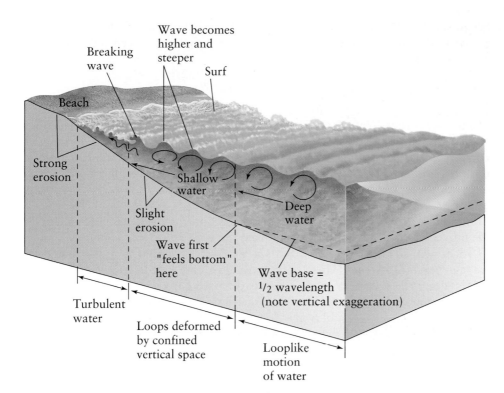

Breaking wave

Wave becomes higher and steeper

Surf

Beach

Strong erosion

Slight erosion

Shallow water

Deep water

Wave first "feels bottom" here

Wave base = ¹/₂ wavelength (note vertical exaggeration)

Turbulent water

Loops deformed by confined vertical space

Looplike motion of water

**Figure 13.16**
**HOW WAVES CHANGE AS THEY NEAR THE SHORE**
Waves change their form as they travel from deep water through shallow water to shore. In the process, the circular loops of water movement found in deep water are flattened as the water becomes shallower and the wave begins to hit the bottom. This drawing has been stretched vertically, and the size of the loops has been exaggerated to emphasize the wave motion.

**Figure 13.17**
**RIDING THE WAVE**
Orienting the board for the best ride, a surfer skims the inside of a breaking wave off the coast of Hawaii.

broad sheet along the bottom and partly in localized narrow channels known as *rip currents*, which create dangerous undertows that can sweep unwary swimmers out to sea.[3]

### Erosion and Transport of Sediment by Waves

Surf is a powerful erosional force because it possesses most of the original energy of the waves that created it. Wave erosion takes place not only at sea level but also below sea level and—especially during storms—above sea level. In the surf zone, rock particles transported by waves are worn down, becoming smoother, rounder, and smaller. At the same time, through continuous rubbing and grinding with these particles, the surf wears down and deepens the bottom and eats into the land. The surf is like a saw cutting horizontally into the land. The energy of the surf is eventually consumed in turbulence, in friction at the bottom, and in the movement of sediment thrown up from the bottom.

Sediment is also brought to the sea by rivers, where it is redistributed by wave-generated currents. These currents build the sediment into distinctive deposits along the shoreline or transport it offshore onto the continental shelves. The movement of sediment is accomplished mainly by two processes, transport by the *longshore current* and *beach drift*.

Most waves reach the shore at an oblique angle (Fig. 13.18A). Part of the force of an incoming wave is oriented perpendicular to the shore; this produces the crashing surf. Another component of the wave motion is oriented parallel to the shore. The parallel component sets up a **longshore current** flowing parallel to the shore within the surf zone. While surf erodes sediment at the shore, the longshore current moves the sediment along in the surf zone.

**longshore current** A current, within the surf zone, that flows parallel to the coast.

Meanwhile, on the exposed beach, incoming waves produce an irregular pattern of water movement along the shore. Because waves generally strike the beach at an angle, the *swash* (uprushing water) of each wave travels obliquely up the beach before gravity pulls the water down (Fig. 13.18B). This zigzag movement of water carries sand and pebbles first up, then down the beach slope. Successive movements of this type gradually cause sediment to be transported along the shore, a process known as **beach drift**. The greater the angle of waves to the shore, the greater the rate of drift. Marked pebbles have been observed to drift along a beach at a rate of more than 800 m/day (2600 ft). When the volume of sand moved by beach drift is added to that moved by longshore currents, the total can be very large.

**beach drift** The movement of particles along a beach as they are driven up and down the beach slope by wave action.

The continual interaction between erosional and depositional forces operating along coasts is occasionally interrupted by exceptional storms that erode cliffs and beaches at rates far greater than the long-term average. After a single storm, cliffs of compact sediment on Cape Cod were observed to have retreated up to 5 m (16.5 ft)—more than 50 times the normal annual rate of retreat. Infrequent bursts of rapid erosion of this type can have a significant impact on coastlines.

In calm weather an exposed beach receives more sediment than it loses and therefore becomes wider. But during storms the increased energy in the surf erodes the exposed part of a beach and makes it narrower. Because storminess is seasonal, changes in beach profiles are more likely to occur at certain times of the year than at others. Along parts of the Pacific coast of North America, winter

---

[3] If you happen to encounter a rip current while you are swimming, don't try to swim *against* the current. Keep a cool head and swim sideways to the current. Even though you may be carried out to deeper water, you will eventually swim free of the rip.

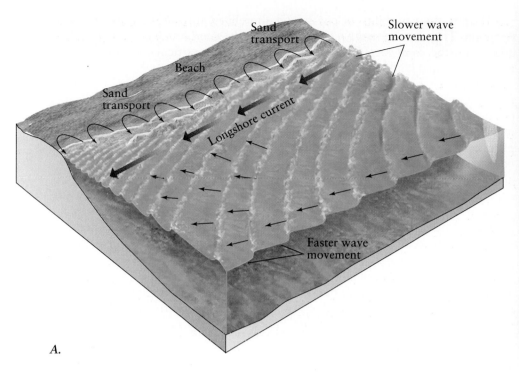

**Figure 13.18**
**LONGSHORE CURRENT AND BEACH DRIFT**
*A.* A longshore current is generated as waves approach a beach at an oblique angle; the current flows parallel to the shore. Such a current can transport considerable amounts of sediment along a coast. *B.* Surf swashes obliquely onto a Brazilian beach and forms a series of arc-shaped cusps. The water rushing onto the beach loses momentum and flows back down the sandy slope in a direction nearly perpendicular to the shoreline. A grain of sand moves along a zigzag path as successive waves reach the shore. The net motion of sand along the beach is called *beach drift.*

storm surf tends to carry away fine sediment, and the remaining coarse material assumes a steep profile. In calm summer weather, fine sediment drifts in and the beach assumes a gentler profile.

## Shorelines and Coastal Landforms

The end result of the constant interplay between erosional and depositional forces along coastlines is a wide variety of shorelines and coastal landforms. The world's coasts do not fall into easily identifiable classes. Their forms depend on (1) the geologic processes at work; (2) the structure and erodability of coastal rocks; and (3) how long these processes have been operating. Changes in sea level can also influence the development of coastal features.

Many coastal and offshore landforms are relics of times when the sea level was either higher or lower than it is now (Fig. 13.19 shows an example of lowered sea level due to rising of the land). Repeated emergence and submergence of coastlines, accompanied by erosion and redeposition of shoreline deposits, has resulted in complex coastal landforms.

Despite the variability of coasts and shorelines, three basic types are most common. They are the *rocky (cliffed) coast*, the *lowland beach and barrier island coast*, and the *coral reef*. Each is characterized by a particular set of erosional and depositional landforms.

*Rocky Coasts*   The most common type of coast, comprising about 80 percent of ocean coasts worldwide, is a rocky or cliffed coast. Seen in profile, the usual elements of a cliffed coast are a wave-cut cliff and wave-cut bench or terrace, both products of erosion. A **wave-cut cliff** is a coastal cliff cut by wave action at the base of a rocky coast. As the upper part of the cliff is undermined, it collapses and the resulting debris is redistributed by waves. An undercut cliff that has not yet collapsed may have a well-developed notch at its base. Below a wave-cut cliff you can often find a *wave-cut bench* or *terrace*, a platform that has been cut across bedrock by the surf. The shoreward parts of some benches are exposed at low tide. If the coast has been uplifted, a wave-cut bench and its sediment cover can be completely exposed.

The rocky character of cliffed coasts may be misleading to people looking for a stable platform on which to build a home. Cliffed shorelines are susceptible to frequent landslides and rock falls as erosion eats away at the base of the cliff. Roads, buildings, and other structures built too close to such cliffs can be damaged or destroyed when sliding occurs.

*Beaches and Barrier Islands*   Beaches are a striking feature of many coasts. Most people think of a beach as the sand surface above the water along a shore. Actually, a **beach** is defined as wave-washed sediment along a coast. A beach thus includes sediment in the surf zone, which is underwater and therefore continually in motion. At low tide, when a large part of a beach is exposed, onshore

**wave-cut cliff** A coastal cliff cut by wave action at the base of a rocky coast.

**beach** Wave-washed sediment along a coast.

**Figure 13.19**
**WAVE-CUT BENCHES**
Two wave-cut benches at Tongue Point, New Zealand. The upper bench is the flat greenish-brown surface. The lower bench is the gray surface just above sea level. Both are former sea floor. They were elevated above sea level by two stages of crustal uplift along this coast.

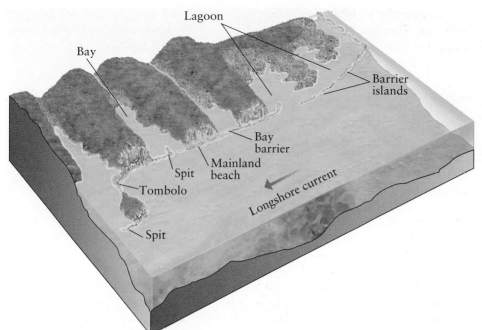

Labels on figure: Lagoon, Bay, Barrier islands, Bay barrier, Mainland beach, Spit, Tombolo, Spit, Longshore current

**Figure 13.20**
**DEPOSITIONAL LANDFORMS ALONG A COAST**
These common landforms are caused by deposition of sediment along a coast. The local direction of beach drift and longshore transport by sediment of the longshore current is toward the free end of the spits.

winds may blow beach sand inland to form belts of coastal dunes. A landform commonly associated with beaches is the **barrier island**, a long, narrow, sandy island lying offshore and parallel to the coast (Fig. 13.20). A barrier island consists of one or more ridges of sand dunes associated with successive shorelines in a region of rising sea level. Barrier islands are found along most lowland coasts. For example, the Atlantic coast of the United States consists mainly of a series of barrier beaches ranging from 15 to 30 km (about 10 to 20 mi) in length and 1.5 to 5 km (1 to 3 mi) in width, located 3 to 30 km (2 to 20 mi) offshore. Sand dunes are typically the highest topographic features along this coastline, which includes Coney Island, New York, and the long chain of islands centered on the Cape Hatteras coast of North Carolina.

During major storms, surf washes across low places on barrier islands and erodes them, cutting inlets that may remain open permanently (Figs. 13.20 and 13.21A). At such times fine sediment is washed between the barrier island and the mainland. In this way the length and shape of barrier islands is always changing. Studies of these processes suggest that island development is closely related to the amount of sediment in the system, the direction and intensity of waves and currents, the shape of the seabed, and the stability of sea level.

Other common depositional landforms associated with beaches include the *spit* (an elongated ridge of sand or gravel that projects from land and ends in open water) and the *tombolo* (a spitlike ridge of sand or gravel that connects an island to the mainland); these are illustrated in Figure 13.20. A well-known example of a large, complex spit is Cape Cod, Massachusetts (Fig. 13.21B). The elongate bay lying inshore from a barrier island or other low, enclosing strip of land (such as a coral reef) is called a *lagoon*. Lagoons are commonly fed by *estuaries*, the wide, fan-shaped mouths of rivers in the tidal zone where fresh and salt water meet. Lagoons and estuaries are important habitats for a wide variety of plants, birds, and animals (Fig. 13.21C). They also play an important role in the protection of mainland shorelines because they serve as buffers against storm waves. Unfortunately, these sensitive environments are particularly susceptible to the impacts of human activities.

**barrier island** A long, narrow, sandy island lying offshore and parallel to a lowland coast.

A.

B.

**Figure 13.21**
**BARRIER ISLANDS, SPITS, AND LAGOONS**
*A.* This is an aerial view of the Outer Banks, Cape Hatteras, North Carolina, a barrier island. *B.* Cape Cod is a complex spit on the Atlantic coast in Massachusetts. Waves and currents rework sediment that has been eroded from the peninsula, forming the south side of Cape Cod Bay, and transport it northward and southward. An eddy carries the sediment around the north point of the spit and into the bay. *C.* This is a high-altitude aerial photograph along the south shore of Long Island, in Nassau and Suffolk counties, New York. A barrier island separates a lagoon from the Atlantic Ocean.

C.

*Coral Reefs*  Many of the world's tropical coastlines consist of limestone **reefs** built by vast colonies of organisms, principally corals, that secrete calcium carbonate (limestone). Reefs are built up very slowly over thousands of years. Each of the tiny coral animals, called *polyps*, deposits a protective layer of calcium carbonate; over time the layers build up, forming a complex reef structure. *Fringing reefs* form coastlines that closely border the adjacent land. *Barrier reefs* are separated from the land by a lagoon, as in the case of the Great Barrier Reef off Queensland, Australia (Fig. 13.22). Reefs are highly productive ecosystems inhabited by a diversity of marine life forms. They also perform an important role in the recycling of nutrients in shallow coastal environments. They provide physical barriers that dissipate the force of waves, protecting the ports, lagoons, and beaches that lie behind them, and they are an important aesthetic and economic resource.

Corals require shallow, clear water in which the temperature remains above 18°C (65°F). Reefs therefore are formed only at or close to sea level and are characteristic of warm, low latitudes. Because of their very specific temperature and light requirements, coral reefs are highly susceptible to damage from human activities as well as from natural causes such as tropical storms.

**reef** A structure composed mainly of the calcareous remains of marine organisms (principally corals).

**Figure 13.22**
**CORAL REEF HOSTS BOUNTIFUL LIFE**
The Great Barrier Reef, a barrier coral reel on the continental shelf of northeastern Australia, is home to one of the world's most diverse groups of marine plants and animals.

## WHAT'S AHEAD

A key point to remember from this chapter is that the Earth system works the way it does because of interactions between the ocean and the atmosphere. The ocean and the atmosphere control the climate and the distribution of climate zones. Climate, in turn, controls the distribution of plants and animals, the kinds of weathering that occur, and the processes of erosion that change the landscape. Weathering and erosion produce the regolith, which in turn is the source of sediment, and sediment is a key element of the rock cycle. Whatever aspect of the Earth you examine, you find that everything and every process is linked to everything else; that is the nature of Earth system.

In the next chapter we examine the effects of the two most pronounced climatic extremes: cold climates that lead to the formation of ice and the effects of glaciation; and hot, dry climates that lead to the formation of deserts and the effects of desert winds. We will see that in many respects the history of the Earth is a story of climatic change, the effects of which are recorded in the geologic record.

## *Chapter Highlights*

**1.** The amount of heat energy from the Sun reaching the Earth's surface is greatest near the equator and least near the poles.

**2.** The composition of **air,** excluding water vapor and aerosols, is 99.96 percent by volume nitrogen, oxygen, and argon.

**3.** Air temperature changes markedly as one moves upward from the surface. As a result, there are four distinct layers in the atmosphere, each with a distinct temperature profile.

**4.** Winds result from the unequal heating of the Earth's surface. This causes convection currents, which move heat from the equator toward the poles. This simple flow of air is disrupted by the distribution of water bodies and land masses, and by the effects of the Earth's rotation. Rotation produces the **Coriolis effect,** which causes winds to veer to the right in the Northern Hemisphere and to the left in the Southern Hemisphere.

**5.** The ocean is at least 4.0 billion years old. The water in the ocean is thought to have condensed from steam produced by primordial volcanic eruptions. Seawater now covers 71 percent of the Earth's surface and is unevenly distributed; the Northern Hemisphere has considerably less ocean area than the Southern Hemisphere.

**6.** Seawater ranges in **salinity** from 33 to 37 per mil.

Local variations are due to freezing and evaporation, which make the sea saltier, and to rain, snow, and river flow, which make the sea less salty.

**7.** The water in the ocean forms layers based on density, which is controlled by temperature and salinity. The surface layer, about 100 m (330 ft) deep, moves in great, slow, wind-driven currents.

**8.** The heat-absorbing capacity of the ocean combines with the atmosphere to mediate the global **climate.**

**9. Tides** are due to the gravitational pull of the Moon and the Sun, combined with the Earth's rotation.

**10.** Waves are caused by wind blowing across the surface of the ocean. The stronger the wind, the farther it blows, and the longer it lasts, the larger the wave.

**11.** Waves shape **beaches** and cause erosion of cliffs and other shoreline features.

**12.** Sediment is transported along the coast by **longshore currents** and by **beach drift.**

**13.** Shorelines are highly variable, but three basic types are most common; rocky (cliffed) coasts, lowland beaches and **barrier island** coasts, and coral **reef** coasts. Shorelines and coastal landforms are shaped by a combination of erosional and depositional processes.

## ▶ *The Language of Geology*

- air 368
- barrier island 391
- beach 390
- beach drift 388

- climate 376
- Coriolis effect 373
- greenhouse effect 371
- longshore current 388

- ozone layer 372
- reef 393
- salinity 378
- surf 386

- tides 384
- wave base 386
- wave-cut cliff 390
- weather 376

## ▶ *Questions for Review*

1. Why do statements about the composition of air exclude water vapor and aerosols?
2. In what layer of the atmosphere do weather-forming processes develop?
3. What are the major factors that control global patterns of windflow?
4. What are the four major ocean basins of the world, and in which hemisphere is each located?
5. What are the four main reasons that the salinity of seawater differs slightly from place to place in the oceans?
6. Explain the origin of surface ocean currents. What is the source of energy for the currents?
7. Most places along the shore see two high tides and two low tides each day. Use Figure 13.13 to explain why this is so.
8. Waves are caused by wind blowing across the water. What factors control the size of the waves?
9. Explain what happens to waves as they move from deep water and approach the shoreline.
10. How are coral reefs formed, and where are they found?

## ▶ *Questions for Thought and Discussion*

1. It is possible that there may once have been microscopic life on Mars or on some of the larger moons of Jupiter. As a member of a team designing tests to be carried out by remote, unmanned landers, what features would you test to see if life as we know it could have existed in the past?
2. Refer to Figures 13.11 and 13.12 and suggest how surface ocean currents might have differed during the early Tertiary, at a time when North and South America were not connected. Do some library research and suggest some tests for your hypothesis.
3. Visit a shoreline. If possible, visit the same spot on the shoreline on two or more occasions. What kinds of coastal landforms do you observe? What kinds of changes can you notice from one visit to the next? (If you don't live near a coast, this might be a good excuse to take a vacation near the ocean.)

For an interactive case study on weather conditions and flooding in the U.S. Midwest, visit our Web site.

# · 14 ·

# DESERTS, GLACIERS AND CLIMATIC CHANGE

The 1996 volcanic eruption at Vatnajökull, Iceland, was one of the most important eruptions of this century, though few but geologists and nearby residents were aware of it. What made this eruption so interesting from a scientific perspective—and so devastating—is that the volcano is located beneath an immense thickness of glacial ice.

When the eruption started in late September, the ice began to melt. At first, the water flowed under the glacier toward Lake Grimsvötn at a rate of 5000 m$^3$/s (well over a million gallons per second). The water level in the lake rose 20 m (more than 60 ft) per day. Then, in early October, a violent explosion in the volcano caused a huge fissure to open in the top of the glacier. Part of the ice cap collapsed into the cauldron, and a column of volcanic ash was emitted. By this time the water pressure under the glacier had built up to the point where the ice cap itself was actually lifted up, and catastrophic flooding—a *glacial outburst*—occurred.

The meltwaters moved with enormous force, causing a large portion of the glacier to move, and leaving huge crevasses in the ice. Preventative works carried out before the flood, such as flood control channels and banks of stones, were of some help. But the force of the flood was so great that solid concrete bridges crumpled like matchsticks. Iceland—a country accustomed to dealing with the forces of nature—had just experienced its greatest and most destructive natural disaster in recent decades.

◆

## In this chapter we discuss

- Types of deserts, both hot and cold
- The processes that characterize deserts and drylands
- Why glaciers form, and how they move and change
- How glaciers shape the landscape
- What processes lead to global climatic change

*Subglacial Eruption*
Vatnajökull volcano in Iceland erupted in 1996, causing extensive melting of the overlying glacier, and devastating glacial outburst flooding.

# Do scientists understand...

*What makes ice ages begin and end?*

*. . . In this chapter you will learn about this complex topic.*

When viewed from space, the Earth looks blue because of the vast expanse of the ocean. However, a polar view of either hemisphere appears largely white because of the extensive cover of snow and ice. In the Northern Hemisphere, much of the ice is a thin sheet floating on the Arctic Ocean. In the Southern Hemisphere, it consists of a vast system of land ice overlying the continent of Antarctica and adjacent islands and seas, as well as floating sea ice extending far beyond the coastline (Fig. 14.1). The water locked up in polar ice is part of the hydrosphere, but it is different from the rest of the water in the hydrosphere because it stays frozen year round. This perennially frozen part of the hydrosphere is referred to as the **cryosphere** (*cryo-* means cold).

These immense expanses of polar ice are linked in important ways to both the ocean and the atmosphere. As we saw in chapter 13, polar ice formation is

**cryosphere** The perennially frozen part of the hydrosphere.

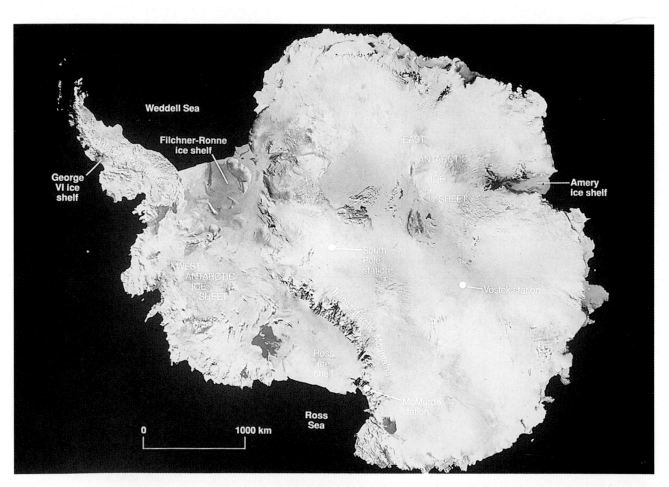

Figure 14.1
**FROZEN DESERT**
Taken from space, this view of the South Pole shows a white, permanently frozen world. The East Antarctic Ice Sheet covers the rocky Antarctic continent. The much smaller West Antarctic Ice Sheet covers a volcanic island arc and the surrounding sea floor. Major ice shelves occupy large coastal bays. The total area of the ice-covered regions of Antarctica is nearly equal to that of Canada and the United States combined.

398

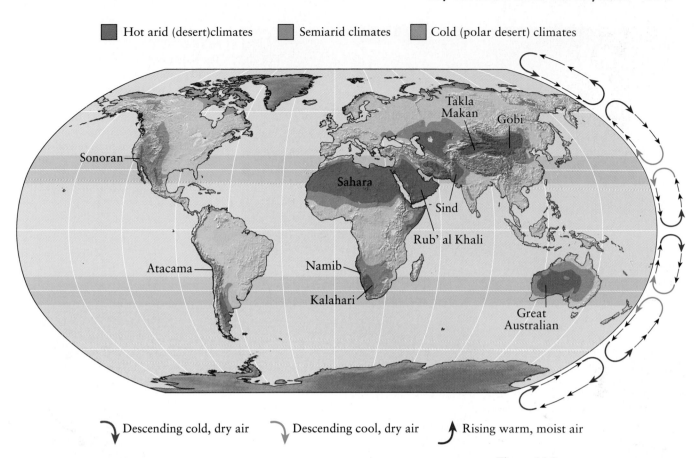

Descending cold, dry air    Descending cool, dry air    Rising warm, moist air

**Figure 14.2**
**THE WORLD'S DESERTS**
This map shows the distribution of arid and semi-arid climates and the major deserts associated with them. Many of the world's great deserts are located where belts of dry air descend along the 30°N and S latitudes. Compare this to Figure 13.7.

one of several interconnected processes that drive deep ocean circulation. It is also an important factor in controlling climate and sea level. Polar ice is puzzling, however. It represents one of the great water reservoirs in the Earth system, containing almost 74 percent of all fresh water. It is therefore surprising to discover that the polar regions are among the driest regions on the Earth, with very low annual precipitation, only a few centimeters per year. The poles are frozen deserts.

What geologic processes characterize these areas of extreme climate? What controls the global distribution of wet and dry regions? What makes the polar regions different from other kinds of deserts? Deserts and ice caps shrink and grow, both seasonally and on geologic time scales; what factors control these and other climatic variations? In this chapter we will try to answer these questions.

# DRY VS. WET: DESERTS AND DRYLANDS

Recall from chapter 13 (Fig. 13.7) that convection in the atmosphere, combined with the Coriolis effect, creates huge belts or cells of rising and falling air masses (Fig. 14.2). This results in three global belts of high rainfall and four belts of low rainfall. The high-rainfall belts are regions of *convergence* where huge, moist air masses meet and flow upward. These belts lie in the equatorial region, resulting in tropical climates, and along the two polar fronts, resulting in the warm-humid (tropical) and cold-humid (polar front) climate zones shown in Figure 13.8. The four global belts of low rainfall are regions of *divergence* where cool, dry air masses move downward and apart. These belts lie in the polar regions and in the

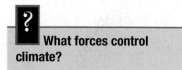

**? What forces control climate?**

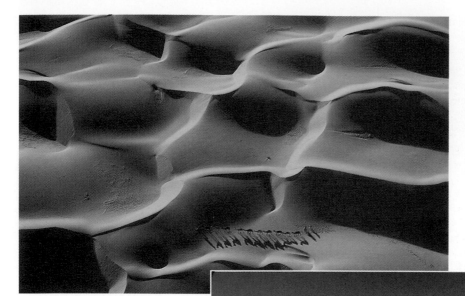

A.

Figure 14.3
HOT DESERTS
The Sahara (*A*) is the greatest of the world's subtropical deserts. Here a camel caravan crosses the desert near Nouakchott, Mauritania. The Namib Desert in southwestern Africa (*B*), shown here where the desert meets the Atlantic Ocean, is typical of coastal deserts. The Gobi Desert (*C*) in Mongolia is an inland continental desert, far from the moisture of the ocean.

B.

C.

subtropical regions along the 30°N and 30°S latitudes. The dry air masses and resulting low annual precipitation in these regions create two belts of dry climate in the subtropical regions, as well as the dry, cold climates of the polar regions.

The effects of the wet and dry belts created by the atmospheric circulation system are most clearly demonstrated by the global distribution of deserts (Fig. 14.2). These regions are not randomly scattered across the globe; rather, they are related to global circulation patterns and to local geographic features. Note, for example, the predominance of large deserts within the belt of dry air near the 30° latitude in both the Southern and Northern Hemispheres.

## Types of Deserts

The word "desert" literally means a deserted (that is, almost uninhabited) region that is nearly devoid of vegetation. However, irrigation has changed the meaning of the word by making many desert regions suitable for agriculture and therefore habitable. As a result, the term **desert** is now generally used in reference to an **arid land**, where annual precipitation is less than 250 mm (10 in). Desert lands total about 25 percent of the land area of the world outside the polar regions (Fig. 14.3). In addition, there is a smaller percentage of *semiarid* land in which the annual rainfall ranges between 250 and 500 mm (10 to 20 in).

Five types of deserts are recognized:

1. *Subtropical deserts.* The most extensive deserts, the Sahara, Kalahari, and Great Australian, are called subtropical because they are associated with the two belts of low rainfall near the 30°N and S latitudes.

2. *Continental interior deserts.* A second type of desert is found in continental interiors far from sources of moisture, where hot summers and cold winters prevail. The Gobi and Takla Makan deserts of central Asia fall into this category. These deserts are formed because air that travels a very long distance over land eventually contains so little water vapor that hardly any is available for precipitation.

3. *Rainshadow deserts.* A third kind of desert is found where a mountain range creates a barrier to the flow of moist air, causing a zone of low precipitation called a *rainshadow* on the downwind side of the range. Mountains do not completely block air, but they do effectively remove most of the moisture from the air. The Cascade Range and Sierra Nevada of the western United States form such barriers and are responsible for the desert regions lying immediately east of these mountains.

4. *Coastal deserts.* These deserts occur locally along the western margins of continents where cold, upwelling seawater cools and stabilizes maritime air flowing onshore, decreasing its ability to form precipitation. As the air encounters the land, the moisture it holds condenses into coastal fogs. The coastal deserts of Peru, Chile, and southwestern Africa are among the driest places on the Earth.

The four kinds of desert mentioned thus far are all hot deserts where rainfall is low and summer temperatures are high. The fifth differs from these in fundamental ways.

5. *Polar deserts.* These vast deserts are found in the polar regions, where precipitation is extremely low owing to the sinking of cold, dry air. Aside from the year-round cold, these deserts differ from hot deserts in one important respect: the surface of a polar desert, unlike that of deserts in warmer latitudes, is often underlain by abundant $H_2O$, nearly all of

**desert** Arid lands, which generally lack vegetation and cannot support a large population.

**arid land** Land where annual precipitation is less than 250 mm (10 in).

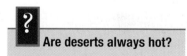
**Are deserts always hot?**

which is ice. The ice accumulates, even though precipitation is very low, because the precipitation is always in the form of snow and it is always too cold for the snow to melt.

Later in the chapter we will take a closer look at the unique environment of the polar desert. For now, let's focus on the geologic processes that characterize the world's hot deserts.

?  **How does wind move things?**

## Wind Erosion

As we learned in chapter 8, wind is an important agent of erosion and sedimentation. Processes related to wind—*eolian* processes—are particularly effective in arid and semiarid regions.

### Wind-Blown Sediment

Sediment carried by the wind tends to be finer than that moved by water or ice. Because the density of air (1.22 kg/m$^3$, or 0.074 lb/ft$^3$) is far less than that of water (1000 kg/m$^3$, or 62.4 lb/ft$^3$), air cannot move as large a particle as water flowing at the same velocity. In most regions with moderate winds, the largest particles that can be lifted in the airstream are grains of sand.

When the wind begins to blow across a bed of sand, the grains roll forward along the ground. This is called **surface creep.** As wind speed increases, grains may be bumped or lifted into the air, where they travel along arc-shaped paths, landing a short distance (a few centimeters) downwind. This process is called **saltation.** Finer particles may be carried aloft, where they encounter faster-moving winds that transport them some distance downwind before they slowly settle to the ground. The very finest dust particles can reach heights of a kilometer (0.6 mi) or more and are swept along in **suspension** as long as the wind keeps blowing. Such fine particles can be carried all the way across the ocean by the wind. Figure 14.4 shows how wind moves sediment.

### Mechanisms of Wind Erosion

Flowing air erodes the land surface in two ways. The first, **abrasion,** results from the impact of wind-driven grains of sand. Airborne particles act like tools, chip-

**surface creep** Sediment transport in which the wind causes grains to roll along the ground.

**saltation** Sediment transport in which particles move forward in a series of short jumps along arc-shaped paths.

**suspension** Sediment transport in which fine grains are lifted by wind currents and carried along above the level of the ground.

**abrasion** Wind erosion in which airborne particles chip small fragments off rocks that stick up from the surface.

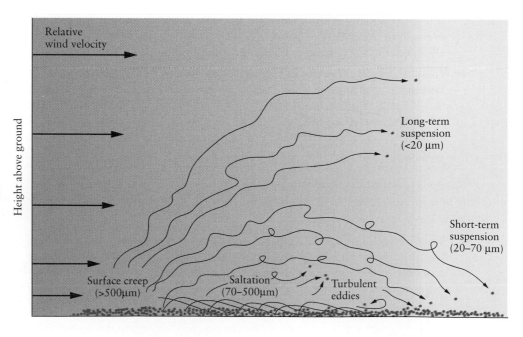

**Figure 14.4
HOW WIND MOVES SEDIMENT**
Moderate winds move the largest sand grains by surface creep. Slightly smaller sand grains move forward by saltation (bouncing). Finer particles are carried aloft where faster wind transports them downwind before they slowly settle to the ground. The very finest dust particles reach greater heights and are swept along in suspension as long as the wind keeps blowing.

# Abrasion by Wind

Figure 14.5

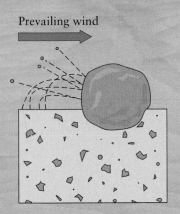

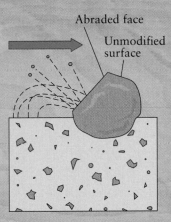

*A.* Ventifacts are formed when wind-blown sand grains pepper the upwind side of a rock that is sticking up above the desert surface.

*B.* The sand grains chip away at the rock, eventually abrading a smooth surface. Antarctica.

*C.* Ventifacts have at least one smooth, abraded surface facing upwind. These ventifacts litter the ground near Lake Vida in Victoria Valley, Antarctica.

ping small fragments off rocks that stick up from the surface. When rocks are abraded in this way they acquire distinctive, curved shapes and a surface polish. A bedrock surface or stone that has been abraded and shaped by wind-blown sediment is called a *ventifact* ("wind artifact") (Fig. 14.5).

The second erosional process, **deflation** (from the Latin word *flare* meaning "to blow away"), occurs when the wind picks up and removes loose particles of sand and dust (Fig. 14.6A). Deflation on a large scale takes place only where there is little or no vegetation and loose particles are fine enough to be picked up by the wind. It is especially severe in deserts but can occur elsewhere during times of drought when no moisture is present to hold soil particles together. Continued deflation sometimes leads to the development of *desert pavement;* most of the fine particles are removed, leaving a continuous pavement-like covering of coarse particles (Fig. 14.6B).

> **deflation** A mechanism of wind erosion in which loose particles of sand and dust are picked up and removed by the wind, leaving only the coarser particles behind.

## Wind Deposits and Desert Landforms

Wind erosion is confined chiefly to arid and semiarid lands. However, deposits of wind-transported sediment are also found in regions that are moist and covered by vegetation. Why is this? Part of the explanation is that the global climate is always changing. As the climate becomes warmer, cooler, wetter, or drier, the locations of different climate zones shift in response to these changes. But wind-blown particles can also be transported far from their place of origin. For example, distinctive reddish dust particles from the Sahara Desert have been identified in the soils of Caribbean islands, in the ice of Alpine glaciers, in deep sea sediments, and in the tropical rainforests of Brazil. When wind-blown particles settle, they sometimes form distinctive eolian deposits, the most important of which are *loess* and *dunes.*

> **?** **Why is wind-blown dust found in tropical rainforests?**

### Loess

Most regolith contains a small proportion of wind-laid dust. Usually, the dust is thoroughly mixed with other sediments and is indistinguishable from them. However, in some regions wind-laid dust is thick and uniform. Known as **loess**

> **loess** A fine, yellow-brown, wind-blown sediment.

# From Deflation to Desert Pavement

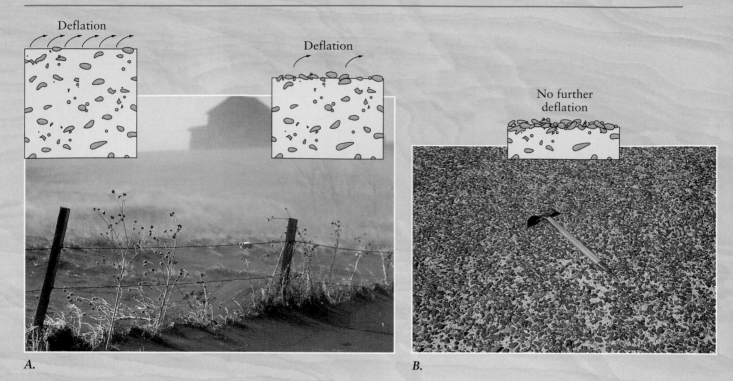

**Figure 14.6**
The drawings show progressive deflation of a sediment with different-sized particles can lead to the formation of desert pavement. *A.* The left-hand photo shows deflation in progress, removing dry soil from a plowed field in eastern Colorado. If the deflation continues long enough, and if the soil contains mixed particle sizes, including gravel, the result can be what you see in (*B*). The right-hand photo shows a desert pavement on the floor of Searles Valley, California. It consists of a layer of gravel, too coarse to be moved by the wind, that covers finer sediment and prevents further deflation.

(the German word for *loose,* pronounced "luhss"), this wind-deposited sediment consists largely of silt but is commonly accompanied by some fine sand and clay. In countries where it is thick and widespread, loess is an important resource because of the productive (though highly erodible) soils that develop on it. The rich agricultural lands of the upper Mississippi Valley, the Columbia Plateau of Washington State, the Loess Plateau of China, and much of Eastern Europe have loess soils that provide food for millions of people. Loess deposits can also provide shelter; in central China caves carved in loess house thousands of families (Fig. 14.7).

## Dunes

**dune** A hill or ridge of sand deposited by winds.

A **dune** is a hill or ridge of sand deposited by winds. Although little is known about how dunes are initially formed, it is likely that a dune develops where some minor surface irregularity or obstacle distorts the flow of air. On encountering an obstacle close to the ground, the wind sweeps over and around it but leaves a pocket of slower-moving air immediately downwind. In this pocket of low wind velocity, sand grains moving with the wind drop out and begin to form a mound. The mound, in turn, influences the flow of air over and around it and may continue to grow into a dune.

A typical dune is asymmetrical, with a gentle windward slope (the side facing toward the wind) and a steep lee face (the side facing away from the wind). Pushed by the wind, sand moves by saltation up the gentle windward slope (Fig. 14.8). When it reaches the top, the sand cascades down the steep lee slope, also

Figure 14.7
**WELCOME HOME**
A cave excavated by hand in thick, compact loess provides a roomy and comfortable home for a family in the loess region near Xian, China.

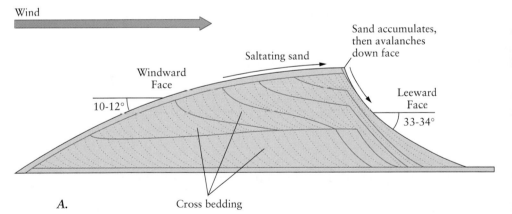

*A.*

Figure 14.8
**HOW SAND DUNES FORM**
*A.* The cross section through a sand dune shows the typical gentle windward slope (facing the wind) and steep slip or leeward face (facing away from the wind). Sand grains saltate up the windward slope to the top of the dune, only to avalanche down the slip face. The black lines inside the dune show old slip faces. *B.* The photo shows these same patterns in ancient sandstones from Utah, which tells us this was once a sandy, windswept desert.

*B.*

called the *slip face*. The slip face is always on the leeward side, so we can tell which way the wind was blowing from the asymmetrical form of the dune. Criss-crossed strata within the dune, called *cross-beds,* are former slip faces.

The sliding sand on the slip face comes to rest at the *angle of repose,* the steepest angle at which loose particles will come to rest. Thus, the dune's slip face lies at the angle of repose. The angle of repose varies for different materials, depending on factors such as the size and angularity of the particles. For dry, medium-sized sand particles it is about 33–34°; the angle is generally steeper for coarser materials such as gravel and gentler for fine materials such as silt.

The type of dune that is formed is controlled by the wind velocity, direction, and variability; the amount of vegetation cover; and the characteristics and availability of the sand. These three factors are shown in the triangle in Figure 14.9, along with photos of common types of dunes. Some types of dunes tend to remain stationary. Others (such as barchan dunes) may migrate over long distances in response to changing environmental conditions. When sand deposits invade nondesert lands, severe degradation and loss of agricultural productivity can result. This is part of the process of *desertification.*

## Desertification

**?**

**How are deserts formed?**

In the region south of the Sahara lies a drought-prone belt of dry grassland known as the Sahel (Arabic for *border*). There the annual rainfall is normally only 100 to 300 mm (4 to 12 in), most of it falling during a single brief rainy season. In the early 1970s the Sahel experienced the worst drought of the century (Fig. 14.10D). For several years in a row the annual rains failed to appear, causing the adjacent desert to spread southward by as much as 150 km (93 mi). The drought extended from the Atlantic to the Indian Ocean, a distance of 6000 km (3700 mi), and affected a population of at least 20 million. The results of the drought were intensified by the fact that between about 1935 and 1970 the human population of the region had doubled and the number of domestic livestock had also increased dramatically. This resulted in severe overgrazing, so the grass cover was devastated by the drought. Many millions of cattle (about 40 percent of the total) died. Millions of people suffered from thirst and starvation. Many died, and mass starvation was alleviated only by worldwide relief efforts.

**desertification** The invasion of desert conditions into nondesert areas.

Such invasion of desert conditions into nondesert areas, referred to as **desertification,** can result from natural environmental changes as well as from human activities. The major signs of desertification are lower water tables, higher levels of salt in water and topsoil, reduction in surface water supplies, unusually high rates of soil erosion, and destruction of vegetation. Although there is evidence of natural desertification in the geologic record, there is increasing concern that human activities can accelerate widespread desertification. Desertification caused by human activities is called *land degradation,* to identify it as anthropogenic (human-caused) rather than natural. Areas that are most susceptible to desertification, whether by natural or anthropogenic causes, are shown in Figure 14.10. Most are semiarid lands adjacent to the world's great deserts. Many are areas where population densities are high and resource use is too intensive to be successfully supported by the dry soils and fragile ecosystems.

As the map in Figure 14.10 shows, drought and desertification are not confined to Africa. During the mid-1930s, many farm families in the southern Great Plains region of the United States abandoned their homes and lands and trekked westward. They were part of the great migration described by John Steinbeck in his award-winning novel *The Grapes of Wrath.* The primary cause of this migration was a severe drought, which caused huge dust storms (Fig. 14.10A). The drought and dust storms devastated the land by destroying crops and burying

# Dunes

## A. BARCHAN DUNES
These crescent-shaped dunes point down-wind across an irrigated field in the Danakil Depression, Ethiopia. Barchans up to 30 m (almost 100 ft) high are formed on hard, flat desert floors where the supply of sand is limited and wind direction is constant. They can migrate many kilometers.

**Dune Triangle**

- E · Longitudinal dunes
- C ·
- A · Crescent dunes
- B · Transverse dunes
- Parabolic dunes
- D ·
- No dunes
- Sand
- Wind
- Vegetation

## B. TRANSVERSE DUNES
An abundant supply of sand allows barchan dunes in China's vast Takla Makan Desert to merge into these continuous, asymmetrical ridges, perpendicular to the strongest wind.

## D. PARABOLIC DUNES
Parabolic dunes and blowouts such as those shown here, are common in vegetated coastal regions. The open end of the U or V shape faces up-wind.

## C. STAR DUNES
Wind blowing from all directions focuses sand into these isolated, stationary, star-shaped dunes in Libya. This radar satellite image reveals massive dunes up to 300 m (almost 1000 ft) high.

**Figure 14.9**

## E. LONGITUDINAL DUNES
These long, narrow ridges, which roughly parallel prevailing winds, cover vast deserts in Africa and Australia.

# Drought and Desertification

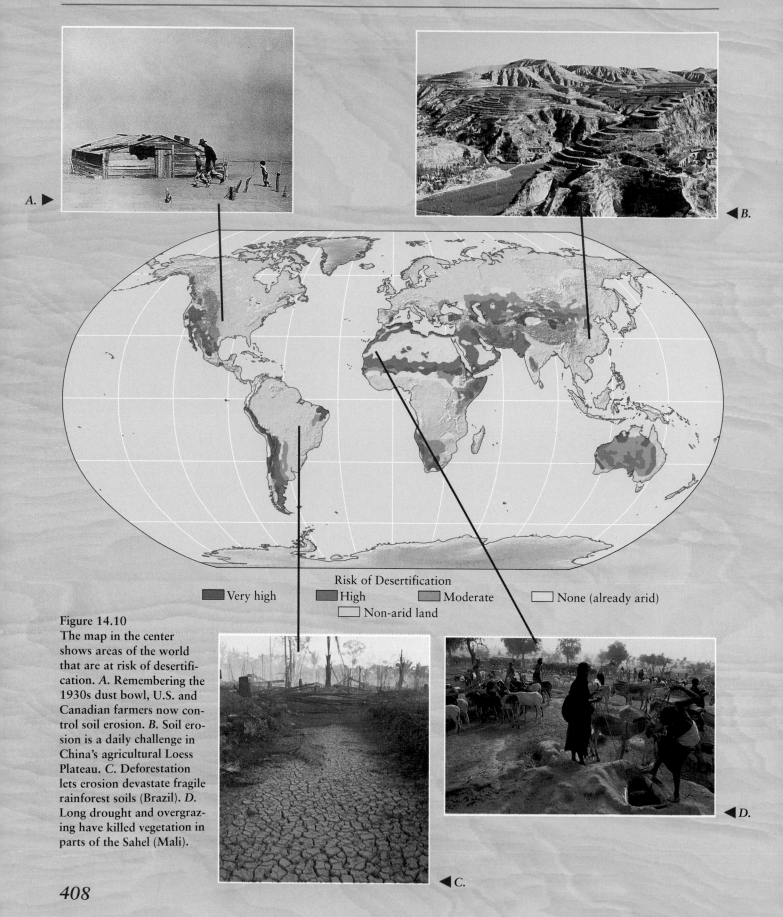

**A.** ▶

◀**B.**

Risk of Desertification

■ Very high ■ High ■ Moderate ☐ None (already arid)
☐ Non-arid land

**Figure 14.10**
The map in the center shows areas of the world that are at risk of desertification. *A.* Remembering the 1930s dust bowl, U.S. and Canadian farmers now control soil erosion. *B.* Soil erosion is a daily challenge in China's agricultural Loess Plateau. *C.* Deforestation lets erosion devastate fragile rainforest soils (Brazil). *D.* Long drought and overgrazing have killed vegetation in parts of the Sahel (Mali).

◀**D.**

◀**C.**

formerly productive fields under drifting sand and dust. The southern plains came to be called the "dust bowl," and we refer to that period as the "dust-bowl years."

The effectiveness of the wind in creating the dust bowl was enhanced by decades of poor land use practices. The grasses that grew on the prairies when the original settlers arrived protected the rich topsoil from wind erosion. However, the grasses were progressively replaced by plowed fields and seasonal grain crops, which left the ground bare and vulnerable for part of the year. Although today the land is still potentially vulnerable in drought years, improved farming practices have reduced the likelihood of similar catastrophes in the future.

How can desertification be halted or even reversed? The answer lies largely in understanding the geologic principles involved and in the application of measures designed to reestablish a natural balance in the affected areas. Elimination of incentives to exploit arid and semiarid lands beyond their natural capacity and long-range planning aimed at minimizing the negative effects of human activity should help. Soil management techniques are widely known and readily available; they just need to be applied more aggressively. These techniques include a range of agricultural approaches (such as crop rotation and terracing of steep slopes), as well as identification and mapping of vulnerable soils, avoidance of heavy use of agricultural chemicals, and reforestation of vulnerable lands. The preservation of productive lands is essential to maintaining the world's food production capacity at the level needed for an increasing population.

# COLD VS. WARM: GLACIERS AND ICE SHEETS

Desertification can be an expression of climatic change. Natural processes involving changes in both precipitation and temperature have caused the Sahara Desert to advance and retreat many times over the past 10,000 years, independent of recent human activities in the region. In the vast deserts of the polar ice sheets, another climate battle is played out. The expansion and shrinking of glaciers and ice sheets, both in the polar regions and in more temperate alpine settings, is an expression of the complex interplay between temperature and precipitation in the global climate system. The existence of glaciers and ice sheets is linked to the interaction of several Earth systems: tectonic forces that produce high, mountainous areas; the ocean, a source of moisture; and the atmosphere, which delivers the moisture to the land in the form of snow. We now turn our attention to these great polar deserts and other parts of the cryosphere.

## Types of Glaciers

Annual snowfall is generally very low in polar regions because the air is too cold to hold much moisture. The small amount of snow that does fall doesn't usually melt, however, because summer temperatures are very low. In areas where the amount of snow that falls each winter is greater than the amount that melts during the following summer, the covering of snow gradually grows thicker. As it accumulates, its increasing weight causes the snow at the bottom to compact into a solid mass of ice. When the accumulating snow and ice become so thick that the pull of gravity causes the frozen mass to move, a glacier is born. We define a **glacier** as a permanent body of ice, consisting largely of recrystallized snow, which shows evidence of movement owing to the pull of gravity.

**glacier** A permanent body of ice, consisting largely of recrystallized snow, that shows evidence of movement owing to the pull of gravity.

Glaciers are obviously cold because they consist of ice and snow. However, when we drill holes through glaciers in a variety of geographic environments, we find a large range in ice temperatures. This allows us to divide glaciers into warm and cold types. The difference between them is important, for it influences the behavior and movement of the ice. Throughout a warm glacier, more commonly called a *temperate glacier*, the ice is near its melting point. In such glaciers, which

# Glaciers and Ice Caps

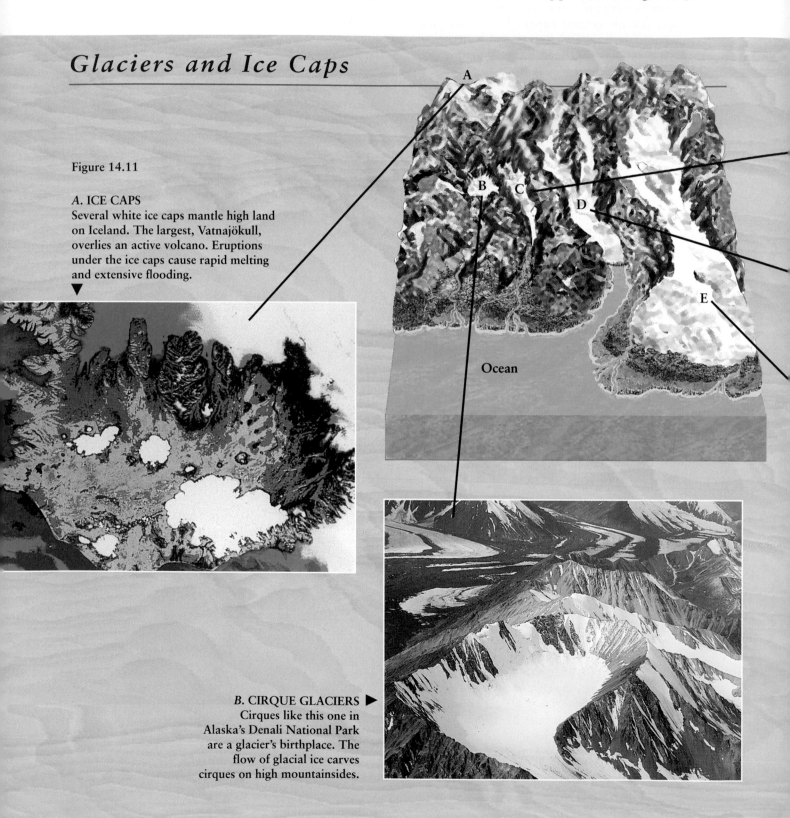

Figure 14.11

*A.* ICE CAPS
Several white ice caps mantle high land on Iceland. The largest, Vatnajökull, overlies an active volcano. Eruptions under the ice caps cause rapid melting and extensive flooding.

Ocean

*B.* CIRQUE GLACIERS ▶
Cirques like this one in Alaska's Denali National Park are a glacier's birthplace. The flow of glacial ice carves cirques on high mountainsides.

are formed in low and middle latitudes, meltwater and ice can exist together at equilibrium. At high latitudes and altitudes, where the mean annual temperature is below freezing, the temperature in a glacier remains low and little or no seasonal melting occurs. Such a cold glacier is commonly called a *polar glacier*.

Both temperate and polar glaciers can vary considerably in shape and size. They have been classified into several types, illustrated in Figures 14.11 and 14.12.

◀ **C. VALLEY GLACIERS**
Ice from cirques flows down existing valleys, becoming valley glaciers. Dozens cover Denali National Park. Mount McKinley, the highest North American peak, lies near the center of the glacial region, from which the valley glaciers radiate.

**D. FJORD GLACIERS** ▶
Icebergs "calve" from the front of this fjord glacier in southwestern Greenland. A fjord is a glacier-carved trough in bedrock that has filled with sea-water as the glacier retreats.

▲ *E.* PIEDMONT GLACIERS
In the Swiss Alps, Gorner Glacier flows from the mountains as a valley glacier and onto lowlands as a large ice lobe, where it becomes a piedmont glacier.

These are the main types of glaciers and ice caps:

1. The smallest, a *cirque glacier,* occupies a protected, bowl-shaped depression called a *cirque,* produced by glacial erosion on a mountainside.

2. A cirque glacier that expands outward and downward into a valley becomes a *valley glacier.* Many high mountain ranges contain glacier systems that include valley glaciers tens of kilometers long.

3. Along some sea coasts in high latitudes, large valley glaciers occupy deep fjords. (A *fjord* is the seaward end of a glacier-carved bedrock trough whose floor lies far below sea level.) These are called *fjord glaciers.*

4. A *piedmont glacier* is a broad lobe of ice that terminates on open slopes beyond a mountain front. Piedmont glaciers are normally fed by one or more large valley glaciers.

5. *Ice caps* cover mountain highlands or low-lying land at high latitudes. They generally flow radially outward from their center.

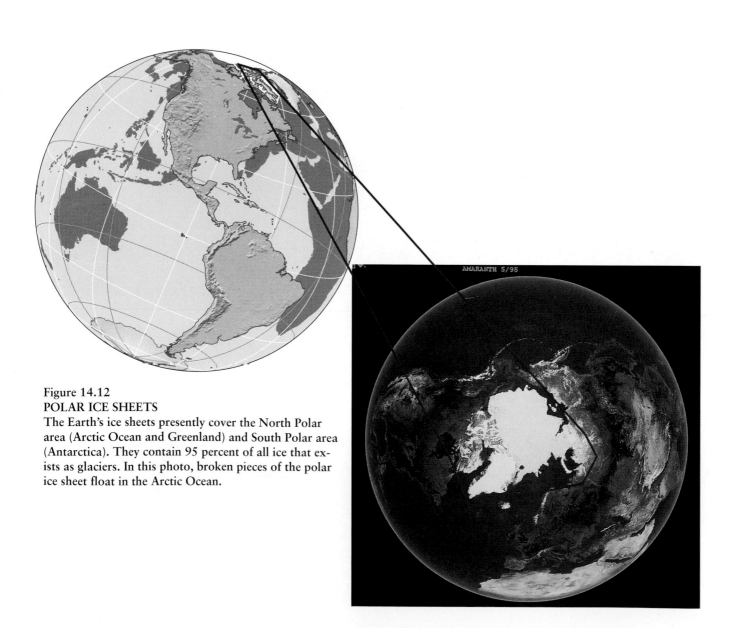

**Figure 14.12**
**POLAR ICE SHEETS**
The Earth's ice sheets presently cover the North Polar area (Arctic Ocean and Greenland) and South Polar area (Antarctica). They contain 95 percent of all ice that exists as glaciers. In this photo, broken pieces of the polar ice sheet float in the Arctic Ocean.

6. Huge continent-sized *ice sheets*, reaching thicknesses of more than 4 km (2.5 mi), overwhelm nearly all the land surface within their margins. Modern ice sheets, which are found only in Greenland and Antarctica, contain about 95 percent of all glacial ice.

7. A great deal of the Earth's ice floats as *sea ice* on the ocean surface.[1] Floating ice is formed when the surface of the ocean cools below the freezing point of seawater. As ice crystals begin to form, a soupy mixture is produced at the ocean surface. In the absence of waves or turbulence, the crystals freeze together to form a continuous cover of ice. Sea-ice growth then proceeds by the addition of ice to the base. Over the course of a year some ice is lost from the surface, but more is added to the base of the sea ice. Floating *ice shelves* hundreds of meters thick occur along the coasts of Antarctica, and smaller ones are found among the Canadian Arctic islands.

## Changing Glaciers

Although we define a glacier as a "permanent" body of ice, in reality glaciers are constantly changing in several ways. The snow that falls on their surfaces gradually changes to ice. Glaciers shrink and grow in response to seasonal changes in temperature and precipitation. The ice in a glacier moves, slowly but surely, under the influence of gravity. The margins of glaciers advance or retreat from year to year and over geologic time, as climatic conditions change. Let's look now at some of these changes.

### How Glaciers Are Formed

Newly fallen snow is very porous and easily penetrated by air. The presence of air in the pore spaces allows the delicate points of each snowflake to evaporate. The resulting water vapor condenses in tiny spaces in the snowflakes, eventually filling them. In this way the fragile ice crystals slowly become smaller, rounder, and denser, and the pore spaces between them disappear (Fig. 14.13). Snow that survives for a year or more becomes more compact as it is buried by successive snowfalls. As the years go by, the snow gradually becomes denser and denser until it is no longer penetrable by air. At this point it has become *glacier ice*.

Further changes take place as the glacier ice is buried deeper and deeper. As snowfall adds to the glacier's thickness, the increasing pressure causes the small grains of glacier ice to grow. This increase in size is similar to what happens when a fine-grained rock recrystallizes as a result of metamorphism in the Earth's crust (chapter 10). Glacier ice is technically a rock, but it has a much lower melting temperature than any other naturally occurring rock, and because of its low

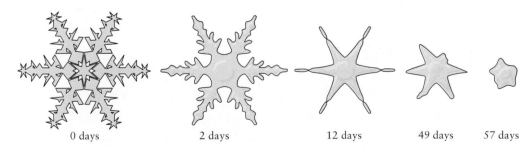

0 days          2 days          12 days          49 days          57 days

Figure 14.13
**FROM SNOW TO ICE**
As a new snowflake is slowly converted into a granule of ice, evaporation and recondensation cause its delicate points to disappear, and it becomes much more compact.

---

[1] Floating sea ice forms from the direct freezing of ocean water. In contrast, *icebergs* are chunks of land ice that have broken off from glaciers that terminate in the ocean.

How does a glacier balance its budget?

density—about 0.9 g/cm³ (56.2 lb/ft³)—it floats in water, which has a density of 1 g/cm³ (62.4 lb/ft³).

## How Glaciers Change in Size

The mass of a glacier is constantly changing as the weather varies from season to season and, over time, as local and global climates change. In a way, a glacier is like a checking account. Instead of being measured in terms of money, the balance of a glacier's account is measured in terms of the amount of snow deposited, mainly through snowfall in the winter, and the amount of snow (and ice) withdrawn, mainly through melting and evaporation during the summer. The additions are collectively called *accumulation* and the losses *ablation* (Fig. 14.14). The total added to the account at the end of a year—the difference between accumulation and ablation—is a measure of the glacier's *mass balance*. The account may have a surplus (a positive balance) or a deficit (a negative balance), or it may hold exactly the same amount at the end of the year as it did at the beginning.

Over a period of years a glacier may gain more mass than it loses. In such cases the glacier's volume increases and its front, or *terminus,* is likely to advance. Conversely, a succession of years in which a glacier's mass balance decreases will cause the terminus to retreat. If no net change in mass occurs, the terminus is likely to remain in the same place.

## How Glaciers Move

As noted earlier, part of the definition of a glacier is that it shows evidence of movement owing to the pull of gravity. But glaciers move very, very slowly. How can we observe this movement? One way to prove that glaciers move is to walk onto a glacier near the end of summer and carefully measure the position of a boulder on its surface relative to a fixed point beyond the glacier's edge. If you

**Figure 14.14**
**GLACIERS HAVE A BUDGET**
This valley glacier has been sliced open so you can see what happens inside. Near the head of the glacier is the accumulation zone, where snow accumulates and turns into new ice. Near the end, or terminus, is the zone of ablation, where snow is lost to evaporation and melting. In the deep parts of the glacier the ice flows, but near the surface it cracks, forming deep crevasses.

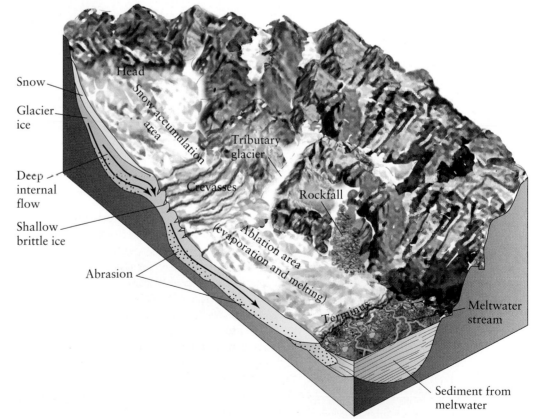

*A.*

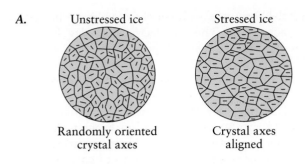

Unstressed ice                    Stressed ice

Randomly oriented            Crystal axes
crystal axes                         aligned

*B.*

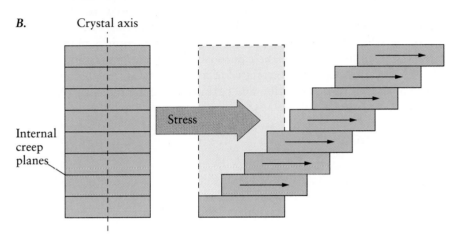

Crystal axis

Internal creep planes

Stress

Figure 14.15
**DEFORMING ICE**
Ice crystals deep within a glacier move by means of internal creep. *A.* Randomly oriented ice crystals are reorganized under stress, until their internal crystal axes and planes are parallel. *B.* Creep along internal planes causes slow deformation, as in a deck of playing cards.

measure the boulder's position again a year later, you will find that it has moved in the downglacier direction. Actually, it is the ice that has moved, carrying the boulder along.

Measurements of surface velocity across a valley glacier show that the uppermost ice in the central part of the glacier moves faster than the ice at the sides. This is similar to what happens to water flowing in a stream (chapter 12). In most glaciers, flow velocities range from a few centimeters to a few meters a day, about the same rate at which groundwater percolates through crustal rocks. Hundreds of years have elapsed since ice that is now exposed at the terminus of a very long glacier fell as snow near the top of its accumulation area.

If we examine glacial ice and the terrain on which it lies, we find clues that tell us that ice moves in two basic ways: by internal flow and by basal sliding across the underlying rock or sediment.

*Internal Flow.* As the weight of overlying snow and ice in a glacier increases, individual ice crystals are subjected to higher and higher stress. Under this stress, ice crystals deep within the glacier are deformed by slow displacement (termed *creep*) along internal crystal planes. The movement is very similar to the way in which cards in a deck of playing cards slide past one another if the deck is pushed from one end (Fig. 14.15). As the compacted, frozen mass moves, the crystal axes of the individual ice crystals are forced into the same orientation and end up with their internal crystal planes oriented in the same direction.

In contrast to the deep parts of a glacier, where ice flows as a result of internal creep, the surface portion has relatively little weight on it and is brittle. When a glacier passes over a change in slope, such as a cliff, the surface ice cracks as tension pulls it apart. When the crack opens up, it forms a *crevasse,* a deep, gaping fissure in the upper surface of a glacier (as shown in Fig. 14.14).

*Basal Sliding.* Sometimes ice at the bottom of a glacier slides across its *bed* (the rocks or sediment on which the glacier rests). This is called *basal sliding*. In some temperate glaciers, meltwater at the base acts as a lubricant. In such glaciers basal sliding may account for up to 90 percent of total observed movement. By contrast, polar glaciers are so cold that they are frozen to their bed; they therefore move primarily by internal flow.

Occasionally, glaciers exhibit an unusual phenomenon marked by rapid movement and dramatic changes in size and form. Such an event is called a *surge.* When a surge occurs, a glacier seems to go berserk. Ice in one part of the glacier begins to move rapidly downglacier, producing a chaos of crevasses and broken pinnacles. Rates of movement have been observed that are up to 100 times those of nonsurging glaciers. In 1993–94 Alaska's Bering Glacier experienced a surge in which ice in a portion of the glacier moved approximately 9 km (5.6 mi) in less than nine months. The causes of surging are poorly understood. In many cases they are believed to be related to a buildup of water pressure at the base of the glacier that reduces friction and permits unusually rapid basal sliding.

## The Glacial Landscape

As glaciers move, they change the landscape by eroding and scraping away material as well as by transporting and depositing material at their ends and along their margins. In changing the surface of the land over which it moves, a glacier acts like a file, a plow, and a sled. As a file, it rasps away firm rock. As a plow, it scrapes up weathered rock and soil and plucks out blocks of bedrock. As a sled, it carries away the load of sediment acquired by plowing and filing, along with rock debris that falls onto it from adjacent slopes. Let's look more closely at the landforms that result from these processes.

**?** **How do glaciers modify the landscape?**

### How Glaciers Sculpt the Land

The base of a glacier is studded with rock fragments of various sizes that are all carried along with the moving ice. When basal sliding occurs, small fragments of rock embedded in the basal ice scrape away at the underlying bedrock and produce long, nearly parallel scratches called *glacial striations.* Larger particles gouge out deeper *glacial grooves* (Fig. 14.16A). Because glacial striations and grooves are aligned parallel to the direction of ice flow, they help geologists reconstruct the flow paths of former glaciers.

In mountainous regions glaciers produce a variety of distinctive landforms, such as bowl-shaped cirques, which are formed through a combination of plucking, frost-wedging, and abrasion. As cirques on opposite sides of a mountain are eroded, their walls meet to form a sharp-crested ridge called an *arête.* When glacial ice moves downward from a cirque, it scours a valley channel with a distinctive U-shaped cross section and a floor that usually lies well below the level of tributary valleys (Fig. 14.16B). The retreat of a glacier can leave behind a terrain full of pits and pockmarks, which subsequently fill with water to form ponds and lakes (Fig. 14.16C). An example of a very small lake formed by glacial processes (partly erosion and partly deposition) is Walden Pond, made famous by the writer Henry Thoreau. Some examples of very large glacially formed lakes are the Great Lakes, Lake Winnipeg, and Great Bear Lake.

### Glacial Deposits

Like streams, glaciers carry a load of sediment particles of various sizes. Unlike a stream, however, a glacier can carry part of its load at its sides and even on its surface. A glacier can carry very large rocks and small fragments side by side

Figure 14.16
GLACIAL SCULPTING
Glaciers plow, file, and scrape the land, creating a variety of erosional land-forms. *A.* A bedrock surface near Findelen Glacier in Switzerland displays grooves and striations etched by rocky debris in the base of the moving glacier when it moved over this site. In the background rises the Matterhorn. It is an example of a *glacial horn,* a landform sculpted by the glaciers that surround it. *B.* The U-shaped glacial trough in the Beartooth Plateau, near Red Lodge, Montana, was carved by an alpine glacier during the Ice Age. *C.* Lake-filled kettles in central Chile were formed at the end of the last glaciation, when debris-covered ice slowly melted away.

*A.* ▲

*B.* ▶

*C.* ▶

◀ *A.*

*B.* ▼

**Figure 14.17**
**GLACIAL DEPOSITS**
*A.* This large erratic boulder was deposited by a glacier in Denali National Park, Arkansas. *B.* The curving ridge of sand and gravel in this photo is an esker in Kettle-Moraine State Park, Wisconsin. The esker marks the bed of an ancient river of meltwater that flowed underneath a continental ice sheet near its margin. *C.* The dark stripes running down the center of the glacier are a medial moraine. This is an aerial view of the Grand Plateau Glacier in the St. Elias Mountains, Glacier Bay National Park. *D.* Lobuche Glacier, which flows out of a high cirque near Mount Everest in the Himalaya, has retreated upslope from a terminal moraine that it deposited over a hundred years ago.

**till** A heterogeneous mixture of finely crushed rock (*rock flour*), sand, pebbles, cobbles, and boulders deposited by a glacier.

without separating them by size or weight. Thus, sediments deposited by a glacier are not sorted or stratified the way stream deposits usually are. This can be seen by examining glacial **till**, a heterogeneous mixture of finely crushed rock called *rock flour,* sand, pebbles, cobbles, and boulders deposited by a glacier. In most cases the boulders and rock fragments in a till are different from the underlying bedrock, indicating that the components of the till were transported to their present site from somewhere else. A glacially deposited rock that is different from the underlying bedrock is called an *erratic,* from the Latin word for wanderer (Fig. 14.17A).

Underneath some large glaciers are flowing streams that carry meltwater and sediment. When such streams emerge from the terminus of the glacier, they may deposit their sediment load, thereby forming a broad, sweeping plain called an *outwash plain.* If the glacier subsequently retreats, the former locations of the streams may be marked by sinuous, curving deposits of stream sediment called *eskers* (Fig. 14.17B). Another common glacial landform is the *drumlin,* a streamlined, elongate hill consisting of glacially deposited sediment that lies parallel to the direction of ice flow. Drumlins are formed beneath glaciers through a combination of sculpting and deposition of glacial till in successive layers.

▲ C.

D. ▶

The boulders, rock fragments, and other sediment carried by the glacier may be deposited along its margins or terminus. A ridge or pile of debris that is being transported or has been deposited along the edges of a glacier is called a **moraine.** A *lateral moraine* is a ridge of sediment along the side of a glacier. If two glaciers converge, they may trap lateral moraines between them, forming a ridge of material that rides along the middle of the ice stream; this is called a *medial moraine* (Fig. 14.17C). At the glacier's terminus, debris accumulates in a *terminal moraine* (Fig. 14.17D). Geologists have used the locations of glacial moraines in the United States and Canada to determine how far the glacial ice cover extended over North America during the last Ice Age.

**moraine** A ridge or pile of debris that is being transported or has been deposited along the edges of a glacier.

## Periglacial Landforms

Areas that are in close proximity to glacial ice are called **periglacial** (*peri* means "near"). In such regions, intense frost action and a large annual range in temperature create a distinctive set of landforms. The most common type of environment in present-day periglacial regions is the treeless *tundra*, with long winters, very short summers, poorly developed soils, and low, scrubby vegetation. A feature typical of tundra regions is **permafrost,** ground that is perennially (year-

**periglacial** Refers to areas that are in close proximity to glacial ice.

**permafrost** Ground that is perennially below the freezing point of water.

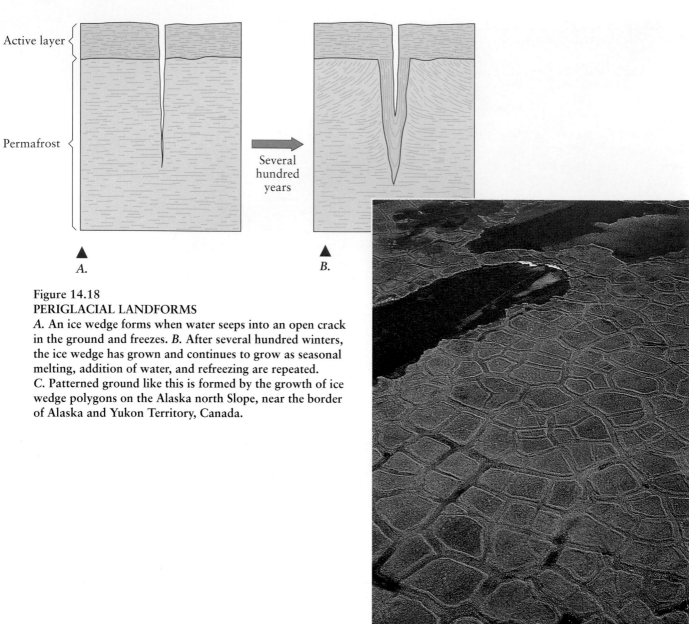

Active layer

Permafrost

Several hundred years

A.

B.

**Figure 14.18**
**PERIGLACIAL LANDFORMS**
*A.* An ice wedge forms when water seeps into an open crack in the ground and freezes. *B.* After several hundred winters, the ice wedge has grown and continues to grow as seasonal melting, addition of water, and refreezing are repeated.
*C.* Patterned ground like this is formed by the growth of ice wedge polygons on the Alaska north Slope, near the border of Alaska and Yukon Territory, Canada.

C. ▶

round) below the freezing point of water. In the tundra, soil moisture is solidly frozen throughout the winter months. During the short summer, however, the ice thaws in a thin layer near the surface, called the *active layer.* The thawing leaves the soil water-saturated and vulnerable to mass wasting and *solifluction,* the downslope movement of water-saturated regolith.

Much of the ice in permafrost is in pore spaces. Another common occurrence is the *ice wedge,* in which surface water enters a crack in the ground and then freezes. The ice-filled crack may open or partially melt during the summer, allowing more water to enter (Fig. 14.18A). Repeated freezing and addition of new ice causes the wedge to thicken, until it becomes as wide as 3 m (10 ft) and as deep as 30 m (50 ft) (Fig. 14.18B). Sometimes ice wedges link together into polygonal shapes characteristic of *patterned ground,* a common landform in periglacial regions (Fig. 14.18C).

# GLOBAL CLIMATIC CHANGE

*For Hot, Cold, Moist, and Dry, four champions fierce,*
*Strive here for mastery . . .*
John Milton (1608–1674)

Last winter may have been colder than the winter before, and last summer may have been wetter than the previous summer. Does this mean that climate is changing? Does it mean that the world is heading for another glaciation or that deserts will spread? Not necessarily. To be identified as a climatic change, average conditions must shift over a span of years. Several years of abnormal weather may not mean that a change is occurring, but trends that persist for a decade or more may signal a shift to a warmer, cooler, wetter, or drier climatic regime.

## The Ice Ages

In the past few million years numerous cycles of cooling and warming have occurred (Fig. 14.19). Periods during which the average temperature at the Earth's surface dropped by several degrees and stayed low long enough for existing ice sheets to grow larger (and new ones to form) are called **glaciations** (or *ice ages, glacial periods, glacial stages,* or *glacial epochs*). Periods between glaciations, when the ice sheets retreated and sea levels rose, are called *interglacials* (or *interglacial stages*).

During the past 1.6 million years—the Pleistocene and Holocene Epochs—the Earth has experienced more than 20 glacial-interglacial cycles. The timing of these cycles has varied, with extreme temperature *minima* (low points) occurring roughly every 100,000 years over the past million years, and every 20,000 to 40,000 years before that. Glaciation has been especially prevalent during the Pleistocene, but it is not a new phenomenon. The rock record contains evidence of glacial ages that occurred as long ago as 2.3 billion years. Today, about 10 percent of the world's land area is covered with glacier ice; of this area, 84 percent lies in the Antarctic region.

### Studying Ancient Climates

How do geologists know what the Earth's climate and surface temperature were like 20,000 or 100,000 or even 100 million years ago? Historical records of temperature and weather have only been kept on a regular basis since about the

**glaciation** Periods during which the average temperature at the Earth's surface dropped by several degrees and stayed low long enough for existing ice sheets to grow larger (and new ones to form).

> **?**
> **How do geologists tell what the climate was like a long time ago?**

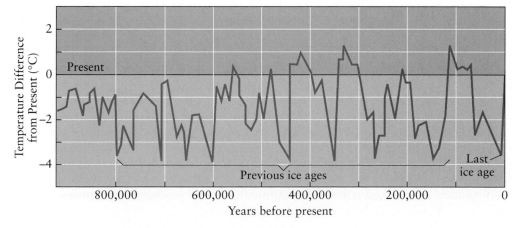

Figure 14.19
**GLOBAL TEMPERATURE CHANGES**
The blue line shows how temperatures have varied from 1 million years ago to the present. Note the cyclical variation between temperature minima (the glacial periods) and temperature maxima (the interglacials).

**The monitoring of global climatic change** has some certainties and many uncertainties. For example, we know that the amount of $CO_2$ in the atmosphere is increasing. We know that the rate of increase has risen sharply since the Industrial Revolution, and we know that $CO_2$ is emitted by the burning of fossil fuels. We also know that $CO_2$ functions as a greenhouse gas in the atmosphere, causing it to become warmer. But there are many things that we don't know with certainty. For example, are warm periods in geologic history always linked with high levels of atmospheric $CO_2$? Do high levels of atmospheric $CO_2$ cause global warming? Alternatively, does a warm climate encourage the release of $CO_2$ into the atmosphere from soils, plants, and water (the chicken-and-egg question)? Is the present high level of atmospheric $CO_2$ unusual in the context of geologic history? Will it cause global warming and, if so, by how much?

One of the ways in which scientists have tried to answer these questions is by looking for actual samples of ancient air in order to find out what the composition of the atmosphere was long ago. Where can they find such samples? Tiny bubbles of ancient air are trapped in ice deep beneath the surfaces of continental ice sheets (Fig. B14.1). Ice cores drilled at Vostok Station in Antarctica contain a record of ancient air going back more than 160,000 years—far longer than any historical climate records. When the ice is melted under controlled conditions in a laboratory, the air samples can be extracted and analyzed.

Analyses of ancient air samples have revealed some important information about past climates. For example, the Vostok samples and similar samples from Greenland show that during glacial ages the atmosphere typically contained far less $CO_2$ than it did during warm interglacial periods. The rapid increase of atmospheric $CO_2$ from preindustrial levels (about 280 ppm by volume, or 0.028%) to present levels (about 355 ppm by volume, or 0.0355%) is unprecedented in the ice-core record and implies that something unusual is happening. Studies like these, which have been carried out using ice from polar glaciers as well as from temperate glaciers, will help Earth scientists develop a detailed history of global climatic change. Then we will have a useful context in which to understand the nature and extent of the climatic changes we seem to be witnessing.

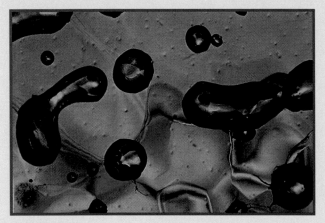

**Figure B14.1**
**ANCIENT AIR**
These air bubbles are trapped in glacier ice. If the ice is melted and the gas collected, the proportion of $CO_2$ and other trace gases can be measured in these samples of the ancient atmosphere.

mid–1800s. Evidence concerning earlier climates are preserved in records that range from tree rings to ice caps, layered sediments and ancient rock strata. Through **paleoclimatology**, the study of ancient climates, geologists employ a variety of techniques to interpret these records.

For example, paleontologists infer past climates from assemblages of fossil plants and animals. Fossilized pollen spores in old bogs and lake bottom sediments have been particularly useful in reconstructing past climatic changes on a relatively fine scale. Sedimentologists and stratigraphers can infer many things about past climates from the nature of the rocks and sediments they study. The sedimentary rock record reveals the past distribution of sediments, their mineral-

**paleoclimatology** The study of ancient climates.

ogy, and the mechanisms of weathering, transport, and deposition. Sediments and sedimentary rocks can also preserve a record of short-term changes in the chemical composition of water and air, such as changes in acidity (pH). Ancient soil horizons, called *paleosols,* which represent former land surfaces, can provide a great deal of information about climate and weather in ancient environments. Geochemists can also learn about the environments in which weathering has occurred by examining the minerals present in paleosols; these enable them to determine the chemical compositions of the ambient air and water at the time that the soil was formed.

Former ground surface temperatures can be determined by studying variations in temperature profiles preserved in sediment layers. Isotope geologists can determine past temperatures from studies of terrestrial and marine sediments and core samples from polar ice. Layered lake bottom sediments and corals can reveal fine seasonal variations. The amount of dust in the atmosphere in earlier periods can be deduced from studies of the distribution of eolian sediments such as loess. Scientists have even obtained samples of ancient air, up to 160,000 years old, trapped in bubbles in both polar and temperate glacial ice. Chemical tests on this air can reveal its composition, and tests on the ice itself can reveal whether global temperatures were relatively warm or cool at the time the ice was deposited (see *Geology Around Us* on page 422).

## The Most Recent Ice Age

The most recent glaciation began about 30,000 years ago. An extensive ice sheet that had formed over eastern Canada spread south and west (Fig. 14.20). At the same time, another great ice sheet that originated in Scandinavia spread southward across northwestern Europe, covering England, France, and Germany. Other large ice sheets grew over arctic regions of North America and Eurasia, including some areas that are now submerged by shallow polar seas, and over the

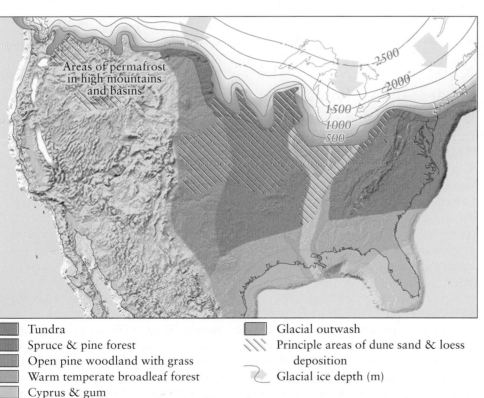

Figure 14.20
THE MOST RECENT ICE AGE. During the most recent glaciation, about 20,000 years ago, so much water was tied up as ice that sea level fell by about 100 m (more than 300 ft). More land was exposed, so coastlines were different from the present coastlines. Most of Canada and parts of the northern United States were covered by ice sheets.

- Tundra
- Spruce & pine forest
- Open pine woodland with grass
- Warm temperate broadleaf forest
- Cyprus & gum
- Diverse plants, determined by altitude, latitude & topography
- Glacial outwash
- Principle areas of dune sand & loess deposition
- Glacial ice depth (m)

mountain ranges of western Canada. The ice sheets in Greenland and Antarctica grew larger and advanced across areas of the surrounding continental shelves. Glaciers also developed in major mountain ranges, including the Alps, Andes, Himalaya, and Rockies, as well as in numerous smaller ranges and on isolated peaks.

Geologists have been able to determine the timing, extent, and nature of the most recent Ice Age (and earlier glaciations as well) on the basis of a variety of evidence, some geologic and some biologic. For example, rocks that were scratched and grooved by glaciers reveal the direction in which the ice was moving. Glacial landforms, moraines in particular, reveal the geographic extent of land ice sheets. The radiocarbon ages of trees that were felled by advancing ice tell geologists when the ice arrived in a given region. Great thicknesses of wind-blown loess were deposited just south of the ice limits during glacial times. These deposits, which contain fossil plants and animals that are characteristic of cold, dry weather, reveal that the Ice Age climate was both colder and dustier than today's climate.

The most recent Ice Age was a time when the great woolly mammoths, mastodons, long-horn bison, and saber-toothed tigers roamed North America. Early humans migrated into North America from Asia, walking across the exposed continental shelf in today's Bering Strait in Alaska. (The land was exposed because sea levels were lower during the Ice Age.) They used stone and wooden tools to hunt the large mammals, perhaps driving some of them to extinction. Plants and animals that were characteristic of cool climates were driven toward regions that today are tropical or subtropical. Huge floating ice sheets like those found in present-day Arctic and Antarctic seas occupied large areas of the Atlantic Ocean. About 10,000 years ago the Earth emerged from this Ice Age. We have now passed the time of maximum warmth in the glacial-interglacial cycle; temperatures peaked in a warm period about 6000 to 7000 years ago, in a period known as the *Holocene optimum*. Since then temperatures have been gradually cooling, with some distinctly cooler fluctuations (such as the so-called *Little Ice Age*, a period of generally cool climates that lasted from about A.D. 1300 to the mid-1800s).

| ? | Were human beings in existence during the most recent Ice Age? |

## What Causes Climatic Change?

What factors cause the climate to warm and cool, producing changes in the Earth's environments? The search for an answer has been difficult because the climate system is very complicated, with many interacting subsystems. Climates also change on several time scales, ranging from decades to many millions of years. Modern impacts of human activities on the climate system are making it increasingly difficult to separate natural from anthropogenic influences. By reconstructing past climates, geologists can determine the range and time scales of climatic variations. Then they can use the data to test the accuracy of computer models that simulate both past and future climatic change. We know that the Earth's climate *will* change, for the geologic record of climatic change is clear. What we lack is a clear view of *how* it will change and at what rate.

### Natural Influences on Climate

Several mechanisms cause natural climatic changes. They involve not only the atmosphere but also the lithosphere, the ocean, and the biosphere, all interacting in a complex way. For example, geographic changes resulting from tectonism—the shifting of continents, the uplift of continental crust and creation of large mountain chains, and the opening or closing of ocean basins—have a significant impact on oceanic and atmospheric circulation and, therefore, on global climate.

Even the Earth's internal processes can affect climate. For example, large, explosive volcanic eruptions sometimes produce vast quantities of dust and tiny aerosol droplets of sulfuric acid, both of which scatter the Sun's rays and cause global cooling.

The characteristics of the Earth's orbit and rotation also play an important role in controlling the timing and cyclicity of climatic variations. The *eccentricity* (departure from circularity) of the Earth's orbit, the *tilt* of the planet's axis of rotation, and the *precession* (wobbling) of the axis all have an impact on *insolation,* the amount of solar radiation reaching the Earth's surface at any given time (Fig. 14.21). These movements all vary on different time scales. The combined effect of all three mechanisms causes insolation to vary by as much as 10 percent in a given location and season. Periodic variations in climate caused by fluctuations in insolation (most pronounced at high latitudes) due to changes in the Earth's orbital and rotational characteristics are called *Milankovitch cycles* after a Croatian mathematician who studied this phenomenon. The combined effect of tilt and precession is roughly correlated with 20,000- to 40,000-year glacial-interglacial cycles. Variations in eccentricity may contribute to the 100,000-year cycles.

All of these influences on climate are tempered by the moderating action of the Earth's atmosphere through the *greenhouse effect* (chapter 13). Greenhouse gases in the atmosphere allow short-wavelength solar radiation to pass through the atmosphere, but absorb outgoing longer-wavelength infrared radiation (heat). Water vapor ($H_2O$) in the atmosphere is the most important natural greenhouse gas, but naturally occurring minor gases such as carbon dioxide ($CO_2$) and methane ($CH_4$) also store heat, providing significant warming. These gases absorb and trap heat in the lower part of the atmosphere, warming the Earth's surface. Without this natural greenhouse warming by the atmosphere, the average temperature at the surface would be much cooler and life on Earth would be far different from what it is today.

> **?** **Why is there so much concern about the greenhouse effect?**

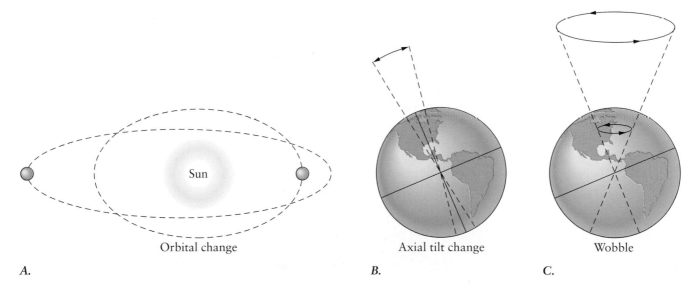

A. Orbital change    B. Axial tilt change    C. Wobble

**Figure 14.21**
THE ASTRONOMIC BASIS FOR CLIMATIC CYCLES
*A.* Over a long period, the Earth's orbit varies, carrying the planet farther from the Sun. *B.* The tilt of the Earth's axis, presently at 23.5 degrees, also varies. *C.* The Earth also wobbles slowly like a spinning top. All of these changes happen on different time scales. When added together, they have a complicated effect on insolation, the amount of Sun reaching the surface of the Earth at any given time.

## Anthropogenic Influences on Climate

As we have seen, climatic fluctuations are a normal part of the functioning of the atmosphere and climate system. During the past decade or so, however, the possibility of accelerated global warming and the role of anthropogenic emissions in the warming process have been subjects of scientific and political debate. We know that greenhouse gases such as carbon dioxide, methane, and other chemicals are emitted into the atmosphere as a result of human activities (especially from the burning of fossil fuels). The rate of emission has been increasing at an exponential rate since the Industrial Revolution. What are the effects of such emissions on global climate? Might they contribute to accelerated global warming?

Historical records of surface temperature have been used to estimate annual average global temperatures over the past 100 years. From these estimates it appears that global temperatures have increased by 0.5°C to 0.6°C (0.9° to 1.1°F) (Fig. 14.22). But is the warming trend real, or is it a statistical anomaly? It is extremely difficult to determine average surface temperatures conclusively on a time scale as short as this, particularly because the existing data from the early part of the past 100 years are insufficient for the task. Will the warming trend continue? Is it a long-term trend or a short-term fluctuation? How hot will the surface temperature of the Earth become? How much of the change is natural, and how much is caused by human actions? And if the global climate does become warmer, what effects will it have on humans and ecosystems?

These questions have not yet been answered conclusively. The Earth's climate system is very complex, and some parts of it are still poorly understood. For example, clouds play a dual role: they warm the surface by trapping infrared radiation, but they also cool the surface by reflecting incoming solar radiation. If the global temperature gets warmer, the rate of evaporation of ocean water will increase, perhaps causing the amount of cloud cover to increase. Will a change in

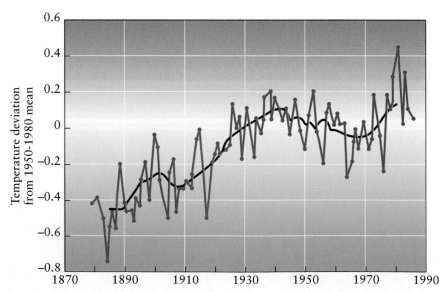

**Figure 14.22**
**100 YEARS OF TEMPERATURE CHANGES**
The data shown on this diagram suggest that global temperatures have warmed by 0.5°C to 0.6°C over the past 100 years. The red spiky curve shows the mean annual temperature, and the blue curve shows a running mean (that is, a statistically smoothed curve). The zero line is the mean temperature for the period 1950–80.

cloud cover have an overall warming effect or a cooling effect? No one knows for certain. Geologists, climatologists, and Earth scientists are working hard to understand how different parts of the Earth system work together to determine global climate and how the system can be affected by human actions. We will learn more about their work in chapter 17.

# WHAT'S AHEAD

Thus far we have been learning about the Earth system and how it functions. In the last part of the book we turn our attention to life on Earth. Life is affected by and, in turn, affects the natural functioning of the Earth system. In chapter 15 we examine the history of life on Earth: how it began and how it has evolved into its present state. People are entirely dependent on the Earth and its resources. Without water, air, and soil, we could not live. Without mineral and energy resources, we could not conduct the daily business of modern society. We will learn more about the Earth's resources and our uses for them in chapter 16. Finally, we will consider the role of geologists and other Earth scientists in the twenty-first century. In chapter 17 we will revisit many of the problems associated with natural hazards and environmental changes, both natural and anthropogenic. Because these changes involve the Earth system, geoscientists may be able to help solve some of the problems.

## *Chapter Highlights*

**1.** In the equatorial region and along the two polar fronts, moist air masses converge and rise, producing belts of high rainfall. In the polar regions and the subtropics, air masses descend and diverge, producing belts of low rainfall. The global distribution of deserts is influenced by the locations of these wet and dry climatic belts.

**2.** The term **desert** refers to **arid** lands where annual rainfall is less than 250 mm (10 in). Five types of deserts have been identified: subtropical, continental interior, rainshadow, coastal, and polar.

**3.** Eolian (wind) erosion is particularly effective in arid and semiarid regions. Wind moves particles through **surface creep, saltation,** and **suspension.** Flowing air erodes the land surface through the processes of **abrasion** and **deflation.**

**4.** **Loess** is wind-deposited sediment that underlies many fertile agricultural regions. **Dunes** are hills or ridges of sand deposited by winds. Common types of dunes are barchan, transverse, star, parabolic, and longitudinal dunes.

**5.** **Desertification** involves the invasion of desert conditions into nondesert lands, which results in land degradation and loss of agricultural productivity. Desertification can be caused by natural environmental changes or by human activities.

**6.** The perennially frozen part of the hydrosphere is called the **cryosphere.**

**7.** A **glacier** is a permanent body of ice that shows evidence of movement. Types of glaciers include cirque glaciers, valley glaciers, fjord glaciers, piedmont glaciers, ice caps, ice sheets, and floating ice shelves.

**8.** Glaciers move by means of internal flow and basal sliding across the underlying rock or sediment. In most glaciers velocities of flow range from a few centimeters to a few meters a day, but occasionally a glacier undergoes a rapid surge. The mass of a glacier may change from season to season and year to year through accumulation and ablation.

**9.** When basal sliding occurs, small fragments of rock embedded in the ice scrape away at the underlying bedrock, producing glacial striations and grooves aligned parallel to the direction of ice flow. In mountainous regions, glaciers produce a variety of distinctive erosional landforms, such as cirques, arêtes, and U-shaped valleys.

**10.** **Till** is an unsorted mixture of rock flour, sand, pebbles, cobbles, and boulders deposited by a glacier. Flowing streams underneath glaciers may deposit sediment in the form of broad, sweeping outwash plains or sinuous eskers. A **moraine** is a ridge or pile of debris being carried along by a glacier or deposited along the edge or terminus of a glacier. Common **periglacial** features include **permafrost,** ice wedges, and patterned ground.

**11.** **Glaciations** are periods during which the global average surface temperature drops by several degrees and stays cool long enough for the polar ice sheets to grow larger and for new ice sheets to form. Warmer periods between glaciations are called interglacials. Glaciations were especially common during the Pleistocene Epoch, but the rock record contains evidence of glacial ages occurring as much as 2.3 billion years ago. The most recent glaciation began about 30,000 years ago and ended about 10,000 years ago.

**12.** The eccentricity, precession, and tilt of the Earth's orbit and rotation are particularly important in controlling the cyclicity of climatic variations because they affect the amount of solar radiation reaching the Earth's surface (insolation) at any given time. Periodic climatic variations caused by fluctuations in insolation at high latitudes that result from changes in the Earth's orbital characteristics are called Milankovitch cycles.

**13.** External influences on climate are moderated by the greenhouse effect. Water vapor ($H_2O$), carbon dioxide ($CO_2$), and methane ($CH_4$) are the most important natural greenhouse gases. During the past decade or so there has been increasing concern about accelerated global warming and the role of anthropogenic emissions of $CO_2$, $CH_4$, and other greenhouse gases.

**14.** Historical records of surface temperatures suggest that over the past 100 years average global temperatures have increased by about 0.5° to 0.6°C. But the Earth's climate system is very complex, and parts of it are still poorly understood. Many questions remain to be answered about global climatic change and the role of human activities in such changes.

## ▶ *The Language of Geology*

- abrasion 402
- arid land 401
- cryosphere 398
- deflation 403
- desert 401
- desertification 406

- dune 404
- glaciation 421
- glacier 409
- loess 403
- moraine 419
- paleoclimatology 422

- periglacial 419
- permafrost 419
- saltation 402
- surface creep 402
- suspension 402
- till 418

## ▶ *Questions for Review*

1. What are the smaller reservoirs that together make up the cryosphere? Is the cryosphere the largest reservoir for fresh water in the hydrosphere? (Return to Fig. 12.3 to find the answer.)
2. Using a map of the world, give an example of each of the five main types of deserts.
3. What is the difference between deflation and abrasion?
4. How does wind differ from water in the way it picks up and moves particles?
5. How can human actions accelerate the natural process of desertification?
6. In what ways are glaciers like streams? In what ways are they different?
7. Is glacier ice a rock? If so, what kind? Go back to chapter 2 and check the definition of "rock" to support your answer. If glacier ice is a rock, does this mean that the individual grains of ice are minerals?
8. Describe the difference between a terminal moraine, a lateral moraine, and a medial moraine.
9. How much of the world's land area is covered by ice today?
10. What is a Milankovitch cycle, and what is the significance of such cycles for global climate?

# ► *Questions for Thought and Discussion*

1. Look at the photograph of sandstones shown in Fig. 14.8B. Can you tell which way the wind was blowing when it produced these dune deposits?

2. Investigate the current status of drought and land degradation in the Sahel. Can you find any information about soil erosion control techniques or other methods that are being used to combat desertification?

3. Do some research on the most recent Ice Age. Do you live in an area that was formerly covered by ice? How thick was the ice? Is there any evidence in the landforms around you to indicate that the area was formerly glaciated?

4. Find out if your city, state, province, or country has set goals for the reduction of $CO_2$ emissions to limit its contribution to global warming. What steps have been taken to meet these goals?

For an interactive case study on global climatic change, visit our Web site.

# The
# Art
# of
# Geology

Nothing in the world
is as soft, as weak, as water;
nothing else can wear away
the hard, the strong,
and remain unaltered

———

*—Lao Tzu*
**Tao Te Ching**

And now the rains had really come, so heavy and persistent that even the village rainmaker no longer claimed to be able to intervene. He could not stop the rain now, just as he would not attempt to start it in the heart of the dry season, without serious danger to his own health. The personal dynamism required to counter the forces of these extremes of weather would be far too great for the human frame. And so nature was not interfered with in the middle of the rainy season. Sometimes it poured down in such thick sheets of water that earth and sky seemed merged in one gray wetness. It was then uncertain whether the low rumbling of Amadiora's thunder came from above or below.

—*Chinua Achebe*
**Things Fall Apart**

*John Warwick Smith,* "Hafod, upper part of the Cascade."

Now, if I wanted to be one of those ponderous scientific people, and "let on" to prove what had occurred in the remote past by what had occurred in a given time in the recent past, or what will occur in the far future by what has occurred in late years, what an opportunity is here! Geology never had such a chance, nor such exact data to argue from! No "development of species," either! Glacial epochs are great things, but they are vague—vague. Please observe.

In the space of one hundred and seventy-six years the Lower Mississippi has shortened itself two hundred and forty-two miles. This is an average of a trifle over one mile and a third per year. Therefore, any calm person, who is not blind or idiotic, can see that in the Old Oolitic Silurian Period, just a million years ago next November, the Lower Mississippi River was upward of one million three hundred miles long, and stuck out over the Gulf of Mexico like a fishing-rod. And by the same token, any person can see that seven hundred and forty-two years from now the Lower Mississippi will be only a mile and three-quarters long, and Cairo and New Orleans will have joined their streets together, and be plodding comfortably along under a single mayor and a mutual board of aldermen. There is something fascinating about science. One gets such wholesale returns of conjecture out of such a trifling investment of fact.

—*Mark Twain*
**Life on the Mississippi**

# PART FIVE

# LIVING ON PLANET EARTH

Throughout much of Earth history, life has been present on this planet. So far, the Earth is the only planet we know of that is hospitable to life. But life also has played a central role in the development of the Earth system as we know it today. In Part V, we close the book with a brief synopsis of the history of life on Earth and the impacts of life on the chemical evolution of the atmosphere and hydrosphere. Today, the human inhabitants of the Earth rely heavily on the material resources of the planet. We use minerals, energy, soil, and water to sustain life and carry out all the activities of modern society. As the human population grows, it becomes even more important that we understand how such resources form, how and where they occur, and how they can be managed to ensure their continued viability. As we move into the twenty-first century, geoscientists will continue to play an important role in providing this knowledge and in enhancing our understanding of the Earth system.

▶ All of these topics and more are addressed in the chapters of Part V, which are

Chapter 15:   A Brief History of Life on Earth

Chapter 16:   Earth Resources

Chapter 17:   The Role of Geoscientists in the Twenty-first Century

*Difficult Farming*

Hard work and human ingenuity have carved fields from the steep slopes of the Cordillera Blanca, near Chavin, Peru.

# A BRIEF HISTORY OF LIFE ON EARTH

A gentle, sneezing Brachiosaurus; a rip-roaring Tyrannosaurus; a very hungry Velociraptor; and a Triceratops with a tummyache. These prehistoric creatures were brought to life by Steven Spielberg in the hit movie *Jurassic Park*. In the movie dinosaur DNA is extracted from blood in a mosquito trapped in amber for millions of years. Through numerous steps that are only vaguely alluded to in the film, the DNA is somehow transfigured into an embryo, which is incubated in a plastic egg until it is ready to hatch. *Voilà!* But could scientists ever really bring extinct animals or plants back to life?

In an eerie case of life mimicking art, during the filming of *Jurassic Park,* scientists extracted DNA from a 40-million-year-old bee and a 25-million-year-old termite, both of which had been fossilized in amber. However, many obstacles would have to be overcome before extinct animals—especially animals as ancient as dinosaurs—could be cloned and brought back to life. DNA deteriorates over time; only a small percentage of the original DNA information would be intact, even if dinosaur blood could be extracted from a fossilized mosquito. Even if most of the essential parts of the DNA could be retrieved, joining the millions (or even billions) of parts together into a meaningful sequence would be next to impossible. Transcribing the DNA onto messenger RNA, creating proteins, and somehow transforming the proteins into a fertilized egg—all of these crucial steps continue to elude scientists.

But just a few years ago, it was thought to be impossible that scientists would ever extract and clone DNA from an extinct animal, clone a mammal, or clone an adult animal. All have now been accomplished. And then came *T. rex . . . ?*

◆

### In this chapter you will learn about

- How life has influenced the chemistry of the ocean and atmosphere
- How life may have originated on the Earth
- The theory of evolution
- The fossil record, and how organisms become fossilized
- How the main groups of organisms have changed and developed since the beginning of the Phanerozoic Eon

▲

*Dinamation*
This *T. Rex* at the American Museum of Natural History seems to be getting ready for dinner.

When and where did life begin, and how has it evolved into its present form? We begin our study of the history of life on the Earth by looking at the environment in which it unfolded. The history of life is closely intertwined with that of the atmosphere–hydrosphere system. Without a hospitable atmosphere and hydrosphere, life as we know it could not survive. Without life, the atmosphere and ocean would not exist in their present forms.

# THE EARTH'S CHANGING ATMOSPHERE AND HYDROSPHERE

Like other Earth systems, the atmosphere has evolved through time (Fig. 15.1). Compared to the Sun, which is representative of the material from which the solar system was formed, the Earth contains less of certain volatile elements, particularly nitrogen, argon, hydrogen, neon, and helium. Portions of these elements were lost when the envelope of gases or *primary atmosphere* that surrounded the early Earth was stripped away by the strong solar wind or by incessant meteorite impacts, or both. Little by little the planet generated a new, *secondary atmosphere* by degassing volatile materials from its interior.

More than 4 billion years ago, in the Hadean Eon (Fig. 15.2), the chemical composition of the atmosphere was very different from what it is now. The atmosphere probably consisted of water vapor, carbon dioxide, and nitrogen, with some sulfur compounds and hydrogen chloride. There was no free oxygen ($O_2$), a necessity for most forms of life on the Earth today. It was also very hot, even though the Sun's *luminosity* (brightness) was lower than it is today. The early atmosphere was composed primarily of greenhouse gases, which trapped heat near the surface. It was too hot for water to exist as a liquid at the surface, so there were no oceans, lakes, or rivers. The atmospheric pressure was also much greater than it is today. Altogether, the early Earth was inhospitable to life as we know it.

## The Role of Geologic Processes

The atmosphere changed over time, and so did the other Earth systems with which it is connected. As the Earth cooled, water vapor in the atmosphere began to condense. It fell as rain and then began to collect in low-lying areas, forming bodies of water on the surface, perhaps as early as 4.4 billion years ago. The early rain was highly acidic because water vapor reacted with the gases in the atmosphere to create acids.[1] The acidic rain reacted with the rocks of the crust, causing chemical weathering. Through reactions with minerals the acidic rainwater was slowly neutralized. Sediments (the products of chemical and physical alteration of the crust) began to form. Thus, the composition of all three subsys-

Have the atmosphere and hydrosphere changed since the Earth was formed?

---

[1] Water vapor + carbon gases = carbonic acid ($H_2CO_3$)
Water vapor + sulfur gases = sulfuric acid ($H_2SO_4$)
Water vapor + nitrogen gases = nitric acid ($HNO_3$)
Water vapor + chlorine gases = hydrochloric acid ($HCl$)

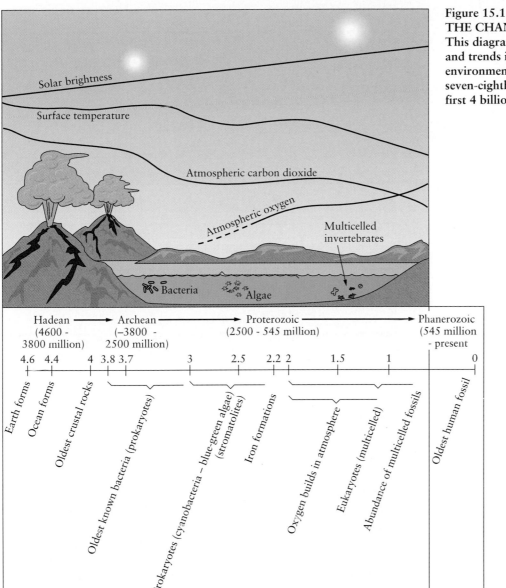

**Figure 15.1**
**THE CHANGING EARTH**
This diagram shows the major events and trends in the history of the surface environment during approximately seven-eighths of the Earth's history (the first 4 billion years or so).

tems—the atmosphere, the hydrosphere, and the lithosphere—began to change as materials were exchanged among them.

Throughout geologic history volcanism has been the main process by which volatile materials have been released from the Earth. Gases are still a major component of volcanic emissions; therefore, the degassing process is still going on, though at a much slower rate. The main chemical constituent of volcanic gases (as much as 97 percent by volume) is water vapor, with varying amounts of nitrogen, carbon dioxide, hydrogen, sulfur dioxide, chlorine, hydrogen sulfide, methane, ammonia, carbon monoxide, and other gases. Overall, the volume of volcanic gases released over the past 4 billion years or so is thought to be large enough and in approximately the right proportions to account for the entire volume of the oceans and atmosphere, with one very important exception: oxygen. The atmosphere is approximately 21 percent oxygen. Where did this abundant oxygen come from?

**?**
**Where did the oxygen in the atmosphere come from?**

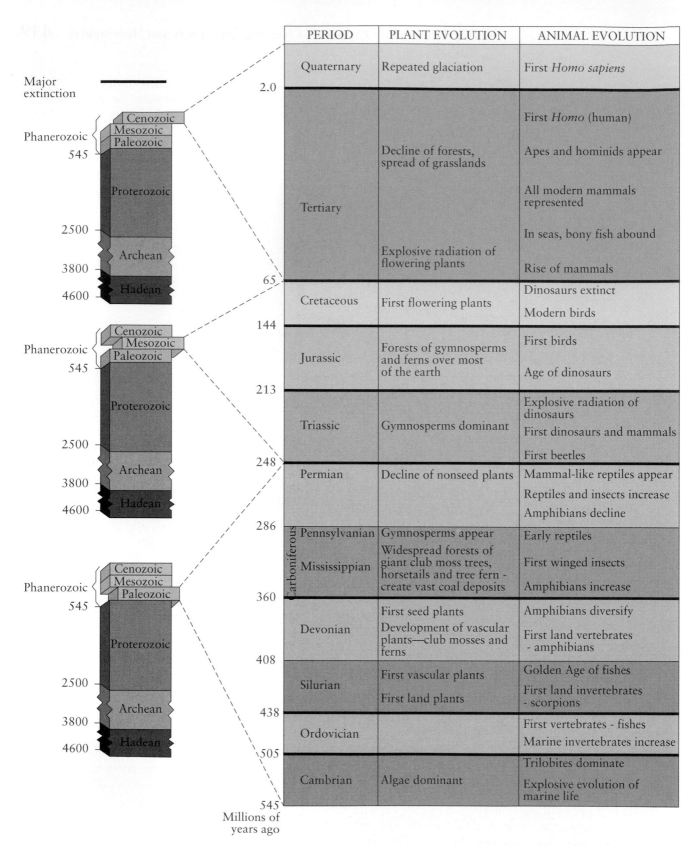

| PERIOD | PLANT EVOLUTION | ANIMAL EVOLUTION |
|---|---|---|
| Quaternary | Repeated glaciation | First *Homo sapiens* |
| Tertiary | Decline of forests, spread of grasslands<br><br>Explosive radiation of flowering plants | First *Homo* (human)<br><br>Apes and hominids appear<br><br>All modern mammals represented<br><br>In seas, bony fish abound<br><br>Rise of mammals |
| Cretaceous | First flowering plants | Dinosaurs extinct<br>Modern birds |
| Jurassic | Forests of gymnosperms and ferns over most of the earth | First birds<br><br>Age of dinosaurs |
| Triassic | Gymnosperms dominant | Explosive radiation of dinosaurs<br>First dinosaurs and mammals<br>First beetles |
| Permian | Decline of nonseed plants | Mammal-like reptiles appear<br>Reptiles and insects increase<br>Amphibians decline |
| Pennsylvanian | Gymnosperms appear | Early reptiles |
| Mississippian | Widespread forests of giant club moss trees, horsetails and tree fern - create vast coal deposits | First winged insects<br><br>Amphibians increase |
| Devonian | First seed plants<br>Development of vascular plants—club mosses and ferns | Amphibians diversify<br><br>First land vertebrates - amphibians |
| Silurian | First vascular plants<br>First land plants | Golden Age of fishes<br><br>First land invertebrates - scorpions |
| Ordovician | | First vertebrates - fishes<br>Marine invertebrates increase |
| Cambrian | Algae dominant | Trilobites dominate<br>Explosive evolution of marine life |

Major extinction

Phanerozoic { Cenozoic / Mesozoic / Paleozoic

545
Proterozoic
2500
Archean
3800
Hadean
4600

2.0
65
144
213
248
286
360
408
438
505
545

Carboniferous

Millions of years ago

**Figure 15.2**
**THE GEOLOGIC TIME SCALE**
The major divisions of geologic time shown here are the result of scientific study using stratigraphy, correlation of fossils, and radiometric dating, as discussed in chapter 3. Most of the major boundaries on the geologic time scale represent major environmental changes. Mass extinctions of species are indicated by red lines; two of the most important are the great Permian extinction (240 million years ago) and the K-T extinction (65 million years ago), in which dinosaurs died out.

*438*

Some oxygen was probably generated in the early atmosphere by the break-down of water molecules into hydrogen and oxygen as a result of interactions with ultraviolet light (through a process called *photodissociation*). This is an important process, but it doesn't come close to accounting for the present high level of oxygen in the atmosphere. Another oxygen-producing process was required, but it had to await the appearance of life.

## The Role of Biologic Processes

Most organisms in the biosphere depend ultimately upon photosynthesis, either directly or indirectly, to obtain food. In **photosynthesis**, light energy is used to cause carbon dioxide to react with water, producing organic substances (carbohydrates) and releasing oxygen (Fig. 15.3). Essentially all of the free oxygen now in the atmosphere originated through photosynthesis.

During the early phase of life, much of the free oxygen produced during photosynthesis combined with iron in ocean water and was deposited in the form of iron oxide-bearing minerals. Evidence of the gradual transition from oxygen-

**photosynthesis** The process whereby plants utilize light energy to cause carbon dioxide to react with water, producing carbohydrates and releasing oxygen.

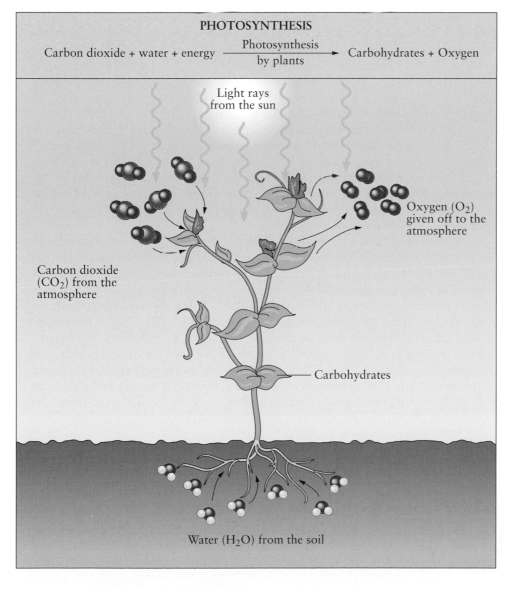

**PHOTOSYNTHESIS**

Carbon dioxide + water + energy $\xrightarrow[\text{by plants}]{\text{Photosynthesis}}$ Carbohydrates + Oxygen

Light rays from the sun

Oxygen ($O_2$) given off to the atmosphere

Carbon dioxide ($CO_2$) from the atmosphere

Carbohydrates

Water ($H_2O$) from the soil

Figure 15.3
**PHOTOSYNTHESIS**
Through photosynthesis, plants combine carbon dioxide ($CO_2$) and water ($H_2O$) to make carbohydrate, a basic food and energy resource. The process also releases a waste gas, oxygen, which humans and animals need to survive.

poor to oxygen-rich ocean water is preserved in seafloor sediments. The minerals in seafloor sedimentary rocks that are more than 2 billion years old are characterized by the presence of *reduced* (oxygen-poor) iron compounds. In rocks that are less than about 1.8 billion years old, *oxidized* (oxygen-rich) compounds predominate. The chemical sediments that were precipitated on the ocean floor during this period reflect the transition through alternating bands of red (oxidized iron) and black (reduced iron) minerals. These rocks are called *banded iron formations*. Since ocean water is in constant contact with the atmosphere and the two systems function together in a state of dynamic equilibrium, the transition to an oxygen-rich atmosphere must also have occurred during this period in the Proterozoic Eon (Figs. 15.1 and 15.2).

Eventually, enough oxygen was created through photosynthesis by early oxygen-producing organisms (probably closely related to modern *Cyanobacteria*) to permit oxygen to build up in the atmosphere. Along with the buildup of molecular oxygen came an increase in ozone levels. When ozone began to function as a screen to filter out harmful ultraviolet radiation, organisms were finally able to survive and flourish in shallow waters and, eventually, on land. This critical stage in the evolution of the atmosphere was reached around 600 million years ago. Sure enough, the fossil record shows an explosion of life forms at that time (the transition from Precambrian time to the Phanerozoic Eon, Fig. 15.2).

The nitrogen content of the atmosphere is also controlled by biologic processes. Nitrogen, the main component of the atmosphere, is a nutrient that is fundamentally important for plant growth. The vast quantity of nitrogen stored in the atmosphere occurs in the form of gaseous $N_2$, which cannot be directly assimilated by plants or animals. Some microorganisms, including some blue-green algae and soil bacteria, have the ability to use $N_2$ directly, in a process called *nitrogen fixation*. Some types of soil bacteria maintain *symbiotic* (mutually dependent and mutually beneficial) relationships with leguminous plants, through which the bacteria supply nitrogen to the plants and the plants, in turn, supply organic compounds to the bacteria. Nitrogen-fixing bacteria and the plants with which they are associated are thus responsible for moving nitrogen from the atmospheric reservoir to the biosphere. In turn, nitrogen is lost to the biosphere and returned to the atmosphere through the process of *denitrification,* in which certain bacteria convert usable nitrogen back into gaseous $N_2$.

The role of life in the chemical evolution of the atmosphere and hydrosphere doesn't stop with the biogeochemical cycles of oxygen and nitrogen. The biosphere has also had a profound impact on the carbon cycle, especially the carbon content of the atmosphere and ocean. The shells of marine organisms are composed primarily of calcium carbonate ($CaCO_3$). These shells provide a storage reservoir for carbon dioxide (note that $CaCO_3 = CaO + CO_2$). When the organisms die, their shells are buried by seafloor sediments. Eventually they are transformed into limestone. Limestone is a long-term reservoir for carbon dioxide, removing it from the atmosphere and hydrosphere. If all the carbon dioxide stored in limestone and other sedimentary rocks were released, there would be as much $CO_2$ in the Earth's atmosphere as there is in the atmosphere of Venus. There, the greenhouse effect runs rampant and the surface temperature averages 480°C (almost 900°F), hot enough to melt lead!

> **[?]** Photosynthesis produces oxygen. In what other ways does life affect the chemistry of the atmosphere?

# EARLY LIFE

Life itself was a major factor in the chemical transformation of the Earth's atmosphere and hydrosphere. Through photosynthesis, early life forms released oxygen, which built up in the atmosphere. Through the removal of carbon dioxide, the greenhouse was brought under control and surface temperatures moderated. But where and how did life begin? How did simple life forms evolve into the complex forms we see today? How were these life forms preserved as fossils in the rock record? We will address each of these questions in turn.

## The Origin of Life

Four steps must have been taken for life to develop from inorganic matter and eventually evolve into the complex forms we know today. They are: (1) chemosynthesis, (2) biosynthesis, (3) replication and reproduction, and (4) metabolism.

### Chemosynthesis

**Chemosynthesis** is the synthesis, from inorganic material, of small organic molecules such as *amino acids*, the basic building blocks of *proteins*. This may have been the first step on the way to life. In 1923 a Russian scientist, Aleksandr Oparin, hypothesized that simple organic compounds may have been synthesized from a primitive atmosphere of ammonia ($NH_3$), methane ($CH_4$), water vapor ($H_2O$), and hydrogen ($H_2$). He proposed that energy for the chemical reactions was supplied by lightning bolts or by ultraviolet radiation from the Sun (Fig. 15.4). Thirty years later an American scientist, Stanley Miller, carried out an ex-

**chemosynthesis** The synthesis, from inorganic material, of organic molecules.

Figure 15.4
THE SPARK OF LIFE?
Is this the type of environment in which the ingredients of life were made? Lightning bolts discharge through volcanic gases over Mount Pinatubo, Philippines. Similar discharges in the early atmosphere may have created organic compounds from which the large molecules needed for life were subsequently created.

periment to check this hypothesis. He passed electric sparks through a mixture of gases and recovered some amino acids and other simple organic compounds. In later experiments, all of the important protein-forming amino acids were synthesized, along with other biologically important compounds.

Subsequent research has exposed a difficulty: It is doubtful that methane, ammonia, and hydrogen were major components of the Earth's primitive atmosphere. As mentioned earlier, it seems likely that the dominant gases in the early atmosphere were carbon dioxide, water vapor, and nitrogen. Recent investigations have therefore focused on the synthesis of organic molecules from $CO_2$, $N_2$, and $H_2O$ with some methane, ammonia, and carbon monoxide.

Some scientists hypothesize that organic molecules may have arrived from some other part of the solar system or even from beyond the solar system. This hypothesis has attracted attention for two reasons. First, astronomers have found small organic molecules in interstellar space. Second, among the many kinds of meteorites that fall on the Earth, one kind, *carbonaceous chondrites,* contains organic molecules; some meteorites may even contain evidence of extraterrestrial life (Fig. 15.5). If interstellar dust, carbonaceous chondrites, or both, fell on the Earth, perhaps they provided the molecules needed for life.

Oparin believed that the organic molecules from which life originated collected as a "soup" in surface waters. The "soup" concept works just as well for organic matter from interstellar space or meteorites as it does for organic matter created by chemosynthesis on the Earth. However, a basic problem is that a high concentration of complex organic molecules would be required. This violates the *second law of thermodynamics,* which basically tells us (in this context) that it would be more energetically favorable for such a mixture of organic compounds to disintegrate into simple parts than to collect into a multitude of complex, organized molecules.

In spite of its problems, most scientists agree with at least some aspects of the "soup" hypothesis. But in what environment was this "soup" formed? Charles

> **?** **What do comets and meteorites have to do with the origin of life on the Earth?**

*A.*

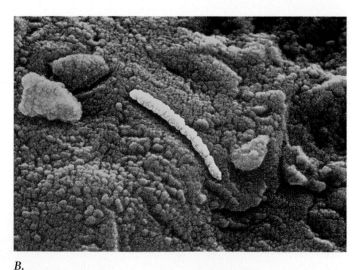

*B.*

Figure 15.5
ORGANIC MATERIAL FROM OUTER SPACE
This meteorite (*A*), originally a piece of the planet Mars, fell to Earth about 16,000 years ago and was found in Antarctica in 1984. In 1996, tiny, tube-like structures (*B*) were discovered within the rock. The structures, shown here in a scanning electron micrograph, are less than 1/100th the diameter of a human hair. Some scientists have interpreted these structures as the fossils of microscopic organisms. The rock itself is about 4.5 billion years old and the microfossils are at least 3.6 billion years old.

Figure 15.6
THE BIRTHPLACE OF LIFE?
Deep-sea life forms abound in the warm waters near this seafloor hydrothermal vent near the Galápagos Islands. "Black smokers" like this one may
have provided an environment conducive to the origin of life on the Earth.

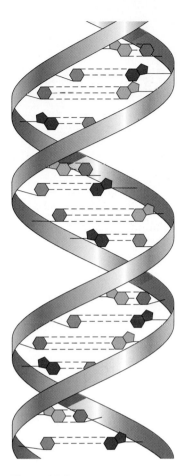

Figure 15.7
DNA: THE PLAN OF LIFE
The two strands of the twisted,
chain-like molecule of DNA are
held together by organic molecules. DNA stores the instructions and information needed for
a living organism to grow.

Darwin envisioned life originating in a "warm little pond" on land. Hydrothermal vents at seafloor spreading centers (called *"black smokers"*) or hot springs
like those at Yellowstone Park could have provided the raw materials and heat
needed for chemosynthesis (Fig. 15.6). Even the bubbles in sea foam could have
provided a site for the collection of early organic material. In fact, the synthesis
of organic molecules may have taken place in numerous locations, only to be
wiped out by meteorite impacts or harsh environmental conditions, before life finally took hold. Wherever the first organic molecules came from, the big problem is in the next steps: How did the early molecules link together to become the
larger "life" molecules? And how did these molecules eventually develop the capability to metabolize and to replicate themselves?

## Biosynthesis, Replication, and Metabolism

*Biosynthesis* is the *polymerization* (linking together) of small organic molecules
to form larger organic molecules (*biopolymers*), including proteins. One problem
with the "soup" hypothesis is that when amino acids polymerize to create proteins, water is eliminated. In an aqueous environment this is difficult, if not impossible, to achieve. However, if amino acids are dehydrated and heated, polymerization can occur. The resulting biopolymers, which contain up to 200 amino
acids, may have been formed when some "soup" along ancient shorelines dried
out and was subsequently heated by solar radiation or volcanic heat.

   Biopolymers can form chains and divide, but they have no mechanism for
replicating themselves. How did these large organic molecules develop such a
mechanism? You may recall from biology classes that the plan of a living organism is encoded in its **DNA** (*deoxyribonucleic acid*), a biopolymer that consists of
two twisted chain-like molecules held together by organic molecules (Fig. 15.7).
The information and instructions stored in the DNA are decoded and executed
by **RNA** (*ribonucleic acid*), a single-strand molecule similar to one-half of a dou-

**DNA** Deoxyribonucleic acid; a
double-chain biopolymer that
contains all the genetic information needed for organisms to
grow and reproduce.

**RNA** Ribonucleic acid; a single-
strand molecule similar to one-
half of a double DNA strand.

**Which came first: RNA or proteins?**

ble DNA strand. Proteins cannot reproduce without RNA because the RNA contains the information required to construct an exact duplicate of the protein molecule.[2]

Where did the first molecules of RNA come from? Simple clay minerals may provide an answer. The atoms in some clay minerals are organized in a crystalline pattern that repeats itself over and over. If a defect appears in the pattern it, too, is repeated over and over again in much the same way that a genetic mutation in modern DNA and RNA is repeated. It is possible that such clay minerals could have served as molecular templates or blueprints, organizing the organic precursors of life into the repetitious linkages that characterize RNA. A problem with the clay hypothesis is that modern proteins and RNA do not contain any structures that can be directly related to clay minerals. If clay did play a role, it must have been very early in the process of RNA development. Once RNA became established, its presence facilitated the synthesis of protein molecules. Protein synthesis eventually became completely dependent on RNA, setting the stage for modern cell replication.

The basic chemistry of these primitive, self-replicating precursors of living organisms was established long before 3.55 billion years ago. But still another crucial step was needed in order for early life forms to take hold: they had to develop the ability to "eat." The earliest organisms probably obtained nutrients by simply absorbing organic compounds from their surroundings. Eventually, organisms had to develop mechanisms for synthesizing and processing chemicals into useable food energy. The set of biochemical reactions through which organisms produce and extract food energy is called *metabolism*. Early organisms slowly evolved an increasingly complex repertoire of metabolic capabilities. One of the earliest of these was the anaerobic (i.e., non-oxygenated) process of *fermentation*. Let's look more closely at the structures of cells and how cells carry out life-supporting processes.

## The Organization of Life

**cell** The basic structural and functional unit of life; a complex grouping of chemical compounds enclosed in a porous membrane.

Organisms carry out the various biochemical processes required for metabolism and replication within the protected chemical environment of the **cell,** the basic structural and functional unit of life. A cell is a complex grouping of chemical compounds enclosed in a *membrane,* a porous wall. The development of the cell membrane was a crucial step in the evolution of life. The membrane separates the materials and chemical reactions that occur inside the cell from the environment outside it. This makes it possible for local organization within the cell to increase. The porous membrane also facilitates the exchange of materials and energy between the cell and its environment.

All organisms are composed of one or more cells. Many bacteria are *unicellular* (one-celled), but most other organisms are *multicellular* (more than one cell). Cells may be small (0.01 mm, or 0.0004 in) or large (a few cm, or even larger in rare cases), but whatever their size, all cells are either *prokaryotic* or *eukaryotic.*

### Prokaryotes and Eukaryotes

**prokaryotic cell** A cell without a well-defined nucleus; refers to single-celled organisms that have no membrane separating their DNA from the cytoplasm.

**Prokaryotic cells** (from the Greek *pro,* meaning "before," and *karyote* meaning "nucleus"—hence "before a nucleus") appear to be the earliest cell forms.

---

[2] Both DNA and RNA are crucial for the replication of modern cells, but most experts agree that there must have been an early phase in which RNA alone was sufficient. The phrase "RNA world" has been coined to describe this phase in the development of life.

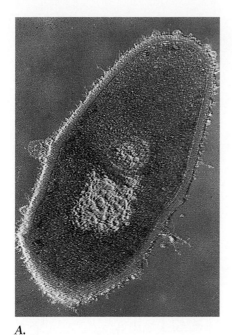

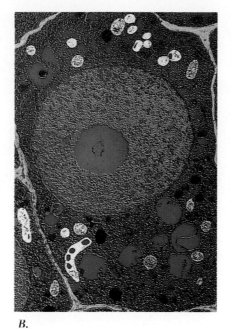

A.                                    B.

Figure 15.8
PROKARYOTES AND
EUKARYOTES
*A.* This is a bacterium, a prokaryotic cell. Its DNA is concentrated
in a poorly defined area that is
not contained by a membrane.
*B.* This is a eukaryotic cell from a
plant root. It has a well-defined,
membrane-bound nucleus. The
cells appear colored because they
have been stained for greater clarity under the microscope.

Prokaryotes house their DNA in a poorly demarcated part of the cell (Fig. 15.8A). The main body of the cell, the *cytoplasm,* lacks distinctly defined areas in which the cell's various functions are carried out. Most important, the portion of the cell that houses the genetic information is not separated from the cytoplasm by a membrane. Present-day bacteria are modern prokaryotes.

**Eukaryotic cells** (from *eu* meaning "good" or "true"—hence, "with a true nucleus") are larger and more complex than prokaryotic cells. Their DNA is housed in a well-defined nucleus that is separated from the cytoplasm by a membrane (Fig. 15.8B). The cytoplasm contains a variety of well-defined parts, called *organelles,* each of which has a specific function. Humans, animals, plants, fungi, and many other living things consist of eukaryotic cells.[3]

Earlier in the chapter we mentioned that early organisms derived their food energy through the anaerobic process of fermentation. These organisms were prokaryotes. Today there are still many organisms that cannot tolerate oxygen and so obtain their energy through anaerobic processes. All organisms that are not anaerobic, including some prokaryotes, are aerobic and obtain energy through *respiration,* which means that they use oxygen to oxidize carbohydrates, creating carbon dioxide, water, and energy. Respiration is more efficient than fermentation at producing energy for a cell, because it releases all of the available energy. Fermentation does not use all the available energy; its end product, alcohol, can still be oxidized. Alcohol is a high-energy compound, which can be used as a fuel; carbon dioxide and water, the products of respiration, cannot.

**eukaryotic cell** A cell that includes a nucleus with a membrane, as well as other membrane-bound organelles.

---

[3] All living organisms that are not prokaryotes belong to one of four kingdoms of eukaryotes: Protista (single-celled and simple multicellular eukaryotes), Fungi (mushrooms, lichens, and their relatives), Animalia (multicellular organisms that obtain their food by consuming other organisms), and Plantae (multicellular, sexually reproducing eukaryotes, that produce their own food and are more complicated than algae). The kingdoms are further subdivided, in a hierarchical manner, into successively narrower categories: Phylum, Class, Order, Family, Genus, and Species. A modern human is classified as follows: Kingdom: Animalia; Phylum: Chordata; Class: Mammalia; Order: Primates; Family: Hominidae; Genus: *Homo;* Species: *Homo sapiens.*

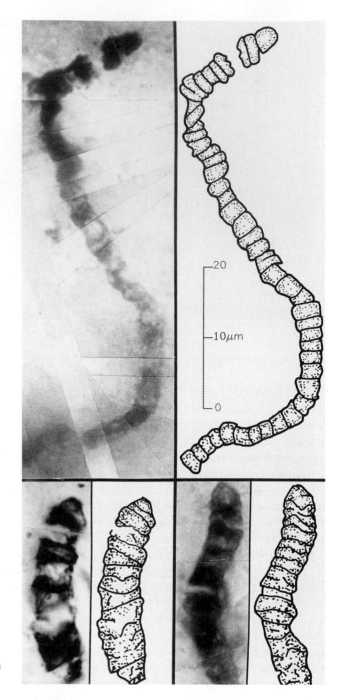

Figure 15.9
OLDEST KNOWN FOSSILS
These are examples of the most ancient fossil prokaryotes ever
found. They came from a 3.5 billion-year-old chert in Western
Australia. Adjacent to each photo is a sketch. The magnification
is indicated by the scale; 10µm = 0.01 mm = 0.0004 in.

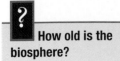

How old is the
biosphere?

## The Age of the Biosphere

The most ancient fossils that have been found are about 3.55 billion years old.[4]
Some of them are remains of microscopic prokaryotes (Fig. 15.9). Others consist
of layers of thin sheets of calcium carbonate that were precipitated from seawater as a result of the action of blue-green photosynthetic bacteria (also prokary-

---

[4] The "chemical signatures" of biological processes have been detected in rocks even older than
this—as much as 3.9 billion years old. This suggests that life originated very soon after the formation of the Earth, even while the planet was still being bombarded by meteorites.

otes). The layered structures, called *stromatolites,* are not the remains of actual organisms, but they provide clear evidence of the presence of organisms. Similar structures are still being formed today (Fig. 15.10).

For at least 2 billion years, through the Hadean and Archean Eons and part of the Proterozoic, the only life on the Earth was prokaryotic. A variety of prokaryotic cells developed (although only one type, cyanobacteria, carried out oxygen-producing photosynthesis). How and where the first simple eukaryotes came into being is a subject of much speculation. We can be reasonably sure that eukaryotes arose from prokaryotes. The chemical pathways in the two classes of cells are so similar that they must be related. Moreover, the organelles in eukaryotes closely resemble those of some prokaryotes. Most authorities believe that organelles were once prokaryotic bacteria and that eukaryotes arose through a process in which larger prokaryotes enclosed smaller cells.

Once the transition to an oxygenated atmosphere had occurred, the emergence of eukaryotes became more likely, for a number of reasons. Prokaryotes need free space around them; crowding interferes with the movement of nutrients and water into and out of the cell. Aerobic eukaryotes are not bothered by crowding, so they can form three-dimensional colonies of cells. Eukaryotes (with the possible exception of the very earliest forms) use oxygen for respiration. This is much more efficient than the anaerobic process of fermentation, so eukaryotes

Figure 15.10
**AN ANCIENT FORM OF LIFE**
These odd-looking bumps are stromatolites, layers of calcium carbonate that are formed in warm, shallow seas by photosynthetic bacteria. These modern stromatolites are forming in the intertidal zone at Shark's Bay, Western Australia. Geologists have found fossil stromatolites more than 1.5 billion years old.

do not require as large a surface area as prokaryotes to facilitate the movement of food and waste. This means that eukaryotic cells can be larger (that is, have a greater volume-to-surface ratio) than prokaryotic cells. Because of their greater efficiency in metabolism, eukaryotes have the energy needed to maintain a more complex cell structure.

## From Simple to Complex

Eukaryotes appeared at least 1.4 billion years ago, in the middle of the Proterozoic Eon. The exact date is not known with certainty, but the fossil evidence clearly shows that by 1 billion years ago they were well established. With the appearance of eukaryotes and the transition to an oxygenated atmosphere, more habitats became available, and many new life forms emerged.

The earliest fossils of larger multicellular organisms appear just at the end of the Proterozoic Eon in rocks that are about 600 million years old. These fossils, which are now known from a number of localities worldwide, were first found in the Ediacara Hills of South Australia and are called the *Ediacara fauna*. Nearly identical animals have since been discovered in rocks of similar age in other parts of the world. The Ediacara fauna lived in quiet marine bays. They were jelly-like animals with no hard parts (Fig. 15.11). The Ediacara animals represent a huge jump in complexity from the first unicellular eukaryotes, which appeared 800 million years earlier. Scientists still do not know much about what happened during those 800 million years.

Figure 15.11
FROM SIMPLE TO COMPLEX:
THE EDIACARA ANIMALS
The Ediacara animals are the most ancient multicelled animals that have ever been found. *A. Mawsonia spriggi* was probably a floating, disc-shaped animal like a jellyfish, 13 cm (5 in) across. *B. Dickinsonia costata* was a worm-like creature, 7.5 cm (3 in) across.

A.

B.

# EVOLUTION AND THE FOSSIL RECORD

What were the mechanisms through which simple single-celled and multicellular organisms eventually diversified into the vast array of organisms that inhabit the Earth today? In the ongoing search for an answer to this question, the most significant contribution was made by Charles Darwin, whose theory of evolution revolutionized biology and paleontology. On December 27, 1831, Darwin departed from England aboard the *H.M.S. Beagle.* He was trained as a clergyman but went along on the voyage as an unpaid naturalist and companion for the captain. When he set sail, Darwin believed in biblical creation and the fixity of species. When he returned, his views had changed considerably, though he still saw the hand of God at work.

Darwin kept scrupulous notes and made paintings and sketches of the plants, animals, and fossils he saw on the voyage. He was particularly impressed with the many species of finches he saw on the Galápagos Islands. The South American mainland, just a short distance away, had only one species of finch. What could account for this difference? Darwin reasoned that long ago finches from the nearby mainland had colonized the islands and had subsequently changed as a result of having to adapt to their new environment. In 1859, many years after his return to England, Darwin published his observations and ideas in a book called *On the Origin of Species by Means of Natural Selection.* He waited a long time to publish his findings because he was concerned about the uproar it might—and did—cause.

In his book, Darwin outlined the theory[5] of **evolution,** which basically says that new species evolve from old species. All present-day organisms are descendants, through a gradual process of adaptation to environmental conditions, of different kinds of organisms that existed in the past. Darwin was not the first to suggest evolution as an explanation for the variety and distribution of species on the Earth. But he was the first to provide such a thorough discussion, supported by clear evidence gathered during his voyage and through subsequent research. Moreover, *On the Origin of Species* was published at a time when scientific thought about the Earth was changing rapidly. The concept of uniformitarianism was widely accepted. There was a general consensus that the Earth was very old.[6] And it had been conclusively shown that some plant and animal species that had once lived were now extinct. Most important, Darwin was the first to propose a reasonable mechanism through which evolution could be achieved. That mechanism was *natural selection.*

## Natural Selection

Darwin (and, almost simultaneously, a young naturalist named Alfred Wallace) proposed that evolution could be achieved through the process of **natural selection,** in which poorly adapted individuals tend to be eliminated from a population. When this happens, there are fewer descendants to inherit the genetic characteristics of the poorly adapted individuals and pass them on to the next generation. All natural populations have individuals with varied characteristics.

> **What was so important about Darwin's finches?**

> **evolution** The changes that species undergo through time, eventually leading to the formation of new species.

> **natural selection** The process whereby individuals that are well-adapted to their environment survive, passing on the favorable characteristics to their offspring.

---

[5] Recall from chapter 4 that a *theory* is a hypothesis that has been tested and supported by experimentation and observation. Plate tectonics is a unifying theory. Evolution is also a unifying theory; it draws together a very large amount of information in a simple, testable model.

[6] In chapter 3, however, we learned that Lord Kelvin's faulty estimate of the age of the Earth caused temporary problems for evolutionary theory. Darwin knew that Kelvin's estimate did not provide enough time for evolution to have occurred.

At any one time and in any one environment, some of these characteristics will be more advantageous than others; they will enable the individual to compete more effectively for scarce resources or to escape predators more easily. These individuals are more likely to survive, and they are likely to have offspring with similar characteristics. Thus, over time an entire population evolves as natural selection favors individuals that are particularly well adapted to their environment. This is informally called "survival of the fittest."

The hypothesis that natural selection could be the mechanism for evolution caused heated debate among Darwin's contemporaries. This was partly because it seemed to contradict the biblical version of creation. Today most scholars accept the idea that religious texts offer spiritual, symbolic explanations for creation that complement, rather than contradict, the scientific explanation (Fig. 15.12). Another problem was that the science of *genetics*—how an individual's

**Figure 15.12**
**IN THE BEGINNING ...**
This Byzantine painting from the twelfth century shows a Biblical interpretation of the origin of life. Light radiating from the hand of God falls on the water-covered Earth.

characteristics are encoded in its genes and passed on to its offspring—was poorly understood in the 1800s. Now we know that inheritable traits are encoded in *genes*, specific protein sets that make up the molecules of DNA. Individuals within a population differ from one another genetically, and it is this variability that makes natural selection possible. New variations are introduced when genetic mutations occur. A *mutation* is a chemical change in genetic traits. Sometimes mutations are unfavorable. Sometimes, however, they provide a competitive edge for the individual in which they occur, and that individual will survive to pass on the favorable trait to a new generation.

Today scientists have refined their understanding of how evolution and natural selection work. Most researchers agree that a population's characteristics may diverge when part of the group is subjected to new environmental conditions. This can happen, for example, if part of a population becomes geographically isolated from the rest. A rising mountain chain or an invasion by the sea might provide a physical barrier that separates two groups. Or some individuals might migrate across a large river or to an island and thus become isolated from the main population. In such cases, the separated group must adjust to a new and different environment. Natural selection will favor those individuals that are best suited to the new environment. Before the separation occurred, the two groups were part of a single **species**, a population of similar individuals that can interbreed and produce fertile offspring. After the separation they may eventually become so different that they can no longer interbreed successfully; in such a case, a new species has developed. The 13 species of finches that Darwin observed on the Galápagos Islands (Fig. 15.13) are a perfect example of this process. Finches that migrated to the islands from the mainland developed differently shaped beaks as they adjusted to their new environments and diets. Some became tree-dwellers and ate insects or berries; others became ground-dwellers and ate seeds or cacti.

The mechanisms of evolution are generally agreed upon, but there is still much discussion about the *rate* of evolutionary change in natural populations.

**species** A population of similar individuals that can interbreed and produce fertile offspring.

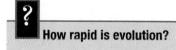

**How rapid is evolution?**

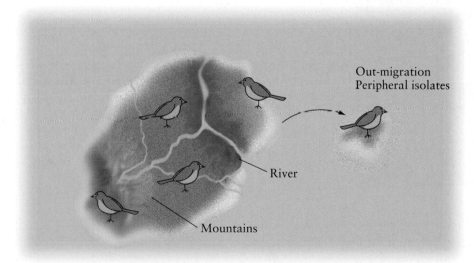

Figure 15.13
**THE BIRTH OF A NEW SPECIES**
As in the case of Darwin's finches, a new species may evolve when part of a population is isolated from the rest by the presence of a natural barrier such as a river or a mountain chain, or by migration to an isolated island, or by the splitting apart or joining of a continent. As the isolated group adjusts to its new environment, different characteristics emerge and are strengthened through natural selection.

**The Galápagos Islands changed the way** we look at the world. After Charles Darwin visited the islands off the coast of Ecuador in 1835, he noted how the different islands kept groups of closely related animals far apart from one another, allowing them to develop into separate species. Those ideas would eventually become part of our shared intellectual heritage with the publication, in 1859, of *On the Origin of Species*.

In this century geologists learned that the Galápagos are no more than 3 million years old, a finding that made the evolutionary pace of these animals seem remarkably fast. Then, in the 1970s, biochemists discovered that this pace was truly astounding: some of the Galápagos animals, they found, have been evolving for at least 10 million years—that's 7 million years of evolution that took place on islands that didn't exist! Studies of DNA from the blood proteins of iguanas showed that all Galápagos iguanas evolved from South American iguanas, but there is only one iguana stock on the mainland (Fig. B15.1). This means that the different iguana species on the islands probably evolved from a common ancestor that rafted over to the islands. But according to the proteins, that ancestral iguana lived at least 10 million years ago, long before the islands were there.

Not until the 1990s did geologists finally figure out how animals could be older than the land beneath their feet: there were older Galápagos Islands, but they were drowned. As discussed in chapter 4, island chains can form when a hot plume of molten rock rises through the Earth's mantle. When it hits the outer crust, it forms a volcano, which eventually builds an island. As a lithospheric plate drifts over the hot plume, new islands are formed one by one, in a chain. Gradually, as the islands age, they begin to erode, and the crust below them sags. Ultimately, they disappear underwater. It appears that this process may be happening in the Galápagos Islands. A team of researchers from Oregon State University, led by geologist David Christie, used radiometric dating to reveal the presence of underwater islands up to 9 million years old, located to the east of existing islands in the Galápagos archipelago.

**Figure B15.1**
**GALÁPAGOS IGUANA**
The different species of iguanas that now inhabit the Galápagos Islands have been evolving for millions of years from a single ancestral species from the mainland of South America.

Ancient iguanas from South America could have come to these islands millions of years ago. They made their home there, and their descendants branched into different species. When the older islands vanished, the animals made the short swim over to a new island. The discovery reminds biologists to take into consideration the changing Earth when they study the evolution of life on islands. "Some biologists used to talk about an 'Atlantis hypothesis' as if it were a bit too imaginative to be real," says Christie, "but it's really not that unusual."

*Source:* Carl Zimmer, "Darwin's Atlantis," *Discover*, January 1993.

Darwin argued that evolution must be very slow, requiring many generations to produce any appreciable change in a population's characteristics. The view that evolutionary change occurs gradually and continuously, so that an ancestral species grades imperceptibly into its descendant species, is called *phyletic gradualism*. On the other hand, a species may persist for a long time but then change very rapidly when a new, favorable trait emerges. (Note that these are not mutually exclusive processes.) The idea that a species (or many species) may go through occasional periods of very rapid change is called *punctuated equilibrium*. Darwin, in fact, suggested that this phenomenon might occur, although he did not give it a name.[7] Proponents of punctuated equilibrium argue that it is supported by the fossil record, because few transitional organisms are preserved as fossils. If the change to a new species occurs very rapidly—over a period of thousands of years, for example, instead of millions of years—it is unlikely that many individuals from the transitional period would be preserved in the fossil record.

## How Fossils Are Formed

In chapter 3 we defined a **fossil** as the remains of a living organism that died and became preserved and incorporated in the Earth's crust. Although fossilization can occur in many ways, it is much more common for a newly deceased organism to be destroyed. If the organism is exposed to running water, air, scavengers, or bacteria, it will decompose or be eaten, and it will be broken into parts and scattered around. Hard parts such as bones, teeth, and shells are less easily destroyed and hence are more likely to be preserved than soft or delicate parts like skin, hair, leaves, flesh, eggshells, or feathers. In any case, for an organism to be preserved as a fossil it must be quickly covered up by a protective layer of sand, mud, or in rare cases tree sap, ice, tar, or volcanic ash.

Sometimes an organism is preserved with little or no alteration. For example, insects many millions of years old have been trapped almost intact in tree sap, which recrystallizes to form the mineral amber (Fig. 15.14). (Hence the premise for the movie *Jurassic Park*.) Organisms can also be preserved if they are trapped in ice or tar, like the ancient wooly mammoths found frozen in Siberia (Fig. 15.15), or the unfortunate prehistoric animals that fell into the La Brea tar pits in Los Angeles, California. Animals may also be preserved by natural *mummification*, in which the soft parts dry and harden before the organism is buried by sediment.

More often, however, the remains of organisms are preserved in an altered state. Bones and other hard parts are sometimes replaced by minerals carried in solution by groundwater (a process called *replacement* or *mineralization*). Wood that has been preserved in this manner is called *petrified wood* (Fig. 15.16). Often the infiltrating minerals fill tiny pores and spaces in bones, teeth, wood, or shells, strengthening and hardening them. This is called *permineralization*. The remains of plants are sometimes preserved by *carbonization*, which occurs when volatile material in the plant evaporates, leaving behind a thin film of carbon.

All of the mechanisms of fossilization that have been mentioned so far preserve actual parts of the organism. Sometimes the organism itself is not preserved

**fossil** The remains of an organism that died and became preserved and incorporated in the Earth's crust.

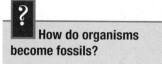

**How do organisms become fossils?**

---

[7] In chapter 11 of *On the Origin of Species* Darwin wrote, "although each species must have passed through numerous transitional stages, it is probable that the periods during which each underwent modification, though many and long as measured by years, have been short in comparison with the periods during which each remained in an unchanged condition."

# Ways to Become a Fossil

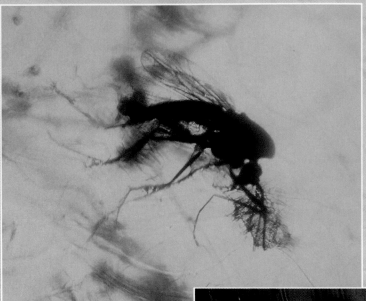

◀ Figure 15.14
**ANCIENT MOSQUITO FOSSILIZED IN AMBER**
Even the delicate legs and wings of this ancient mosquito are preserved in its casing of amber, which is fossilized tree resin. The mosquito is of Eocene to Oligocene age (more than 24 million years old).

Figure 15.15 ▶
**FROZEN IN TIME**
This young wooly mammoth was preserved almost intact, including its fur, skin and flesh, by being frozen whole.

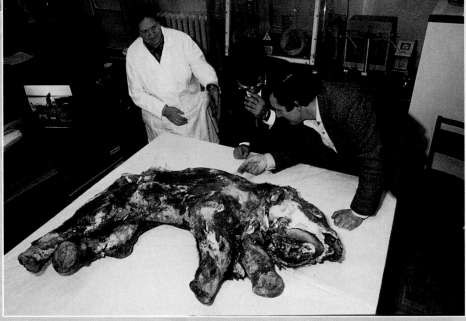

**trace fossil** The fossilized evidence of life processes of an organism, such as tracks, burrows, or footprints.

but leaves an imprint or *mold* in the soft sediment that covered it. Molds can reveal fine details long after the organism itself has been destroyed. Dinosaurs and birds also leave nests, which may even contain the remnants of eggshells (Fig. 15.17). It is also possible for an organism to leave behind other evidence of their presence, called **trace fossils**. For example, worms often leave burrows or borings. Footprints are a common form of trace fossil (Fig. 15.18). Some prehistoric animals even left behind feces, called *coprolites* when fossilized, that provide clues about their characteristics, habits, and diets.

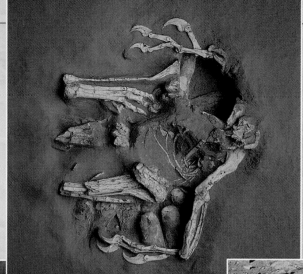

Figure 15.17 ▶
**NESTING MOTHER**
This 2 m (8 ft) long fossilized dinosaur (Oviraptor) was found curled protectively around a nest containing at least 20 eggs. This is considered to be the first proof that dinosaurs cared for their young.

Figure 15.18
**DINOSAURS WALKED HERE**
More than 65 million years ago, dinosaurs left their tracks here in the soft, red mud that has now turned into rock. Evidence of an animal's presence, such as tracks or burrows, are caled *trace fossils*.
▼

◀
Figure 15.16
**PETRIFIED WOOD**
Over many millennia this log in Petrified Forest National Park, Arizona, has been mineralized. Although their woody texture is preserved, the logs in this forest now consist entirely of minerals.

# LIFE IN THE PHANEROZOIC EON

Much of the history of life is the history of unicellular and multicellular organisms, that is, microbial life. Indeed, the geologic record of life from its origin until the end of Precambrian time—almost 3 billion years—is dominated by **microfossils**, fossils so small that they must be studied under a microscope. The study of such fossils is called *micropaleontology*. About 600 million years ago, with the

**microfossil** A fossil so small that it must be studied under a microscope.

455

appearance of larger multicellular animals like the Ediacara fauna, life forms began to diversify very rapidly. We now take a whirlwind tour through the major events in the history of life during Phanerozoic time.

## The Cambrian Explosion

**?**

**Why was there an explosion of biodiversity during the Cambrian?**

The Phanerozoic Eon, which starts with the Cambrian Period, was a time of incredible diversification (Fig. 15.19). Why was this so? One hypothesis is that sexual reproduction, which developed with the eukaryotes, caused the Cambrian explosion of biological diversity. Another hypothesis is that before that time there was too little oxygen in the atmosphere to support the metabolism of larger organisms. As noted earlier, the increase of ozone ($O_3$) in the atmosphere may have shielded Cambrian life forms from harmful ultraviolet radiation. The rising oxygen content of the atmosphere may also have affected the biochemistry of calcium phosphate and calcium carbonate, the two most important skeleton- and shell-building components.

Whatever the reasons, a great many changes occurred about 550 million years ago in the early Cambrian. Compact animals were evolving to replace the soft-bodied, jelly-like organisms of Ediacara times: trilobites (Fig. 15.20), mollusks (clams and sea snails), and echinoderms (sea urchins). All of these (except trilobites) are types that have persisted up to the present. These animals were all equipped with gills, filters, efficient guts, a circulatory system, and other characteristics of more advanced life forms. The Cambrian also saw the development of skeletons, both internal and external. Skeletons gave many organisms a selective advantage, protecting them against predators, against drying out, against being injured in turbulent water, and so on. However, many soft-bodied creatures also persisted in the Cambrian. This can be seen in the fossils of the Burgess Shale (Fig. 15.21), a beautiful collection of soft-bodied animals and plants that were covered by black muds in Cambrian times and eventually preserved as fossils in

**Figure 15.19
EXPLODING BIOLOGICAL DIVERSITY**
This diagram shows how biological diversity has increased throughout geologic time. Note the huge increase in diversity associated with the Cambrian explosion at the beginning of the Phanerozoic Eon.

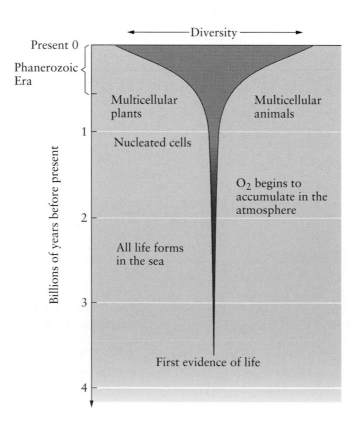

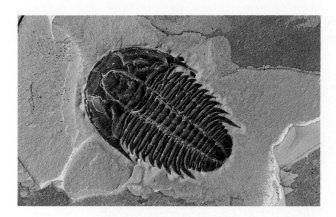

**Figure 15.20**
**TRILOBITE, KING OF THE CAMBRIAN SEAS**
This is a fossil trilobite from the Cambrian Period. Trilobites were one of the first animals to develop a hard covering, presumably as a defense against predators. This sample was collected in Utah.

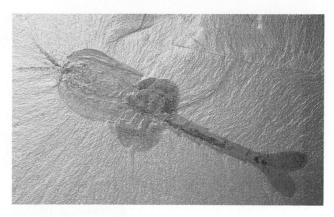

**Figure 15.21**
**SOFT-BODIED FAUNA**
**OF THE BURGESS SHALE**
This specimen of *Waptia fiedensis*, an arthropod, is one of many soft-bodied life forms preserved in the Burgess Shale of British Columbia. The specimen is about 3 cm (1.25 in) in length.

shale near Field, British Columbia. The rich diversity and extraordinary life forms of the Cambrian marine environment are almost unparalleled in the history of the Earth.

## From Sea to Land

The great proliferation of life in the Cambrian was confined to the sea. Successful organisms diversified and flourished; unsuccessful ones disappeared. By 500 million years ago, the main kinds of structural organization for animal life had been established. The one big step that remained was to leave the sea and occupy the land. Eventually all kingdoms of life took that step.

The requirements for life on land are the same for all organisms. Here are the most important ones:

1. *Structural support.* This is needed because, whereas aquatic organisms are buoyed up by water, on land organisms must contend with gravity.

2. *An internal aquatic environment,* with a plumbing system giving it access to all parts of the organism and methods of conserving water against loss to the surrounding atmosphere.

3. *Methods of exchanging gases with air* instead of with water.

4. *A moist environment for the reproductive system.* This is essential for all sexually reproducing organisms.

Let's examine how the first land organisms—plants—met these requirements.

## Plants

Scientists believe that land plants evolved from green algae more than 600 million years ago (Fig. 15.22). Eventually, vascular plants evolved, with structural support from stems and limbs (requirement 1). They also have a *vascular system* (requirement 2), a set of channels through which water and dissolved elements are transferred from the roots to the leaves. Requirement 3 (gas exchange) occurs by diffusion and is controlled by adjustable openings in the leaves called *stomata* (from the Greek for mouth). When carbon dioxide pressure inside the leaf is

**Figure 15.22**
**GREEN ALGAE: THE**
**EARLIEST LIFE ON LAND?**
Scientists believe that land plants probably evolved from green algae more than 600 million years ago. This photo of algae growing in a stream shows what early life may have looked like on land in late Precambrian time.

high, the stomata open; when it is low, they close. The stomata also close when the plant is short of water, thereby protecting it from drying out.

The earliest land plants were seedless (Fig. 15.23). Mosses, ferns, and ground pine are modern examples of seedless plants. Many of the adult forms of seedless plants can tolerate some drought. During dry spells, they release spores that lie dormant until moist conditions return. But all plants rely on moisture for the sexual phase of their reproductive cycles. Without moisture the reproductive cells have no medium in which to reach each other and fuse, so fertilization does not occur. Consequently, seedless plants have never been able to survive in places that lack a dependable supply of moisture for at least part of the growing season. Seedless plants reached their peak in the Mississippian and Pennsylvanian Periods, when they dominated the vast forests on the tropical floodplains and deltas of North America, Europe, and Asia. The remains of these plants produced the fossil fuel (coal) that was essential to the Industrial Revolution two centuries ago.

By the Middle Devonian a few plants were already on the way to meeting requirement 4—providing their own moist environment to facilitate sexual reproduction. The plants that evolved at that time were the **gymnosperms** ("naked-seed plants"), which included *Glossopteris*.[8] The female cell of a gymnosperm is attached to the vascular system and therefore has a supply of moisture. The male cell is carried in a pollen grain with a waxy coating. When the two fuse, a seed results. The seed provides moisture and nutrients that sustain the growth of the young plant until it can support itself through photosynthesis. This allowed the vascular plants to survive in other habitats besides swampy lowlands. Naked-seed plants survive today; gingkos (Fig. 15.24) and conifers are examples.

Gymnosperms were very successful. Freed from their original swampy habitat, they did not have to compete with the great seedless trees of the coal forests. Instead, they established themselves in the drier uplands of the new supercontinent of Pangaea. By the end of the Pennsylvanian Period, they had spread over most of the world. But gymnosperms have one important drawback. The male cell-carrier, the pollen, is spread through the air. What chance does a pollen grain in the air have of finding a female cell? The odds are against it. To ensure success, therefore, gymnosperms have to make huge amounts of pollen. Flowering plants (**angiosperms,** or enclosed-seed plants) solved the problem of the distribution of pollen (Fig. 15.25). For a small incentive (nectar or a share of the pollen), insects

**gymnosperm** Naked-seed plant.

**angiosperm** Flowering plant, or enclosed-seed plant.

---

[8] Do you recall the role played by *Glossopteris* in Wegener's early discussions concerning continental drift? See chapter 4.

# Plants: Ancient and Modern

Figure 15.23
SEEDLESS PLANTS
*A.* This fossil fern, preserved in shale, is about 350 million years old. *B.* This modern fern, called a long beech fern (*Thelypteris phegopteris*), shows the spores (small dark spots) on the underside of the frond (leaf).

A.

B.

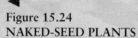

Figure 15.24
NAKED-SEED PLANTS
Naked-seed plants developed from the seedless plants late in the Devonian Period. These are leaves of modern and fossil gingkos. The fossil is from North Dakota. Gingkos are long-lived relics of the ancient family of naked-seed plants.

Figure 15.25
FLOWERING PLANTS
Plants evolved flowers and fruits as an incentive for insects to do the work of distributing pollen. This 15-million-year-old fossil found in Idaho is shown next to its modern equivalent, the sweet gum fruit.

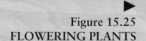

459

will deliver the pollen from one flower to another or from one part of the plant to another. Birds and other animals also help out the angiosperms by eating their seed-bearing fruits. The seeds are distributed throughout the animal's territory in its feces. Angiosperms evolved after gymnosperms, but by the end of the Cretaceous Period the angiosperms had become the dominant land plants. Their life cycle is not significantly different from that of gymnosperms, but they have developed close relationships with animals: insects for pollination, and birds and *quadrupeds* (four-footed animals) for seed dispersal. Many flowering plants, also rely on the wind to help disperse their seeds and pollen.

The last frontier for plants—the dry steppes, savannas, and prairies—were not colonized until the Tertiary Period, when grasses evolved. This was also the time when animal life on land had culminated in the great grazing herds of the high plains of all continents except Antarctica. To see how these animals evolved, we must return to the Paleozoic Era and retrace the first, tentative steps of early animal life.

## Animals

The first major expansion of multicellular marine animals occurred in the last few million years of Precambrian time, at the end of the Proterozoic Eon. Let's look at the various pathways by which these simple animals diversified into the many different types of animals that now occupy both land and sea.

### Arthropods and Insects

**?** **Which creatures were the first to move from sea to land?**

Among the creatures in the Cambrian seas were many that belong to the phylum *Arthropoda,* so called because of their jointed legs. Modern arthropods include crabs, spiders, centipedes, and insects, members of the most diverse phylum on the Earth. They were the first creatures to make the change from sea to land.

With a few exceptions, early arthropods were quite small and light. They were covered with a hard shell of *chitin* (a fingernail-like material). Thus, they were well adapted for life on land in regard to structural support and water conservation. The first to go on land were probably Silurian centipedes and millipedes. Insects were abundant by the Mississippian Period and included dragonflies with a wing span of up to 60 cm (24 in) (Fig. 15.26). For all their success as land creatures, however, the arthropods have very primitive respiratory and vas-

**Figure 15.26**
**ANCIENT INSECT**
This ancient dragonfly from the early Cretaceous Period (about 140 million years ago) was fossilized in the rocks of the Santana Formation, Brazil. The fossil is about 7 cm (3 in) in length.

cular systems. For example, insects breathe through tiny tubes that penetrate the outer coating. This mode of respiration severely limits the size of an organism and is the reason why most insects are small.

The arthropods have an open vascular system. That is, their "blood" is simply body fluid bathing the internal organs; it does not circulate in closed vessels. The fluid is generally kept in motion by a sluggish "heart" that is little more than a contracting tube. At first it seems odd that these primitive animals could have diversified into more than a million terrestrial species. Obviously, the arthropods' simple vascular system is effective. And it's close to indestructible; whoever heard of a cockroach having a heart attack?

## Fishes and Amphibians

Among the fossils of the Burgess Shale is a small, inconspicuous fossil called *Pikaia*. *Pikaia* is a *chordate* because it has a *notochord*, a cartilaginous rod running along the back of the body. (Humans are also chordates. We have a notochord as embryos; later it is replaced by the backbone.) *Pikaia* and other Cambrian fish were jawless, probably feeding on organic matter dredged from the seafloor. In the jawless fish, we can see the first important stage in the development of *vertebrates*, animals that possess backbones.

> **?**
> **What do humans have in common with fossils from the Burgess Shale?**

Jawed fish came next. With jaws came a great burst of diversification. As can be imagined, the possession of jaws allowed fish to move into a wide range of ecological niches that had been vacant until then. The original jawless fish, only a few centimeters long, were quickly joined by larger fish. These included 9-m (29.5-ft) armored carnivores, sharks and other cartilaginous fish, and the huge order of ray-finned fish that are familiar to us as game and food fish.

The first fish to venture onto land, in the Devonian Period, may have been a member of an obscure order called *Crossopterygii*, or lobe-finned fish. (Interestingly, crossopterygians were thought to be extinct until a living example called *coelacanthus* was found in the Indian Ocean in 1938.) Crossopterygians had several features that could have enabled them to make the transition to land. Their lobe-like fins, for example, contained all the elements of a quadruped limb. They had internal nostrils, a characteristic of air-breathing animals. As fish, they already had developed a vascular system that was adequate for life on land.

However, recent DNA studies of modern coelacanths, amphibians, and *Dipnoi* (African, Australian, and South American lungfish) suggest that lungfish, rather than lobe-finned fish, may have been the ancestors of the first land-dwelling amphibians. The lungs used by modern lungfish to take occasional gulps of air during periods of drought were developed long before the first amphibians appeared. Presumably it would have been a relatively simple adaptation for these lungs to become full-time suppliers of air for amphibians in a terrestrial environment.

Amphibians never developed an effective method for conserving water. To this day they retain permeable skins, which is one reason they have never become wholly independent of aquatic environments. Although amphibians have been successful for many millions of years, they have never met the reproductive requirement for life on land. In most amphibian species, the female lays her eggs in water, the male fertilizes them there after a courtship ritual, and the young are fish-like when first hatched (e.g., tadpoles). Like the seedless plants, the amphibians, with one foot on the land, have remained tied to the water for breeding. Although some became quite large (2 to 3 m, about 3 yd), they never diversified much after the Devonian Period. One branch went on to become reptiles. Of the rest, those that survive are frogs, toads, newts, salamanders, and limbless water "snakes," which have returned to life in the water.

**Figure 15.27**
**ARCHAEOPTERYX**
The skeleton and teeth of *Archaeopteryx* were very similar to those of dinosaurs. The most unusual feature of this fossil is the very detailed impressions of feathers surrounding the wing bones.

> **?**
> **Which came first, the chicken or the egg?**

## Reptiles, Birds, and Mammals

Reptiles freed themselves from the water by evolving an egg that could be incubated outside of the adult, and by developing a watertight skin. These two "inventions" enable them to occupy terrestrial niches that the amphibians had missed because of their need to live near water. The amniotic egg did for reptilian diversity what jaws did for diversity in fishes. Originating in the Mississippian and Pennsylvanian coal swamps, by the Jurassic Period the reptiles had moved over the land, up into the air, and back to the water. They had also produced two orders of dinosaurs (the largest quadrupeds ever to walk the Earth) and given rise to two new vertebrate classes, mammals and birds.

Birds first appeared near the end of the Jurassic Period. An early example of a bird, *Archaeopteryx* ("ancient wing"), would have been classified as a dinosaur were it not for the discovery that it had feathers (Fig. 15.27). Prior to *Archaeopteryx,* vertebrate animals made the transition into the air in the form of pterosaurs—flying reptiles with long wings and tails.

In many ways, mammals are better equipped to live on the land than were the great reptiles. However, it is difficult to pick out a single mammalian "invention" comparable to the jaws of fish or the reptilian egg. The mammalian advantage may simply be a set of interdependent improvements managed by a larger brain and supported by a faster metabolism. Mammals, mostly quadrupeds, are adapted to a faster and more versatile life than the reptiles could ever have led. The placental uterus is sometimes regarded as the key to mammalian success, but it's really only a piece of equipment required by the intricacy of the fetus. By comparing brain-to-body weight ratios in archaic and modern reptiles and mammals, it can be shown that increase in mammalian brain size is a continuing process, whereas in reptiles brain size has not increased; the ratios in modern reptiles do not differ significantly from those in archaic ones. What is perhaps more significant is that if something (a meteorite? climatic change?) had not wiped out the great reptiles of the end of the Cretaceous Period, we might not even be here to think about them.

## The Human Family

The family of humans, Hominidae, did not descend from the modern ape family, Pongidae. Instead, the two families probably diverged from an earlier apelike family. Unfortunately, the fossil record for both modern humans and modern apes is poor, and many transitional forms are missing. The oldest known genus

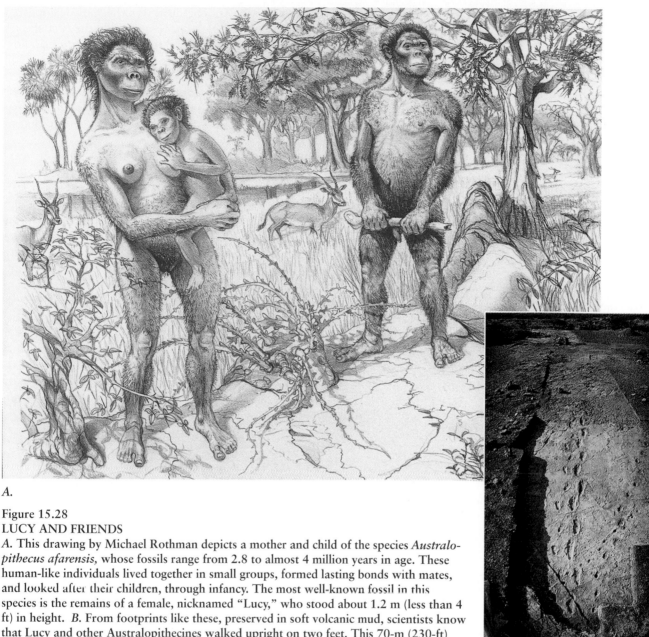

*A.*

**Figure 15.28**
**LUCY AND FRIENDS**
*A.* This drawing by Michael Rothman depicts a mother and child of the species *Australopithecus afarensis,* whose fossils range from 2.8 to almost 4 million years in age. These human-like individuals lived together in small groups, formed lasting bonds with mates, and looked after their children, through infancy. The most well-known fossil in this species is the remains of a female, nicknamed "Lucy," who stood about 1.2 m (less than 4 ft) in height.  *B.* From footprints like these, preserved in soft volcanic mud, scientists know that Lucy and other Australopithecines walked upright on two feet. This 70-m (230-ft) trail includes the footprints of two adults and possibly a child, stepping in the footprints of one of the adults. The prints to the right are those of an extinct three-toed horse.

*B.*

of the Hominidae (or **hominid**) family is *Ardipithecus*, dating back to 4.4 million years. The first hominid family that was clearly bipedal (walked upright) was *Australopithecus*, best represented by *Australopithecus afarensis*, of which the famous "Lucy" is an example (Fig. 15.28A). These hominids were only about 1.2 m (under 4 ft) in height but had a brain capacity larger than that of chimpanzees. Fossils of this species range from about 3.9 million to 3.0 million years in age, a span of almost a million years. From the shape of its pelvis and from footprints left in soft volcanic mud more than 3 million years ago (Fig. 15.28B), we know that these individuals walked upright, though their skulls looked more apelike than human.

Other species of *Australopithecus*—*Australopithecus africanus, aethiopicus, boisei,* and *robustus*—inherited the African continent in succession, each over-

**hominid** A member of the family of humans, the Hominidae.

lapping in time with the preceding species. The australopithecines disappeared altogether about 1.1 million years ago. *Homo erectus,* probably the first species of our own genus (*Homo*), was more widely traveled than *Australopithecus.* Fossils of *Homo erectus* dating back about 1.8 million years have been found in Africa, Europe, China ("Peking man"), and Java ("Java man"). Even earlier, a problematic species sometimes called *Homo habilis* used stone tools. Since toolmaking is the distinguishing feature of the genus *Homo,* some experts include *habilis* in this genus; others argue that the skull of this species is more like that of the australopithecines.

*Homo erectus* disappeared around 300,000 years ago. By 230,000 years ago, *Homo neandertalensis*—"Neandertal man"—had appeared. Unfortunately, the fossil record between 400,000 and 100,000 years ago is poor, so the transition from *Homo erectus* to Neandertal is not well understood. We do know from burial sites that Neandertal people practiced some form of religion. On the basis of similarities in teeth and brain size (slightly larger than our own), some experts argue that Neandertal was part of our own species; thus, they label it *Homo sapiens neandertalensis.* However, recent studies suggest that the DNA of Neandertal is different from our own; thus, *Homo sapiens* may not be a direct descendant of Neandertal. Neandertals disappeared about 30,000 years ago; they were replaced rather suddenly by the biologically modern *Cro-Magnon* people, the first indisputable example of our own species, *Homo sapiens.*

Did the Cro-Magnon evolve from the Neandertal, or were they a distinct species? Both peoples lived in Europe for about 5000 years before the disappearance of the Neandertal. Did they interbreed? Were the Neandertal wiped out by the Cro-Magnon? These and many other questions await answers, as paleontologists continue to search for clues in the hominid fossil record.

## Mass Extinctions

**mass extinction** A catastrophic episode in which many species become extinct within a geologically short time.

Embedded in the fossil record is a story of adaptation and recovery following catastrophic episodes in which many species become extinct within a geologically short time. Such episodes are called **mass extinctions.** Most people are aware that the dinosaurs became extinct about 65 million years ago, at the boundary between the Cretaceous (K) and Tertiary (T) Periods. But many are not aware that other animal and plant species were also affected. Approximately one quarter of all known animal families living at the time, including marine and land-dwelling species, became extinct at the end of the Cretaceous Period. This mass disappearance of species is clearly evident in the fossil record. It is the reason that early paleontologists selected this particular stratigraphic horizon to represent a major boundary in the geologic time scale.

The great K-T extinction is not unique, nor was it the most dramatic of such occurrences. There have been at least 5 and possibly as many as 12 mass extinctions during the past 250 million years (Fig. 15.2). The most devastating of these occurred 245 million years ago at the end of the Permian Period, when as many as 96 percent of all species died out. Another great extinction occurred at the end of the Triassic Period, and several earlier extinctions in the Paleozoic Era affected marine organisms.

What causes mass extinctions? Some observational evidence suggests that the great extinction at the end of the Cretaceous Period may have been caused by a giant meteorite impact. If an extraterrestrial body such as a meteorite or a comet 10 km (6 mi) in diameter struck the Earth, it would cause massive environmental devastation. The effects could include earthquakes, tsunamis, widespread fires, acid rain, atmospheric particulates that might cause global darkness, and intense climatic changes. Evidence for these and other effects has been found in the stratigraphic horizon that marks the K-T boundary. Throughout the world the boundary is also marked by a thin layer of clay that is rich in the element iridium (Ir). This is consistent with an influx of extraterrestrial material because meteorites contain a great deal of Ir compared to the amount contained in terrestrial rocks.

It is possible that a meteorite impact caused the K-T extinction, but the causes of other major extinctions are not as clear. Many scientists feel that some extinctions—particularly the great marine extinctions of the Paleozoic Era—were more likely caused by climatic or other environmental changes than by catastrophic events such as meteorite impacts. The study of mass extinctions seems particularly relevant today; the present rate of species extinctions from anthropogenic causes is rivaled only by the greatest mass extinctions in Earth history.

# WHAT'S AHEAD

Now that we have introduced life into the Earth system, we are almost at the end of our exploration of the *Geology Today*. In this chapter you have learned how closely interrelated are the histories of all parts of the Earth system. The atmosphere, hydrosphere, and lithosphere are inextricably linked together in their physical and chemical evolution. The biosphere, too, is part of this integrated system. Life has been, and will continue to be, a profound influence on the functioning of the entire Earth system.

We humans remain closely dependent on the Earth system for all of our needs. In turn, we continue to affect the physical and chemical characteristics of all parts of the system, but at a greater rate and magnitude than ever before. In short, humans have become a geologic force to be reckoned with. In the next chapter we will learn more about the mineral and energy resources that supply the needs of modern society. Finally, in the last chapter of the book, we will examine some of the impacts of human activities on the Earth system. The expertise of geoscientists can help us manage these impacts and meet our continuing need for Earth resources and for protection from the effects of natural hazards.

## Chapter Highlights

**1.** The history of life is intertwined with that of the atmosphere–hydrosphere system. Both temperature and pressure were high in the early atmosphere, which consisted primarily of water vapor, carbon dioxide, and nitrogen; there was no free oxygen ($O_2$). As early as 4.4 billion years ago, the Earth cooled enough for surface water to begin to collect.

**2.** Volcanic emissions can account for the entire volume of the oceans and atmosphere, with the exception of oxygen. Most of the free oxygen in the atmosphere originated through the process of **photosynthesis**. Life (particularly shelled organisms) also influenced the chemical composition of the atmosphere by providing a reservoir for carbon dioxide, which brought the greenhouse effect under control and moderated surface temperatures.

**3. Chemosynthesis** of amino acids, the basic building blocks of proteins, may have been the first step on the way to life. Early organic molecules may have collected in an organic "soup," possibly in a warm pond, at hydrothermal vents, or even on bubbles in sea spray. Drying and heating may have facilitated the polymerization of small organic molecules to form proteins. Metabolism began with the absorption of nutrients from the surrounding environment, eventually evolving into more complex biochemical processes.

**4.** The fundamental structural and organizational unit of life is the **cell**. There are two kinds of cells: **prokaryotic** and **eukaryotic**. Through the Hadean and Archean Eons and part of the Proterozoic, the only life on the Earth was prokaryotic. Eukaryotes arose from prokaryotes around 1.4 billion years ago.

5. The most ancient fossils that have been found are about 3.55 billion years old. The earliest known fossils of large multicellular organisms are in 600-million-year-old rocks from the end of the Proterozoic Eon.

6. The theory of **evolution** states that all present-day organisms are descendants of different kinds of organisms that existed in the past. The mechanism for evolution is **natural selection,** in which poorly adapted individuals tend to be eliminated from a population.

7. A population's characteristics diverge when part of the group is subjected to new environmental conditions. This can happen if part of the population becomes isolated from the rest. After the separation, the two groups may eventually become so different that a new **species** develops.

8. There are many mechanisms through which an organism can become a **fossil.** Organisms may be fossilized essentially intact as a result of preservation in tree sap, ice, or tar, or through mummification. More commonly, organisms are preserved in an altered state through replacement, permineralization, or carbonization. Organisms may also leave behind evidence, such as molds or trace fossils, without having any body parts preserved.

9. There is evidence in the fossil record of numerous **mass extinctions** of species. The most devastating was the Permian extinction, in which 96 percent of all species died out. Some mass extinctions may have been caused by catastrophic events such as meteorite impacts; others may have been caused by environmental changes.

10. Biological diversity exploded in the Cambrian Period. It was promoted by the introduction of gills, filters, efficient guts, circulatory systems, and internal and external skeletons.

11. To succeed on land, organisms require structural support; an internal aquatic environment and devices for conserving water; a means of exchanging gases with air; and a moist environment for the reproductive system.

12. The earliest vascular land plants were seedless plants, which reached their peak in the Mississippian and Pennsylvanian Periods. By the Middle Devonian, the **gymnosperms** had evolved. Flowering plants (**angiosperms**) solved the problem of pollen and seed distribution by enticing insects and animals to disperse them.

13. The arthropods are members of the most diverse phylum on the Earth. They include crabs, spiders, centipedes, and insects. Arthropods were the first creatures to make the change from sea to land. They have a simple but effective vascular system.

14. The first fish appeared in the Cambrian and were jawless. The development of jaws was accompanied by a great burst of diversification. Fish first ventured onto land in the Devonian, eventually giving rise to the amphibians. Amphibians are still dependent on aquatic environments.

15. Reptiles freed themselves from reliance on water by evolving a watertight skin and an egg that could be incubated outside of the animal and away from an aquatic environment. By the Jurassic Period, reptiles had produced two orders of dinosaurs and given rise to two new vertebrate classes, mammals and birds.

16. The oldest known genus of the Hominidae (or **hominid**) family is *Ardipithecus,* succeeded by *Australopithecus.* The australopithecines disappeared about 1.1 million years ago. Fossils of *Homo erectus,* the first species of our own genus, have been found in Africa, Europe, China, and Java and date back about 1.8 million years. *Homo erectus* disappeared around 400,000 years ago, giving way to the Neandertals. About 30,000 years ago the Neandertals were replaced by the Cro-Magnon, the first indisputable example of our own species, *Homo sapiens.* There are still many unanswered questions concerning the ancestry of modern humans.

# ▶ *The Language of Geology*

- angiosperm 458
- cell 444
- chemosynthesis 441
- DNA 443
- eukaryotic cell 445
- evolution 449
- fossil 453
- gymnosperm 458
- hominid 463
- mass extinction 464
- microfossil 455
- natural selection 449
- photosynthesis 439
- prokaryotic cell 444
- RNA 443
- species 451
- trace fossil 454

# ► Questions for Review

1. How did life affect the chemical composition of the atmosphere through geologic time?
2. What is the Oparin "soup" hypothesis? In what kinds of environments might early organic molecules have collected to form an organic "soup"?
3. What are the two basic kinds of cells, and what are the differences between them?
4. How old are the oldest fossils that have ever been found?
5. What are the common processes through which organisms become fossilized? Which parts of an organism are most likely to be preserved?
6. Why did biological diversity explode during the Cambrian Period?
7. What features must organisms have if they are to be successful on land?
8. Describe the important developments in the evolution of land plants.
9. What was the first type of organism to make the transition from sea to land? What was the first type of animal to make the transition?
10. Were there ever two or more hominid species on the Earth at the same time?

# ► Questions for Thought and Discussion

1. Recall from chapter 1 that the Earth and Venus are so similar in size and overall composition that they are almost "twins." Why did these two planets evolve so differently? Why is the Earth's atmosphere rich in oxygen and poor in $CO_2$ compared to that of Venus? What would happen to ocean water (and then to the atmosphere) if the Earth were just a little bit closer to the Sun?
2. *Archaeopteryx* is a classic example of a transitional species in the fossil record. Investigate the characteristics of *Archaeopteryx* and its dinosaur predecessors. In what ways were they like and unlike modern birds?
3. What do you think would have happened to mammals, which were mostly small, unobtrusive, rat-like creatures during the Mesozoic Era, if the K-T extinction had not wiped out the dinosaurs?
4. What do you think would have happened if two hominid species, such as the Neandertal and Cro-Magnon peoples, encountered one another? Would they have fought? Would they have interbred? (Can two different species interbreed successfully and produce fertile offspring?)

For an interactive case study on habitat and biodiversity in urban ecosystems, visit our Web site.

# · 16 ·

# EARTH RESOURCES

**Gold is nearly indestructible;** it doesn't tarnish, corrode, or dissolve in common liquids, and it is so valuable that people take good care of any they own. As a result, almost all the gold that has ever been mined is still in circulation. Gold is the ultimate recyclable metal. The gold from Cleopatra's bracelets may now reside in a tooth filling or a modern wedding band.

Because gold is so durable, we know with reasonable accuracy how much has been produced since mining started about 17,000 years ago; the total to the end of 1997 is about 140,000 tons. Because gold is a very dense metal, if 140,000 tons were stacked on the playing area of a football field, the pile would be only 1.6 m (5.3 ft) high. This doesn't seem like much gold after 17,000 years of hard work.

By comparison, consider copper, a technologically much more important metal than gold. In just one year, 1997, 10 million tons of copper were mined. Ten million tons of copper stacked on the playing area of a football field would make a pile 2488 m (8158 ft) high! Gold may seem romantic, but considering its main uses—jewelery, coins, and bullion in bank vaults—you have to wonder if it has been worth all the labor and hardship.

............◆............

## In this chapter you will discover

- The difference between renewable and nonrenewable resources
- The importance of natural resources to modern society
- Where our energy supplies come from
- How, where, and why mineral deposits form

*For the Love of Gold*

Gold miners, Serra Pellada, Brazil, in the 1980s. Each miner was awarded a small square claim by the government. The square sides visible in the picture indicate the size of a claim (about 3 m, or 10 ft). As mining became deeper, the ore had to be carried out on miner's backs. Accidents, injury, and death were common.

**Earth resources** Useful things that are extracted from the Earth.

**Earth resources** are the useful things that we get from the Earth. They include air, water, building materials, metals, fertilizers, oil, coal, gemstones, and many other materials.[1] Three kinds of Earth resources have been discussed in previous chapters; they are soil, water, and air. Air is everywhere around us, and soil and water are widespread. All living creatures use these instinctively. Only humans have learned to exploit energy and mineral resources, which are less widely distributed.

# RENEWABLE AND NONREWABLE RESOURCES

Resources can be either *renewable* or *nonrenewable*. A *renewable resource* can be replenished or regenerated. For example, even though we may consume a food crop each season, a new crop grows during the following season. A layer of soil lost to erosion will eventually be regenerated by the physical, chemical, and biological processes of soil formation. Groundwater drawn from wells may eventually be replenished by rainwater. But what does "eventually" mean? Some resources take a *very* long time to regenerate—longer than humans are willing or able to wait. For example, it may take hundreds of years for groundwater supplies to be replenished and many thousands of years for a 10-inch layer of arable soil to form from bare rock. So we must modify our definition to say that a **renewable resource** is one that can be renewed *on a human time scale*.

> **If an average human life span is 75 years, how many life spans must pass before a large mineral deposit is formed?**

**renewable resource** A resource that can be replenished or regenerated on the scale of a human lifetime.

**nonrenewable resource** A resource that cannot be replenished or regenerated on the scale of a human lifetime.

Resources such as coal, oil, copper, iron, gold, and fertilizers are mined from mineral deposits. Some mineral deposits are known to be forming today, but the rate of formation is exceedingly slow. For example, it may take 600,000 years for a large copper deposit to be formed. From a human point of view, all mineral resources are one-crop resources, and the Earth's supply of those "crops" is fixed. We therefore define a **nonrenewable resource** as one that cannot be replenished or regenerated on a human time scale.

## Natural Resources and Human History

> **What was the first nonrenewable resource used by early humans?**

Our hunter-gatherer ancestors lived entirely on renewable resources—the animals they could catch and the plants they could gather. But several million years ago they crossed one of the great thresholds of human development—they picked up suitably shaped stones to use in hunting and became the first and only species to routinely use nonrenewable resources. Before long they discovered that some stones, such as flint, chert, and obsidian, are more easily shaped into sharp spear points than other stones (Fig. 16.1). Because the best stones could be found only in a few places, our ancestors began trading various kinds of stone. Then they started gathering and trading salt. Hunting communities can meet their dietary needs for salt by eating meat, but after humans began farming, their diets were based on grain, and they needed extra salt. We don't know when or where salt mining started, or when or where the mining of flint, chert, and obsidian started,

---

[1] The broader term *natural resources* includes these Earth resources, plus biological resources: plants and animals.

470

Figure 16.1
**SHARP STONES: THE FIRST NONRENEWABLE RE-SOURCES**
This is a collection of Clovis spear points made of chert and quartz from the Lehner Mammoth Kill site, Hereford, Arizona. The largest spear points are about 7.5 cm (3 in) long.

but long before the beginning of recorded history the land was crisscrossed by trade routes for the exchange of nonrenewable resources.

Insofar as we know, metals were first used about 17,000 years ago. The first metals used were copper and gold, both of which can be found in nature as pure metals (Fig. 16.2). Natural metallic copper is rare, however, and eventually other sources of copper were needed. About 6000 years ago our ancestors learned how to extract copper from certain minerals through a process called smelting. By 5000 years ago they had learned how to smelt lead, tin, zinc, silver, and other metals, as well as how to mix metals to make alloys such as bronze (copper + tin) and pewter (tin + lead + copper). Smelting of iron is more difficult than smelting of copper and was not achieved until about 3300 years ago.

The first fuels to be used as sources of energy were wood and animal dung. The first people to use oil instead of wood were the Babylonians, who lived in what is now Iraq about 4500 years ago. They used oil from natural seeps in the

Figure 16.2
**EARLY USES OF METAL**
Gold, copper, and silver were the first metals used by humans because they can be found in nature in their metallic form. This gold cup, approximately 4000 years old, was made for Queen Pu-alu of Sumer in what today is Iraq. The cup stands approximately 13 cm (5 in) high.

valleys of the Tigris and Euphrates rivers. The first people to mine coal and drill for natural gas were the Chinese, beginning about 3100 years ago. Thus, by the time the Roman empire came into existence in Europe about 2500 years ago, humans had come to depend on a very wide range of natural resources—not just fuels and metals but also processed materials such as cements, plasters, glasses, porcelains, and pottery. The list of materials mined, processed, and used by humans has grown steadily since then. Today we have uses for almost all the naturally occurring chemical elements, and more than 200 kinds of minerals are mined and used.

## Natural Resources and Modern Society

**?** How would people survive without mineral and energy resources?

Can you imagine a world without machines? In the modern world of 1998 there are almost 6 billion human inhabitants, and the population is growing larger by almost 100 million bodies each year. Such a huge growing population couldn't operate without machines. As the *Geology Around Us* essay in chapter 2 ("Minerals in Everyday Life") makes clear, each of us uses—directly or indirectly—a very large amount of material derived from nonrenewable resources. Without them we could not build planes, cars, televisions, or computers. We could not distribute electrical power or build tractors to till the fields and produce food. Machines are used for manufacturing, transportation, and communication. The metals needed to build machines, as well as the fuels needed to run them, are nonrenewable resources extracted from the Earth. Without these resources, industry would collapse and living standards would deteriorate dramatically.

Imagine what life would be like if we ran out of energy resources and we had to rely entirely on human muscle power. If a healthy adult rides an exercise bike that drives an electrical generator hooked to a light bulb, the best an average person can do in a nonstop eight-hour day of pedaling is to keep a 75-watt bulb burning. In North America the same amount of electricity can be purchased from a power company for about 10 cents. Viewed in this way, we can see that human muscle power is really puny. Our ancestors realized this a long time ago, and over many thousands of years they found ways to supplement muscle power. First, they domesticated beasts of burden such as horses, oxen, camels, elephants, and llamas. Then they learned to make sails and use wind power, build dams and use water power, and convert the heat energy of wood, oil, and coal into mechanical power. Today we use supplementary energy in every part of our lives, from food production and transportation to housing and recreation. North Americans are the world's biggest energy users. An average North American uses energy, directly or indirectly, at a rate of 10,000 watts—equivalent to 133 75-watt light bulbs burning continuously. Energy use in many developing countries is much lower, so the worldwide average rate of energy use is only 1600 watts per person. To put this into perspective, consider this comparison: If we had to rely strictly on muscle power, it would take 65 people working continuously in eight-hour shifts to produce enough energy for the average person, but it would take 400 people to produce the energy for just one North American!

Clearly, a tremendous amount of material and energy is needed to keep human societies going. How else could food be delivered from distant farms to densely packed cities? How else could we operate in cold climates? This raises the question of the Earth's carrying capacity, which we can ask in the following way: Are there enough natural resources to raise the living standards of all peoples to the levels they desire, and just how many people will there be? The answers to these questions are not clear. But one thing is clear: as the population grows, we will need more and more resources. In chapter 17 we will look more closely at the role of geologists in finding and managing new resources.

The issues discussed in the preceding pages are social issues involving human dependence on Earth resources. We turn next to the practical issues of how energy and mineral resources are formed and where they are found.

# ENERGY RESOURCES

As noted earlier, there are both renewable and nonrenewable sources of energy. Examples of renewable energy sources are fuelwood, wind, and water power. Examples of nonrenewable energy sources include coal, oil, natural gas, and nuclear energy. Everywhere in the world, even in the least developed countries, nonrenewable sources supply at least half of the energy used (Fig. 16.3). The main nonrenewable source of energy is fossil fuels.

## Fossil Fuels

A **fossil fuel** is organic matter trapped in sediment or sedimentary rock. The organic matter, the remains of either plants or animals, has undergone various changes during and after burial. The kind of sediment, the kind of organic matter, and the kinds of post-burial changes that occur determine which type of fossil fuel is formed. The principal fossil fuels are peat, coal, oil, and natural gas.

All fossil fuels derived their energy from the Sun. The energy-trapping mechanism was *photosynthesis* (shown in Fig. 15.3), in which plants absorb the Sun's

**fossil fuel** Organic matter trapped in sediment or sedimentary rock; the principal fossil fuels are peat, coal, oil, and natural gas.

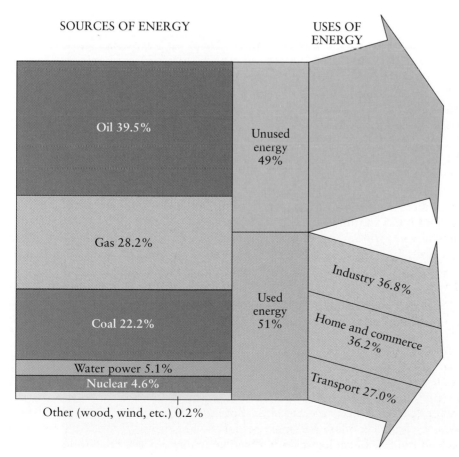

Figure 16.3
**ENERGY SOURCES AND USES**
Fossil fuels (oil, gas, coal) account for 90 percent of the energy used in the United States. The large amount of lost energy—nearly half, as the upper-right arrows show—arises both from inefficiencies in energy use and from the fact that the laws of thermodynamics put limits on the efficiency of any engine, furnace, or other device, so that only some energy can be employed usefully.

SOURCES OF ENERGY

USES OF ENERGY

Oil 39.5%

Gas 28.2%

Coal 22.2%

Water power 5.1%

Nuclear 4.6%

Other (wood, wind, etc.) 0.2%

Unused energy 49%

Used energy 51%

Industry 36.8%

Home and commerce 36.2%

Transport 27.0%

? From what do fossil fuels obtain their energy?

energy and use it to combine water ($H_2O$) and carbon dioxide ($CO_2$), creating oxygen and organic compounds called *carbohydrates*. When plant matter decays or burns, the photosynthetic reaction is reversed—oxygen from the air combines with carbohydrates to form $H_2O$ and $CO_2$, and the solar energy trapped in the organic matter is released as heat. Any organic matter that escapes decay and is buried in sediment becomes part of a vast reservoir of stored solar energy. The total amount of organic matter stored in sediments and sedimentary rocks is less than 1 percent of all the organic matter that has been formed by plants and animals. Over geologic time, however, the amount of stored organic matter has become very large.

Let us now look at how three kinds of stored organic matter—coal, oil and natural gas— are formed.

## Peat and Coal

Organic matter that accumulates on land comes from trees, bushes, and grasses. These land plants are rich in organic compounds that tend to remain solid. In water-saturated places such as swamps and bogs, the remains accumulate to form **peat**, a loose aggregate of plant remains with a carbon content of about 60 percent. Peat is a biogenic sediment. (Fig. 16.4).

The accumulation of plant remains to form peat is the initial stage in the formation of the black, combustible sedimentary rock we call **coal.** As layers of peat formed millions of years ago were compressed and heated by being buried under more and more overlying sediment, water and gaseous compounds such as carbon dioxide ($CO_2$) and methane ($CH_4$) escaped. The loss of these constituents left a higher proportion of carbon in the residue, and carbon is what we want for

**peat** A loose aggregate of plant remains with a carbon content of about 60 percent, which typically accumulates in water-saturated places such as bogs and swamps; a biogenic sediment.

**coal** A black, combustible sedimentary rock that is formed by compression and heating of layers of peat buried by overlying sediments.

**Figure 16.4**
**PEAT, THE STARTING POINT OF COAL**
A peat cutter harvests peat from a bog in western Ireland. When dried, peat provides fuel for heat and cooking. Peat is higher in energy than firewood, but lower than coal, because it is in the process of changing from plant matter to coal. If the peat cutter could wait a few million years, he would harvest much higher-energy coal.

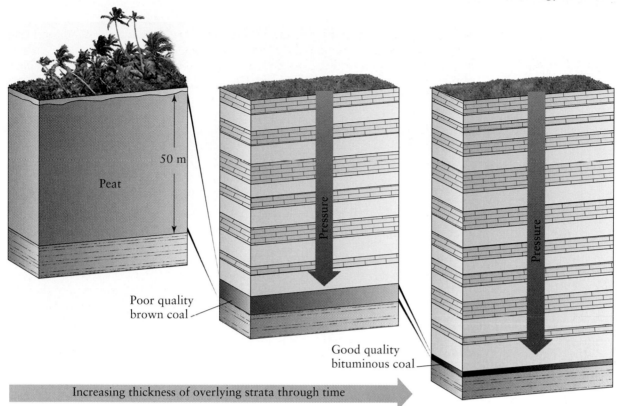

50 m

Peat

Poor quality
brown coal

Pressure

Good quality
bituminous coal

Pressure

Increasing thickness of overlying strata through time

**Figure 16.5**
**COALIFICATION**
Plant matter in peat is slowly converted into coal. This *coalification* happens as a result
of the decomposition of plants, increased pressure and temperature as overlying sediments
build up, and the passage of time—millions of years. By the time a layer of peat has been
converted into bituminous coal, its thickness has been reduced by 90 percent and the
proportion of carbon raised from 60 to 80 percent. Thus the desired heat source—the
burnable element, carbon—becomes concentrated.

burning. The higher the carbon content, the greater the *rank* of the coal. Coal of
lowest rank is called *lignite*, while *anthracite* is the name for coal of highest rank.
Much of the world's coal is *bituminous coal*, intermediate between lignite and
anthracite (Fig. 16.5). The conversion of peat (a sediment) into coal (a rock) is
called *coalification*. Lower ranks of coal are sedimentary rocks, but anthracite is
so changed from its original form that it is considered a metamorphic rock. Coal
occurs in strata (miners call them *seams*) along with other sedimentary rocks,
mainly shales and sandstones; anthracite occurs in slates—low-grade metamorphic rocks. Most coal seams are 0.5 to 3 m (1.5 to 10 ft thick), although some
reach thicknesses of more than 30 m (100 ft) (Fig. 16.6). Coal seams tend to
occur in groups. For example, 60 seams of bituminous coal have been found in
western Pennsylvania. A handful of them are thick enough and of sufficient quality to mine.

Peat has been formed more or less continuously since land plants first appeared on the Earth about 450 million years ago. The size of peat swamps has
varied greatly, however, and therefore the amount of coal formed has also varied.
By far the largest amounts of peat swamp formation occurred during the warm
Carboniferous and Permian Periods, 360 to 245 million years ago. The great coal
seams of Europe and eastern North America were formed from peat deposited at
this time, when the swamp plants were different from today—giant ferns and

**Figure 16.6**
**MINING A COAL SEAM**
This photo shows the Peabody Coal Mine on the Navajo Reservation at Black Mesa, Arizona. The flat-lying bed of coal is mined by removing the overlying strata prior to digging out the coal.

scale trees (*gymnosperms*). A second great period of peat formation peaked during the Cretaceous Period, 144 to 66.4 million years ago. During this period the plants of peat swamps were flowering plants (*angiosperms*), much like those found in swamps today.

The luxuriant growth needed to form thick and extensive coal seams is most likely to occur in a tropical or semitropical climate. This implies either that the global climate was warmer when the plant matter of today's coal deposits originally accumulated or that the swamps were all located in the tropics. Probably both conditions were involved. Coal deposits that were formed in warm low-latitude environments but are now in frigid polar lands provide compelling evidence for the slow drift of continents over great distances; the many places where coal is found in North America today provide an excellent example (Fig. 16.7).

Today peat formation occurs in wetlands such as the Okeefenokee Swamp in Florida and the Great Dismal Swamp in Virginia and North Carolina (Fig. 16.8). The Great Dismal Swamp, one of the largest modern peat swamps, contains an average thickness of 2 m (6.5 ft) of peat. However, unless this swamp lasts many millions of years, not enough peat will accumulate in the Great Dismal Swamp to produce a coal seam as thick as some of the seams in the Appalachians and the Rockies.

### Petroleum

**petroleum** Gaseous, liquid, and semisolid naturally occurring substances that consist chiefly of hydrocarbon compounds.

In the ocean, microscopic phytoplankton (tiny floating plants) and bacteria (simple, single-celled organisms) are the principal sources of organic matter trapped in sediment. Over long periods of time, such organic matter has been trapped in clay-rich sediment that has been slowly converted into shale. During this process, organic compounds were transformed into *oil* and *natural gas*. These are the main forms of **petroleum**—gaseous, liquid, and semisolid naturally occurring substances that consist chiefly of hydrocarbon compounds. Liquid petroleum (from the Latin words *petra*, rock, and *oleum,* oil) is referred to as *crude oil* when it emerges from the ground. From this state it must be distilled and refined and thereby separated into the various petrochemicals that we actually use, such as gasoline.

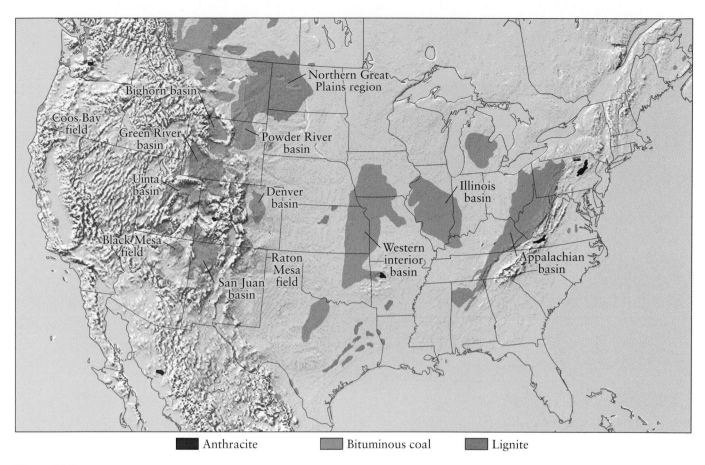

Anthracite     Bituminous coal     Lignite

Figure 16.7
COAL, AN ABUNDANT FOSSIL FUEL
Coal seams are widespread in the United States.

Figure 16.8
MODERN COAL SWAMP
This is the Great Dismal Swamp in North Carolina, a modern example of a peat swamp. It has the necessary slow-flowing, nearly stagnant water and abundant vegetation. Given millions of years, the plants will accumulate and become buried beneath other sediments (sand, silt, clay). Under the right conditions, the plants will form coal and the sediments will form sandstone, siltstone, and shale.

**Figure 16.9**
**THE FIRST OIL WELL IN THE UNITED STATES**
This is a reconstruction of the original engine shed of the world's first commercially productive oil well, drilled by Edwin Drake in 1859. It is shown at Drake's Well Museum in Titusville, Pennsylvania.

**oil** The liquid form of petroleum, a naturally occurring substance composed mainly of hydrocarbon compounds.

**natural gas** A naturally occurring hydrocarbon (a form of petroleum) that is gaseous at ordinary pressures and temperatures.

? **What do petroleum and plastic have in common?**

Oil—the liquid form of petroleum—first came into widespread use in about 1847, when a merchant in Pittsburgh started bottling and selling oil from natural seeps to be used as a lubricant. Five years later a Canadian chemist discovered that heating and distilling oil yields *kerosene*, a liquid that could be used in lamps. Kerosene lamps quickly replaced the then widely used whale oil lamps and greatly reduced the use of candles in household lighting. Soon wells were being dug by hand near Oil Springs, Ontario, to produce oil. In 1856, using the same hand-digging process, workers in Romania were producing 2000 barrels a year. In 1858, the first well drilled for oil was put down in southern Ontario. The following year, on August 27, 1859, in the first successful well drilled for oil, at Titusville, Pennsylvania, oil-bearing strata were encountered at a depth of 21 m (70 ft). Soon up to 35 barrels of oil[2] were being pumped each day (Fig. 16.9).

The modern use of **natural gas** (naturally occurring hydrocarbons that are gaseous at ordinary temperatures and pressures) started with an accidental discovery at Fredonia, New York, in 1821. A water well drilled in that year produced not only water but also bubbles of a mysterious gas. The gas was accidentally ignited and produced a spectacular flame. A new well was drilled on the same site, and wooden pipes were installed to carry the gas to a nearby hotel where it was used in 66 gaslights. By 1872 natural gas was being piped as far as 40 km (25 mi) from its source.

Today petroleum products are used for a wide variety of purposes in addition to their main use as fuel for heating and operating vehicles and machines. Different components of petroleum are separated and used in fertilizers, lubricants, asphalt, and in the manufacture of an array of synthetic materials including plastics and synthetic fabrics.

---

[2] A barrel is equivalent to 42 U.S. gallons of oil, about 160 liters.

Deposits of petroleum are nearly always found in association with sedimentary rocks that formed in a marine (ocean) environment. When marine microorganisms die, their remains settle to the bottom and collect in the fine seafloor mud, where they start to decay. The decay process quickly uses up any oxygen that is present. The remaining organic material is then preserved and covered with more layers of mud and decaying organisms. Throughout the Phanerozoic Eon, there have been many marine basins where muddy sediments with partially decayed organic material have been subjected to increasing heat and pressure with continued burial. Heat and pressure initiate a series of complex physical and chemical changes, called *maturation,* that break down the organic material and turn it into liquid and gaseous hydrocarbon compounds. Meanwhile, the muddy sediment itself is turned into a shale. A shale in which organic material has been converted into oil and natural gas is referred to as a *source rock.*

Long ago geologists realized that oil and gas are formed in one rock formation and travel, or *migrate,* to another rock formation at a later time. The gas and small droplets of oil in the source rock are eventually squeezed out by the weight of overlying rocks and, being light, slowly migrate toward the surface. For accumulation to occur, however, the migrating oil and gas must encounter a rock formation with lots of open pore spaces through which they can flow. Thus, oil accumulation requires rock with a high proportion of large interconnected pore spaces, called a *reservoir rock.* The rocks with the most suitable pore spaces are sandstones and limestones, and therefore these are the most common types of reservoir rocks.

Most of the petroleum formed in sediments eventually makes its way to the surface. It is estimated that no more than 0.1 percent of all the organic material that was originally buried in sediments is eventually trapped in reservoir rocks. It is not surprising, therefore, that the greatest proportion of oil and gas pools are found in rock that is no more than 2.5 million years old. This does not mean that older rocks produced less petroleum, just that oil in older rocks has had more time in which to escape. Sometimes an impermeable rock gets in the way of the migrating oil and gas and prevents it from going any farther. Such a formation is called a *cap rock.* The most common type of cap rock is shale.

A geologic situation that includes a source rock that contributes organic material, a reservoir rock that allows for the accumulation of oil, and a cap rock that stops the migration is referred to as a **petroleum trap** (or a *hydrocarbon trap*) (Fig. 16.10). Both water and natural gas are associated with oil in petroleum traps. All three fluids migrate out of the source rock and into the reservoir rock. Because water is the densest of the three fluids, it ends up on the bottom, with the oil in the middle and the gas, which is lightest, on top.

## Tar Sands and Oil Shales

Oil that is exceedingly viscous, will not flow easily, and cannot be pumped is called *tar.* **Tar sands** are sands in which all the pores are filled by dense, viscous, asphalt-like oil. They are found not only in sand but also in a variety of sedimentary rocks and unconsolidated sediments.

The largest known occurrence of tar sands is in Alberta, Canada, where the Athabasca tar sand covers about 5000 km² (about 2000 mi²) (Fig. 16.11). It contains as much as 600 billion barrels of tar. Similar deposits almost as large have been found in Venezuela and in the former Soviet Union.

Another potential source of petroleum is a waxlike organic substance called *kerogen,* which is found in certain very fine-grained sedimentary rocks such as shales. If burial temperatures are not high enough to form oil and natural gas, kerogen will be formed instead. If the kerogen is mined and heated, it breaks

> **?** **Why is oil mostly found in rocks that are geologically "young"?**

**petroleum trap** A source rock to contribute organic material, a reservoir rock to allow for the accumulation of oil, and a cap rock to stop migration.

**tar sand** A sediment or sedimentary rock in which the pores are filled by dense, viscous, asphalt-like oil.

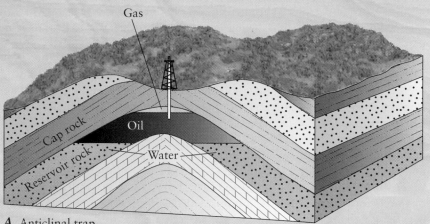

**A.** Anticlinal trap

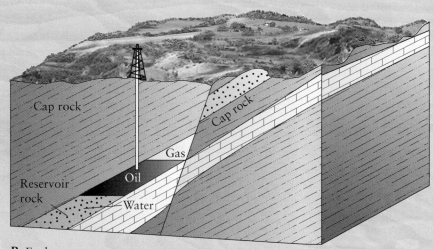

**B.** Fault trap

**Figure 16.10**
A petroleum trap requires the presence of a source rock, reservoir rock to contain the oil or gas, and cap rock. Common types of petroleum traps are formed by *(A)* anticlinal folds; *(B)* faults; *(C)* unconformities; *(D)* stratigraphic thinning and pinch-out of a reservoir rock; *(E)* salt domes distorting strata; and *(F)* fossil coral reefs.

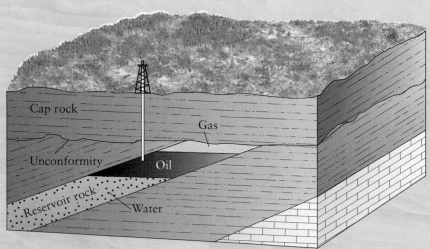

**C.** Stratigraphic trap

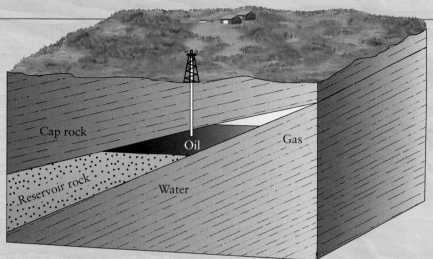

**D.** Pinch-out

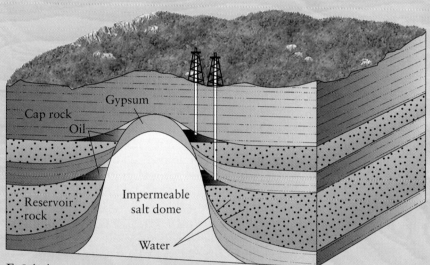

**E.** Salt dome

**F.** Reef

481

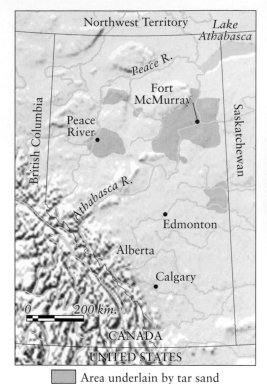

Area underlain by tar sand

*A.*

*B.*

**Figure 16.11**
**COOKING TAR FROM A SANDSTONE**
*A.* Part of the province of Alberta, Canada, is underlain by the Athabasca tar sand. *B.* The tar-cemented Athabasca sandstone is mined and then heated by hot water and steam in order to soften the tar and break the bond between the tar and sand. The photograph shows the rotating conditioning drums in which tar-sand, hot water, and steam are mixed.

**oil shale** A fine-grained sedimentary rock with a high content of kerogen, a waxy organic substance.

down and forms oil and gas. To be considered an energy resource, the kerogen in an **oil shale** must yield more energy than the amount needed to mine and heat it.

The world's largest deposit of rich oil shale formed millions of years ago, during the Eocene Epoch, in large, shallow lakes in Colorado, Wyoming, and Utah. In three of them a series of organic-rich sediments was deposited, eventually forming the Green River oil shales (Fig. 16.12). The U.S. Geological Survey

**Figure 16.12**
**OIL SHALE IN COLORADO**
The Green River oil shale contains the equivalent of 2000 billion barrels of recoverable oil, it is estimated. Like tar sands, the shale is mined and "cooked" to recover crude oil. The photo shows a Colorado canyon where stream erosion has exposed beds of oil shale.

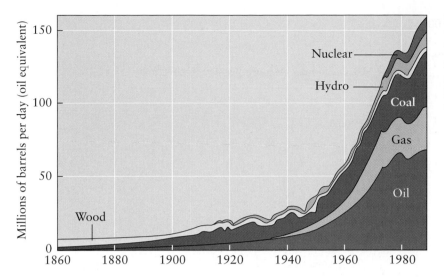

Figure 16.13
WORLD ENERGY USE
The history of world energy consumption shows oil's rise to predominance in the current energy mix. Also note that all fuels have increased in consumption except fuelwood.

estimates that these shale deposits, if mined and processed, would yield about 2000 billion barrels of oil.

## Use of Fossil Fuels

Figure 16.13 shows the history of energy consumption and the changing character of the world's energy "mix," including the dramatic increase in demand for energy after World War II and the subsequent increase in dependence on oil. The shift from coal to oil was driven by new technologies, primarily the internal combustion engine, which required fuel in liquid form. The ability to discover new fossil fuel deposits to meet these increasing demands was greatly improved by the advent of new technologies for exploration, including seismic studies, satellite imagery, computer modeling, and digital recording methods.

Despite our successes in increasing supplies of fossil fuels, we cannot ignore the fact that fossil fuels are nonrenewable resources. What happens when we reach the limit? Figure 16.14 presents projections of world energy production

**?** **Will there always be enough energy to meet the world's needs?**

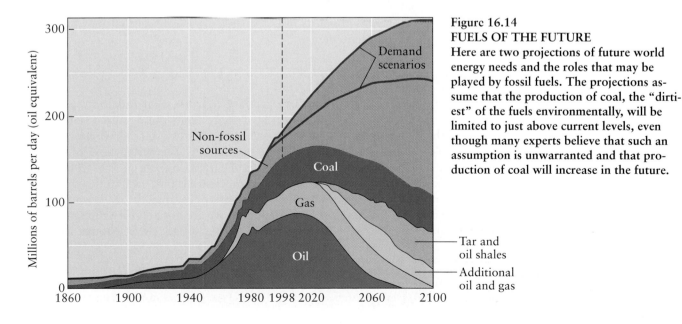

Figure 16.14
FUELS OF THE FUTURE
Here are two projections of future world energy needs and the roles that may be played by fossil fuels. The projections assume that the production of coal, the "dirtiest" of the fuels environmentally, will be limited to just above current levels, even though many experts believe that such an assumption is unwarranted and that production of coal will increase in the future.

**Figure 16.15**
**HOW MUCH FOSSIL FUEL?**
This diagram shows the estimated amounts of fossil fuel still in the ground and the amounts considered to be recoverable. Flowing oil means oil that flows into a well and can be easily pumped out. In all oil wells, half or more of the oil remains trapped in small pores and cannot be pumped out. Note that the unit used is a barrel of oil, or oil equivalent. For comparison, 0.22 ton of coal = 1 barrel of oil.

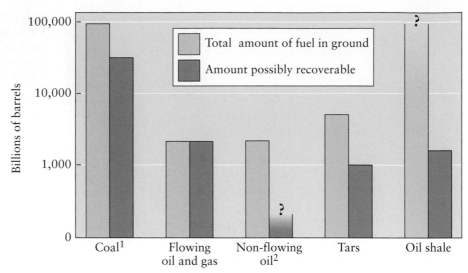

[1] The unit of comparison is a barrel of oil: 0.22 ton coal = 1 barrel of oil.
[2] Half or more of the oil in a producing oil field remains trapped in small pores and cannot be pumped out. It is not known how much, if any of this non-flowing oil can be recovered.

into the twenty-first century. A range of possible scenarios is shown. It is assumed that the demand for energy will eventually level off as a result of increased efficiency and conservation. The projections also assume an increased contribution from so-called *unconventional* oil and gas reserves, including tar sands and oil shales. Coal, considered the "dirtiest" and therefore least environmentally acceptable of the fossil fuels, is limited to just over current levels of production in these projections. To fulfill these demand scenarios, it will be necessary to place much greater emphasis on other sources of energy besides fossil fuels.

Realistically, it is highly unlikely that the demand for fossil fuels will decrease substantially or that an alternative fuel source will supplant the need for oil. Are supplies of fossil fuels adequate to meet future demands? If we use a barrel of oil as our unit of measurement, we can compare quantities of all fossil fuels (Fig. 16.15). Comparing the estimated amounts of fossil fuels remaining, it is apparent that coal is the most abundant and accessible fossil fuel and that despite environmental concerns about its use, it is the only fossil fuel that may have the capacity to meet long-term demand.

## Other Energy Sources

The Earth's energy comes from three primary sources: solar radiation, geothermal heat, and tidal energy (Fig. 16.16). Energy from these sources circulates through the many pathways and reservoirs of the Earth system, driving processes such as photosynthesis and atmospheric circulation. The actual flow of energy across the Earth's surface, all of which could be used (in theory), is 174,000 terawatts.[3] (A terawatt is a trillion watts.) We noted earlier that each of the world's 6 billion people uses energy at an average rate of approximately 1600 watts for a total rate of 9.6 terawatts. It is clear, then, that we are not about to run out of energy in an absolute sense. The question is whether we are clever enough to learn

---

[3] Remember that a watt is a unit that measures the rate at which energy is used. To get an amount of energy, multiply the use rate in watts by the time. For example, the unit used on bills from local power companies is the kilowatt hour, meaning the use of energy at rate of 1000 watts for 1 hour.

how to use alternative sources of energy rather than relying so heavily on nonrenewable fossil fuels.

The major alternative energy sources are solar, biomass (fuel wood and animal waste), wind, wave, tidal, hydroelectric, nuclear, and geothermal power. Some of these (especially biomass) are renewable energy sources. Others (especially nuclear power) are technically nonrenewable, but the amount of energy available is so large that they are in effect inexhaustible. Tidal energy, solar energy, and other energy sources that derive their energy directly or indirectly from the Sun (such as wind and waves) will not be exhausted as long as the Sun–Moon–Earth system continues to exist.

## Sun, Wind, and Water Power

*Solar energy* reaches the Earth from the Sun at a rate more than ten thousand times greater than that at which humans use energy from all sources. We already put some of the Sun's energy to work in greenhouses and solar homes, but the total amount used in these ways is small. Direct solar energy is best suited to supplying heat at or below the boiling point of water, which makes it especially useful for such applications as home heating and heating of water for home use. Converting solar energy directly into electricity is a major technological challenge. It can be done using photovoltaic cells, which have been around for decades. So far the costs of these devices are too high and their efficiency is too low for most uses, although they are widely used in small calculators, radios, and the like. Photovoltaic technology is constantly improving, however, and the cost of energy generated in this manner is decreasing; eventually, this technology may be low enough in cost for widespread use.

**?** **In how many ways does solar energy work in the Earth system?**

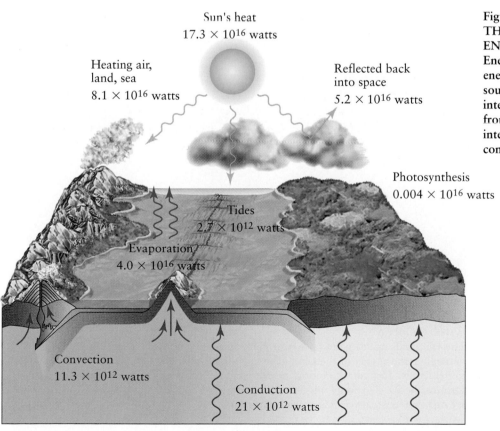

Sun's heat
$17.3 \times 10^{16}$ watts

Heating air,
land, sea
$8.1 \times 10^{16}$ watts

Reflected back
into space
$5.2 \times 10^{16}$ watts

Photosynthesis
$0.004 \times 10^{16}$ watts

Tides
$2.7 \times 10^{12}$ watts

Evaporation
$4.0 \times 10^{16}$ watts

Convection
$11.3 \times 10^{12}$ watts

Conduction
$21 \times 10^{12}$ watts

**Figure 16.16**
**THE EARTH'S ENERGY BUDGET**
Energy comes into the Earth's energy budget from three main sources: external (solar radiation), internal (geothermal energy), and from Earth–Moon–Sun tidal interactions, a very small component.

**biomass energy** Any form of energy that is derived more or less directly from plant life, including fuel wood, peat, animal dung, and agricultural wastes.

**?** **Can wind and water ever meet our energy needs?**

Solar energy is sometimes stored for long periods in the form of fossil fuels. But living plant matter also contains stored solar energy. Any form of energy that is derived more or less directly from the Earth's plant life is **biomass energy.** Biomass in the form of fuel wood was the dominant source of energy until the end of the nineteenth century, when it was displaced by coal. Biomass fuels—primarily fuel wood, peat, animal dung, and agricultural wastes—are still widely used throughout the world. Use of wood as a fuel is greatest in less developed countries, where the cost of fossil fuel is very high in relation to income.

Winds, as we learned in chapter 13, are an indirect form of solar energy. For thousands of years wind power has been used as a source of power for ships and windmills. Today, huge windmill "farms" are being erected in particularly windy places (Fig. 16.17). In Denmark, about 3000 wind turbines supply electricity throughout this breezy coastal country. Although there are some problems with windmill technologies, it seems very likely that windmills will soon be cost competitive with coal-burning electrical power plants. Unfortunately, much wind energy is contained in very high-altitude winds. Steady surface winds can provide only about 10 percent of the amount of energy now used by humans. Therefore, wind power may be locally important but probably will not become globally significant.

Waves created by winds blowing over the ocean are another indirect form of solar energy. As you will see if you watch any coastline during a storm, waves contain an enormous amount of energy. For centuries wave power has been used to ring bells and blow whistles that serve as navigational aids, but so far no one has discovered how to tap waves as a source of power on a large scale.

Tides are another water-related energy source. Their energy comes from the Earth's rotation and from gravitational interactions with the Moon and the Sun.

**Figure 16.17**
**WIND ENERGY**
The windmills at this wind farm near Palm Springs, California, generate electricity by harnessing the energy of the wind. The fan blades resist air flow, so they turn in reaction to the winds. This rotary motion in each windmill turns an electrical generator, transforming the wind's energy into "clean" (pollution-free) electricity. Wind farms can operate only where steady winds prevail year-round.

As discussed in chapter 13, this gravitational pull raises tidal bulges in the ocean, one on either side of the Earth. As the Earth rotates, each coastline "runs into" these bulges, experiencing two high tides and two low tides every day. If a dam is constructed across the mouth of a bay so that water is trapped at high tide, the water can be released at low tide and used to drive a turbine. The use of tidal energy is not entirely new; a site in Britain dating from A.D. 1170 still uses tidal power to run a mill. In many places around the world, the difference in water height between high and low tide makes it feasible to develop tidal power, but, like wind power, tidal power can satisfy only a small fraction of human energy needs and thus can never be important beyond a local scale.

**Hydroelectric energy** is the only form of water-derived power that currently meets a significant portion of the world's energy needs. Hydroelectric power is generated from the energy of a flowing stream of water; it is primarily gravitational energy. In order to convert the power of flowing water into electricity, it is necessary to build dams (Fig. 16.18). The flowing water is used to run turbines, which convert the energy into electricity. The total recoverable energy from the water flowing in all of the world's streams is estimated to be equivalent to the energy obtained by burning 15 billion barrels of oil per year. Thus, even if all the potential hydropower in the world were developed, it could not meet all of today's energy needs. Another problem is that reservoirs eventually fill up with silt, so even though water power is inexhaustible, dams and reservoirs have limited lifetimes. We have to conclude that hydroelectric power is very important for countries with large rivers and suitable dam sites, such as Canada, but it has limited potential for global development.

> **hydroelectric energy** The energy from a flowing stream of water, used to run turbines and generate electricity.

## Nuclear Power

**Nuclear energy** is the heat energy produced by controlled *nuclear reactions;* that is, reactions in which atoms of one species of chemical element are transformed into atoms of another species by nuclear change. Nuclear energy can be generated in two ways: by splitting heavy atoms into lighter atoms, a process called *fission,* or by combining two light atoms to make a heavier atom, a process called *fusion.*

> **nuclear energy** The heat energy produced by controlled nuclear reactions.

**Figure 16.18**
**HYDROPOWER**
The Hoover Dam on the Colorado River in Arizona and Nevada impounds water, creating Lake Mead. Water from the lake is released through the hydroelectric power station (lower left, below the dam), turning turbines that generate electricity. As long as the Colorado flows and the dam does not silt severely, power will continue to be generated here.

Some atoms are naturally unstable and decay by giving off nuclear particles and energy. This, as we discussed in chapter 3, is called radioactivity, and the energy given off by natural radioactivity is the main source of the Earth's internal heat. Three of the species of radioactive atoms that keep the Earth hot are uranium–235, uranium–238, and thorium–232. They can be mined and used to obtain nuclear energy by fission.

When a fissionable atom is hit by a neutron, it not only releases heat and fragments that form new, lighter elements, but it also ejects some neutrons from its nucleus. If these neutrons strike other fissionable atoms, they split too, creat-

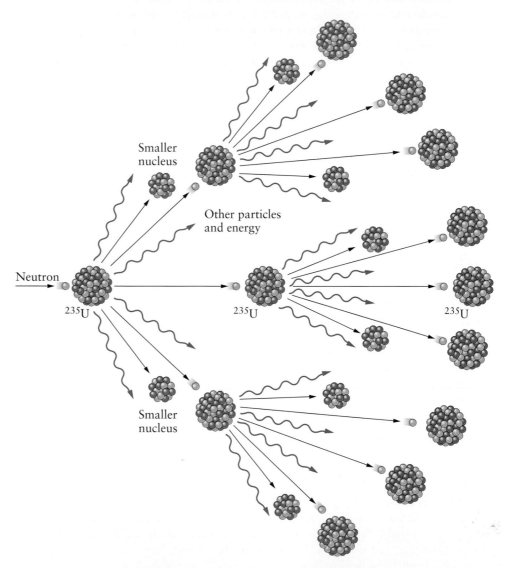

**Figure 16.19**
**ATOMIC ENERGY**
When the nucleus of a uranium–235 atom is hit by a neutron (left), the atom splits (fissions) into two smaller, lighter atoms of different elements, giving off a great deal of heat energy in the process. The fission process also releases other neutrons (black arrows) which, in turn, hit other uranium 235 atoms, making them split in the *chain reaction* shown here. This process releases tremendous heat energy, which is used to generate steam, which turns turbines that generate electricity. This is the basic operating principle of a nuclear power plant.

ing a continuous *chain reaction* (Fig. 16.19). When such a chain reaction proceeds without control, an atomic explosion occurs. But if the reaction is controlled, the heat can be harnessed to do helpful work, like generating electricity. The fissioning of just 1 gram of uranium–235 produces as much heat as the burning of 13.7 barrels of oil. There is enough recoverable uranium and thorium in the crust so that nuclear energy is essentially limitless.

Nuclear power plants use the heat energy from fission to produce steam that can drive turbines and generate electricity. Approximately 17 percent of the world's electricity is derived from nuclear power plants. In France, more than half of all the electrical power comes from nuclear plants; the proportion is rising sharply in some other European countries and in Japan. The reason for the increase is that Japan and most European countries do not have enough fossil fuel supplies to be self-sufficient. Nuclear power is considered—even by many environmentalists—to be a clean source of energy because it does not produce harmful emissions under normal conditions, although accidents can release toxic gases, as happened at the Chernobyl disaster in the former Soviet Union. The main problem plaguing the nuclear power industry is that fission generates highly radioactive waste that must be isolated from the biosphere and the hydrosphere for many thousands of years. This presents a serious and complex disposal problem that has not been resolved and probably won't be for many years.

In principle, nuclear fusion—the joining together or fusing of two small atoms to create a single larger atom, with an attendant release of heat energy—is another potential source of nuclear power. Nuclear fusion is fueled by *deuterium*, a heavy isotope of hydrogen, of which the Earth has a virtually endless supply in the form of water. The primary byproduct of nuclear fusion would be helium, a nontoxic, chemically inert gas.

All of this conjures up images of a cheap, clean power source with a virtually endless fuel supply. So why are we not using nuclear fusion to produce energy? Fusion is the nuclear process that occurs in the cores of stars, the process responsible for the tremendous amount of heat energy generated by the Sun. But that is the problem. For nuclei to fuse, the temperature must be similar to that in the central part of a star—millions of degrees. In 1989 a furor arose when two researchers announced that they had induced nuclei to fuse at room temperature, a process called *cold fusion*. Controlled fusion at room temperature would probably solve all our energy needs forever, but unfortunately, the work of those researchers could not be replicated in other laboratories and cold fusion seems to be a utopian dream. Routine use of fusion power remains a distant goal.

**? Why did the claim to have harnessed "cold fusion" create a lot of excitement?**

### Geothermal Power

The Earth's internal energy, **geothermal energy** (Fig. 16.20), has been utilized for more than 50 years where it is easily tapped, in New Zealand, Italy, Iceland, and more recently in the United States. The people of Iceland, for example, have found many ways to harness geothermal heat to combat their country's cold climate. Using water warmed by hot volcanic rocks, they heat their houses, grow tomatoes in hothouses, and swim year-round in naturally heated pools. Icelanders also use volcanically produced steam to generate most of their electricity.

*Hydrothermal reservoirs* are underground systems of hot water or steam that circulate in fractured or porous rocks. These reservoirs are near the surface because the sources of heat are shallow magma chambers or thick layers of recently erupted volcanic rocks. To be used efficiently, hydrothermal reservoirs should be 200°C (about 400°F) or hotter, and this temperature must be reached within 3 km (2 mi) of the surface. Therefore, most of the world's hydrothermal reservoirs

**geothermal energy** The Earth's internal energy.

Figure 16.20
**HEAT ENERGY FROM THE GROUND**
Steam in rock fractures flows up a well to a power plant where it spins turbines, generating electricity without polluting the atmosphere. After use, waste water and any remaining steam are pumped back underground again.

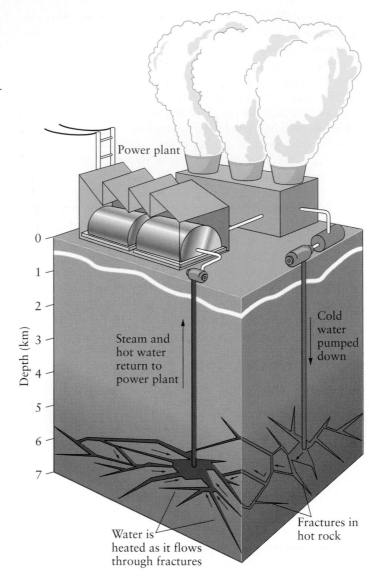

Power plant

Depth (km)

0
1
2
3
4
5
6
7

Steam and hot water return to power plant

Cold water pumped down

Fractures in hot rock

Water is heated as it flows through fractures

are close to the margins of tectonic plates, where volcanic activity has occurred recently and where hot rocks or magma can be found close to the surface. The Geysers geothermal installation in northern California—the largest producer of geothermal power in the world—is a hydrothermal reservoir system

Another type of geothermal deposit is *hot, dry rocks*. In experiments carried out in the Jemez Mountains in New Mexico, scientists drilled deep holes into the hot rocks near the edge of an extinct (but still hot) volcano. They shattered the hot rock with explosives, creating an artificial reservoir, and pumped water through it to produce steam that could run turbines. A major difficulty with these tests was that water did not flow uniformly through the hot rock. Instead, it mostly followed the larger cracks (through which it could flow more easily), and the rocks that lined them soon cooled. Hot, dry rocks are much more abundant than hydrothermal reservoirs; if the problems encountered in the Jemez Mountains experiments could be overcome, geothermal energy might someday play a major role in meeting human energy needs.

# MINERAL RESOURCES

The number and diversity of minerals and rocks that provide materials used by humans are so great that it is almost impossible to classify them. Nearly every kind of common rock and mineral can be used for something (Table 16.1), but the economic reality is that we mine about 200 different minerals and about a dozen types of rock. Those that are most valuable also tend to be rare. However, it is convenient to group mineral resources according to how they are used (Table 16.2). There are two broad groups of mineral resources: *metallic* and *nonmetallic*. Metallic minerals are mined specifically for the metals that can be extracted from them. Examples are zinc, a valuable plating metal that we refine from the mineral *sphalerite* (ZnS), and lead, used in automobile batteries, which we refine from the mineral *galena* (PbS). Nonmetallic minerals are mined for their properties as minerals, not for the metals they contain. Examples are salt, gypsum, and clay.

The branch of geology that is concerned with discovering new supplies of useful minerals is *economic geology*. A *mineral deposit* is a local concentration of

**TABLE 16.1    Some Mineral Products Used in Everyday Life**

| | |
|---|---|
| Building materials | Sand, gravel, stone, brick (clay), cement, steel, tar (asphalt) |
| Plumbing and wiring | Iron and steel, copper, lead, tin, cement, asbestos, glass, tile (clay), plastic |
| Paint | Mineral pigments (e.g., iron, zinc, titanium), and fillers (e.g., talc, mica, asbestos) |
| Appliances | Iron, copper, many rare metals |
| Food | Grown with mineral fertilizers: processed, packaged, and delivered by machines made of metal |

**TABLE 16.2    Nonrenewable Mineral Substances**

| *METALS* | |
|---|---|
| Abundant metals | Iron, aluminum, magnesium, manganese, titanium, silicon |
| Scarce and rare metals | Copper, lead, zinc, nickel, chromium, gold, silver, tin, tungsten, mercury, molybdenum, uranium, platinum, and many others |
| *NONMETALS* | |
| Used for chemicals | Sodium chloride (halite), sodium carbonate, borax, calcium fluoride (fluorite) |
| Used for fertilizers | Calcium phosphate (apatite), potassium chloride, sulfur, calcium carbonate (limestone), sodium nitrate |
| Used for building | Gypsum (for plaster), limestone (for cement), clay (for brick and tile), asbestos, sand, gravel, crushed rock, shale (for brickmaking and for cement) |
| Used for jewelry | Diamond, corundum (sapphire and ruby), garnet, amethyst (quartz), beryl (emerald), and many others |
| Used for ceramics | Clay, feldspar, quartz |
| Used for abrasives | Diamond, garnet, corundum, pumice, quartz |

**ore** A deposit from which one or more minerals can be extracted profitably.

? **What is the difference between a mineral deposit and an ore deposit?**

a given mineral. Economic geologists seek mineral deposits from which the desired minerals can be recovered least expensively. In distinguishing between profitable and unprofitable mineral deposits, we use the word **ore,** meaning a deposit from which one or more minerals can be extracted profitably. All ores are mineral deposits, but not all mineral deposits are ores. Whether or not a given mineral deposit is an ore is determined by how much it costs to extract the mineral and by how much people are prepared to pay for it—in other words, can it be mined profitably? Factors to consider are the level of concentration, or *grade,* of the desired substance, the size of the deposit, how deeply it is buried, ease of access to a road or railroad, the cost of the labor required to extract the substance, the demand for the commodity, the cost of meeting any environmental restrictions, and costs of reclamation after mining has ceased.

## Finding and Assessing Mineral Resources

Mineral resources are nonrenewable. Therefore, they can be exhausted by mining. Moreover, economically exploitable minerals are localized in distinct areas within the Earth's crust. The uneven distribution of exploitable deposits is the main reason that no nation, not even the United States or Russia, is self-sufficient in mineral supplies. For example, the United States has little or no aluminum, manganese, or vanadium deposits and must import most of what is used.

It is difficult to assess the quantity of a given material available in a country and even more difficult to anticipate whether new deposits will be discovered there. As a result, it is exceedingly difficult to predict production and supplies over a period of years. A country that can meet its needs for a given mineral substance today may find that it must import the substance in the future.

For example, England was once able to meet most of its own mineral needs, but today it can no longer do so. A little more than a century ago, England was a great mining nation, producing and exporting such materials as tin, copper, tungsten, lead, and iron. Today, the known deposits of those minerals have been exhausted. The pattern followed by England—a pattern of intensive mining followed by depletion of the resource, declining production and exports, and increasing dependence on imports—can be applied in any local or regional context to estimate the remaining effective lifetime of a given mine or mineral resource.

Over the past few decades there has been a slow but steady shift of mineral exploration and production away from the industrialized nations and toward the less developed parts of the world, such as New Guinea and Peru. In the industrialized world, the geologic locations that are most favorable for conventional mineral exploration have already been prospected, assessed, and in some cases mined and depleted. This doesn't mean that there are no more mineral deposits to be found; it just means that geologists will have to look harder, develop new exploration techniques, and learn to look for mineral deposits in unconventional locations.

## How Mineral Deposits Are Formed

To a certain extent, the distribution of mineral resources can be linked to specific tectonic environments. This permits economic geologists to identify locations where specific types of mineral deposits are most likely to be found (Fig. 16.21).

For a mineral deposit to have formed, one or more processes must have locally enriched the rock in one or more minerals. Minerals can become concentrated as a result of these six types of processes:

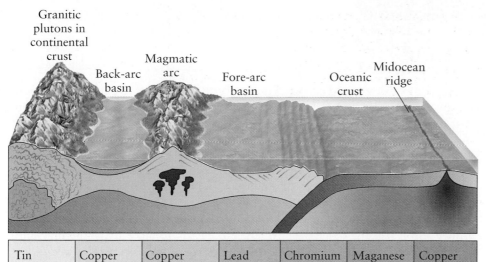

Figure 16.21
**PLATE TECTONICS AND MINERALIZATION**
Plate tectonic structures control where many minerals are concentrated into economic deposits.

| Tin Tungsten Bismuth Copper | Copper Zinc Gold Chromium | Copper Gold Silver Tin Lead Mercury Molybdenum | Lead Zinc Copper | Chromium | Maganese Cobalt Nickel | Copper Zinc |
|---|---|---|---|---|---|---|

1. Hot water solutions flowing through fractures and pores in rocks, which produces *hydrothermal mineral deposits.*

2. Metamorphic recrystallization, which produces *metamorphic mineral deposits.*

3. Crystallization within a body of magma, which produces *magmatic mineral deposits.*

4. Precipitation from lake water or seawater, which produces *sedimentary mineral deposits.*

5. The action of waves or currents in flowing surface water, which produces *placer deposits.*

6. Weathering processes, which produce *residual mineral deposits.*

We will look at each of these processes, with examples from different parts of the world.

*Hydrothermal Mineral Deposits* Many famous mines, such as those at Butte, Montana, and Grass Valley, California, contain ores that were formed when minerals were deposited from *hydrothermal solutions*—hot, water-rich fluids (*hydro* is the Greek word for water, and *thermal* is the Greek word for heat). Hydrothermal reservoirs of the kind used for geothermal energy, discussed earlier in this chapter, are one source of hydrothermal solutions but, as discussed in the *Geology Around Us* essay "Hydrothermal Mineral Deposits Forming Today," there are others. Hydrothermal solutions deposit their mineral loads in cracks, forming mineral *veins* (Fig. 16.22). Hydrothermal mineral deposits are the primary sources of many important metals that you use daily: copper (electric wiring), lead (car batteries), zinc (plating steel), tin (solder), molybdenum (toughening steel), tungsten (filaments in light bulbs), gold (jewelery), and silver (photographic film).

## Hydrothermal Mineral Deposits Forming Today

**Geologists wondered for more than a century** how to tell an ore-forming hydrothermal solution from a non-ore-forming one. Then, starting in 1962, a 16-year period of discoveries changed everyone's ideas about mineral deposits.

The first discovery was accidental. Drillers seeking oil and gas in the Imperial Valley of southern California were astonished in 1962 when their drills tapped a 320°C (610°F) brine at a depth of 1.5 km (about 0.9 mi). As the brine flowed upward, it cooled and precipitated minerals it had been carrying in solution. Over three months, the well deposited 8 tons of siliceous scale containing 20 percent copper and 8 percent silver. The drillers had clearly found a hydrothermal solution that could, under suitable flow conditions, form a rich mineral deposit.

The Imperial Valley is a sediment-filled graben covering the join between the Pacific and North American plates, where the mid-ocean ridge passes under North America (Fig. B16.1). Volcanism is the source of heat for the brine solution discovered in 1962. These brines provided the first unambiguous evidence that hydrothermal solutions can leach metals from ordinary sediments.

Almost before geologists had a chance to absorb the significance of the Imperial Valley discovery, a second remarkable find was announced. In 1964, oceanographers discovered a series of hot, dense brine pools at the bottom of the Red Sea. The brines are trapped in the graben formed over the spreading center between the Arabian and African plates (Fig. B16.2). They are so much more salty, and therefore more dense, than seawater, that they remain ponded in the graben even though they are as hot

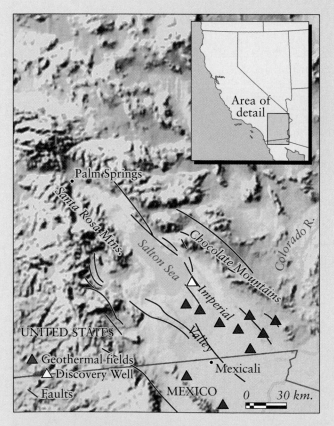

**Figure B16.1**
**THE FIRST DISCOVERY OF A HYDROTHERMAL SOLUTION**
The Imperial Valley of California is a graben flanked by the Chocolate Mountains on the east and the Santa Rosa Mountains on the west. A present-day, ore-depositing hydrothermal solution was discovered in a well drilled at the southern end of the Salton Sea. Places where hot spring activity is known and where other hydrothermal solutions may be present at depth are marked by triangles.

*Metamorphic Mineral Deposits* Hydrothermal solutions are sometimes associated with contact metamorphism. (Recall that this is the alteration of minerals and rocks that are adjacent to bodies of very hot rock or magma.) Regional metamorphism—the alteration of large areas of rock as a result of the high temperatures and pressures associated with the tectonic activity of mountain build-as 60°C (140°F). Many such brine pools have now been discovered on the floor of the Red Sea.

The Red Sea brines rise up the normal faults associated with the graben and, like the Imperial Valley brines, have evolved to their present compositions through reactions with the enclosing rocks. The Red Sea brine discovery was surprising, but even more surprising was the discovery that sediments at the bottom of the pools contained ore minerals such as chalcopyrite, galena, and sphalerite. In other words, the oceanographers had discovered a modern sedimentary mineral deposit in the process for formation.

The third remarkable discovery was really a series of discoveries that commenced in 1978. Scientists using deep-diving submarines made a series of dives on the mid-ocean ridge in the Pacific Ocean, at 21° N latitude. To their amazement, they found 300°C (570°F) hot springs emerging from the sea floor 2500 m (6200 ft) below sea level. Around the hot springs lay a blanket of sulfide minerals. The submariners watched a modern hydrothermal ore deposit forming before their eyes.

Each of the discovery sites—Imperial Valley, Red Sea, and 21° N—is on a spreading center, so there is no doubt that the deposits are forming as a result of magmatic activity associated with plate tectonics. Soon the hunt was on to see if seafloor deposits could be found associated with subduction zones. In 1989, a joint German-Japanese oceanographic expedition to the western Pacific discovered the first modern subduction-related deposits, and since then several other subduction-related modern ore deposits have been found. No longer are geologists limited to speculating about how hydrothermal mineral deposits *might* have formed. Today they can study them as they grow.

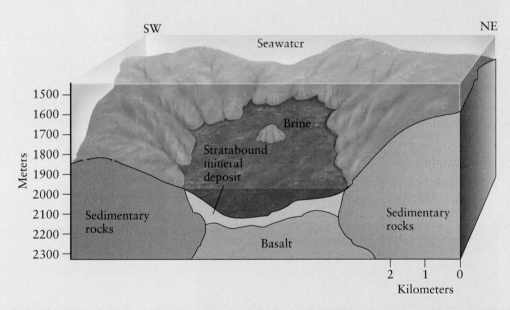

Figure B16.2
**HYDROTHERMAL SOLUTION IN THE RED SEA**
The topography at the bottom of the Red Sea in a region called the Atlantis II brine pool, after the oceanographic vessel from which the discovery was made. Hot, dense brines rise up normal faults that flank the Red Sea graben, forming pools of hot, dense hydrothermal solutions from which copper and zinc minerals precipitate.

ing—can also concentrate minerals and thus create metamorphic mineral deposits. Responding to stress during regional metamorphism, minerals in a rock may become concentrated and separated into distinct bands or layers (Fig. 16.23). Examples are micas, asbestos, graphite, and some gemstones. An important metamorphic process is recrystallization, which usually produces coarser-

Figure 16.22
GOLD
A geologist inspects a quartz vein that was deposited by a hydro-thermal solution and is being mined for the gold it contains at Mary Nevin Mine, Cripple Creek, Colorado.

grained, interlocking aggregates. For example, the pure, white marbles of Carrara, Italy, which are highly valued by sculptors, began as porous aggregates of sea shells and other forms of calcium carbonate. Through metamorphic recrystallization, all traces of the original shells were obliterated, and the minerals were transformed into seamless crystalline marbles composed entirely of interlocking grains of calcite.

Figure 16.23
ORE FORMED BY METAMORPHISM
Ore of the Tem-Piute Mine, Nevada. White is calcite, purple is fluorite. Ore minerals visible are sphalerite (brown, lower left), pyrite (gold) and scheelite ($CaWO_4$), the sugary pale brown mineral upper left and lower right. Scheelite is an important tungsten ore mineral.

*Magmatic Mineral Deposits*  How magmas can form valuable ores is interesting. When a large chamber of basaltic magma crystallizes, one of the first minerals formed is chromite. Chromite is the main ore mineral of chromium, used to make tough steel and to beautify car parts by plating. The chromite crystals, which are denser than the magma around them, settle to the bottom of the magma chamber, like particles of sediment settling in water (Fig. 16.24A). This process can produce almost pure layers of chromite (Fig. 16.24B).

Another product from magma is the diamond-rich rock *kimberlite* (mentioned in chapters 5 and 11). Kimberlites are long, pipelike bodies of igneous rock that formed from magma that originated deep in the mantle—150 km (about 100 mi) or more. How and why kimberlite magma is formed remains a puzzle, but it rises explosively upward, punching a circular hole in the crust and transporting broken fragments of mantle rock (called *xenoliths*) as it goes. One mineral brought up by kimberlites is diamond, a high-pressure carbon mineral that is formed only at depths greater than 150 km.

**Figure 16.24**
**A MAGMATIC ORE DEPOSIT**
*A.* Chromite, the main ore mineral of chromium, crystallizes from a magma and, because it is denser than the magma, sinks to the bottom and accumulates in a process called crystal settling. *B.* Layers of pure chromite (black) enclosed in layers of plagioclase, settled out during the crystallization of the Bushveld Igneous Complex. This unusually fine outcrop is located at Dwars River in South Africa.

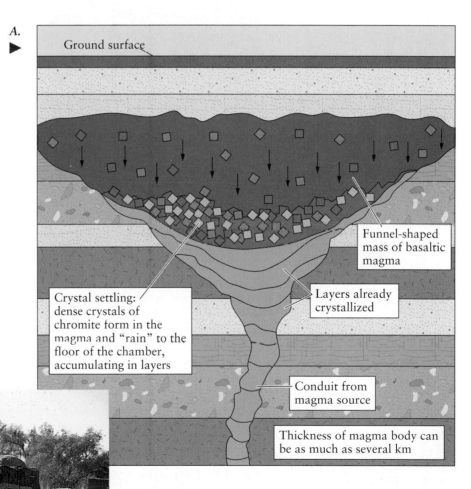

A.
Ground surface

Funnel-shaped mass of basaltic magma

Crystal settling: dense crystals of chromite form in the magma and "rain" to the floor of the chamber, accumulating in layers

Layers already crystallized

Conduit from magma source

Thickness of magma body can be as much as several km

B.

**Figure 16.25**
**DESERT SALTS**
Present-day evaporite salts encrust the floor of Death Valley in California. It happens like this: a shallow lake is created during rainy periods. The water evaporates and the lake dries up, forcing salts to crystallize out of the brine. Polygonal fractures are formed as the drying sediment contracts.

**?** **What kind of ore deposit do aluminum cans come from?**

*Sedimentary Mineral Deposits* Sedimentary mineral deposits are formed when substances in solution precipitate from lake or seawater. One cause of precipitation is evaporation, which leaves layers of salts in *evaporite deposits* (Fig. 16.25). Household products like sodium carbonate and borax come from deposits that were formed by the evaporation of lake waters. When seawater evaporates, it can leave behind gypsum, halite (salt), and a variety of potassium salts that are used in fertilizers.

Another cause of precipitation is biochemical reactions in seawater. These precipitate the mineral apatite, the main source of phosphate fertilizers. In sedimentary rocks more than about 2 billion years old, there are unusual iron-rich sediments called banded-iron formations. These are thought to be ancient biochemical precipitates from seawater.

*Placer Deposits* Moving water sorts light and heavy mineral grains. A heavy mineral deposit formed in this way is called a *placer*. The flowing water may be a stream or the longshore current in the ocean. The heavy minerals that typically become concentrated in this way are gold and platinum, diamonds, chromite, and minerals containing zirconium and titanium.

*Residual Mineral Deposits* Weathering occurs because newly exposed rock is not chemically stable when it comes into contact with rainwater and the atmosphere. Chemical weathering, especially in tropical climates, can concentrate minerals by removing the more soluble ones. The soluble minerals are dissolved and carried downward by the infiltrating water. This leaves behind a residual concentration of less soluble minerals. A common result is *laterite,* a hard, highly weathered soil that is rich in insoluble minerals. In laterite, less soluble iron minerals can be concentrated, turning lateritic soils a reddish-yellow color. They contain so much iron that some laterites are mined for iron and sometimes for nickel.

Figure 16.26
BAUXITE (ALUMINUM ORE)
Bauxite from Weipa, in Queensland, Australia, is the main ore of aluminum. It forms by leaching more soluble minerals from rocks near the surface under tropical conditions, leaving a residue of less soluble aluminum-rich, minerals behind. The photograph shows rounded masses of aluminum hydroxide (gibbsite) in a matrix of iron and aluminum hydroxides. The sample is about 10 cm (4 in) across.

Iron-rich laterite is by far the most common kind of residual mineral deposit. But the most important to people is *bauxite*, the source of aluminum ore. Bauxites are most commonly found in the tropics (Fig. 16.26). Where bauxites are found in temperate climates today— in France, China, Hungary, and Arkansas— we can infer that the climate was tropical millions of years ago when the bauxites formed. The explanation is that the bauxites were indeed formed in the tropics and that we find them in places with temperate climates today because plate tectonic motions have moved them there.

# WHAT'S AHEAD

Throughout this book we have learned about the Earth system and its many components. We have answered a lot of questions about how the Earth functions, but at the same time we have raised many new questions. There is still much to be learned about the Earth system and the interrelationships among its subsystems. We humans are affected by geologic processes every day. Some of these processes, such as mountain building and climatic change, are slow, while others, such as earthquakes and landslides, are very rapid. We need to understand how the Earth functioned in the past, so that we can figure out where we are going—where valuable mineral deposits may lie hidden, or how sea levels might change, or where earthquakes and volcanic eruptions will occur. We also must better understand how our actions will affect the functioning of the natural Earth system. In the final chapter we will look more closely at the role of geoscientists in the twenty-first century.

## Chapter Highlights

**1. Earth resources** can be either **renewable** on a human time scale, or **nonrenewable**. Most of the mineral and energy sources on which humans rely are nonrenewable. Exponential growth of the human population is now increasing by 100 million people each year.

**2. Fossil fuels** are the organic remains of plants and animals trapped in sediments and sedimentary rocks. They derived their energy from the Sun via the process of photosynthesis. The kind of fuel that formed depended on the kind of sediment, the kind of organic remains, and the changes that occurred in the organic matter after burial.

**3. Peat,** the accumulated remains of plants in swamps and bogs, is the first stage in the formation of **coal.** There have been two great periods of coal formation in the Earth's history, one during the Carboniferous and Permian Periods and the other in the Cretaceous Period.

**4. Petroleum** is formed from the organic matter of microscopic plants and bacteria trapped in marine shales. Heat converted the organic matter into **oil** and **natural gas,** the main forms of petroleum. Once formed, petroleum was squeezed out of a source rock and migrated through porous reservoir rocks such as sandstones and limestones, where it may have been caught by cap rocks in petroleum traps of various kinds.

**5.** Tar is oil that is so viscous that it does not flow. **Tar sands** can be mined in order to recover the tar. The solid organic matter in **oil shales,** called kerogen, can be heated and converted into oil and gas.

**6.** Solar, wind, wave, and tidal power are renewable energy sources but are all of limited usefulness for a variety of technical reasons. **Biomass energy** is derived more or less directly from the Earth's plant life. **Hydroelectric energy** is generated from the energy of running water. It is an important renewable source of energy, but it is insufficient to meet all the current needs of human societies.

**7. Nuclear energy** is not renewable but is so vast that it is essentially limitless. Fission energy is derived from the controlled splitting of heavy radioactive isotopes of uranium and thorium into two or more lighter isotopes. Fusion energy comes from the joining of two light isotopes to form a heavier one. Controlled fission has been attained, but controlled fusion has not.

**8. Geothermal energy** comes from the Earth's internal heat. It is important only in places where reservoirs of hot water or steam can be found at shallow depth, principally in volcanic regions.

**9.** Mineral resources are nonrenewable and are unevenly distributed as a consequence of the rock cycle and plate tectonics. Exploration geologists seek useful minerals in the form of **ores,** deposits from which one or more minerals can be extracted profitably.

**10.** There are six types of mineral deposits based on the geologic processes that concentrate the minerals: hydrothermal, metamorphic, magmatic, sedimentary, placer, and residual deposits.

## ▶ The Language of Geology

biomass energy 486
coal 474
Earth resource 470
fossil fuel 473
geothermal energy 489
hydroelectric energy 487

natural gas 478
nonrenewable resource 470
nuclear energy 487
oil 478
oil shale 482
ore 492

peat 474
petroleum 476
petroleum trap 479
renewable resource 470
tar sand 479

## ► Questions for Review

1. What is the difference between renewable and nonrenewable resources?
2. Which of the following resources are renewable and which ones are nonrenewable: soil, rainwater, oil, forest timber, gold, gypsum?
3. What is photosynthesis, and what role does it play in the formation of fossil fuels?
4. Describe the changes that occur as peat is transformed into bituminous coal.
5. Since land plants evolved about 450 million years ago, there have been two great periods of coal formation. When were they, and what hypothesis can you offer to explain why so much coal was formed in these two periods?

6. What steps are necessary for the formation of petroleum?
7. Name five different renewable energy sources. Why are they not major sources of power today?
8. There are two ways of generating nuclear power; what are they?
9. What are the six processes by which minerals become concentrated into mineral deposits?
10. What processes are thought to have formed deposits of each of the following: copper, salt, chromium, bauxite, apatite?

## ► Questions for Thought and Discussion

1. Oil production in the United States can satisfy only half of the country's needs; the rest is imported. If imports were cut off, what changes would you expect to occur in your lifestyle?
2. In attempting to seek energy from sources other than oil or gas, what might be the best options for the United States? Would the same options be valid for Canada? for Japan?
3. Many hydrothermal mineral deposits of copper, gold, silver, and other metals have been found in the countries bordering on the Pacific Ocean. Can you offer an explanation for this remarkable concentration? If you were part of a team of exploration geologists looking for large copper deposits, where would you focus your search?
4. Given that we are now dependent on nonrenewable sources of energy and minerals and that the global population continues to increase in number, how do you think the world can or will adjust in the future?

For an interactive case study on energy resources and economic geology, visit our Web site.

# · 17 ·

# THE ROLE OF GEOSCIENTISTS IN THE 21ST CENTURY

The idea that the Earth is an integrated set of systems is not new. In 1863, the great mineralogist James Dana wrote in his *Manual of Geology*:

> The Earth ... has not only its comprehensive system of growth, in which strata have been added to strata, continents and seas defined, mountains reared, and valleys, rivers, and plains formed, all in orderly plan, but also a system of currents in its oceans and atmosphere, the Earth's circulating system; its equally worldwide system in the distribution of heat, light, moisture, and magnetism, plants and animals; its system of secular variations ... in climate and meteorological phenomena. In these characteristics the sphere before us is an individual, as much so as a crystal or a tree. . . .

Scientists have returned to this concept of the Earth as an integrated "individual" because it is an approach that makes sense in today's interconnected world. Still, much has changed since Dana's time. Throughout the 1800s and much of the 1900s, Earth scientists worked to build a comprehensive understanding of the Earth through painstaking collection and analysis of information about each distinct part of the system. This type of intensely focused analysis is still important, but now Earth scientists have tools that permit them to study the Earth system as an integrated whole.

One of the best-known of these new tools is a group of instrument-bearing satellites known as the Earth Observing System, or EOS, which were designed to study the atmosphere-ocean-land system from space. Using the instruments aboard the EOS satellites, NASA scientists will provide information concerning climatic change; atmospheric chemistry, including ozone; natural hazards, such as earthquakes, volcanoes, floods, and hurricanes; and changes in land use and land cover. EOS is a good example of how Earth science has changed and will continue to change in the 21st century.

◆

## In this chapter you will learn about

- The types of jobs held by geoscientists today
- The role of geoscientists in furthering our understanding of the Earth system
- The role of geoscientists in reducing the risks from natural hazards
- The role of geoscientists in the sustainable management of resources

*EOS: Earth Observing System*

NASA's Earth Observing System (EOS) satellites were designed to facilitate the study of the atmosphere-ocean-land system from space.

# In our rapidly changing world...

### How will geoscientists help shape the twenty-first century?

*... We discuss this question in this final chapter of the book.*

With this chapter we come to the end of our exploration of geology and the Earth system. We hope you have gained a basic understanding of how the Earth works and how geologic processes and cycles affect each of us every day. In Part One of the book we introduced you to a three-ringed diagram illustrating these cycles and their interconnections. Figure 1.19 had a lot of blank spaces; we used the information in subsequent chapters to fill those spaces little by little. In Part

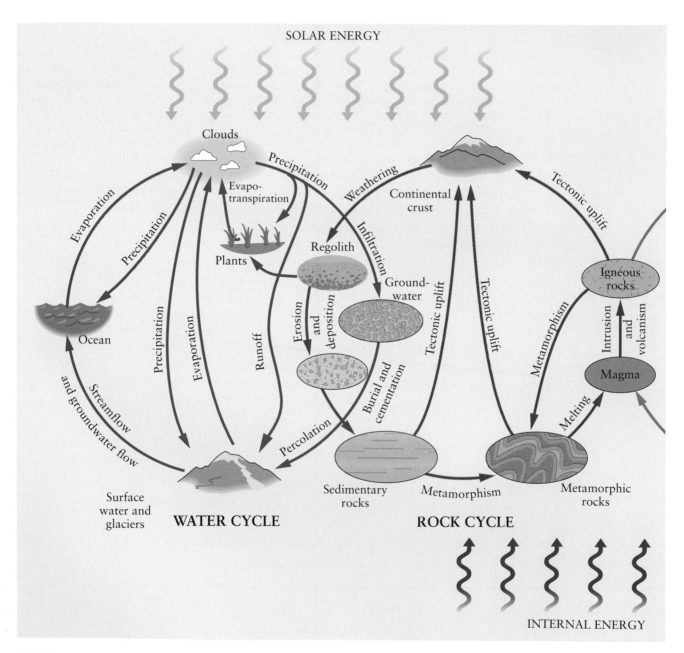

Two, we explored the internal Earth processes that characterize the tectonic cycle (Fig. 4.20). The chapters of Part Three focused on the center ring of the diagram, which illustrates the crustal processes of the rock cycle (Fig. 7.1). Finally, in Part Four, we filled in the blanks remaining in the lefthand ring diagram, the water cycle (Fig. 12.1), which operates in the dynamic environment of the Earth's surface and near-surface. Now we return, one last time, to our three-ringed diagram (Fig. 17.1). Study the diagram for a moment; can you see how each cycle is distinct, yet interconnected with the others? We live on the surface of the Earth, yet we are fundamentally influenced by processes in each of the three cycles.

Through these explorations of Earth processes and materials, we hope that the preceding chapters also have given you an idea of the scope of work done by

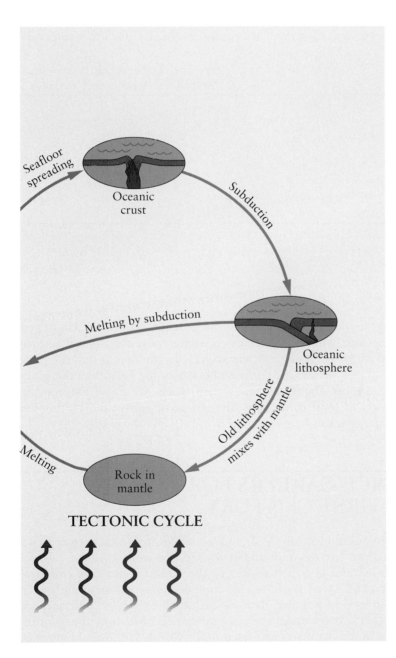

**Figure 17.1**
**THE EARTH SYSTEM**
Here is the Earth system diagram we first introduced in chapter 1. It consists of the three interconnected cycles that comprise the Earth system: the tectonic cycle, the rock cycle, and the hydrologic cycle. We have now filled in the blank spaces on the diagram with the names of all the various geologic processes that we have studied throughout the book.

**TABLE 17.1    Employment of Geoscientists**

A. This table shows what percentage of geoscientists is employed by each sector of the economy.

| Sector | Percentage of Geoscientists Employed in Each Sector |
| --- | --- |
| 4-yr. college or university | 20.2 |
| Other educational institutions | 2.8 |
| Business/industry | 46.7 |
| Self-employed | 4.7 |
| Nonprofit | 1.9 |
| Federal government | 15.9 |
| State/local government | 7.8 |

B. This table shows what percentages of geoscientists are employed in various areas of specialization.

| Specialization | Percentage of Geoscientists Employed in Different Areas of Specialization |
| --- | --- |
| Atmospheric and space scientists | 10.2 |
| Geologists | 64.9 |
| Oceanographers | 3.3 |
| Astronomers | 2.9 |
| Postsecondary earth, environmental, and marine science teachers | 12.4 |
| Mining and geological engineers | 6.3 |

*Source:* From the American Geological Institute home page, at *http://www.agiweb.org/*. The data were gathered as part of the National Science Foundation's annual survey of college graduates, 1993, and reflect conditions as of April 15, 1993.

**geoscience** Earth science.

**Earth science** The broad field of scientific study that includes geology, oceanography, atmospheric science, and other areas of specialization concerning the Earth and its processes and materials.

geologists. Geology, in the strict sense, is just one part of the broader field of **geoscience,** or **Earth science,** which also includes oceanography, atmospheric science, and other areas of study. What all geoscientists have in common is that the Earth is their laboratory. Geoscientists are curious about how the Earth functions, and they try to find out more about it through their work. We finish the book by looking more closely at some of the jobs done by geoscientists and by thinking about the role of geoscientists in a changing world.

# GEOSCIENCE CAREERS FOR THE TWENTY-FIRST CENTURY

The American Geological Institute (AGI), in conjunction with other scientific societies, has carried out surveys and studies to find out what kinds of jobs are done by geoscientists and in which sectors of the economy they are employed (Table 17.1 and Fig. B17.1). They have also compiled information about typical salaries for a given level of training, data for each field by gender and race, and statistics about other aspects of geoscience employment. Much of the information in this section of the chapter comes from AGI's World Wide Web home page, where you can find more information about jobs and careers in the geosciences. The address is <*http://www.agiweb.org/*>.

## What Do Geoscientists Do?

The basic role of geoscientists is to gather and interpret data about the Earth in order to increase our understanding and improve the quality of human life. Geoscientists investigate the materials, processes, and history of the Earth. They study and help predict the effects of natural hazards such as volcanic eruptions, earthquakes, floods, and landslides. Geoscientists provide information to be used in solving problems and establishing policies for resource management, environmental protection, and public health, safety, and welfare. There are many areas of specialization in the geosciences; some are described in the box *Geology Around Us*.

## Where Do Geoscientists Work?

Geoscientists may be found sampling the deep ocean floor or collecting rock specimens on the Moon (Fig. 17.2). But the work of most geoscientists is more "down to Earth." They are explorers for and extractors of new mineral or hydrocarbon resources; they are consultants on engineering or environmental problems. They work as researchers, teachers, writers, editors, museum curators. They often divide their time among work in the field, the laboratory, and the office. Generally speaking, their work includes a mix of indoor and outdoor duties and travel.

Field work usually consists of preparing geologic maps, collecting samples, and making measurements that will be analyzed in the laboratory (Fig. 17.3). For example, rock or fossil samples may be X-rayed, studied under a polarizing or electron microscope, or analyzed for chemical content. Geoscientists may also conduct experiments or design computer models to test theories about geologic phenomena. In the office, they integrate field and laboratory data and write reports that include maps and diagrams illustrating the results of their studies. Such maps may pinpoint areas where ores, coal, oil, natural gas, or underground water may be found, or indicate subsurface conditions at construction sites.

## The Job Outlook

The employment outlook in the geosciences, as in any profession, varies with economic conditions. The outlook is good for the early twenty-first century. Dwindling energy, mineral, and water resources, increasing environmental concerns, hazard assessment, and global issues such as rising sea levels present new challenges to organizations that employ geoscientists. Career opportunities in environmental geoscience are increasing along with the growing need to maintain natural environments and meet society's demands for resources. Salaries vary, of course, depending on level of training and type of employment.

Most Earth scientists are employed in industries related to oil and gas, mining and minerals, and water resources. Some geoscientists are self-employed geologic consultants or work with consulting firms. Most consulting geologists have had prior experience in industry, teaching, or research. Many of the geoscientists in the United States work for the federal government or for state governments. Most work for the U.S. Geological Survey, but others work for the Department of Energy, the Forest Service, the Bureau of Land Management, the National Aeronautics and Space Administration, the U.S. Army Corps of Engineers, or

# GEOLOGY AROUND US

## Areas of Specialization in Geoscience

Figure B17.1

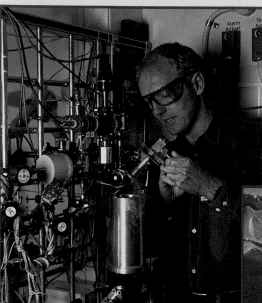

◄ A. Climatologist extracting samples of ancient air from Antarctic ice cores.

▼ B. Glaciologists measuring depth and flow in a glacial stream.

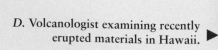

▲ C. Marine geologist studying corals.

D. Volcanologist examining recently erupted materials in Hawaii. ▶

## Areas of Specialization in Geoscience

| | |
|---|---|
| *Atmospheric scientists* and *climatologists* | study the Earth's wind, weather, and climate patterns, and the physics and chemistry of the atmosphere |
| *Economic geologists* and *mining geologists* | explore for and develop geologic materials that have profitable uses |
| *Engineering geologists* | investigate geologic factors that affect engineered structures such as bridges, buildings, airports, and dams |
| *Environmental geologists* | work to solve problems caused by pollution, waste disposal, and urban development, and hazards such as flooding and erosion |
| *Geochemists* | investigate the nature and distribution of chemical elements in rocks and minerals |
| *Geochronologists* | determine the age of rocks by measuring the products of decay of radioactive elements, thus helping to reconstruct the Earth's history (*chron* means "time") |
| *Geomorphologists* | investigate the nature, origin, and development of present landforms and their relationship to underlying structures (*morph* means "shape") |
| *Geophysicists* | study the Earth's interior and its magnetic, electric, and gravitational fields |
| *Glaciologists* | study the physical properties and movement of glaciers and ice sheets and the nature of glacial deposits |
| *Hydrologists* and *hydrogeologists* | study the abundance, distribution, and quality of groundwater and surface water (*hydro* means "water") |
| *Limnologists* | study the characteristics of ponds, lakes, and other inland surface waters (*limn* means "pool") |
| *Marine geologists* and *oceanographers* | investigate the oceans and continental shelves |
| *Mineralogists* | study the formation, composition, and properties of minerals |
| *Paleontologists* | study fossils in order to understand past life forms and how they have changed through time, and to reconstruct past environments (*paleo* means "ancient") |
| *Petroleum geologists* | are involved in exploration for and production of oil and natural gas |
| *Petrologists* | determine the origin and genesis of rocks by analyzing mineral or grain relationships (*petra* means "rock") |
| *Planetary geologists* | study the Moon and planets in order to understand how the solar system evolved |
| *Sedimentologists* | study sedimentary rocks and the processes of sediment formation, transportation, and deposition |
| *Seismologists* | study the location and force of earthquakes and trace the behavior of earthquake waves |
| *Stratigraphers* | investigate the time and space relationships of layered rocks and their fossil and mineral content (*strata* means "layers") |
| *Structural geologists* | study deformation, fracturing, and folding in the Earth's crust |
| *Volcanologists* | investigate volcanoes and volcanic phenomena |

*Based on* American Geological Institute, *Careers in the Geosciences,* Alexandria, VA.

# Geologists in the Field

**Figure 17.2**
Much of the work of geology is carried out through field studies. A. Planetary geologist Jack Schmidt's field area has only been visited by a handful of people. Schmidt selected samples of lunar rocks to bring back for laboratory study on Earth. B. Volcanologists measure the temperature and composition of gases and lavas during a volcanic eruption. C. Structural geologists and other field geologists use instruments such as this compass to determine the orientations of folded and faulted rocks.

state geologic surveys. In Canada, geoscientists are similarly employed at the Geological Survey of Canada, Environment Canada, federal and provincial parks and ministries, provincial geological surveys, and conservation authorities.

## Geoscientists and Public Policy

**? What is the role of geoscientists in public policy-making?**

Geology affects our lives every day in innumerable ways, from the siting of a waste disposal facility to the opening of a mine, the eruption of a volcano, the flooding of a river, or the price of oil. Many geoscientists feel that in coming

# Geologists in the Lab

A.

B. ▲

C. ▲

**Figure 17.3**
Work in the laboratory furthers our understanding of natural systems. *A.* Geologist and Apollo astronaut Jack Schmidt and lunar sample curator James Gooding examine some of the lunar samples brought back to Earth by Schmidt and his colleagues. *B.* Seismologists in Indonesia monitor seismometers for earthquake activity. *C.* This illustrator is working on a scientifically accurate reconstruction of an Ordovician seafloor community.

decades it will be increasingly important to gain a better understanding of the Earth system and to communicate this knowledge to decision makers and policy makers more effectively. In California, for example, geologists have influenced public policies concerning construction and land use. They did so by informing policy makers about landslides and earthquakes, both of which are common in some parts of California. Through more rigorous building codes and zoning laws, lives have been saved and damage has been prevented.

Geologic processes have a profound influence on human interests and activities. But do geologists themselves bring a special perspective to the public policy

**?**
How do public and private decision makers deal with uncertainty?

arena? According to Mohamed El-Ashry, a geologist who is the former environmental director of the World Bank:

> *Geologists understand a long-term perspective. We deal in millions and billions of years. We know how long it takes to renew resources, that it won't be in human lifetimes. . . . With our knowledge as Earth scientists, we can help guide policy toward broad, long-range economic development. We can help, with our knowledge, to improve decision-making, not only in this country but on an international scale.*[1]

Increasingly, policy makers must deal with situations that are not cut-and-dried, that involve scientific uncertainty. Will the global climate change? If so, when and how much, and what will the environmental impacts be? Will the Mississippi River flood next season? If so, how high will the water go and how fast will the currents flow? Will deforestation in the Amazon lead to soil erosion and land degradation? If so, how can it be stopped? Will fossil fuel or strategic mineral resources be depleted in the near future? If so, when? Can we find replacements? There are no easy answers to these questions. They are characterized by scientific uncertainty, coupled with uncertainty about the continued impacts of human activities.

Most policy makers are uncomfortable with uncertainty. It is easier to make decisions on the basis of known facts. But geologists are used to dealing with uncertainty. They must find the tiniest clues to discover the early history of the Earth's crust or to figure out how dinosaurs cared for their young. They use indirect methods to study parts of the Earth that are beyond their reach, such as the core, the mantle, and the depths of the ocean. Perhaps geoscientists, with their unique experience in dealing with scientific uncertainty, can bring new perspectives to policy making.

We can group the contributions of geoscientists into three broad categories: (1) developing a clearer understanding of Earth resources and the environmental impacts of their exploitation; (2) improving our understanding of geologic hazards and their impacts; and (3) enhancing our knowledge of the natural Earth system. Let's look more closely at each of these categories.

# UNDERSTANDING EARTH RESOURCES

Geologists have the task of locating new deposits of the energy and mineral resources on which modern societies depend. As the human population grows, the demand for such resources is growing even faster. But many Earth resources are nonrenewable and therefore limited, as discussed in chapter 16. Will geologists always be able to find sufficient resources to meet society's needs? And if nonrenewable resources are limited, how many people can the planet support?

## Population and Carrying Capacity

The answers depend partly on how fast the human population is growing and partly on the rate of per capita resource use. When conditions are favorable, populations of living organisms, including humans, tend to grow *exponentially.*

---

[1] Mary Beck Desmond, "Earth Scientist at the International Development Front," *Geotimes,* March 1992, pp. 21–23.

One bacterium divides into two, which then divide into four. The four bacteria become 8, then 16, then 32, and then 64. After 10 divisions there are over 1000 bacteria. After 20 divisions there are over 1 million! This kind of growth, in which a population increases by a constant percentage in a given unit of time, is called **exponential growth.**

Throughout human history, even in prehistoric times when our ancestors started using nonrenewable resources, the world's human population has been characterized by spurts of exponential growth (Fig. 17.4). The population seems to have increased dramatically after each of the three major "revolutions": the Neolithic Revolution (in which stone tools and agriculture were first used), the Industrial Revolution, and the modern Medical-Technological Revolution. These revolutions made life much easier, allowing population numbers to explode, followed by a leveling-off period. Today the world's population is increasing by 1.7 percent per year. This will amount to about 100 million new people this year. As a result, the population has doubled since 1960 and will reach 6 billion by the year 2000. Can the Earth sustain a population of 6 billion?

The maximum population that can be supported by a system without causing permanent damage to the system is its **carrying capacity.** Numerous efforts have been made to figure out the Earth's carrying capacity with respect to a large human population. On the basis of mineral and energy resources and biological productivity, different studies have suggested that the planet is capable of sustaining a population of 2 billion, or 6 billion, or 12 billion, or even 100 billion.

Why are these estimates so different? Part of the reason is that different human populations consume different amounts of resources. For example, on a per capita basis Americans use about four times as much steel and 23 times as much aluminum as people in Mexico. The average Japanese consumes nine times as much steel as the average Chinese. Canadians routinely top the list in per capita energy consumption. In terms of overall use of nonrenewable resources, the average Swiss consumes as much as 40 Somalis. The point is not that the lifestyle of the Swiss or the Mexicans or the Americans or the Canadians is right or wrong. Nor is it that everyone's living standard should be lowered to a minimum in order to increase the Earth's carrying capacity. Rather, the point is that there are many choices to be made, many resource issues to consider, many ways to use resources wisely, and many ways to do more with less.

**exponential growth** Growth in which a population increases by a constant percentage in a given unit of time.

**?** **If the human population continues to grow exponentially, how large might it be by the year 2100?**

**carrying capacity** The maximum population that can be sustained by a system, without causing permanent damage to the system.

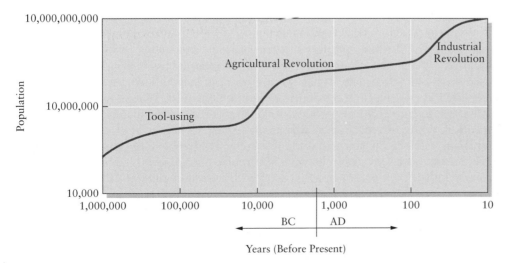

Figure 17.4
**POPULATION EXPLOSIONS**
Human population has grown in exponential surges after each of the great "revolutions": the Neolithic Revolution (start of toolmaking), the Agricultural Revolution (origin of agriculture) and the modern Industrial (Medical-Technological) Revolution.

## Issues in Resource Management

Most living resources are renewable—*if* they are managed properly. On the other hand, most solid Earth resources are nonrenewable. Some resources, such as soil and groundwater resources, are renewable, but people may not be willing or able to wait long enough for a depleted resource to be replenished. Increasingly, the guiding principle in the management of both renewable and nonrenewable resources is **sustainable development,** that is, cautious, planned utilization of Earth resources to meet current needs without degrading ecosystems or jeopardizing the future availability of those resources.

The sustainable management of renewable resources must ensure that the rate at which the resource is being used does not exceed that at which it is replenished or regenerated. This means that fish should not be harvested faster than new fish are hatching; trees should not be cut down faster than new trees can grow; and water should not be withdrawn from wells faster than water can replenish underground storage reservoirs. If a renewable resource becomes too severely depleted, it may never regenerate, no matter how long we wait.

This is reasonably obvious in the case of *living* resources like fish, for which there is often a critical level below which the population simply cannot survive. Fishing industries in many parts of the world, including the eastern coast of Canada, have collapsed (both economically and biologically) because fish were harvested beyond the critical level for the fish population in those areas. In some cases a population can be renewed artificially. Seedlings can be planted; fish can be spawned in fish farms and then transferred to rivers. However, there is always the possibility that biological diversity, either of the species or of the ecosystem as a whole, will be lost. For example, populations of fish that are grown in captivity quickly lose their genetic diversity unless their stocks are replenished by fish from the wild. The ability of a fish population to regenerate itself may also be reduced by other environmental factors, both anthropogenic and natural. These include accelerated erosion caused by logging, agriculture, or mining; reduced water supply or quality due to the withdrawal or contamination of water; and climatic fluctuations.

It is also possible for *nonliving* renewable Earth resources to become depleted to the point at which they cannot regenerate naturally. For example, a groundwater reservoir can be replenished by rainfall, although the process may take many years or even centuries. However, if too much water is withdrawn from an underground aquifer, the particles in the rocks or sediments of the aquifer may become compacted. When this happens, the aquifer can never be replenished, no matter how much it rains and no matter how long we wait.

Nonrenewable resource management is different from renewable resource management. The key feature of nonrenewable resources is this: the more we use, the less remains. Nonrenewable resources must be treated with respect and used wisely and sparingly. When we use up nonrenewable resources, we do so not just for ourselves but for future generations as well. Wherever possible, we must seek to extend the availability of these resources through conservation, substitution, reuse, or recycling. A wise management strategy will also look toward replacing reliance on nonrenewable resources with reliance on renewable or inexhaustible ones. For example, we might replace nonrenewable fossil fuels with solar, geothermal, tidal, or biomass energy (discussed in chapter 16).

## Will We Run Out?

Are there enough mineral and energy resources available to raise the living standards of all peoples to the levels they desire? The answer to this question is not

---

**sustainable development** The cautious, planned utilization of Earth resources to meet current needs without degrading ecosystems or jeopardizing the future availability of those resources.

---

**?** **Why are renewable and nonrenewable resources managed in different ways?**

clear. Many geologists are concerned that nonrenewable resources may be inadequate and that shortages will eventually hamper development. Unfortunately, there is no sure way to know exactly how much of any mineral or energy resource remains to be found. Throughout the history of energy and mineral exploration, geologists have continually refined and improved the techniques used to locate new deposits, as discussed in chapter 16. Much ingenuity has been expended in bringing mineral and energy resource production to its present state. Today the known and easily exploitable deposits are fast being depleted while demand continues to grow. We can be certain that even more ingenuity will be needed to extend mineral resources into the future while limiting the negative environmental impacts of resource extraction.

Whatever the Earth's carrying capacity may turn out to be, we apparently haven't yet reached its limits. The human population continues to grow, and we continue to find, process, and make use of Earth resources in a wide variety of ways. Perhaps the only firm conclusion we can reach is that while the Earth's carrying capacity is very large, it is not infinite. It is likely that the Earth's mineral and energy resources can be extended far into the next century through careful use and conservation practices, new extraction and processing technologies, and new discoveries by geologists and geophysicists. In the long run, however—in the twenty-first century and beyond—our descendants will have to manage Earth resources very differently from how we and our parents and grandparents have managed them. As the global community prepares for the future, the role of geoscientists in policy making and education will become more and more crucial.

# UNDERSTANDING GEOLOGIC HAZARDS

Geologic processes that we call "hazardous" have existed throughout the Earth's history. Even the most destructive geologic events are part of the normal functioning of this dynamic planet. To a great extent, they have made the planet habitable. Earthquakes and volcanic eruptions, for example, are part of the system of plate tectonics that has formed the continents and shaped the landscape. Wind and water cause flooding, landslides, and windstorms, but they also replenish the soil and sustain life. Knowledge about geologic processes and their hazards should play an integral role in the planning of human activities. Before we address the challenges inherent in assessing geologic hazards and risks, let's clarify the meanings of these terms.

## Hazards and Their Impacts

The terms *natural hazard* and *geologic hazard* encompass a wide variety of geologic circumstances, materials, processes, and events. **Geologic hazards** include earthquakes, volcanic eruptions, floods, landslides, and other geologic processes. The term **natural hazard** is slightly more comprehensive; it includes biologic and meteorologic processes and events such as locust infestations, wildfires, tornadoes, and hurricanes, as well as geologic hazards.

Some natural hazards are catastrophic events that strike quickly but with devastating effects. For example, there is a small but real risk of a large meteorite impact, like the one thought to be related to the extinction of the dinosaurs (chapter 15). (In fact, it is now known that a large asteroid will pass quite close to the Earth in the year 2020). Events that strike quickly and with little warning, such as meteorite impacts, earthquakes, or flash floods, are called **rapid onset hazards.** Other hazardous processes operate more slowly. Droughts, for example,

**geologic hazard** Earthquakes, volcanic eruptions, floods, landslides, and other geologic processes that may be damaging to human interests.

**natural hazard** Naturally occurring hazardous processes or events, including locust infestations, wildfires, tornadoes, and hurricanes, as well as geologic hazards.

**rapid onset hazard** Hazardous events that strike quickly and with little warning, such as meteorite impacts, earthquakes, or flash floods.

# Geologic Hazards

A. ▲

B. ▲

**Figure 17.5**
Geologists can contribute to the mitigation of damages from natural disasters. *A.* This highway collapsed during the 1995 Kobe, Japan earthquake. *B.* In 1997, the Soufrière Hills volcanic on the islands of Monserrat erupted. The resulting ash falls, lava flows, and volcanic mud flows virtually wiped out the town of Plymouth off the map. *C.* The Red River flooded in the spring of 1997, leaving huge areas of North Dakota (shown here near Fargo), Minnesota and Manitoba under water. *D.* This Florida town was virtually leveled by Hurricane Andrew's 200 km/h winds in 1992.

**?**
**When do we call something a "hazard"?**

**technological hazards** The hazards associated with exposure to potentially harmful substances, such as radon (a naturally occurring radioactive soil gas), mercury, asbestos fibers, or coal dust.

can last ten years or more. The socioeconomic impacts of extended droughts are caused by the cumulative effects of season after season of below-average rainfall.

In general, natural processes are called "hazardous" only when they threaten human life, health, or interests, either directly or indirectly (Fig. 17.5). In other words, we take an *anthropocentric* (human-centered) approach to the study and management of natural and geologic hazards. This is just human nature; we are naturally concerned with the protection of human life and property. But this approach has important implications because it can lead to a style of hazard management in which geologic processes are cast as the "enemy" and efforts are made to manipulate the environment into submission. A somewhat different approach focuses on improving scientific understanding of geologic processes and their triggering mechanisms. This provides a foundation for better preparation and decision making in natural hazard management.

A different type of hazard, sometimes referred to as **technological hazards,** is associated with everyday exposure to hazardous substances, such as radon (a naturally occurring radioactive soil gas), mercury, asbestos fibers, or coal dust. Still another type of anthropogenic hazard arises from pollution and degradation of the natural environment.

The effects of hazardous events or processes are classified as either primary, secondary, or tertiary. *Primary effects* are those that result from the impacts of

C. ▲

D. ▶

the event itself: water damage from a flood; wind damage from a cyclone; or the collapse of a building as a result of ground motion during an earthquake. *Secondary effects* result from hazardous processes that are associated with, but not directly caused by, the main event. Examples include forest fires touched off by lava flows; house fires caused by gas lines breaking during an earthquake; or disruption of water and sewage services by a flood. *Tertiary effects* are long-term or even permanent. These might include loss of wildlife habitat or permanent changes in a river channel from flooding; regional or global climatic changes and crop losses after a major volcanic eruption; or permanent changes in topography as a result of an earthquake.

## Vulnerability

Exactly how vulnerable are we to natural hazards? During the past two decades, as many as 3 million lives have been lost as a direct result of hazardous events. Another 800 million people have suffered the loss of property or health. A United Nations committee estimated that by the end of the 1990s the Earth will have experienced tens of thousands of damaging landslides and earthquakes; a million thunderstorms; 100,000 floods; and several hundred to several thousand tropical cyclones and hurricanes, tsunamis, droughts, and volcanic eruptions.

**?** **Why are natural disasters so devastating to less developed countries?**

According to the World Bank, natural disasters cause about US$40 billion each year in physical damage. Windstorms, floods, and earthquakes alone cost about US$18.8 million *per day!*

The concept of *vulnerability* encompasses not only the physical effects of a natural hazard but also the status of people and property in the area. A complicated web of factors can increase vulnerability to natural hazards, especially catastrophic events. Aside from the simple fact of living in a hazardous area, vulnerability can also depend on population density, construction styles, and building codes (or sometimes the lack of them). Scientific understanding and public education and awareness are very important. The existence of an early warning system and effective lines of communication, as well as the availability and readiness of emergency personnel, can have a crucial effect on response times. Some cultural factors (such as unwillingness to leave the home at certain times) can also influence public response to warnings. During the eruptions of Mount Pinatubo in the Philippines in 1991, government officials showed video footage of devastating volcanic eruptions in order to convince reluctant residents that they should leave their tribal lands.

These factors help explain why less developed countries are especially vulnerable to natural hazards. Something as simple as an inoperative telephone line—an everyday reality in many developing countries—can cause an entire early warning system to fail. The actual dollar value of property damage from an event such as an earthquake or a flood may be higher in an industrialized country, but the *relative* value of monetary losses is *much* greater, on average, in developing countries. Poverty itself contributes to increased vulnerability: inadequate housing and high population densities on environmentally sensitive lands can lead to much higher losses of life in a natural disaster. For example, many of the poorest people in Bangladesh are landless and are forced by circumstance to farm the shifting lowlands of the Brahmaputra River delta. Population densities are high, communication systems are poor, the country is subjected to many tropical cyclones, and the low-lying delta lands are extremely susceptible to flooding. This combination of factors renders the poor in Bangladesh particularly vulnerable to the effects of natural hazards.

Human intervention in natural systems can increase vulnerability. This can happen in two basic ways: (1) by leading to the development and habitation of lands that are susceptible to hazards, such as floodplains and areas underlain by "quick" clays; and (2) by increasing the severity or frequency of natural hazards. For example, overly intensive agriculture may accelerate erosion of sensitive soils. Mining of groundwater can trigger land subsidence. Anthropogenic global warming may lead to a higher frequency and intensity of tropical cyclones. And modification of river channels may lead to floods of unexpected magnitude.

## Assessing Hazards and Risks

The terms *hazard assessment* and *risk assessment* are often used interchangeably, but they have different meanings. **Hazard assessment** involves asking questions such as: "How often can we expect a very large event to occur?" and "When an event does occur, what will the effects be like?" Specifically, hazard assessment consists of:

- Determining when and where hazardous events have occurred in the past.
- Determining the severity of physical effects generated by events of a given magnitude.
- Determining how often we can expect damaging events to occur in a particular place.

**hazard assessment** Determining when, where, and how frequently hazardous events have occurred in the past and the nature of the physical effects of the event in a particular location, and portraying the information in a form that can be used by decision makers.

- Determining what a particular event would be like if it were to occur now.

- Portraying this information in a form that can be used by decision makers.

The information provided by hazard assessment is used by many people. Political leaders use it to make decisions about evacuation and funding. Emergency personnel make decisions about response procedures. Planners and engineers make decisions about land use, zoning, and building codes. Often the results of hazard assessment are portrayed in the form of maps to make the information as accessible and understandable as possible (Fig. 17.6).

**Risk assessment** differs from hazard assessment because it looks at the economic losses, deaths and injuries, and loss of services that are likely to occur when a specific event—such as an earthquake—strikes a given area. Risk assessment starts by establishing the *probability* that a hazardous event of a particular magnitude will occur within a given time period. It then takes into account factors such as:

- The locations of buildings, facilities, and emergency systems in the community.

- Their potential exposure to the physical effects of the hazard.

- The community's vulnerability—that is, the potential loss of life, injury, or damage—when subjected to these specific physical events.

Thus, risk assessment incorporates social and economic factors as well as the scientific factors involved in hazard assessment. Hazard assessment describes the effects associated with a particular event. Risk assessment focuses more on the likely amount of damage and the actions that can be taken to reduce vulnerability. Risk assessment involves asking questions such as: "When a situation of this type exists or an event of this type occurs, what will the damage to the community be like?" and "Even though the likelihood of an event of this magnitude is small, would the consequences be unacceptably severe?"

Risk is sometimes stated in probabilities. For example, it has been estimated that smoking 1.4 cigarettes, having a single chest X-ray, or being exposed to

> **?** **Who makes use of the results of hazard assessments and risk assessments?**

**risk assessment** Establishing the probability that a hazardous event will occur within a given period and estimating its impact, taking into account the potential exposure and vulnerability of the community.

**Figure 17.6**
**HAZARD ASSESSMENT**
This figure shows a landslide susceptibility map and recommended land use policies for the Congress Springs area near San Francisco, California. The results of hazard assessment are often portrayed in map format so that they can be understood and used by local planners and decision makers.

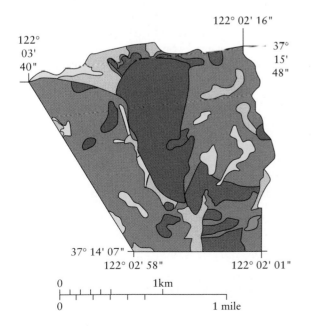

| Relative stability | Map area | Geologic conditions | Recommended land use | | |
|---|---|---|---|---|---|
| | | | Houses | Roads Public | Private |
| Most stable ↑ | | Flat to gentle slopes; subject to shallow sliding, soil creep, and settlement | ✓ | ✓ | ✓ |
| | | Gentle to moderately steep slopes in older stabilized landslide debris; subject to settlement, soil creep, and shallow and deep landsliding | ✓ | ✓ | ✓ |
| | | Steep to very steep slopes; subject to mass-wasting by soil creep, slumping and rock fall | ✓ | ✓ | ✓ |
| | | Gentle to very steep slopes in unstable material subject to sliding, slumping and soil creep | ✗ | ✗ | ✗ |
| | | Moving shallow (>10 ft) landslide | ✗ | ✗ | ✗ |
| Least stable | | Moving deep landslide, subject to rapid failure | ✗ | ✗ | ✗ |

earthquake hazards by living in Southern California for seven months all carry the same statistical risk, increasing the chance of an untimely death by approximately 1 in a million. Alternatively, risk can be stated in terms of cost (damages and injuries expressed in dollar value lost). In either case, risk assessment can help both decision makers and scientists compare and evaluate hazards, set priorities, and decide where to focus attention and resources.

## Predicting Hazards

Geoscientists are often called upon to make predictions about hazardous situations. A **prediction** is a statement of probability that is based on scientific understanding and observations. Prediction requires the monitoring of geologic processes that could generate a hazardous event. Such monitoring usually focuses on identifying anomalous phenomena that might be *precursors,* unusual or unexpected physical changes that might be leading up to a sudden, catastrophic event.

Sometimes the term **forecast** is used synonymously with the term *prediction.* In the prediction of floods or hurricanes, forecasting generally refers to short-term prediction (days or hours ahead) of the specific magnitude and time of occurrence of an event. In earthquake prediction, on the other hand, the term *forecast* is generally used to refer to a long-term, nonspecific statement of probability. Before the Loma Prieta (San Francisco) earthquake of October 17, 1989 (the "World Series" earthquake), the U.S. Geological Survey issued a forecast stating that there was a 50 percent probability of a large earthquake occurring along the San Andreas fault in that region within 30 years. This was a relatively nonspecific, long-range forecast (a prediction) based on the general scientific understanding of seismicity and the geologic setting of the area.[2]

In chapter 6 we discussed the monitoring of precursor phenomena in the context of volcanic eruptions. When observations of precursor phenomena have accumulated to the point where they signal the imminent occurrence of a hazardous event, a specific warning can be given. For example, when Mount Pinatubo erupted violently in 1991, scientists observed a variety of precursor events. The most important were changes in the number, intensity, and locations of local earthquakes and in the quantity and composition of gases emitted from the volcano. Careful monitoring allowed scientists to predict the eruption with great accuracy, saving many thousands of lives.

The issuance of a warning is the final step in preparing a community to deal with a hazardous event. As one expert put it, an **early warning** is "a public declaration that the normal routines of life should be altered for a period of time to deal with the danger posed by the imminent event."[3] Successful warnings depend on timeliness, effective communications and public information systems, and credible sources. If a warning is issued prematurely or irresponsibly and the event does not occur as predicted, the results can be disastrous. Like the boy who cried "wolf," the scientist may not get a chance to regain credibility, and public response to a *real* threat may be too slow. This conflict was portrayed with reasonable accuracy in the movie *Dante's Peak,* in which U.S. Geological Survey scientists agonized over whether or not to issue an early warning of a volcanic eruption. In a town heavily dependent on tourism, even the suggestion of a catastrophe like a volcanic eruption could cause an economic disaster. In this particu-

**prediction** A statement of probability that is based on scientific understanding and observations.

**forecast** A prediction. Sometimes refers to short-term predictions (as in weather forecasting), and sometimes refers to long-term, nonspecific predictions (as in earthquake forecasting).

**?** **Why are scientists so cautious about issuing predictions and early warnings?**

**early warning** A public declaration that the normal routines of life should be altered in order to deal with the danger posed by an imminent hazardous event.

---

[2] The threat of a major earthquake in the San Francisco area is still very real. After the Loma Prieta earthquake, the U.S. Geological Survey forecast a 67 percent probability of a major earthquake in this area within the next 30 years.

[3] Walter W. Hays, "Hazard and Risk Assessments in the United States," *Episodes,* vol. 14, no. 1, March 1991. Much of the information in the section "Assessing Hazards and Risks" was based on this useful article.

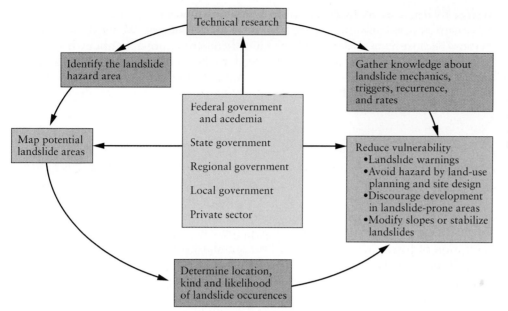

**Figure 17.7**
**HAZARD MITIGATION**
This diagram shows an integrated system designed to reduce landslide hazards. Through such a system, geologists can contribute essential information to all levels of government, the private sector, and emergency response personnel.

lar story, the geologists' decision not to issue an early warning was the wrong one for the unfortunate residents of Dante's Peak.

## The Role of Geoscientists in Hazard Reduction

Some natural hazards, such as major meteorite impacts, may be impossible to predict within any useful time frame (the 1998 summer "disaster" movies *Deep Impact* and *Armageddon* were based on this idea). Although we know that such an event may occur one day, there isn't much we can do to decrease the risk (given current knowledge, technologies, and funding). However, we are faced daily with a wide variety of natural hazards to which we *can* adapt by making certain choices and taking certain actions to prepare ourselves or to decrease our vulnerability. What role should geoscientists play in shaping the public's response to natural hazards?

Ideally, scientific understanding can contribute to a system in which geoscientists cooperate with public and private decision makers to reduce susceptibility to natural hazards. An example is given in Figure 17.7. In this integrated approach, scientific and technical research contributes an understanding of landslide mechanisms and the delineation of the hazard in time and space. This understanding then leads to the development of hazard and risk reduction strategies and landslide hazard maps. All levels of government, academic institutions, and private organizations are involved. Because so many natural disasters are associated with geologic processes, geoscientists have important roles to play in assessing, predicting, and preventing or limiting the damage associated with hazardous events.

# UNDERSTANDING THE CHANGING EARTH SYSTEM

We know that the Earth is changing all the time and that change is a natural part of the functioning of the Earth system. In this book we have investigated natural changes on a wide variety of time scales: from the seconds it takes to release the energy of an earthquake to the billions of years it took for the atmosphere–ocean

anthropogenic Caused by humans.

system to reach its present state (Fig. 17.8). An important part of the work of geoscientists is to help us understand these natural changes. We need this information because it is useful to understand as much as possible about how the Earth system functions. But we also need to be able to distinguish natural changes from **anthropogenic,** or human-induced, changes. Geologic observations of natural changes provide a baseline against which we can measure the impacts of recent anthropogenic changes.

## Anthropogenic Impacts

Over many millennia humans have changed the Earth's natural landscapes. We have built villages and cities, converted forests into agricultural land, and dammed and diverted streams (Fig. 17.9). Beginning in the nineteenth century, ever-greater amounts of mineral and energy resources were needed to fuel the industrial technologies of increasingly populous societies. In particular, the exploitation of fossil fuels greatly raised the standard of living for most people. In spite of the many benefits involved, however, the exploitation of our planet's rich natural resources has not been without cost. In many parts of the world severe environmental deterioration is occurring. In addition to scarring and poisoning the land, we have polluted the oceans and groundwater. Natural and human activities are intertwined, and it is increasingly apparent that *Homo sapiens* has become a major factor in global geologic change.

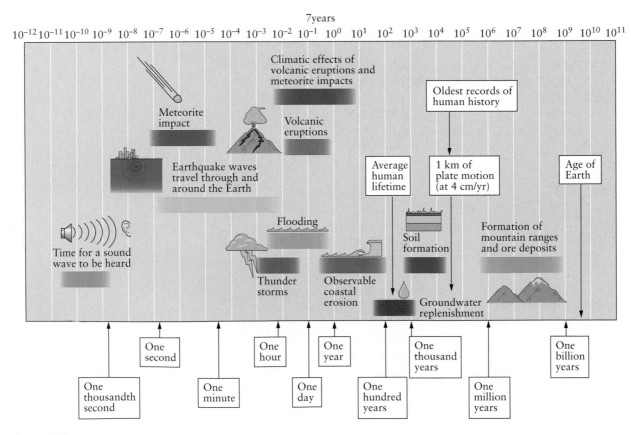

**Figure 17.8**
**TIME SCALES OF GEOLOGIC PROCESSES**
Change in the natural environment happens on many time scales. The things that are important to us socially and politically are measured on a scale ranging from days to years. The duration of geologic processes ranges from seconds (an earthquake or a meteorite impact) to hundreds or thousands of years (soil formation or groundwater renewal) to hundreds of millions of years (the formation of an ore deposit or the uplifting of a mountain range).

Figure 17.9
**PEOPLE CHANGING THE LANDSCAPE**
Throughout human history, people have altered the natural landscape. However, the scale of human interventions has become enormous. This city, on the upper Yangtze River in China, will be flooded when the huge Three Gorges Dam is completed. The water will rise as high as the tower on the lefthand side of the photo.

The massive impact of humanity on the Earth's surface is due partly to consumption-oriented lifestyles and partly to the sheer numbers in the "human herd." Consider the quantities of mineral resources we take from the ground and use to make automobiles, build roads, and produce the myriad other objects we use. On average, 8 metric tons of material a year are dug up and utilized for every man, woman, and child on the Earth. In 1997 there were almost 6 billion people alive on the Earth, which means that the total amount of Earth material dug up that year was almost 50 billion tons (*not* including material moved for such purposes as construction or dredging). If you compare this enormous figure with the 16.5 billion tons of dissolved and suspended matter carried to the sea each year by the world's rivers, you can get an idea of the magnitude of the geologic impact of human activities.[4]

## Uniformitarianism Revisited

In chapter 1 we introduced the concept of *uniformitarianism,* the idea that geologic processes that we observe today have operated in a similar manner throughout much of the Earth's history. We can learn about how ancient rocks were formed by comparing their characteristics to those of rocks and sediments being formed today by similar processes. This principle guides much scientific investigation in geology as well as in other sciences. Uniformitarianism even accounts for infrequent, catastrophic events like meteorite impacts, provided that we step back and take a very long view of their role as natural occurrences in Earth history.

We realize now, however, that although many geologic processes have operated throughout the Earth's history, their rates and magnitudes may have changed. Some of these changes are completely natural. For example, the rate of meteorite impacts was originally very high but decreased over time to the present low rate of impact. The magnitudes of geologic processes have also varied at different times. For example, satellite imagery provides evidence of catastrophic

---

[4] Even the latter figure—16.5 billion tons per year of sediment carried to the sea by rivers—may be two to three times greater than it was before humans began altering the Earth's surface through land-clearing activities such as agriculture and mining.

Figure 17.10
PAST CATASTROPHES
Some geologic processes have changed in magnitude. The Channeled Scablands of eastern Washington were created by catastrophic floods that happened when ice dams broke. Giant ripples formed as raging floodwaters swept around a bend of the Columbia River. The ripples are up to several meters high, and their crests are as much as 100 meters (more than 300 ft) apart. Flooding on this scale is unknown in recorded history.

flooding in eastern Washington state (Fig. 17.10). This curious terrain, known as the Channeled Scablands, was formed when a huge, glacially impounded lake was suddenly released by the melting of ice that had dammed the lake's outlet. When the glaciers began to retreat, the ice dam failed, releasing as much as 2500 km³ (600 km³) of water all at once. This process occurred as many as 54 times. The catastrophic flooding created landforms such as huge current ripples and massive piles of gravel containing huge boulders. Flooding of this magnitude is unknown in recorded history.

Species extinction is another example of a natural process whose rate has changed dramatically at different times. Throughout geologic history species have come and gone; but every now and then a mass extinction occurs. For example, at the end of the Permian Period, 245 million years ago, 96 percent of all species then alive died out in a very short span of time (geologically speaking). In

**?** Does uniformitarianism explain the situation of endangered species today?

Figure 17.11
PAST ENVIRONMENTAL CHANGES
Geologists use many techniques to reconstruct past environmental changes. Some methods, such as the study of sedimentary rocks and polar ice cores, can help geologists understand changes millions of years in the past. Other methods, such as weather records, can reveal changes of brief duration, but these methods do not reach very far into the past.

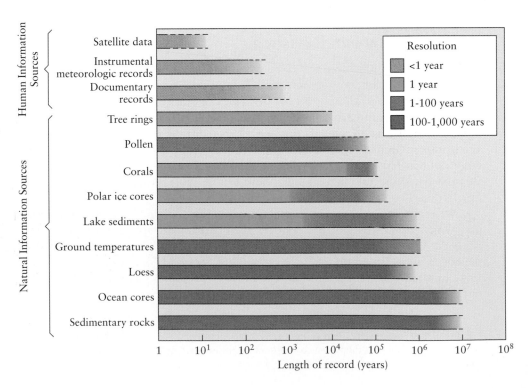

chapter 15 we discussed such mass extinctions as a normal part of the evolution of life, consistent with uniformitarianism. But what about today's endangered species? Through the destruction of habitat and other results of human activities, we are causing a mass extinction of species greater than any that has occurred within the past 65 million years. Even though mass extinctions have occurred at various times in the Earth's history, we must acknowledge that this particular mass extinction has very different causes.

The rate of global climatic change is also being altered by human actions. Short- and long-term climatic fluctuations are a normal part of atmospheric evolution. During the past decade or so, however, the possibility of accelerated global warming and the role of anthropogenic emissions in the warming process have occupied an important place in both scientific and political debates. But it is very difficult to separate the anthropogenic component of global warming from natural climatic fluctuations. Weather and climate changes that can be measured on the scale of a human lifetime are just tiny squiggles compared to the broad, complex changes that characterize the Earth's climate over geologic time.

The only way to understand human-induced climatic change is first to unravel the complex background of natural climatic change. Geologists use a variety of information sources to do this (Fig. 17.11). Geologic information about natural climatic change thus provides a baseline against which the rate and magnitude of anthropogenic changes can be evaluated. Against the geologic background of natural climatic change, scientists are gaining a fuller understanding of the causes and effects of rapid, extreme shifts in climate.

# WHAT'S AHEAD

In response to the changing needs of society, geology, too, is changing. As a science, geology is becoming less descriptive and more quantitative. The use of three-dimensional computer modeling, remote sensing, and other sophisticated analytical techniques has become routine. Geology is becoming more broadly based, bringing together the contributions of many different sciences to formulate a more holistic understanding of the Earth system. The science of geology will continue to evolve and grow along with the needs of human societies.

What's ahead for you? If you are planning to become a geologist, we hope you have gained the scientific foundation you need in order to pursue your career successfully. We wish you the best of luck with the many fascinating opportunities and challenges that await you. Perhaps you took this course not for career purposes but out of personal interest or to meet a degree requirement. If so, we hope the book has helped you become more aware of the geologic nature of our environment and the role of geoscientists. We hope that all of our readers have emerged better informed about our home planet and better prepared to make decisions about the geologic processes that affect our lives on a daily basis.

## Chapter Highlights

**1.** Geology is part of the broader field of **geoscience, or Earth science.** Geoscientists gather and interpret data about the Earth. Geologic information can be used to solve problems and establish policies for resource management, natural hazard mitigation, environmental protection, and public health, safety, and welfare. With their experience in dealing with long time scales and scientific uncertainty, geoscientists can bring a unique perspective to policymaking.

**2.** Geoscientists work in the field (mapping, collecting samples, making measurements), in the lab (studying and analyzing samples, conducting experiments, designing computer models), and in the office (integrating field and laboratory data, writing reports). Most Earth scientists are employed by industries related to oil and gas, mining and minerals, and water resources.

**3.** The human population is characterized by **exponential growth.** As the population grows and living standards improve, the demand for Earth resources grows even more quickly. **Carrying capacity** is the maximum population that a system can support without permanent damage. It is difficult to calculate the Earth's carrying capacity with respect to humans, partly because different human populations consume different amounts of resources.

**4.** **Sustainable development** is the cautious, planned utilization of Earth resources to meet current needs without degrading ecosystems or jeopardizing the future availability of those resources. The primary concern in the sustainable management of renewable resources is to ensure that the rate at which the resource is utilized does not exceed that at which it is replenished or regenerated. Wise management of nonrenewable resources will seek to extend their availability through conservation, substitution, reuse, or recycling. It will also look toward replacing reliance on nonrenewable resources with reliance on renewable ones.

**5.** Hazardous processes have existed throughout the Earth's history. **Geologic hazards** include earthquakes, volcanic eruptions, floods, landslides, and other geologic processes. **Natural hazards** include phenomena such as locust infestations, wildfires, and tornadoes. **Rapid onset hazards** strike quickly and with little warning. **Technological hazards** are associated with exposure to hazardous substances, usually through some aspect of their use in the built environment.

**6.** Primary effects of a hazard are those that result from the event itself. Secondary effects result from processes that are associated with, but not directly caused by, the main event. Tertiary effects are long term or even permanent.

**7.** Many factors affect vulnerability to natural hazards. They include population density, construction styles, building codes, scientific understanding, public education and awareness, the existence of an early warning system, effective lines of communication, and the availability and readiness of emergency personnel.

**8.** **Hazard assessment** focuses on determining the frequency and severity of physical effects associated with a given event and presenting this information in a form that can be used by decision makers. **Risk assessment** incorporates social and economic considerations, in addition to the scientific factors involved in hazard assessment, to establish probabilities and place dollar values on expected losses.

**9.** A **prediction** is a statement of probability that is based on scientific understanding and observations. The term **forecast** can be used to refer to either short-term, specific predictions (as in weather forecasting) or long-term, nonspecific predictions (as in earthquake forecasting). An **early warning** is a public declaration that routines should be altered in order to deal with the danger posed by an imminent event.

**10.** It is important to distinguish natural changes from **anthropogenic** changes in the Earth system. Observations of natural changes provide a baseline against which to measure the impacts of recent anthropogenic changes such as accelerated erosion and global warming.

**11.** According to the principle of uniformitarianism, the geologic processes that we observe today have operated throughout much of the Earth's history. However, the rates and magnitudes of geologic processes sometimes change.

# ▶ *The Language of Geology*

## ► *Questions for Review*

1. What is exponential growth?
2. The world's population has doubled since 1960. At its present growth rate of 1.7 percent per year, in what year will the population have doubled again?
3. Why is it so difficult to calculate the Earth's carrying capacity with respect to a human population? (Try to think of other reasons, besides the ones given in the chapter.)
4. How does the management of renewable resources differ from that of nonrenewable resources?
5. What is the difference between a geologic hazard and a natural hazard? Give examples to illustrate your answer.

6. What are some examples of rapid onset hazards (other than those given in the chapter)?
7. Why are less developed countries especially vulnerable to natural hazards?
8. What are the differences among prediction, forecast, and early warning?
9. Give examples of primary, secondary, and tertiary effects of hazards (other than those given in the chapter).
10. Why do we need to look back into geologic history in order to understand environmental changes that are happening today?

## ► *Questions for Thought and Discussion*

1. The home page of the American Geological Institute (AGI) offers many links with sources that provide information about careers in geology. Check it out: <*http://www.agiweb.org/*>.
2. Investigate current projections for the lifetime of some important Earth resources, such as oil, gas, coal, and iron. Are we likely to run out of any of these crucial resources within your lifetime?
3. Many of the most significant scientific outcomes of the Apollo space program had to do with the geology of the Moon and the early Earth. Geology has also played a key role in other space missions, notably the recent Mars Pathfinder mission. Find out about some of the most important geologic findings associated with these missions. One good place to start is the home page of Planetary Science

Research and Discoveries. The address is <*http://www.soest.hawaii.edu/PSRdiscoveries/*>.
4. Given that we are now dependent on nonrenewable sources of energy and minerals and that the world's population continues to increase, how do you think human societies can or will adjust in the future?
5. Some people think sustainable development is not a useful concept, partly because it may be impossible to implement—or even to define—in the case of nonrenewable resources. Others think it is an extremely important concept, if only because it makes us think about the needs of future generations in planning resource management strategies. What do you think?

## *Virtual Internship: You Be the Geologist*

You can become a "virtual geologist" by trying out the internships on your *Geosciences in Action* CD-ROM.

For interactive case studies on the role of geoscientists in forecasting risks, monitoring environmental change, influencing policy decisions, and promoting sustainable development, visit our Web site.

# APPENDIX

# · A ·

## UNITS AND THEIR CONVERSIONS

## ABOUT SI UNITS

Regardless of the field of specialization, all scientists use the same units and scales of measurement. They do so to avoid confusion and the possibility that mistakes can creep in when data are converted from one system of units, or one scale, to another. By international agreement the SI units are used by all, and they are the units used in this text. SI is the abbreviation of Système International d'U- nités (in English, the International System of Units).

Some of the SI units are likely to be familiar, some unfamiliar. The SI unit of length is the meter (m), of area the square meter ($m^2$), and of volume the cubic meter ($m^3$). The SI unit of mass is the kilogram (kg), and of time the second (s). The other SI units used in this book can be defined in terms of these basic units. Three important ones are:

1. The newton (N), a unit of force defined as that force needed to acceler- ate a mass of 1 kg by 1 m/s²; hence 1 N = 1 kg·m/s². (The period between kg and m indicates multiplication.)

2. The joule (J), a unit of energy or work, defined as the work done when a force of 1 newton is displaced a distance of 1 meter; hence 1 J = 1 N·m. One important form of energy so far as the Earth is concerned is heat. The outward flow of the Earth's internal heat is measured in terms of the number of joules flowing outward from each square centimeter each sec- ond; thus, the unit of heat flow is J/cm²/s.

3. The pascal (Pa), a unit of pressure defined as a force of 1 newton applied across an area of 1 square meter; hence 1 Pa = 1 N/m². The pascal is a numerically small unit. Atmospheric pressure, for example (15 lb/in²), is 101,300 Pa. Pressure within the Earth reaches millions or billions of pas- cals. For convenience, Earth scientists sometimes use 1 million pascals (megapascal, or MPa) as a unit.

Temperature is a measure of the internal kinetic energy (expressed as movement) of the atoms and molecules in a body. In the SI system, temperature is measured on the Kelvin scale (K). The temperature intervals on the Kelvin scale are arbitrary, and they are the same as the intervals on the more familiar Celsius scale (°C). The difference between the two scales is that the Celsius scale selects 100°C as the temperature at which water boils at sea level, and 0°C as the freezing temperature of water at sea level. Zero degrees Kelvin, on the other hand, is absolute zero, the temperature at which all atomic and molecular motions cease. Thus, 0°C is equal to 273.15 K, and 100°C is 373.15 K. The temperatures of processes on and within the Earth tend to be at or above 273.15 K. Despite the inconsistency, earth scientists still use the Celsius scale when geological processes are discussed.

Appendix A provides a table of conversion from older units to Standard International (SI) units.

# PREFIXES FOR MULTIPLES AND SUBMULTIPLES

When very large or very small numbers have to be expressed, a standard set of prefixes is used in conjunction with the SI units. Some prefixes are probably already familiar; an example is the centimeter (which is one hundredth of a meter, or $10^{-2}$ m). The standard prefixes are

| | | |
|---|---|---|
| tera | $1,000,000,000,000$ | $=10^{12}$ |
| giga | $1,000,000,000$ | $=10^{9}$ |
| mega | $1,000,000$ | $=10^{6}$ |
| kilo | $1,000$ | $=10^{3}$ |
| hecto | $100$ | $=10^{2}$ |
| deka | $10$ | $=10$ |
| deci | $0.1$ | $=10^{-1}$ |
| centi | $0.01$ | $=10^{-2}$ |
| milli | $0.001$ | $=10^{-3}$ |
| micro | $0.000001$ | $=10^{-6}$ |
| nano | $0.000000001$ | $=10^{-9}$ |
| pico | $0.000000000001$ | $=10^{-12}$ |

One measure used commonly in geology is the nanometer (nm), a unit by which the sizes of atoms are measured; 1 nanometer is equal to $10^{-9}$ meter.

# COMMONLY USED UNITS OF MEASURE

## Length

### Metric Measure

| | |
|---|---|
| 1 kilometer (km) | = 1000 meters (m) |
| 1 meter (m) | = 100 centimeters (cm) |
| 1 centimeter (cm) | = 10 millimeters (mm) |

| | |
|---|---|
| 1 millimeter (mm) | = 1000 micrometers (μm) (formerly called microns) |
| 1 micrometer (μm) | = 0.001 millimeter (mm) |
| 1 angstrom (Å) | = $10^{-8}$ centimeters (cm) |

### Nonmetric Measure

| | |
|---|---|
| 1 mile (mi) | = 5280 feet (ft) = 1760 yards (yd) |
| 1 yard (yd) | = 3 feet (ft) |
| 1 fathom (fath) | = 6 feet (ft) |

### Conversions

| | |
|---|---|
| 1 kilometer (km) | = 0.6214 mile (mi) |
| 1 meter (m) | = 1.094 yards (yd) = 3.281 feet (ft) |
| 1 centimeter (cm) | = 0.3937 inch (in) |
| 1 millimeter (mm) | = 0.0394 inch (in) |
| 1 mile (mi) | = 1.609 kilometers (km) |
| 1 yard (yd) | = 0.9144 meter (m) |
| 1 foot (ft) | = 0.3048 meter (m) |
| 1 inch (in) | = 2.54 centimeters (cm) |
| 1 inch (in) | = 25.4 millimeters (mm) |
| 1 fathom (fath) | = 1.8288 meters (m) |

## Area

### Metric Measure

| | |
|---|---|
| 1 square kilometers ($km^2$) | = 1,000,000 square meters ($m^2$) |
| | = 100 hectares (ha) |
| 1 square meter ($m^2$) | = 10,000 square centimeters ($cm^2$) |
| 1 hectare (ha) | = 10,000 square meters ($m^2$) |

### Nonmetric Measure

| | |
|---|---|
| 1 square mile ($mi^2$) | = 640 acres (ac) |
| 1 acre (ac) | = 4840 square yards ($yd^2$) |
| 1 square foot ($ft^2$) | = 144 square inches ($in^2$) |

### Conversions

| | |
|---|---|
| 1 square kilometer ($km^2$) | = 0.386 square mile ($mi^2$) |
| 1 hectare (ha) | = 2.471 acres (ac) |
| 1 square meter ($m^2$) | = 1.196 square yards ($yd^2$) |
| | = 10.764 square feet ($ft^2$) |
| 1 square centimeter ($cm^2$) | = 0.155 square inch ($in^2$) |
| 1 square mile ($mi^2$) | = 2.59 square kilometers ($km^2$) |
| 1 acre (ac) | = 0.4047 hectare (ha) |
| 1 square yard ($yd^2$) | = 0.836 square meter ($m^2$) |
| 1 square foot ($ft^2$) | = 0.0929 square meter ($m^2$) |
| 1 square inch ($in^2$) | = 6.4516 square centimeter ($cm^2$) |

## Volume

### Metric Measure

| | |
|---|---|
| 1 cubic meter ($m^3$) | = 1,000,000 cubic centimeters ($cm^3$) |
| 1 liter (l) | = 1000 milliliters (ml) |
| | = 0.001 cubic meter ($m^3$) |
| 1 centiliter (cl) | = 10 milliliters (ml) |
| 1 milliliter (ml) | = 1 cubic centimeter ($cm^3$) |

### Nonmetric Measure

| | |
|---|---|
| 1 cubic yard ($yd^3$) | = 27 cubic feet ($ft^3$) |
| 1 cubic foot ($ft^3$) | = 1728 cubic inches ($in^3$) |
| 1 barrel (oil) (bbl) | = 42 gallons (U.S.) (gal) |

### Conversions

| | |
|---|---|
| 1 cubic kilometer ($km^3$) | = 0.24 cubic miles ($mi^3$) |
| 1 cubic meter ($m^3$) | = 264.2 gallons (U.S.) (gal) |
| | = 35.314 cubic feet ($ft^3$) |
| 1 liter (l) | = 1.057 quarts (U.S.) (qt) |
| | = 33.815 ounces (U.S. fluid) (fl. oz.) |
| 1 cubic centimeter ($cm^3$) | = 0.0610 cubic inch ($in^3$) |
| 1 cubic mile ($mi^3$) | = 4.168 cubic kilometers ($km^3$) |
| 1 acre-foot (ac-ft) | = 1233.46 cubic meters ($m^3$) |
| 1 cubic yard ($yd^3$) | = 0.7646 cubic meter ($m^3$) |
| 1 cubic foot ($ft^3$) | = 0.0283 cubic meter ($m^3$) |
| 1 cubic inch ($in^3$) | = 16.39 cubic centimeters ($cm^3$) |
| 1 gallon (gal) | = 3.784 liters (l) |

## Mass

### Metric Measure

| | |
|---|---|
| 1000 kilograms (kg) | = 1 metric ton (also called a tonne) (m.t) |
| 1 kilogram (kg) | = 1000 grams (g) |

### Nonmetric Measure

| | |
|---|---|
| 1 short ton (sh.t) | = 2000 pounds (lb) |
| 1 long ton (l.t) | = 2240 pounds (lb) |
| 1 pound (avoirdupois) (lb) | = 16 ounces (avoirdupois) (oz) = 7000 grains (gr) |
| 1 ounce (avoirdupois) (oz) | = 437.5 grains (gr) |
| 1 pound (Troy) (Tr. lb) | = 12 ounces (Troy) (Tr. oz) |
| 1 ounce (Troy) (Tr. oz) | = 20 pennyweight (dwt) |

## Conversions

| | |
|---|---|
| 1 metric ton (m.t) | = 2205 pounds (avoirdupois) (lb) |
| 1 kilogram (kg) | = 2.205 pounds (avoirdupois) (lb) |
| 1 gram (g) | = 0.03527 ounce (avoirdupois) (oz) = 0.03215 ounce (Troy) (Tr. oz) = 15,432 grains (gr) |
| 1 pound (lb) | = 0.4536 kilogram (kg) |
| 1 ounce (avoirdupois) (oz) | = 28.35 grams (g) |
| 1 ounce (avoirdupois) (oz) | = 1.097 ounces (Troy) (Tr. oz) |

## Pressure

| | |
|---|---|
| 1 pascal (Pa) | = 1 newton/square meter (N/m$^2$) |
| 1 kilogram/square centimeter (kg/cm$^2$) | = 0.96784 atmosphere (atm) = 14.2233 pounds/square inch (lb/in$^2$) = 0.98067 bar |
| 1 bar | = 0.98692 atmosphere (atm) = 10$^5$ pascals (Pa) = 1.02 kilograms/square centimeter (kg/cm$^2$) |

# Energy and Power

## Energy

| | |
|---|---|
| 1 joule ( J) | = 1 newton meter (N.m) |
| | = $2.390 \times 10^{-1}$ calorie (cal) |
| | = $9.47 \times 10^{-4}$ British thermal unit (Btu) |
| | = $2.78 \times 10^{-7}$ kilowatt-hour (kWh) |
| 1 calorie (cal) | = 4.184 joule ( J) |
| | = $3.968 \times 10^{-3}$ British thermal unit (Btu) |
| | = $1.16 \times 10^{-6}$ kilowatt-hour (kWh) |
| 1 British thermal unit (Btu) | = 1055.87 joules ( J) |
| | = 252.19 calories (cal) |
| | = $2.928 \times 10^{-4}$ kilowatt-hour (kWh) |
| 1 kilowatt hour | = $3.6 \times 10^6$ joules ( J) |
| | = $8.60 \times 10^5$ calories (cal) |
| | = $3.41 \times 10^3$ British thermal units (Btu) |

## Power (energy per unit time)

| | |
|---|---|
| 1 watt (W) | = 1 joule per second ( J/s) |
| | = 3.4129 Btu/h |
| | = $1.341 \times 10^{-3}$ horsepower (hp) |
| | = 14.34 calories per minute (cal/min) |
| 1 horsepower (hp) | = $7.46 \times 10^2$ watts (W) |

### *Temperature*

To change from Fahrenheit (F) to Celsius (C)

$$°C = \frac{(°F - 32°)}{1.8}$$

To change from Celsius (C) to Fahrenheit (F)

$$°F = (°C \times 1.8) + 32°$$

To change from Celsius (C) to Kelvin (K)

$$K = °C + 273.15$$

To change from Fahrenheit (F) to Kelvin (K)

$$K = \frac{(°F - 32°)}{1.8} + 273.15$$

# APPENDIX
## · B ·

# TABLES OF THE CHEMICAL ELEMENTS AND NATURALLY OCCURRING ISOTOPES

TABLE B.1   The Periodic Table of the Elements

Key:
- Atomic number (top)
- Name
- Symbol (large)
- Approximate atomic weight (to nearest whole number) (bottom)

Example: 11 / Sodium / Na / 23

| Group IA | IIA | IIIB | IVB | VB | VIB | VIIB | VIII | VIII | VIII | IB | IIB | IIIA | IVA | VA | VIA | VIIA | VIII |
|---|---|---|---|---|---|---|---|---|---|---|---|---|---|---|---|---|---|
| 1 Hydrogen **H** 1 | | | | | | | | | | | | | | | | | 2 Helium **He** 4 |
| 3 Lithium **Li** 7 | 4 Beryllium **Be** 9 | | | | | | | | | | | 5 Boron **B** 11 | 6 Carbon **C** 12 | 7 Nitrogen **N** 14 | 8 Oxygen **O** 16 | 9 Flourine **F** 19 | 10 Neon **Ne** 20 |
| 11 Sodium **Na** 23 | 12 Magnesium **Mg** 24 | | | | | | | | | | | 13 Aluminum **Al** 27 | 14 Silicon **Si** 28 | 15 Phosphorus **P** 31 | 16 Sulfur **S** 32 | 17 Chlorine **Cl** 35 | 18 Argon **Ar** 40 |
| 19 Potassium **K** 39 | 20 Calcium **Ca** 40 | 21 Scandium **Sc** 45 | 22 Titanium **Ti** 48 | 23 Vanadium **V** 51 | 24 Chromium **Cr** 52 | 25 Manganese **Mn** 55 | 26 Iron **Fe** 56 | 27 Cobalt **Co** 59 | 28 Nickel **Ni** 59 | 29 Copper **Cu** 64 | 30 Zinc **Zn** 65 | 31 Gallium **Ga** 70 | 32 Germanium **Ge** 73 | 33 Arsenic **As** 75 | 34 Selenium **Se** 79 | 35 Bromine **Br** 80 | 36 Krypton **Kr** 84 |
| 37 Rubidium **Rb** 85 | 38 Strontium **Sr** 88 | 39 Yttrium **Y** 89 | 40 Zirconium **Zr** 91 | 41 Niobium **Nb** 93 | 42 Molybdenum **Mo** 96 | 43 Technetium **Tc** 99 | 44 Ruthenium **Ru** 101 | 45 Rhodium **Rh** 103 | 46 Palladium **Pd** 106 | 47 Silver **Ag** 108 | 48 Cadmium **Cd** 112 | 49 Indium **In** 115 | 50 Tin **Sn** 119 | 51 Antimony **Sb** 122 | 52 Tellurium **Te** 128 | 53 Iodine **I** 127 | 54 Xenon **Xe** 131 |
| 55 Cesium **Cs** 133 | 56 Barium **Ba** 137 | 57 *Lanthanum **La** 139 | 72 Hafnium **Hf** 178 | 73 Tantalum **Ta** 139 | 74 Wolfran (tungsten) **W** 184 | 75 Rhenium **Re** 186 | 76 Osmium **Os** 190 | 77 Iridium **Ir** 192 | 78 Platinum **Pt** 195 | 79 Gold **Au** 197 | 80 Mercury **Hg** 201 | 81 Thallium **Tl** 204 | 82 Lead **Pb** 207 | 83 Bismuth **Bi** 209 | 84 Polonium **Po** 210 | 85 Astatine **At** 210 | 86 Radon **Rn** 222 |
| 87 Francium **Fr** 223 | 88 Radum **Ra** 226 | 89 **Actinium **Ac** 227 | 104 Unnil-quadium **Unq** 266 | 105 Unil-pentium **Unp** 262 | 106 Unil-hexium **Unh** 263 | 107 Unil-septium **Uns** 262 | 108 Uniloctium **Uno** 265 | 109 Unilennium **Une** 266 | | | | | | | | | |

*Lanthanide series

| 58 Cerium **Ce** 140 | 59 Prase-odymium **Pr** 141 | 60 Neodymium **Nd** 144 | 61 Promethium **Pm** 147 | 62 Samarium **Sm** 150 | 63 Europium **Eu** 152 | 64 Gadolinium **Gd** 157 | 65 Terbium **Tb** | 66 Dysprosium **Dy** 163 | 67 Holmium **Ho** 165 | 68 Erbium **Er** 167 | 69 Thulium **Tm** 169 | 70 Ytterbium **Yb** 173 | 71 Lutetium **Lu** 175 |
|---|---|---|---|---|---|---|---|---|---|---|---|---|---|

**Actinide series

| 90 Thorium **Th** 232 | 91 Protactinium **Pa** 231 | 92 Uranium **U** 238 | 93 Neptunium **Np** 237 | 94 Plutonium **Pu** 242 | 95 Americium **Am** 243 | 96 Curium **Cm** 247 | 97 Berkellium **Bk** 247 | 98 Californium **Cf** 251 | 99 Einsteinium **Es** 254 | 100 Fermium **Fm** 253 | 101 Mendelevium **Md** 256 | 102 Nobelium **No** 254 | 103 Lawrencium **Lr** 257 |
|---|---|---|---|---|---|---|---|---|---|---|---|---|---|

TABLE B.2   Alphabetical List of the Elements

| Element | Symbol | Atomic Number | Crustal Abundance, Weight Percent | Element | Symbol | Atomic Number | Crustal Abundance, Weight Percent |
|---------|--------|---------------|-----------------------------------|---------|--------|---------------|-----------------------------------|
| Actinium | Ac | 89 | Human-made | Mercury | Hg | 80 | 0.000002 |
| Aluminum | Al | 13 | 8.00 | Molybdenum | Mo | 42 | 0.00012 |
| Americium | Am | 95 | Human-made | Neodymium | Nd | 60 | 0.0044 |
| Antimony | Sb | 51 | 0.00002 | Neon | Ne | 10 | Not known |
| Argon | Ar | 18 | Not known | Neptunium | Np | 93 | Human-made |
| Arsenic | As | 33 | 0.00020 | Nickel | Ni | 28 | 0.0072 |
| Astatine | At | 85 | Human-made | Niobium | Nb | 41 | 0.0020 |
| Barium | Ba | 56 | 0.0380 | Nitrogen | N | 7 | 0.0020 |
| Berkelium | Bk | 97 | Human-made | Nobelium | No | 102 | Human-made |
| Beryllium | Be | 4 | 0.00020 | Osmium | Os | 76 | 0.00000002 |
| Bismuth | Bi | 83 | 0.0000004 | Oxygen[b] | O | 8 | 45.2 |
| Boron | B | 5 | 0.0007 | Palladium | Pd | 46 | 0.0000003 |
| Bromine | Br | 35 | 0.00040 | Phosphorus | P | 15 | 0.1010 |
| Cadmium | Cd | 48 | 0.000018 | Platinum | Pt | 78 | 0.0000005 |
| Calcium | Ca | 20 | 5.06 | Plutonium | Pu | 94 | Human-made |
| Californium | Cf | 98 | Human-made | Polonium | Po | 84 | Footnote[d] |
| Carbon[a] | C | 6 | 0.02 | Potassium | K | 19 | 1.68 |
| Cerium | Ce | 58 | 0.0083 | Praseodymium | Pr | 59 | 0.0013 |
| Cesium | Cs | 55 | 0.00016 | Promethium | Pm | 61 | Human-made |
| Chlorine | Cl | 17 | 0.0190 | Protactinium | Pa | 91 | Footnote[d] |
| Chromium | Cr | 24 | 0.0096 | Radium | Ra | 88 | Footnote[d] |
| Cobalt | Co | 27 | 0.0028 | Radon | Rn | 86 | Footnote[d] |
| Copper | Cu | 29 | 0.0058 | Rhenium | Re | 75 | 0.00000004 |
| Curium | Cm | 96 | Human-made | Rhodium[c] | Rh | 45 | 0.00000001 |
| Dysprosium | Dy | 66 | 0.00085 | Rubidium | Rb | 37 | 0.0070 |
| Einsteinium | Es | 99 | Human-made | Ruthenium[c] | Ru | 44 | 0.00000001 |
| Erbium | Er | 68 | 0.00036 | Samarium | Sm | 62 | 0.00077 |
| Europium | Eu | 63 | 0.00022 | Scandium | Sc | 21 | 0.0022 |
| Fermium | Fm | 100 | Human-made | Selenium | Se | 34 | 0.000005 |
| Fluorine | F | 9 | 0.0460 | Silicon | Si | 14 | 27.20 |
| Francium | Fr | 87 | Human-made | Silver | Ag | 47 | 0.000008 |
| Gadolinium | Gd | 64 | 0.00063 | Sodium | Na | 11 | 2.32 |
| Gallium | Ga | 31 | 0.0017 | Srontium | Sr | 38 | 0.0450 |
| Germanium | Ge | 32 | 0.00013 | Sulfur | S | 16 | 0.030 |
| Gold | Au | 79 | 0.0000002 | Tantalum | Ta | 73 | 0.00024 |
| Hafnium | Hf | 72 | 0.0004 | Technetium | Tc | 43 | Human-made |
| Helium | He | 2 | Not known | Tellurium[c] | Te | 52 | 0.000001 |
| Holmium | Ho | 67 | 0.00016 | Terbium | Tb | 65 | 0.00010 |
| Hydrogen[b] | H | 1 | 0.14 | Thallium | Tl | 81 | 0.000047 |
| Indium | In | 49 | 0.00002 | Thorium | Th | 90 | 0.00058 |
| Iodine | I | 53 | 0.00005 | Thulium | Tm | 69 | 0.000052 |
| Iridium | Ir | 77 | 0.00000002 | Tin | Sn | 50 | 0.00015 |
| Iron | Fe | 26 | 5.80 | Titanium | Ti | 22 | 0.86 |
| Krypton | Kr | 36 | Not known | Tungsten | W | 74 | 0.00010 |
| Lanthanum | La | 57 | 0.0050 | Unnilennium | Une | 109 | Human-made |
| Lawrencium | Lw | 103 | Human-made | Unnilhexium | Unh | 106 | Human-made |
| Lead | Pb | 82 | 0.0010 | Unniloctium | Uno | 108 | Human-made |
| Lithium | Li | 3 | 0.0020 | Unnilpentium | Unp | 105 | Human-made |
| Lutetium | Lu | 71 | 0.000080 | Unnilquadium | Unq | 104 | Human-made |
| Magnesium | Mg | 12 | 2.77 | Unnilseptium | Uns | 107 | Human-made |
| Manganese | Mn | 25 | 0.100 | Uranium | U | 92 | 0.00016 |
| Mendelevium | Md | 101 | Human-made | Vanadium | V | 23 | 0.0170 |

TABLE B.2    (Continued)

| Element | Symbol | Atomic Number | Crustal Abundance, Weight Percent | Element | Symbol | Atomic Number | Crustal Abundance, Weight Percent |
|---------|--------|---------------|-----------------------------------|---------|--------|---------------|-----------------------------------|
| Xenon | Xe | 54 | Not known | Zinc | Zn | 30 | 0.0082 |
| Ytterbium | Yb | 70 | 0.00034 | Zirconium | Zr | 40 | 0.0140 |
| Yttrium | Y | 39 | 0.0035 | | | | |

*Source:* After K. K. Turekian, 1969.

[a]Estimate from S. R. Taylor (1964).

[b]Analyses of crustal rocks do not usually include separate determinations for hydrogen and oxygen. Both combine in essentially constant proportions with other elements, so abundances can be calculated.

[c]Estimates are uncertain and have a very low reliability.

[d]Elements formed by decay of uranium and thorium. The daughter products are radioactive with such short half-lives that crustal accumulations are too low to be measured accurately.

TABLE B.3    Naturally Occurring Elements Listed in Order of Atomic Number, Together with the Naturally Occurring Isotopes of Each Element, Listed in Order of Mass Number

| Atomic Number[a] | Name | Symbol | Mass Numbers[b] of Natural Isotopes |
|------------------|------|--------|-------------------------------------|
| 1 | Hydrogen | H | 1, 2, [3][c] |
| 2 | Helium | He | 3, 4 |
| 3 | Lithium | Li | 6, 7 |
| 4 | Beryllium | Be | 9, [10] |
| 5 | Boron | B | 10, 11 |
| 6 | Carbon | C | 12, 13, [14] |
| 7 | Nitrogen | N | 14, 15 |
| 8 | Oxygen | O | 16, 17, 18 |
| 9 | Fluorine | F | 19 |
| 10 | Neon | Ne | 20, 21, 22 |
| 11 | Sodium | Na | 23 |
| 12 | Magnesium | Mg | 24, 25, 26 |
| 13 | Aluminum | Al | 27 |
| 14 | Silicon | Si | 28, 29, 30 |
| 15 | Phosphorus | P | 31 |
| 16 | Sulfur | S | 32, 33, 34, 36 |
| 17 | Chlorine | Cl | 35, 37 |
| 18 | Argon | A | 36, 38, 40 |
| 19 | Potassium | K | 39, [40], 41 |
| 20 | Calcium | Ca | 40, 42, 43, 44, 46, [48] |
| 21 | Scandium | Sc | 45 |
| 22 | Titanium | Ti | 46, 47, 48, 49, 50 |
| 23 | Vanadium | V | [50], 51 |
| 24 | Chromium | Cr | 50, 52, 53, 54 |
| 25 | Manganese | Mn | 55 |
| 26 | Iron | Fe | 54, 56, 57, 58 |

**TABLE B.3**    *(Continued)*

| Atomic Number[a] | Name | Symbol | Mass Numbers[b] of Natural Isotopes |
|---|---|---|---|
| 27 | Cobalt | Co | 59 |
| 28 | Nickel | Ni | 58, 60, 61, 62, 64 |
| 29 | Copper | Cu | 63, 65 |
| 30 | Zinc | Zn | 64, 66, 67, 68, 70 |
| 31 | Gallium | Ga | 69, 71 |
| 32 | Germanium | Ge | 70, 72, 73, 74, 76 |
| 33 | Arsenic | As | 75 |
| 34 | Selenium | Se | 74, 76, 77, 80, 82 |
| 35 | Bromine | Br | 79, 81 |
| 36 | Krypton | Kr | 78, 80, 82, 83, 84, 86 |
| 37 | Rubidium | Rb | 85, 87 |
| 38 | Strontium | Sr | 84, 86, 87, 88 |
| 39 | Yttrium | Y | 89 |
| 40 | Zirconium | Zr | 90, 91, 92, 94, 96 |
| 41 | Niobium | Nb | 93 |
| 42 | Molybdenum | Mo | 92, 94, 95, 96, 97, 98, 100 |
| 44 | Ruthenium | Ru | 96, 98, 99, 100, 101, 102, 104 |
| 45 | Rhodium | Rh | 103 |
| 46 | Palladium | Pd | 102, 104, 105, 106, 108, 110 |
| 47 | Silver | Ag | 107, 109 |
| 48 | Cadmium | Cd | 106, 108, 110, 111, 112, 113, 114, 116 |
| 49 | Indium | In | 113, 115 |
| 50 | Tin | Sn | 112, 114, 115, 116, 117, 118, 119, 120, 122, 124 |
| 51 | Antimony | Sb | 121, 123 |
| 52 | Tellurium | Te | 120, 122, 123, 124, 125, 126, 128, 130 |
| 53 | Iodine | I | 127 |
| 54 | Xenon | Xe | 124, 126, 128, 129, 130, 131, 132, 134, 136 |
| 55 | Cesium | Cs | 133 |
| 56 | Barium | Ba | 130, 132, 134, 135, 136, 137, 138 |
| 57 | Lanthanum | La | 138, 139 |
| 58 | Cerium | Ce | 136, 138, 140, 142 |
| 59 | Praseodymium | Pr | 141 |
| 60 | Neodymium | Nd | 142, 143, 144, 145, 146, 148, 150 |
| 62 | Samarium | Sm | 144, 147, 148, 149, 150, 152, 154 |
| 63 | Europium | Eu | 151, 153 |
| 64 | Gadolinium | Gd | 152, 154, 155, 156, 157, 158, 160 |
| 65 | Terbium | Tb | 159 |
| 66 | Dysprosium | Dy | 156, 158, 160, 161, 162, 163, 164 |
| 67 | Holmium | Ho | 165 |
| 68 | Erbium | Er | 162, 166, 167, 168, 170 |
| 69 | Thulium | Tm | 169 |
| 70 | Ytterbium | Yb | 168, 170, 171, 172, 173, 174, 176 |
| 71 | Lutetium | Lu | 175, 176 |

**TABLE B.3**    *(Continued)*

| Atomic Number[a] | Name | Symbol | Mass Numbers[b] of Natural Isotopes |
|---|---|---|---|
| 72 | Hafnium | Hf | 174, 176, 177, 178, 179, 180 |
| 73 | Tantalum | Ta | 180, 181 |
| 74 | Tungsten | W | 180, 182, 183, 184, 186 |
| 75 | Rhenium | Re | 185, 187 |
| 76 | Osmium | Os | 184, 186, 187, 188, 189, 190, 192 |
| 77 | Iridium | Ir | 191, 193 |
| 78 | Platinum | Pt | 190, 192, 195, 196, 198 |
| 79 | Gold | Au | 197 |
| 80 | Mercury | Hg | 196, 198, 199, 200, 201, 202, 204 |
| 81 | Thallium | Tl | 203, 205 |
| 82 | Lead | Pb | 204, 206, 207, 208 |
| 83 | Bismuth | Bi | 209 |
| 84 | Polonium | Po | 210 |
| 86 | Radon | Rn | 222 |
| 88 | Radium | Ra | 226 |
| 90 | Thorium | Th | 232 |
| 91 | Protactinium | Pa | 231 |
| 92 | Uranium | U | 234, 235, 238 |

[a]Atomic number = number of protons.

[b]Mass number = protons + neutrons.

[c] ☐ indicates isotope is radioactive.

# APPENDIX
## · C ·

# TABLES OF THE PROPERTIES OF SELECTED COMMON MINERALS

**TABLE C.1**   Properties of the Common Minerals with Metallic Luster

| Mineral | Chemical Composition | Form and Habit | Cleavage | Hardness / Specific Gravity | Other Properties | Most Distinctive Properties |
|---|---|---|---|---|---|---|
| Chalcopyrite | $CuFeS_2$ | Massive or granular. | None. Uneven fracture. | 3.5–4 / 4.2 | Golden yellow to brassy yellow. Dark green to black streak. | Streak. Hardness distinguishes from pyrite. |
| Chromite | $FeGr_2O_4$ | Massive or granular. | None. Uneven fracture. | 5.5 / 4.6 | Iron black to brownish black. Dark brown streak. | Streak and lack of magnetism distinguishes from magnetite. |
| Copper | $Cu$ | Massive, twisted leaves and wires. | None. Can be cut with a knife. | 2.5–3 / 9 | Copper color but commonly stained green. | Color, specific gravity, malleable. |
| Galena | $PbS$ | Cubic crystals, coarse or fine-grained granular masses. | Perfect in three directions at right angles. | 2.5 / 7.6 | Lead-gray color. Gray to gray-black streak. | Cleavage and streak. |
| Gold | $Au$ | Small irregular grains. | None. Malleable. | 2.5 / 19.3 | Gold color. Can be flattened without breakage. | Color, specific gravity, malleability. |
| Hematite | $Fe_2O_3$ | Massive, granular, micaceous | Uneven fracture. | 5–6 / 5 | Reddish-brown, gray to black. Reddish-brown streak. | Streak, hardness. |

**TABLE C.1**    *(Continued)*

| Mineral | Chemical Composition | Form and Habit | Cleavage | Hardness / Specific Gravity | Other Properties | Most Distinctive Properties |
|---------|---------------------|----------------|----------|------------------------------|------------------|------------------------------|
| Limonite (*Goethite* is most common.) | A complex mixture of minerals, mainly hydrous iron oxides. | Massive, coatings, botryoidal crusts, earthy masses. | None. | 1–5.5 / 3.5–4 | Yellow, brown, black, yellowish-brown streak. | Streak. |
| Magnetite | $Fe_3O_4$ | Massive, granular. Crystals have octahedral shape. | None. Uneven fracture. | 5.5–6.5 / 5 | Black. Black streak. Strongly attracted to a magnet. | Streak, magnetism |
| Pyrite ("Fool's gold") | $FeS_2$ | Cubic crystals with striated faces. Massive. | None. Uneven fracture. | 6–6.5 / 5.2 | Pale brass-yellow, darker if tarnished. Greenish-black streak. | Streak. Hardness distinguishes from chalcopyrite. Not malleable, which distinguishes from gold. |
| Rutile | $TiO_2$ | Slender, prismatic crystals or granular masses. | Good in one direction. Conchoidal fracture in others. | 6–6.5 / 4.2 | Reddish-brown (common), black (rare). Brownish streak. Adamantine luster. | Luster, habit, hardness. |
| Sphalerite | ZnS | Fine to coarse granular masses. Tetrahedron shaped crystals. | Perfect in six directions. | 3.5–4 / 4 | Yellowish-brown to black. White to yellowish-brown streak. Resinous luster. | Cleavage, hardness, luster. |
| Uraninite | $UO_2$ to $U_3O_8$ | Massive, with botryoidal forms. Rare crystals with cubic shapes. | None. Uneven fracture. | 5–6 / 6.5–10 | Black to dark brown. Streak black to dark brown. Dull luster. | Luster and specific gravity distinguish from magnetite. Streak distinguishes from ilmenite and hematite. |

**TABLE C.2   Properties of Rock-Forming Minerals with Nonmetallic Luster**

| Mineral | Chemical Composition | Form and Habit | Cleavage | Hardness / Specific Gravity | | Other Properties | Most Distinctive Properties |
|---|---|---|---|---|---|---|---|
| Amphiboles. (A complex family of minerals, *Hornblende* is most common.) | $X_2Y_5Si_8O_{22}$ $(OH)_2$ where $X$ = Ca, Na; $Y$ = Mg, Fe, Al. | Long, six-sided crystals; also fibers and irregular grains | Two; intersecting at 56° and 124° | 5–6 | 2.9– 3.8 | Common in metamorphic and igneous rocks. *Hornblende* is dark green to black; *actinolite*, green; *tremolite*, white. | Cleavage, habit. |
| Andalusite | $Al_2SiO_5$ | Long crystals, often square in cross-section. | Weak, parallel to length of crystal. | 7.5 | 3.2 | Found in metamorphic rocks. Often flesh-colored. | Hardness, form. |
| Anhydrite | $CaSO_4$ | Crystals are rare. Irregular grains or fibers. | Three, at right angles. | 3 | 2.9 | Alters to gypsum. Pearly luster, white or colorless. | Cleavage, hardness. |
| Apatite | $Ca_5(PO_4)_3$ $(F, OH, Cl)$ | Granular masses. Perfect six-sided crystals. | Poor. One direction. | 5 | 3.2 | Green, brown, blue, or white. Common in many kinds of rocks in small amounts. | Hardness, form. |
| Aragonite | $CaCO_3$ | Massive, or slender, needle-like crystals. | Poor. Two directions. | 3.5 | 2.9 | Colorless or white. Effervesces with dilute HCl. | Effervescence with acid. Poor cleavage distinguishes from calcite. |
| Asbestos | | | See Serpentine | | | | |
| Augite | | | See Pyroxene | | | | |
| Biotite | | | See Mica | | | | |
| Calcite | $CaCO_3$ | Tapering crystals and granular masses. | Three perfect; at oblique angles to give a rhomb-shaped fragment. | 3 | 2.7 | Colorless or white. Effervesces with dilute HCl. | Cleavage, effervescence with acid. |
| Chlorite | $(Mg, Fe)_5(Al, Fe)_2$ $Si_3O_{10}(OH)_8$ | Flaky masses of minute scales. | One perfect; parallel to flakes. | 2– 2.5 | 2.6– 2.9 | Common in metamorphic rocks. Light to dark green. Greasy luster. | Cleavage—flakes not elastic, distinguishes from mica. Color. |
| Dolomite | $CaMg(CO_3)_2$ | Crystals with rhomb-shaped faces. Granular masses. | Perfect in three directions as in calcite. | 3.5 | 2.8 | White or gray. Does not effervesce in cold, dilute HCl unless powdered. Pearly luster. | Cleavage. Lack of effervescence with acid. |

**TABLE C.2**  *(Continued)*

| Mineral | Chemical Composition | Form and Habit | Cleavage | Hardness / Specific Gravity | Other Properties | Most Distinctive Properties |
|---|---|---|---|---|---|---|
| Epidote | Complex silicate of Ca, Fe and Al | Small elongate crystals. Fibrous. | One perfect, one poor. | 6–7 / 3.4 | Yellowish-green to dark green. Common in metamorphic rocks. | Habit, color. Hardness distinguishes from chlorite. |
| Feldspars: Potassium feldspar (*orthoclase* is a common variety) | $KAlSi_3O_8$ | Prism-shaped crystals, granular masses. | Two perfect, at right angles. | 6 / 2.6 | Common mineral. Pink, white, or gray in color. | Color, cleavage. |
| *Plagioclase* | $NaAlSi_3O_8$ (albite) and $CaAl_2Si_2O_8$ (anorthite) and all compositions between. | Irregular grains, cleavable masses. Tabular crystals. | Two perfect, not quite at right angles. | 6–6.5 / 2.6–2.7 | White to dark gray. Cleavage planes may show fine parallel striations. | Cleavage. Striations on cleavage planes will distinguish from potassium feldspar. |
| Fluorite | $CaF_2$ | Cubic crystals, granular masses. | Perfect in four directions. | 4 / 3.2 | Colorless, bluish green. Always an accessory mineral. | Hardness cleavage, does not effervesce with acid. |
| Garnets | $X_3Y_2(SiO_4)_3$; X = Ca, Mg, Fe, Mn; Y = Al, Fe, Ti, Cr. | Perfect crystals with 12 or 24 sides. Granular masses. | None. Uneven fracture. | 6.5–7.5 / 3.5–4.3 | Common in metamorphic rocks. Red, brown, yellowish-green, black. | Crystals, hardness, no cleavage. |
| Graphite | C | Scaly masses. | One, perfect. Forms slippery flakes. | 1–2 / 2.2 | Metamorphic rocks. Black with metallic to dull luster. | Cleavage, color. Marks paper. |
| Gypsum | $CaSO_4 \cdot 2H_2O$ | Elongate or tabular crystals. Fibrous and earthy masses. | One, perfect. Flakes bend but are not elastic. | 2 / 2.3 | Vitreous to pearly luster. Colorless. | Hardness, cleavage. |
| Halite | NaCl | Cubic crystals. | Perfect to give cubes. | 2.5 / 2.2 | Tastes salty. Colorless, blue. | Taste, cleavage. |
| Hornblende | | | See Amphibole | | | |
| Kaolinite | $Al_2Si_2O_5(OH)_4$ | Soft, earthy masses. Submicroscopic crystals. | One, perfect. | 2–2.5 / 2.6 | White, yellowish. Plastic when wet; emits clayey odor. Dull luster. | Feel, plasticity, odor. |

TABLE C.2  *(Continued)*

| Mineral | Chemical Composition | Form and Habit | Cleavage | Hardness / Specific Gravity | Other Properties | Most Distinctive Properties |
|---|---|---|---|---|---|---|
| Kyanite | $Al_2SiO_5$ | Bladed crystals. | One perfect. One imperfect. | 4.5 parallel to blade, 7 across blade / 3.6 | Blue, white, gray Common in metamorphic rocks. | Variable hardness, distinguishes from sillimanite. Color. |
| Mica: *Biotite* | $K(Mg, Fe)_3 AlSi_3O_{10} (OH)_2$ | Irregular masses of flakes. | One, perfect. | 2.5–3 / 2.8–3.2 | Common in igneous and metamorphic rocks. Black, brown, dark green. | Cleavage, color. Flakes are elastic. |
| *Muscovite* | $KAl_3Si_3O_{10} (OH)_2$ | Thin flakes. | One, perfect. | 2–2.5 / 2.7 | Common in igneous and metamorphic rocks. Colorless, pale green or brown. | Cleavage, color. Flakes are elastic. |
| Olivine | $(Mg, Fe)_2 SiO_4$ | Small grains, granular masses. | None. Conchoidal fracture. | 6.5–7 / 3.2–4.3 | Igneous rocks. Olive green to yellowish-green. | Color, fracture, habit. |
| Orthoclase | | | See Feldspar | | | |
| Plagioclase | | | See Feldspar | | | |
| Pyroxene (A complex family of minerals. *Augite* is most common.) | $XY(SiO_3)_2$ $X = Y = Ca,$ Mg. Fe | 8-sided stubby crystals. Granular masses. | Two, perfect, nearly at right angles. | 5–6 / 3.2–3.9 | Igneous and metamorphic rocks. *Augite,* dark green to black; other varieties white to green. | Cleavage |
| Quartz | $SiO_2$ | 6-sided crystals, granular masses. | None. Conchoidal fracture | 7 / 2.6 | Colorless, white, gray, but may have any color, depending on impurities. Vitreous to greasy luster. | Form, fracture, striations across crystal faces at right angles to long dimension. |
| Serpentine (Fibrous variety is *asbestos*) | $Mg_3Si_2O_5 (OH)_4$ | Platy or fibrous. | One, perfect. | 2.5–5 / 2.2–2.6 | Light to dark green. Smooth, greasy feel. | Habit, hardness. |
| Sillimanite | $Al_2SiO_5$ | Long needle-like crystals, fibers. | Breaks irregularly, except in fibrous variety | 6–7 / 3.2 | White, gray. Metamorphic rocks. | Hardness distinguishes from kyanite. Habit. |
| Talc | $Mg_3Si_4O_{10} (OH)_2$ | Small scales, compact masses. | One, perfect. | 1 / 2.6–2.8 | Feels slippery. Pearly luster. White to greenish. | Hardness, luster, feel, cleavage. |

**TABLE C.2**   *(Continued)*

| Mineral | Chemical Composition | Form and Habit | Cleavage | Hardness / Specific Gravity | | Other Properties | Most Distinctive Properties |
|---------|---------------------|----------------|----------|---------|---|------------------|------------------------------|
| Tourmaline | Complex silicate of B, Al, Na, Ca, Fe, Li and Mg. | Elongate crystals, commonly with triangular cross section. | None. | 7–7.5 | 3–3.3 | Black, brown, red, pink, green, blue, and yellow. An accessory mineral in many rocks. | Habit. |

**TABLE C.3**   Properties of Some Common Gemstones

| Mineral and Variety | Composition | Form and Habit | Cleavage | Hardness / Specific Gravity | | Other Properties | Most Distinctive Properties |
|---------------------|-------------|----------------|----------|---------|---|------------------|------------------------------|
| Beryl: *Aquamarine* (blue) *Emerald* (green) *Golden beryl* (golden-yellow) | $Be_3Al_2Si_6O_{18}$ | Six-sided, elongate crystals common. | Weak. | 7.5–8 | 2.75 | Bluish green, green, yellow, white, colorless. Common in pegmatites. | Form. Distinguished from apatite by its hardness. |
| Corundum: *Ruby* (red) *Sapphire* (blue) | $Al_2O_3$ | Six-sided, barrel-shaped crystals. | None, but breaks easily across the crystal. | 9 | 4 | Brown, pink, red, blue, colorless. Common in metamorphic rocks. Star sapphire is opalescent with a six-sided light spot showing. | Hardness. |
| Diamond | C | Octahedron-shaped crystals. | Perfect, parallel to faces of octahedron. | 10 | 3.5 | Colorless, yellow; rarely red, orange, green, blue or black. | Hardness, cleavage. |
| Garnet: *Almandite* (red) *Grossularite* (green, cinnamon-brown) *Andradite* (variety *demantoid* is green) | A rock-forming mineral—See Table C.2 | | | | | | |

TABLE C.3 *(Continued)*

| Mineral and Variety | Composition | Form and Habit | Cleavage | Hardness / Specific Gravity | Other Properties | Most Distinctive Properties |
|---|---|---|---|---|---|---|
| Opal (A mineraloid) | $SiO_2 \cdot nH_2O$ | Massive, thin coating. Amorphous. | None. Conchoidal fracture. | 5–6 / 2–2.2 | Colorless, white, yellow, red, brown, green, gray, opalescent. | Hardness, color, form. |
| Quartz: (1) Coarse crystals *Amethyst* (violet) *Cairngorm* (brown) *Citrine* (yellow) *Rock crystal* (colorless) *Rose quartz* (pink) (2) Fine-grained *Agate* (banded, many colors) *Chalcedony* (brown, gray) *Heliotrope* (green) *Jasper* (red) | A rock-forming mineral—See Table C.2 | | | | | |
| Topaz | $Al_2SiO_4(OH, F)_2$ | Prism-shaped crystals, granular masses. | One, perfect. | 8 / 3.5 | Colorless, yellow, blue, brown. | Hardness, form, color. |
| Tourmaline | A rock-forming mineral—See Table C.2 | | | | | |
| Zircon | $ZrSiO_4$ | Four-sided elongate crystals, square in cross-section. | None | 7.5 / 4.7 | Brown, red, green, blue, black. | Habit, hardness. |

# · D ·

# BOWEN'S REACTION SERIES

As discussed briefly in Chapter 6, geochemist N. L. Bowen first recognized the importance of magmatic differentiation (the formation of many different types of rocks from a single starting magma composition) by the process of fractional crystallization. Recall that when fractional crystallization occurs, crystals become separated from the magma they are crystallizing from. This results in a rock and a magma that are different in composition—different from each other and different from the starting composition of the magma as well. Because basaltic magma is far more common than either rhyolitic or andesitic magma, Bowen suggested that all other magmas may be derived from basaltic magma by magmatic differentiation. At least in theory, Bowen argued, a single magma could crystallize into basalt (or gabbro), andesite (or diorite), and rhyolite (or granite) as a result of fractional crystallization.

## CONTINUOUS REACTION SERIES

Bowen carried out experiments to test his hypothesis. He knew that plagioclase feldspars that crystallize from basaltic magmas are calcium-rich. (The most Ca-rich plagioclase feldspar is called *anorthite*.) On the other hand, the plagioclase feldspars that crystallize from rhyolitic magma are sodium-rich (*albite* is the most Na-rich plagioclase feldspar). Andesitic magma, he observed, tends to crystallize plagioclase feldspars of intermediate composition. Bowen's experiments showed that the composition of the first plagioclase to crystallize from a basaltic magma was anorthitic in composition. But the composition of the plagioclase changed, becoming more albitic as crystallization proceeded. Bowen referred to this change as a *continuous reaction series,* by which he meant that the composition of the mineral in the cooling magma changed continuously (reacting chemically with the magma), even though its crystal structure remained unchanged.

Here's how and why this chemical change happens. When a very Ca-rich anorthite crystallizes from a basaltic magma, it uses up some of the calcium from the magma, leaving the magma itself slightly depleted in calcium. The next crystal of plagioclase that crystallizes from the melt, therefore, will be slightly less Ca-rich than the first (and the temperature at which it crystallizes will be slightly lower as well, because the magma is cooling). The next plagioclase crystal, in turn, will be even less Ca-rich (more Na-rich), and so it continues as crystallization progresses (as shown in Fig. D.1). In a continuous reaction series, all of the plagioclase crystals—even the ones formed earliest in the crystallization process—continually change and adjust their compositions as the magma cools. They do this in order to stay in *chemical equilibrium* with the remaining melt.

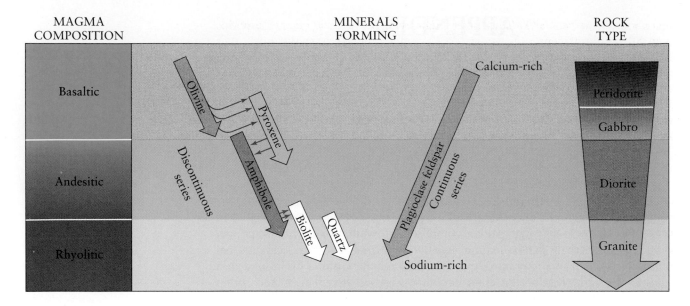

| MAGMA COMPOSITION | MINERALS FORMING | ROCK TYPE |

**FIGURE D.1**
**Bowen's Reaction Series**
The earliest minerals to crystallize from a magma of basaltic composition are Ca-rich plagioclase feldspar (anorthite) and olivine. As cooling and crystallization proceed, the Ca-rich plagioclase reacts with the residual melt and continually changes its composition, becoming more sodium-rich. This is called a *continuous reaction series.* Meanwhile, the early-crystallized olivine reacts with the melt to form pyroxene. Pyroxene, in turn, reacts to form amphibole, and amphibole reacts to form biotite. The composition of the remaining melt becomes increasingly silica-rich, and eventually the final small fraction of melt has the composition of a rhyolitic magma.

One piece of evidence that strongly supports the existence of a continuous reaction series in plagioclase feldspars is the presence of *zoned* feldspar crystals in igneous rocks. A crystal is said to be zoned if it has one chemical composition in its core and a different composition around its rim. In order for a crystallizing plagioclase to change its composition and maintain chemical equilibrium with the surrounding melt, some ions have to move out of the crystal and others have to move in. The speed at which the change can occur is controlled by the rate at which ions can move through the magma to get to the site of the growing crystal, and the rate at which ions can move into and out of the plagioclase crystal structure. Such movement of ions, called *diffusion,* is exceedingly slow. Therefore, equilibrium is rarely attained because the cooling and crystallization rates of magmas are typically faster than diffusion rates. As a result, zoned crystals of plagioclase (Fig. D.2A) are formed. The early-crystallized anorthitic cores can't change their compositions quickly enough, so they become surrounded (or *mantled*) by later-crystallizing albitic rims. Here's the important part: If an albite-rich partial melt were somehow separated from zoned, early-crystallized anorthite-rich crystals, the result would be an albite-rich magma and an anorthite-rich rock—in other words, magmatic differentiation by fractional crystallization.

## DISCONTINUOUS REACTION SERIES

Through his experiments, Bowen also identified several sequences of compositional changes involving other minerals besides the plagioclase feldspars. One of the earliest minerals to form in a cooling basaltic magma is olivine. Olivine contains about 40 percent SiO₂ (silica) by weight, whereas basaltic magma contains about 50 weight percent SiO₂. When the silica-poor olivine crystallizes, it leaves the remaining magma a little richer in silica. Eventually, the olivine is no longer in chemical equilibrium with the melt; the olivine reacts chemically with the melt to form a more silica-rich mineral, pyroxene. Sometimes we find rocks that seem to have been "arrested" in the process of undergoing this reaction; we can see the original core of early-crystallized olivine. mantled by a rim of pyroxene that was forming as a result of the chemical reaction between olivine and melt (Fig. D.2B). The pyroxene, in turn, can eventually react with the melt to form amphibole, and the amphibole can react with the melt to form an even more silica-rich mineral, biotite. Such a series of chemical reactions, in which early formed minerals form entirely new minerals through reaction with the melt, is called a *discontinuous reaction series* (Fig. D.1). If such reactions are interrupted, and low-silica crystals become separated from higher-silica residual melt, magmatic differentiation can occur. If the process of crystal-melt separation happens repeatedly, the composition of the residual melt will become increasingly silica-rich; eventually, the final small fraction of melt will be very silica-rich, with the composition of a rhyolitic magma.

Together, the continuous and discontinuous reactions identified by Bowen through his experiments are known as *Bowen's reaction series* (Fig. D.1). We now know that the process of fractional crystallization rarely goes to the extremes shown in this diagram. Therefore, the answer to the question that Bowen originally investigated—whether large volumes of rhyolitic magma can be formed from basaltic magma by fractional crystallization—is "no." But careful study of almost any igneous rock will reveal evidence that fractional crystallization has occurred at some point in its formation. Magmatic differentiation, fractional crystallization, and Bowen's reaction series play a crucial role in the formation of the great variety of igneous rocks we find on this planet.

A.

B.

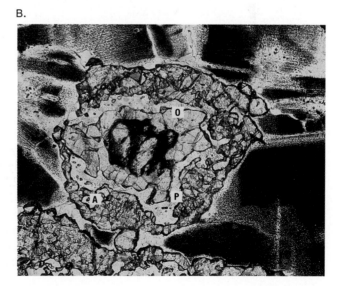

**FIGURE D.2**
**Reaction Textures**
These mineral textures illustrate Bowen's reaction series. A. This zoned plagioclase crystal in an andesite shows evidence of continuous reaction. The core of the crystal, which crystallized first, is anorthitic (Ca-rich) in composition. It is surrounded by a series of bands of increasingly albitic (Na-rich) composition. The grain is about 2 mm long, photographed under a petrographic microscope with crossed polarizers. B. This grain of olivine (O) in a gabbro is surrounded by reactions rims of pyroxene (P) and amphibole (A). This texture demonstrates a discontinuous reaction series. The diameter of the outer rim is 3 mm.

# SYMBOLS COMMONLY USED ON GEOLOGIC MAPS

SYMBOL                      EXPLANATION

Strike and dip of strata

Strike of vertical strata

Horizontal strata, no strike, dip=0

Strike and dip of foliation in metamorphic rocks

Strike of vertical foliation

Anticline; arrows show directions of dip away from axis

Syncline; arrows show directions of dip toward axis

Anticline; showing direction and angle of plunge

Syncline; showing direction and angle of plunge

Normal fault; hachures on downthrown side

Reverse fault; arrow shows direction of dip, hachures on downthrown side

Dip of fault surface; D, downthrown side; U, upthrown side

Directions of relative horizontal movement along a fault

Low-angle thrust fault; barbs on upper block

A.

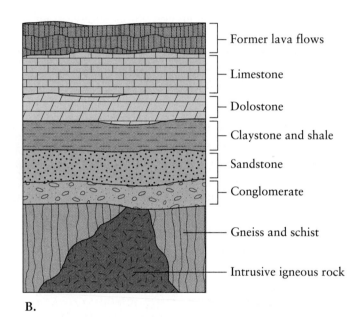

— Former lava flows

— Limestone

— Dolostone

— Claystone and shale

— Sandstone

— Conglomerate

— Gneiss and schist

— Intrusive igneous rock

B.

Representative patterns commonly, but not universally, used to show kinds of rock in geologic maps and cross sections.

# • GLOSSARY •

*Numbers in parentheses at the end of each definition indicate the page or pages on which a term is defined and used as a key term.*

**abrasion** A mechanism of eolian (wind) erosion in which airborne particles chip small fragments off rocks that stick up from the surface. When a rock is abraded, it acquires a distinctive, curved shape and a surface polish; such a rock is called a *ventifact*. (402)

**absolute age** The age of a rock in years (cf. *relative age*). (60)

**air** The gaseous envelope that surrounds the Earth. (368)

**alluvium** Sediment that has been deposited by a stream in fairly recent times. (338)

**andesite** An intermediate-silica volcanic rock, consisting primarily of abundant feldspar with some mafic minerals. (161)

**angiosperm** Flowering plant, or enclosed-seed plant. (458)

**anorogenic granite** Granitic batholiths that form in areas of the crust not related to an orogen. (312)

**anthropogenic** Caused by humans. (522)

**anticline** A fold in the form of an arch, with the rock strata convex upward. (254)

**aquifer** A body of water-saturated, porous and permeable rock or regolith. An aquifer in which the water is free to rise to its natural level is an *unconfined aquifer*; the water in an unconfined aquifer is in contact with the atmosphere, through the porosity of the overlying rocks or sediments. A *confined aquifer* is overlain by impermeable rock units, called *aquicludes,* and its water is held down by the pressure of the overlying impermeable unit. (358)

**Archean Eon** The eon that follows the Hadean Eon and precedes the Proterozoic Eon. (65)

**arid land** Land where annual precipitation is less than 250 mm (10 in). (401)

**asthenosphere** A weak layer within the mantle of the Earth, just below the lithosphere. (12; 144)

**atmosphere** The envelope of gases that surrounds the Earth. (11)

**atom** The smallest individual particle that retains the distinctive properties of a given chemical element. (32)

**banded iron formation** A type of chemical sedimentary rock formed on the sea floor from iron-rich siliceous sediments. All known banded iron formations are at least 1.8 billion years old. (227)

**barrier island** A long, narrow, sandy island lying offshore and parallel to a lowland coast. (391)

**basalt** A mafic volcanic rock with low silica content, consisting primarily of plagioclase feldspar, pyroxene, and olivine. Basalt is the dominant rock type in the oceanic crust. (161)

**batholith** The largest kind of pluton; an irregularly shaped igneous body that cuts across the layering of the rock into which it intrudes and has at least 100 km$^2$ (40 mi$^2$) of surface exposure. (165)

**beach** Wave-washed sediment along a coast. (390)

**beach drift** The movement of particles along a beach as they are driven up and down the beach slope by wave action. (388)

**bed** A *stratum*, or layer, with a succession of sedimentary rock strata. (219)

**bedding** The layered arrangement of strata in a body of sediment or sedimentary rock. (219)

**bed load** The part of the total sediment load of a stream that is moved along the bottom or *bed* of the stream. Bed load typically consists of coarse particles, which move by saltation. (207)

**bedding surface** The top or bottom surface of a rock layer or bed. (219)

**biogenic sediment** Sediment that is composed mainly of the remains of plants and animals. (219)

**biogeochemical cycle** A group of models that describe the movement through the Earth system of chemicals that are essential to life, including carbon, oxygen, nitrogen, sulfur, and phosphorus. (21)

**biomass energy** Any form of energy that is derived more or less directly from the Earth's plant life, including fuel wood, peat, animal dung, and agricultural wastes. (486)

**biosphere** The totality of living matter on the Earth. (11)

**body wave** Any seismic wave that travels outward from the point of origin of an earthquake and has the capacity to travel through the interior of the Earth. (130)

**bonding** The force that holds the atoms together in a chemical compound. (34)

**brittle** Deformation in which the rock fractures or cracks, instead of flowing or bending. (244)

**burial metamorphism** Metamorphism that occurs as a result of the burial of sediments to depths of 10 km (6.2 mi) or more in deep sedimentary basins. Burial metamorphism, the first stage of metamorphism after diagenesis, begins at about 150°C (300°F). It involves mainly chemical recrystallization, and grades into regional metamorphism as temperature and pressure increase. (282)

**carrying capacity** The maximum population that can be sustained by a system, without causing permanent damage to the system. (513)

**cave** Underground open space. (356)

**cell** The basic structural and functional unit of life; a complex grouping of chemical compounds enclosed in a porous membrane. (444)

**cementation** The process whereby substances dissolved in pore water in a sediment precipitate out and cement the sediment grains together. Calcium carbonate, silica, and iron hydroxide are common sediments. Cementation is one of the steps in the process of lithification. (225)

**Cenozoic Era** The most recent era of the Phanerozoic Eon. (65)

**channel** The passageway of a stream. (335)

**chemical sediment** Sediment formed by the precipitation of minerals dissolved in lakewater or seawater. (219)

**chemical weathering** The decomposition of rocks and minerals as a result of chemical and biochemical reactions, producing new minerals that are stable at the Earth's surface. Most chemical weathering happens through *hydrolysis,* that is, through chemical reactions involving water. Some examples of proccsses that cause chemical weathering are ion exchange, solution, and oxidation. (192)

**chemosynthesis** The synthesis, from inorganic material, of organic molecules. (441)

**chert** A biogenic sediment composed of extremely tiny particles of quartz. Bedded chert commonly results from the accumulation of tiny aquatic organisms that contain siliceous particles. (228)

**clast** Fragment; the individual particles in a clastic sediment. (220)

**clastic sediment** Sediment formed from the loose rock and mineral debris produced by weathering and erosion; also known as *detrital sediment.* (219)

**cleavage** The tendency of a mineral to break in preferred directions. (41)

**climate** Weather patterns averaged over a long time. (376)

**coal** A black, combustible sedimentary rock formed by compression and heating of layers of peat buried by overlying sediments. The burial causes water and gaseous compounds to be expelled, and the proportion of the carbon content of the residue increases. The highest grade of coal, anthracite, is considered to be a metamorphic rock. (228; 474)

**collision zone** Where one continent meets another continent along a convergent plate margin. (102)

**compaction** Reduction of pore space in a sediment in response to the weight of overlying sediment accumulation. Compaction forces grains closer together and causes water to be expelled from the sediment; it is the first step in the process of lithification. (225)

**compound** (chemical) A combination of atoms of one or more elements with atoms of another element in a specific ratio. (34)

**compression** Stress that acts in a direction perpendicular to and toward a surface; compressional stress squeezes rocks, shortening them or decreasing their volume. (244)

**compressional wave** One of two types of seismic body wave, consisting of alternating pulses of compression and expansion acting in the direction in which the wave is traveling; cf. *shear wave.* (130)

**condensation** The process by which water changes from a vapor into a liquid or solid. (331)

**conglomerate** A clastic sedimentary rock consisting of large fragments in a finer-grained matrix. In the fragments, the rock is called a *breccia.* (226)

**contact metamorphism** Metamorphism that occurs adjacent to bodies of hot magma that have intruded into cooler rocks. Contact metamorphism involves mainly chemical recrystallization in response to a pronounced increase in temperature and to the involvement of chemically active fluids released by the magma. (282)

**continental crust** The crustal rocks that form the continents and continental shelves; continental crust is thick and low in density compared to oceanic crust. (13)

**continental drift** The slow, lateral movement of continents across the surface of the Earth. (89)

**continental volcanic arc** A chain of andesitic volcanoes formed along a continental margin, where oceanic crust is subducting under continental crust (an ocean-continent plate boundary). (311)

**contour line** Lines of equal elevation on a topographic map. (260)

**convection** A mechanism of heat transfer in which hot, less dense materials rise upward, being replaced by cold, more dense, downward-flowing material to create a convection current. (107)

**convergent margin** A boundary along which two plates come together; the three types of convergent margins are ocean-ocean, ocean-continent, and continent-continent margins. (102)

**core** The innermost compositional layer of the Earth, consisting primarily of metallic iron and nickel. (10; 145)

**Coriolis effect** The phenomenon whereby anything that moves freely with respect to the rotating Earth is caused to veer off-course, towards the right in the Northern Hemisphere and towards the left in the Southern Hemisphere. (373)

**correlation** A method of equating relative ages in successions of strata from two or more different places. (63)

**craton** A portion of the continental crust that has been stable and free from deformation for a very long time (at least a billion years). A craton is in a state of isostatic balance. Cratons form the cores of all continents. (302)

**creep** The imperceptibly slow downslope granular flow of regolith. (212)

**cross bedding** Sedimentary strata (beds) that are inclined with respect to a thicker stratum within which they occur. (222)

**crust** The outermost compositional layer of the Earth; the crust is very thin and low in density compared to the Earth as a whole. (10; 142)

**cryosphere** The perennially frozen part of the hydrosphere. (398)

**crystal** Any solid body with an internal crystal structure, that may be expressed in the form of planar growth surfaces. (39)

**crystal form** The geometric arrangement of planar faces that bound a crystal. (39)

**crystalline** Having an internal crystal structure, that is, a geometric pattern of atoms. (38)

**crystallization** The process whereby crystals form and grow in a cooling magma (or lava). (156)

**deflation** A mechanism of eolian (wind) erosion in which loose particles of sand and dust are picked up and removed by the wind, leaving only the coarser particles behind. Continued deflation may lead to the formation of *desert pavement.* (403)

**deformation** The change in shape or volume of a rock in response to squeezing, stretching, shearing, or any other kind of tectonic force. (242)

**delta** A sedimentary deposit, commonly triangular or fan-shaped, that forms where a stream enters a standing body of water. As the stream loses velocity, its transporting ability suddenly decreases and its sediment load is deposited along the coastline to form the delta. Thus, deltas consist of stream sediment that is not carried away by currents, waves, or tides. (234)

**density** Mass per unit volume. (44)

**deposition** The laying down of sediment by a transporting medium (usually air, water, or ice). (336)

**desert** Arid lands, which generally lack vegetation and cannot support a large population. (401)

**desertification** The invasion of desert conditions into nondesert areas. (406)

**diagenesis** The chemical, physical, and biological changes undergone by a sediment from the moment it is deposited up to, during, and immediately after lithification. (225)

**dike** A sheetlike body of igneous rock that cuts across layering or contacts in the rock into which it intrudes. (168)

**diorite** An intermediate-silica igneous rock, the plutonic equivalent of andesite. (163)

**dip** The angle between a tilted surface and a horizontal plane. (249)

**discharge,** *first meaning:* The amount of water (stream width x stream depth x velocity of flow) passing by a point on a channel's bank during a unit of time. (336) *second meaning:* The process by which subsurface water leaves the saturated zone and becomes surface water. (354)

**divergent margin** A boundary along which two plates diverge (move apart from one another); also called *rifting* or *spreading centers.* (102)

**divide** The topographic high that separates adjacent drainage basins. (340)

**DNA** Deoxyribonucleic acid; a biopolymer that consists of two twisted chain-like molecules held together by organic molecules. The genetic material for all organisms (except viruses), it stores the information on how to make proteins. (443)

**dolostone** A biogenic or bioclastic rock composed predominantly of the mineral dolomite. (228)

**drainage basin** The total area from which water flows into a stream. In general, the greater a stream's annual discharge, the larger its drainage basin. (340)

**ductile deformation** A type of permanent deformation in a rock (or other solid) that has been stressed beyond its elastic limit. Rocks that deform in a ductile manner usually change their shape by flowing or bending. Ductile deformation, also called *plastic* deformation, occurs when the *elastic limit* has been exceeded. (244)

**dune** A hill or ridge of sand deposited by winds. (404)

**early warning** A public declaration that the normal routines of life should be altered for a period of time to deal with the danger posed by an imminent hazardous event. (520)

**Earth resources** Useful things that are extracted from the Earth. (470)

**Earth science** The broad field of scientific study that includes geology, oceanography, atmospheric science, and other areas of specialization; essentially a synonym for *geoscience.* (506)

**Earth system science** The application of a systems approach to the study of the Earth and the relationships among its component spheres. (17)

**elastic deformation** A nonpermanent change in the volume or shape of any solid, including rocks. When the stress is removed, the solid returns to its original shape and size. (244)

**elastic rebound theory** A theory of earthquakes and fault motion that states that energy is stored in bodies of rock subjected to stress along a fault. When the blocks on either side of the fault slip, the stored energy is released suddenly in the form of an earthquake, and the rocks rebound to assume their original shapes. (126)

**element** (chemical) The most fundamental substance into which matter can be separated by chemical means. (32)

**eolian sediment** Sediment that is carried and deposited by the wind. Examples of eolian sediments include loess and sand dunes. (234)

**eon** The four major time divisions of the geologic time scale. (65)

**epicenter** The point on the Earth's surface directly above an earthquake's focus. (132)

**epoch** A unit of geologic time that is shorter than a period. (66)

**era** A unit of geologic time that is shorter than an eon and longer than a period. (65)

**erosion** The wearing away of bedrock and transport of loosened particles. (190; 336)

**estuary** A semi-enclosed body of coastal water in which seawater is diluted with fresh water. (234)

**eukaryotic cell** A cell that includes a nucleus with a membrane, as well as other membrane-bound organelles. (445)

**evaporation** The process by which water changes from a liquid into a vapor. (331)

**evaporite** A chemical sediment that forms when lake water or sea water evaporates, leaving behind salt deposits. (226)

**evolution** The changes that species undergo through time, eventually leading to the formation of new species. The theory that all present-day organisms are descendants, through a gradual process of adaptation to environmental conditions, of different kinds of organisms that existed in the past. (449)

**exponential growth** Growth in which a population increases by a constant percentage in a given unit of time. (513)

**fall** A type of slope failure that involves a sudden, vertical, or nearly vertical drop of rock fragments or debris. (212)

**fault** A fracture in the crust along which movement has occurred. (102; 249)

**flood** An event in which a water body overflows its banks. (344)

**floodplain** The relatively flat valley floor adjacent to a stream channel, which is inundated when the stream overflows its banks. (338)

**flow** Any mass-wasting process that involves a flowing motion of regolith in which the pores are filled with water and/or air. (212)

**focus** The point or area within the Earth where the energy of an earthquake is first released. (130)

**fold** A bend or warp in a layered rock. (253)

**foliation** A planar metamorphic rock texture, that is, a rock texture in which elongate mineral grains like micas tend to be aligned, and the rock tends to split into thin, leaf-like flakes. (276)

**forecast** A prediction. Sometimes the term *forecast* refers to short-term predictions (as in weather forecasting), and sometimes it refers to long-term, nonspecific predictions (as in earthquake forecasting). (520)

**formation** A unit of rock that can be mapped on the basis of rock type and recognizable boundaries, or *geologic contacts,* with other rock units. (261)

**fossil** The remains of an organism that died and became preserved and incorporated into sediments. (60; 453)

**fossil fuel** Organic matter that has been trapped in sediment or sedimentary rock and has undergone changes during and after burial. The principal fossil fuels are peat, coal, oil, and natural gas. (473)

**fractional crystallization** The set of processes whereby crystals and liquids can become separated from one another during crystallization, preventing them from reacting chemically with one another or with the liquid from which they are growing. (164)

**fractionation** A process in which a melt can be separated from the remaining solid material during the course of melting; also known as *fractional melting.* (155)

**fracture** Any kind of crack or break in a rock. (249)

**frost heaving** The uplift of surface rock and regolith as a result of the freezing of subsurface water to form ice. (212)

**frost wedging** The freezing of ice in a confined opening within a rock, causing the rock to be forced apart. A type of mechanical weathering. (193)

**gabbro** A low-silica, mafic igneous rock, the plutonic equivalent of basalt. (163)

**geochemistry** The use of chemical techniques to study Earth materials and processes. (121)

**Geologic Column** The succession of all known strata, fitted together in relative chronological order, on the basis of their fossils or other evidence of relative age. (64)

**geologic cross section** A diagram showing geologic features that occur underground, on a plane that intersects the ground surface. (264)

**geologic hazard** Earthquakes, volcanic eruptions, floods, landslides, and other geologic processes that may be damaging to human interests. (515)

**geologic map** A map that shows the locations and orientations of rock units and structural features. (257)

**geologist** A scientist who studies the Earth. (4)

**geology** The scientific study of the Earth. (4)

**geophysics** The application of physics to the study of the Earth. (121)

**geoscience** Earth science. (506)

**geothermal energy** The Earth's internal energy. (489)

**geothermal gradient** The rate at which temperature increases with depth in the Earth. (151)

**glaciation** Periods during which the average temperature at the Earth's surface dropped by several degrees and stayed low long enough for existing ice sheets to grow larger (and new ones to form). Also called *ice ages, glacial periods, glacial stages,* or *glacial epochs.* (421)

**glacier** A body of ice that is large enough to persist from year to year and within which there is evidence of movement due to the pull of gravity. (208; 409)

**gneiss** A high-grade metamorphic rock with coarse grains and pronounced foliation, in which bands of micaceous minerals have segregated from bands of quartz and feldspar. (287)

**graded bed** Individual beds of sediment or sedimentary rock in which the coarsest clasts are concentrated at the bottom of the bed, grading up to the finest clasts at the top. Graded beds are formed when water containing clasts of mixed sizes slows down, depositing the coarsest (heaviest) clasts first, followed by successively finer clasts. (222)

**granite** A high-silica, felsic igneous rock, the plutonic equivalent of rhyolite, consisting primarily of quartz, feldspar, and mica. (163)

**gravity anomaly** A measurement of the Earth's gravity that differs from the calculated value of gravitational force for that location; a greater-than-expected pull is a *positive anomaly,* while a less-than-expected pull is a *negative anomaly.* (122)

**greenhouse effect** The process through which long-wavelength (infrared) heat energy is absorbed by gases in the atmosphere, thereby warming the surface of the Earth. (371)

**groundwater** Subsurface water, that is, the water contained in spaces within bedrock and regolith. Most groundwater occurs within a depth of 750 m (2460 ft) below ground. (352)

**gymnosperm** Naked-seed plant. (458)

**habit** The distinctive shape of a particular mineral. (40)

**Hadean Eon** The earliest eon in Earth history. (65)

**half-life** The time needed for a radioactive material to lose half its radioactivity. (70)

**hardness** A mineral's resistance to scratching. (42)

**hazard assessment** The process of determining when, where, and how frequently hazardous events have occurred in the past, determining the nature of the physical effects of the event in a particular location, and portraying information in a form that can be used by decision makers. (518)

**high-grade** (metamorphic rocks) Rocks that are metamorphosed under temperature and pressure conditions higher than about 400°C (750°F) and 4000 atm (a depth of about 15 km, or 9 mi), ranging up to the onset of melting. (274)

**historical geology** A branch of geology that is concerned with geologic events that occurred in the past, the establishment of a chronology of Earth history, and the study of organisms that lived long ago. (4)

**hominid** A member of the family of humans, the Hominidae. (463)

**humus** Partially decayed organic matter in soil. (200)

**hydroelectric energy** The energy from a flowing stream of water, used to run turbines and generate electricity. (487)

**hydrologic cycle** The set of interconnected reservoirs and processes whereby water moves around in the Earth system. (20; 330)

**hydrosphere** The part of the Earth system that contains the ocean, lakes, and streams; underground water; and snow and ice, including glaciers. (11; 329)

**igneous rock** A rock formed by the cooling and consolidation of magma. (52; 156)

**impact crater** A crater, or excavated depression, formed on a planetary surface by the collision of a large extraterrestrial object (a meteorite). On the Earth, a meteorite must have appreciable size in order to punch through the atmosphere without breaking up, hit the ground, and excavate a crater. On planets with no atmosphere, however, even tiny meteorite fragments can make impact craters. (297)

**infiltration** The process by which water works its way into the ground, passing through small openings and channels in the soil and other surface materials. (331)

**ion** An atom that has an overall excess positive or negative electrical charge because of the loss or addition of one or more electrons. (33)

**isostasy** The flotational balance of the lithosphere on the asthenosphere. (301)

**isotope** Atoms with the same atomic number but different mass numbers. (33)

**joint** A fracture in a rock, along which no appreciable movement has occurred. (192; 253)

**jovian planets** The "gas giant" outer planets: Jupiter, Saturn, Uranus, and Neptune. (6)

**lake** A water body that has an upper surface exposed to the atmosphere and has no appreciable gradient (that is, the surface is relatively flat). *Ponds* (small, shallow lakes) and *wetlands* (areas of poor surface drainage, such as marshes and swamps, which may contain standing water) are included under the general definition of a lake. (343)

**laterite** A soil that has been strongly *leached,* that is, most of the original substances have been removed in solution. Laterites form in equatorial and tropical lands that have both high rainfall and high temperature. They are composed largely of iron and aluminum oxides and hydroxides, because these substances are relatively insoluble. Named for the Latin word *latere,* brick. (202)

**lava** Magma that reaches the surface in a molten state. (52; 155)

**limestone** A biogenic of bioclastic rock that consists primarily of the mineral calcite. Many limestones are composed of lithified shell fragments from marine organisms. (228)

**lithification** The group of processes by which newly deposited, loose sediment is slowly transformed into sedimentary rock. (225)

**lithosphere** The tough, rocky, outermost part of the Earth, comprising the crust and the uppermost part of the mantle, and defined on the basis of strength rather than composition. (12; 98; 144)

**load** The material carried along by a stream. The load has three parts: *bed load,* particles that move along the bottom; *suspended load,* particles and organic debris that are suspended in the water; and *dissolved load,* dissolved substances that are a product of rock weathering. (336)

**loess** A fine, yellow-brown, wind-blown (eolian) sediment. Loess consists of windblown dust of Pleistocene age, transported from desert surfaces, glacial sediments, and glacial stream deposits at times of ice-sheet retreat. (234; 403)

**longshore current** A current, within the surf zone, that flows parallel to the coast. (388)

**low-grade** (metamorphic rocks) Rocks that are metamorphosed under temperature and pressure conditions from the region of diagenesis up to about 400°C (750°F) and 4000 atm (a depth of about 15 km, or 9 mi). (274)

**luster** The quality and intensity of light reflected from a mineral. (43)

**magma** Molten rock under the ground (often containing crystals and dissolved gases). (52; 155)

**magmatic differentiation** The processes whereby a single magma of a given composition can crystallize into many different kinds of igneous rock. (163)

**magnetic reversal** An event in which the Earth's magnetic polarity reverses itself. (74)

**mantle** The middle compositional layer of the Earth, between the core and the crust. (10; 143)

**mantle plume** A long, thin body of hot rock rising through the mantle, which may cause a volcanic outpouring or *hot spot* at the surface. (108)

**marble** The product of metamorphism and recrystallization of limestone. (289)

**mass extinction** A catastrophic episode in which many species become extinct within a geologically short time. (464)

**mass wasting** The downslope movement of regolith and/or bedrock masses due to the pull of gravity. (206)

**mechanical weathering** The breakdown of solid rock into fragments by physical processes, with no change in chemical composition. Some examples of processes that cause mechanical weathering are frost wedging, crystallization of salt crystals, heating by forest fires, and penetration by plant roots. (192)

**Mercalli intensity scale** A scale of earthquake intensity, based on descriptions of the vibrations that people felt, saw, and heard, and on the extent of damage to buildings during earthquakes; the scale ranges from I (not felt except under favorable circumstances) to XII (waves seen on ground surface, practically all works of construction destroyed or severely damaged). (133)

**mesosphere** The mantle between the bottom of the asthenosphere and the core-mantle boundary. (144)

**Mesozoic Era** The middle era of the Phanerozoic Eon. (65)

**metamorphic facies** The set of metamorphic mineral assemblages that are formed in rocks of different composition under given temperature and stress conditions. (285)

**metamorphic rock** A rock whose original sedimentary or igneous form and mineralogy have been changed as a result of high temperature, high pressure, or both. (53)

**metamorphism** The mineralogical, chemical, and structural adjustments of solid rocks to physical and chemical conditions at depths below the region of sedimentation and diagenesis. (272)

**metasomatism** The process whereby the chemical composition of a rock is altered by the addition or removal of material by solution in fluids. (290)

**meteorite** A fragment of extraterrestrial material that falls to Earth. (9)

**microfossil** A fossil so small that it must be studied under a microscope. (455)

**migmatite** A high-grade metamorphic rock in which a small amount of melting has occurred, and pockets of trapped melt have cooled and solidified within the rock. Thus, a migmatite is a composite rock—primarily metamorphic, but with small pockets of igneous material. (280)

**mineral** A naturally formed, solid, inorganic substance with a specific chemical composition and a characteristic crystal structure. (37)

**Moho** The boundary between the crust and the mantle (short for *Mohorovicic discontinuity*). (143)

**molecule** The smallest unit that has all the properties of a particular compound. (34)

**moment magnitude** A measure of the strength of an earthquake based on field observations of the area over which the fault ruptured, the physical characteristics of the Earth materials involved, and the average amount of ground displacement. (134)

**moraine** A ridge or pile of debris that is being transported or has been deposited along the edges of a glacier. (419)

**MORB** An acronym for **mid-ocean ridge b**asalt. MORB is the magma that rises at seafloor spreading centers and solidifies to form oceanic crust; it varies little in composition around the world. (304)

**natural gas** A naturally occurring hydrocarbon (a form of petroleum) that is gaseous at ordinary pressures and temperatures. (478)

**natural hazard** Naturally occurring hazardous processes or events, including locust infestations, wildfires, tornadoes, and hurricanes, as well as geologic hazards. (515)

**natural selection** The process whereby certain individuals within a population have characteristics that enable them to compete more effectively for scarce resources or to escape predators more easily ("survival of the fittest"). These individuals are more likely to survive, and they are likely to have offspring with similar characteristics. Thus, over time an entire population evolves as natural selection favors individuals that are particularly well adapted to their environment. (449)

**nonrenewable resource** A resource that cannot be replenished or regenerated on the scale of a human lifetime. (470)

**normal fault** A type of fault in which the block of rock on top of the fault surface, the *hanging wall* block, appears to have moved downward relative to the block on the bottom, the *footwall* block. Normal faults are caused by tensional stress, that is, stress that stretches or pulls apart the crust, lengthening and thinning it. (250)

**nuclear energy** The heat energy produced by controlled nuclear reactions, that is, reactions in which atoms of one species of chemical element are transformed into atoms of another species by nuclear change. The energy produced by nuclear power plants today comes from fission, the splitting of heavy atoms into lighter atoms. Fusion energy, produced by combining two light atoms to make a heavier atom, is theoretically possible but has never been achieved at ordinary temperatures and pressures. (487)

**oceanic crust** The crustal rocks that underlie the world's deep ocean basins, consisting primarily of basalt; oceanic crust is thin and dense compared to continental crust. (13)

**oil** The liquid form of petroleum, a naturally occurring substance composed mainly of hydrocarbon compounds. (478)

**oil shale** A fine-grained sedimentary rock with a high content of kerogen, a waxy organic substance. (482)

**ophiolite** Fragments of oceanic crust found on continents. Ophiolites are greenish, serpentine-dominated fragments of metamorphosed oceanic crust basalt that have been caught up in the plate collision process. (304)

**ore** A deposit from which one or more minerals can be extracted profitably. (492)

**orogen** An elongate region of the crust that has been intensely deformed and metamorphosed during a continental collision. (300)

**orogenic granite** Huge batholiths of granite that form in orogens as a result of wet partial melting at the base of the continental crust, where the crust has been thickened as a result of a plate tectonic collision, that is, in an orogen. The resulting magma is viscous and moves upward so slowly that most of it cools and crystallizes underground. (311)

**oxide mineral** A mineral that contains the $O^{2-}$ anion. (46)

**ozone layer** A zone in the stratosphere in which ozone ($O_3$) is concentrated. (372)

**P wave** Primary wave, a compressional seismic body wave; the fastest of the seismic waves. (130)

**paleoclimatology** The study of ancient climates. (422)

**paleomagnetism** The magnetism taken on by a rock long ago, recording the direction of the Earth's magnetic poles at the time of the rock's formation. (74)

**paleontology** The study of fossils and the record of ancient life on Earth. (62)

**paleoseismology** The study of prehistoric earthquakes. (138)

**Paleozoic Era** The earliest era of the Phanerozoic Eon. (65)

**partial melt** A mixture of molten material and solid material. (155)

**peat** A loose aggregate of plant remains with a carbon content of about 60 percent, which typically accumulates in water-saturated places such as bogs and swamps; a biogenic sediment. (228; 474)

**pedalfer** A soil that is rich in clays, plus aluminum and iron hydroxides. Pedalfers form in regions with moderate rainfall and temperate climates, and have well-developed A, B, and C horizons, and sometimes a distinct E horizon. Named from the Greek *pedon*, soil, and *al* and *fer* for *aluminum* and *ferrum* (Latin for iron). (202)

**pedocal** A soil that is rich in calcium carbonate and sometimes in other soluble minerals such as gypsum. Pedocals form in regions with dry climates where soil water tends to evaporate, causing the precipitation of soluble salts near the base of the B horizon. Named from the Greek *pedon*, soil, and the Latin name for calcium (*calx*). (202)

**percolation** The process by which groundwater seeps downward under the influence of gravity. (354)

**periglacial** Refers to areas that are in close proximity to glacial ice. (419)

**period** A unit of geologic time that is shorter than an era and longer than an epoch. (65)

**permafrost** Ground that is perennially below the freezing point of water. (419)

**permeability** A measure of how easily a solid allows fluids to pass through it. (353)

**petroleum** Gaseous, liquid, and semisolid naturally occurring substances that consist chiefly of hydrocarbon compounds. The main forms of petroleum are oil and natural gas. (476)

**petroleum trap** A geologic situation that includes a source rock that contributes organic material, a reservoir rock that allows for the accumulation of oil, and a cap rock that stops the migration. Also known as a *hydrocarbon trap*. (479)

**Phanerozoic Eon** The most recent eon in Earth history. (65)

**photosynthesis** The process whereby plants utilize light energy to cause carbon dioxide to react with water, producing carbohydrates and releasing oxygen. (439)

**phyllite** A metamorphic rock with pronounced foliation, produced by continuing metamorphism of slate. (286)

**physical geology** A branch of geology that is concerned with the processes that operate at or beneath the surface of the Earth and the materials on which those processes operate. (4)

**plate tectonics** The group of processes in which large fragments (plates) of lithosphere move horizontally across the surface of the Earth, and through their movements and interactions generate earthquakes, volcanism, mountain-building, and other geologic processes. (13; 99)

**plateau basalt** A vast, flat sheet of volcanic rock that forms when continental crust passes over a plume-related mantle hot spot, causing a large volume of very fluid basaltic lava to pour onto the continent. Also known as *flood basalt*. (311)

**pluton** An intrusive igneous rock body. (165)

**plutonic rock** An igneous rock formed from magma that cooled and solidified underground; also called an *intrusive rock*. (160)

**polarity** (of magnetic field) The north-south directionality of the Earth's magnetic field. (74)

**porosity** The percentage of the total volume of a body of rock or regolith that consists of open spaces, or pores. (353)

**Precambrian** Geologic time prior to the beginning of the Phanerozoic Eon, that is, prior to 570 million years ago. (66)

**precipitation** The process by which water that has condensed in the atmosphere falls as rain, snow, or hail on the land or ocean. (331)

**prediction** A statement of probability that is based on scientific understanding and observations. Prediction usually involves the monitoring of geologic processes that could generate a hazardous event. (520)

**pressure** The force acting on a surface, per unit area; used when the forces acting on the surface of a body are the same in all directions. Pressure is a type of stress. (243)

**prokaryotic cell** A cell without a well-defined nucleus; refers to single-celled organisms that have no membrane separating their DNA from the cytoplasm. (444)

**Proterozoic Eon** The most recent eon of Precambrian time; the eon that follows the Archean Eon and precedes the Phanerozoic Eon. (65)

**pyroclastic rock** A rock formed from the consolidation of *pyroclasts*, fragments of rock ejected during a volcanic eruption. (172)

**quartzite** The product of metamorphism, recrystallization, and silica cementation of sandstone. (289)

**radioactivity** The process by which an element transforms spontaneously into another isotope of the same element or another element. (69)

**radiometric dating** The use of naturally occurring radioactive isotopes to determine the time of formation (that is, the absolute age) of minerals or rocks. (71)

**rapid onset hazard** Hazardous events that strike quickly and with little warning, such as meteorite impacts, earthquakes, or flash floods. (515)

**recharge** Replenishment of groundwater, which occurs when rainfall and snowmelt infiltrate the ground and percolate downward to the saturated zone. (354)

**recrystallization** The process whereby minerals in an accumulating sediment form new, more stable minerals. Like cementation, recrystallization acts to hold together the grains in a sedimentary rock. (225)

**reef** A structure composed mainly of the calcareous remains of marine organisms (principally corals). (393)

**refraction** The bending of a wave as it passes from one material into another material of differing density. (141)

**regional metamorphism** Metamorphism of an extensive area of the crust, as a result of the stresses and high temperatures associated with plate convergence, collision, and subduction. (283)

**regolith** The loose layer of broken rock and mineral fragments that covers the Earth's surface. (190)

**relative age** The age of a rock, fossil, or other geologic feature measured relative to another feature rather than in years (cf. *absolute age*). (58)

**renewable resource** A resource that can be replenished or regenerated on the scale of a human lifetime. (470)

**reverse fault** A type of fault in which the block on top of the fault plane (the *hanging wall* block) is pushed over the block underneath the fault plane (the *footwall* block). Reverse faults are caused by compressional stress, that is, horizontal stress that pushes blocks of crust together, shortening and thickening the crust. (250)

**rhyolite** A felsic, high-silica volcanic rock, consisting largely of quartz, feldspar, and mica. Rhyolite is often very fine-grained or glassy in texture, so individual minerals may not be identifiable. (161)

**Richter magnitude scale** A scale of earthquake intensity based on the recorded heights, or *amplitudes*, of the seismic waves. (133)

**rift valley** A valley that has formed along a rift, that is, along a divergent plate boundary or spreading center. A continental rift valley may eventually become a passive continental margin, or the rifting may cease and the valley become a failed rift. (317)

**risk assessment** The process of establishing the probability that a hazardous event will occur within a given period and estimating its impact, taking into account the potential exposure and vulnerability of the community. (519)

**river** A stream with a considerable volume and a well-defined channel. (335)

**RNA** Ribonucleic acid; a single-strand molecule similar to one-half of a double DNA strand. (443)

**rock** A naturally formed, coherent aggregate of minerals or solid material such as natural glass or organic matter. (52)

**rock cycle** A model that describes all of the various crustal processes by which rock is formed, modified, transported, decomposed, and reformed. (21)

**S wave** Secondary wave, a seismic body wave that travels with a shearing motion. (131)

**salinity** Saltiness. (378)

**saltation** A mechanism of sediment transport in which particles move forward in a series of short jumps along arc-shaped paths. (206; 402)

**sandstone** A medium-grained clastic sedimentary rock, in which the clasts are typically dominated by quartz grains. If appreciable feldspar is present, the rock is called arkose. If rock fragments are also present, the term *graywacke* is used. (226)

**scale** The ratio between a distance on a map and the corresponding distance on the ground surface that is being mapped. A scale shows the amount by which the size of objects or distances shown on the map has been reduced. (258)

**schist** A coarse-grained metamorphic rock with pronounced schistosity; between phyllite and gneiss in metamorphic grade. (287)

**schistosity** Foliation in coarse-grained metamorphic rocks. (277)

**seafloor spreading** The processes through which the sea floor splits apart along a midocean ridge, the oceanic crust on either side moves away from the ridge, and magma wells up along the ridge to form new oceanic crust. (98)

**sediment** Regolith particles that have been transported in suspension by water, wind, or ice and then deposited. (52)

**sedimentary facies** Any sediment (or sedimentary rock) that can be distinguished from another sediment that accumulated at the same time, but in a different depositional environment. One facies may be distinguished from another by differences in grain size, grain shape, stratification, color, chemical composition, depositional structures, or fossils. (232)

**sedimentary rock** Any rock formed by chemical precipitation from water at the Earth's surface, or by the cementation of sediment. (53)

**seismic discontinuity** A boundary inside the Earth, where the velocities of seismic waves change abruptly. Seismic discontinuities may represent boundaries between layers of differing chemical composition, or boundaries between layers of the same chemical composition but differing physical properties. (142)

**seismic wave** A vibrational energy wave that travels outward from an earthquake's source, causing an elastic disturbance in the rocks through which it passes. (130)

**seismogram** The record made by a seismograph. (130)

**seismograph** An instrument based on the principle of inertia that detects and records vibrations of the Earth resulting from earthquakes (and other causes, such as explosions). (128)

**seismology** The scientific study of earthquakes and seismic waves. (128)

**shale** A very fine-grained clastic sedimentary rock, consisting primarily of clay and silt particles. Shales, which typically break into sheet-like fragments, are distinguished from mudstones, which break into blocky fragments. (226)

**shear** Stress that acts parallel to a surface. (244)

**shear wave** One of two types of seismic body waves that is propagated through materials in a series of shearing movements perpendicular to the direction in which the wave is traveling; cf. *compressional wave*. (130)

**shield volcano** A broad, flat volcano with gently sloping sides, built of successive fluid lava flows. (168)

**silicate mineral** A mineral that contains the silicate anion $(SiO_4)^{4-}$. (47)

**sill** A tabular body of intrusive igneous rock, parallel to the layering of the rocks into which it intrudes. (168)

**siltstone** A rock that consists primarily of silt-sized particles (rock or mineral fragments with diameters in the range of 0.002 to 0.05 mm). (226)

**slate** The product of low-grade metamorphism of shale; a fine-grained, micaceous rock with slaty cleavage, formed under conditions of regional or burial metamorphism. (286)

**slaty cleavage** A property of low-grade, fine-grained metamorphic rocks, in which foliation produces the tendency to break into flat, platelike fragments. (277)

**slide** A type of slope failure that involves rapid displacement of a mass of rock or regolith down a steep or slippery slope, along an essentially planar surface. (212)

**slump** A type of slope failure that involves *rotational* movement of rock or regolith, that is, downward and outward movement along a curved surface. (212)

**soil** A special kind of sediment, consisting of loose particles that have been altered by biological processes to form a material that can support rooted plants; the uppermost part of the regolith. (53; 200)

**soil horizon** One of a succession of zones or layers within a soil profile, each of which has distinct physical, chemical, and biological characteristics. (200)

**soil profile** The sequence of soil horizons from the surface down to the underlying bedrock. (200)

**solar system** The Sun and the group of objects in orbit around it. (5)

**species** A population of similar individuals that can interbreed and produce fertile offspring. (451)

**spring** A flow of groundwater that emerges naturally at the ground surface. (354)

**stock** An irregularly shaped pluton with surface exposure less than 100 km$^2$ (40 mi$^2$). A stock may be associated with a batholith, or it may be the top of a partly eroded batholith. (165)

**strain** A change in shape or volume of a rock in response to stress. (243)

**stratigraphy** The study of strata, that is, the study of rock layers and layering. (58)

**stratovolcano** A steep-sided volcano, constructed of solidifed lava flows interlayered with pyroclastic material; also called a *composite volcano*. (173)

**stratum** (plural = **strata**) A distinct layer of sediment that accumulated at the Earth's surface (also applied to the rocks that form from layers of sediment). (58)

**streak** A thin layer of powdered mineral made by rubbing a specimen on a nonglazed fragment of porcelain called a streak plate. (44)

**stream** A body of water that flows downslope along a clearly defined natural passageway, transporting particles and dissolved substances. (335)

**streamflow** That part of surface runoff that travels in stream channels. (335)

**stress** The force acting on a surface, per unit area. (243)

**strike** The orientation of the line of intersection between a rock layer and a horizontal plane; also the orientation of a linear feature such as a fold axis or a fault. (248)

**strike-slip fault** A type of fault, caused by shear stress, along which the movement of the fault blocks is mainly horizontal and parallel to the strike of the fault. (251)

**structural geology** The study of stress and strain, the processes that cause them, and the types of rock structures and deformation that result from them. (247)

**subduction zone** A boundary along which one lithospheric plate descends into the mantle beneath another plate. (102)

**surf** The "broken," turbulent water found between a line of breakers and the shore. (386)

**surface creep** A mechanism of sediment transport in which the wind causes grains to roll along the ground. (402)

**surface runoff** Precipitation that drains off over the land or in stream channels. (331)

**surface wave** A seismic wave that travels along the surface of the Earth (or along a boundary near the surface). (130)

**suspended load** The part of the total sediment load of a stream that is carried along in suspension, that is, supported by water currents above the level of the stream bed. (207)

**suspension** A mechanism of sediment transport in which fine grains are lifted by wind (or water) currents and carried along above the level of the ground (or the stream bed). (402)

**sustainable development** The cautious, planned utilization of Earth resources to meet current needs without degrading ecosystems or jeopardizing the future availability of those resources. (514)

**syncline** A fold in the form of a trough, with the rock strata concave upward. (254)

**system** Any portion of the universe that can be separated from the rest of the universe for the purpose of observing changes. (16)

**tar sand** A sediment or sedimentary rock in which the pores are filled by dense, viscous, asphaltlike oil. (479)

**technological hazard** The hazard associated with exposure to potentially harmful substances, such as radon (a naturally occurring radioactive soil gas), mercury, asbestos fibers, or coal dust. (516)

**tectonic cycle** A model that describes the movements and interactions of lithospheric plates, the internal processes that drive plate motion, and the types of rock and rock formations that develop as a result of tectonic movement and interactions. (21)

**tension** Stress that acts in a direction perpendicular to and away from a surface; tensional stress pulls or stretches rocks, sometimes causing the volume to increase. (243)

**terrestrial planets** The inner planets of the solar system: Mercury, Venus, Earth, and Mars. (6)

**texture** The size, shape, and arrangement of mineral grains that give a rock its overall appearance. (156)

**thrust fault** A low-angle (that is, shallowly dipping) reverse fault. (250)

**tides** The cycles of regular rise and fall in the level of water in the ocean and other large bodies of water. Tides result from the gravitational attraction of the Moon and (to a lesser extent) the Sun, which causes water to bulge upward wherever the gravitational attraction is strongest. As the Earth rotates, the continents "run into" these bulges, creating high tides. (384)

**till** A heterogeneous mixture of finely crushed rock (*rock flour*), sand, pebbles, cobbles, and boulders deposited by a glacier. (418)

**topographic map** A map that uses contour lines to show the shape and relief of the ground surface, and the location and elevation of surface features. (259)

**trace fossil** The fossilized evidence of life processes of an organism, such as tracks, burrows, or footprints. (454)

**transform fault margin** A fracture in the lithosphere where two plates slide past each other. (102)

**transpiration** The process by which water taken up by the roots of plants passes directly into the atmosphere. (331)

**turbidity current** A turbulent, gravity-driven flow consisting of a dilute mixture of sediment and water; essentially, an underwater landslide. Turbidity currents typically originate on a continental shelf and rush swiftly down the continental slope, depositing a graded layer of sediment (a *tillite*) on the ocean floor. (236)

**unconformity** A break or gap in a stratigraphic sequence, marking the absence of part of the geologic record. (60)

**uniformitarianism** A fundamental principle of geology, which states that the processes operating in Earth systems today have operated in a similar manner throughout much of geologic time; the concept that "the present is the key to the past." (22)

**varve** A pair of rhythmic layers (alternating coarse and fine layers) of sediment deposited in still water over the course of one year. The layering results from seasonal variations in glacial lakes. (222)

**viscosity** The degree to which a substance offers resistance to flow; the more viscous a substance, the less fluid it is. (156)

**volcanic glass** A volcanic rock that formed from a very rapidly cooled lava. Rapid cooling prohibits the formation and growth of crystals, resulting in a glassy, noncrystalline texture. (158)

**volcanic island arc** A chain of islands, each of which is an andesitic stratovolcano formed along a plate subduction edge. Volcanic island arcs form from the wet partial melting of mantle rocks in areas where oceanic crust is subducting beneath another oceanic plate (an ocean-ocean plate boundary). The volcanoes occur in chains parallel to the deep ocean trench that marks the boundary between the two plates. (310)

**volcanic rock** An igneous rock formed from lava and other volcanic materials that cooled and solidified at or near the surface of the Earth; also called an *extrusive rock*. (158)

**volcanic sediment** A special kind of clastic sediment, in which all of the clasts are volcanic in origin. (222)

**volcano** A vent through which lava, pyroclastic material, and gases are erupted. (168)

**water table** The top surface of the saturated zone. (352)

**wave base** The effective lower limit of wave movement (and, by extension, the lower limit of erosion by the bottoms of waves). (386)

**wave-cut cliff** A coastal cliff cut by wave action at the base of a rocky coast. (390)

**weather** Local atmospheric conditions at any given time. (376)

**weathering** The chemical and physical breakdown of rock exposed to air, moisture, and organic matter. (190)

**xenolith** A fragment of rock that is ripped from its source region and carried to the surface by magma; a "foreign" inclusion in an igneous rock. (121)

# · PHOTO CREDITS ·

## The Art of Geology
Page 82: William K. Hartmann, "Earth in Space."

Page 184: Katsushika Hokusai, "Fuji in Clear Water." British Museum, London/Bridgeman Art Library, London/SuperStock, Inc.

Page 323: Thomas Moran, "Grand Canyon of the Yellowstone," 1872. National Museum of American Art, Washington, D.C./Art Resource, NY. Lent by the U.S. Department of the Interior, Office of the Secretary.

Page 430: John Warwick Smith, "Hafod, Upper Part of the Cascade." Courtesy Yale Center for British Art, Paul Mellon Collection.

## Part 1
Opener: ©Yann Arthus-Bertrand/Photo Researchers.

## Chapter 1
Opener: Satellite imagery from EROS Data Center, USGS; Mosaic assembled by National Geographic Society. Figure 1.1: ©Krafft-Explorer/Photo Researchers. Figure 1.2: ©Louis Psihoyos/Matrix International. Figure 1.3: NRSC Ltd./Science Photo Library/Photo Researchers. Figure 1.5: Courtesy NASA. Figure 1.7: ©Breck Kent. Figure 1.8: Courtesy Canada Centre for Remote Sensing Department of Natural Resources Canada. Figure 1.10: Courtesy J.D. Griggs, USGS. Figure 1.11: ©Tom Van Sant/The Geosphere Project. Figure 1.12: Courtesy Earth Satellite Corporation. Figure 1.17: ©Tom Bean/DRK Photo. Figure 1.19a: ©John S. Shelton. Figure 1.19b: François Gohier/Ardea London. Figure 1.20: ©Alberto Garcia/SABA. Box 1.1: Courtesy Dr. Baerbel Lucchitta/USGS Box 1.2 (left): ©World Perspectives. Box 1.2 (right): Courtesy NASA.

## Chapter 2
Opener: ©Thomas J. Abercrombie/National Geographic Society. Figure 2.8: Michael Hochella. Figure 2.9: Courtesy William Sacco. Figure 2.10: ©Jacques Delacour/Okapia 1988/Photo Researchers. Figures 2.11 and 2.13: Courtesy William Sacco. Figure 2.12: ©Breck

Kent/Animals Animals/Earth Scenes. Figure 2.14: Brian J. Skinner. Figures 2.15a and 2.16: Courtesy William Sacco. Figures 2.15b and 2.21b,c: Courtesy The Natural History Museum, London. Figure 2.20c: ©Boltin Picture Library. Figures 2.20d,e-g and h: ©Breck P. Kent. Figures 2.21a and 2.22: ©Breck Kent/Animals Animals/Earth Scenes. Table 2.1: Brian J. Skinner.

## Chapter 3
Opener: ©Walter Imber. Figures 3.1a and 3.2: ©Jeff Gnass. Figure 3.1b: ©Helmut Gritscher/Peter Arnold, Inc. Figure 3.3: Landform Slides. Figure 3.5a: ©Wardene Weisser/Bruce Coleman, Inc. Figure 3.5b: ©Breck Kent/Animals Animals/Earth Scenes. Figure 3.9a: ©Michael Rothman. Figures 3.9b,c: ©Breck Kent. Figure 3.10a: ©Wardene Weissner/Bruce Coleman, Inc. Figure 3.10b: ©Ken Lucas/Visuals Unlimited. Figure 3.17: ©Sam Bowring.

## Part 2
Opener: ©Barbara Cushman Rowell/DRK Photo.

## Part 3
Opener: ©Mark Newman/Photo Researchers.

## Chapter 4
Opener: Mountain Light Photography, Inc. Figure 4.6b: Mark C. Burnett/Photo Researchers. Figure 4.7a: ©Ken Lucas/Visuals Unlimited. Figure 4.8: Courtesy The Natural History Museum, London. Figure 4.14a: ©John Downer/Oxford Scientific Films. Figure 4.14b: ©Edward Mendell/Royal Geographical Society. Figure 4.14c: NASA/Corbis. Figure 4.14d: Mountain Light Photography, Inc. Figure 4.14e: ©David Parker/Science Photo Library/Photo Researchers. Figure 4.21: Marie Tharp and Bruce Heezen. Box 4.1: ©Greg Vaughn/Pacific Stock.

## Chapter 5
Opener: ©Oshihara/Sipa Press. Figure 5.1: Courtesy Ocean Drilling Program. Figure 5.2b: Brian J. Skinner. Figure 5.2c: ©Dane

Penland/Smithsonian Institution. Figure 5.3a: Courtesy Russ Hemley, Carnegie Institution of Washington. Figure 5.3b: Photo by J. Shu; courtesy Russ Hemley, Carnegie Institution of Washington. Figure 5.6: ©John S. Shelton. Figure 5.7: ©Kevin Schafer/Tom Stack & Associates. Figure 5.16: ©Corbis-Bettmann. Figure 5.17a: Courtesy California Historical Society Library. Figure 5.17b: Courtesy M.L. Fuller, U.S. Geological Survey. Figure 5.18: ©Mark Downey/Gamma Liaison. Figure 5.19: ©George Plafker, U.S Geological Survey. Figure 5.20: Courtesy NOAA/NGDC. Figure 5.21: ©Lysaght/Gamma Liaison. Figure 5.24b: Richard Megna/Fundamental Photographs. Box 5.1: ©Rick Rickman/Matirx .

## Chapter 6
Opener: ©Kevin Schafer. Figure 6.1: ©Science VU-ASIS/Visuals Unlimited. Figure 6.3: ©Krafft Explorer/Photo Researchers. Figures 6.7 and 6.8: Courtesy J.D. Griggs, U.S. Geological Survey. Figure 6.9: ©Dr. Lee Tepley/Moonlight Productions. Figure 6.10a: Courtesy William Sacco. Figure 6.10b: ©Tony Waltham. Figure 6.10c: ©Fred Hirschmann. Figure 6.11a: Brian J. Skinner. Figure 6.11b,c: Courtesy William Sacco. Figure 6.12a: ©Fred Hirschmann. Figure 6.12b: ©William Felger/Grant Heilman Photography. Figure 6.13d: ©Craig Johnson. Figure 6.14c: ©Richard Packwood/Oxford Scientific Films. Figure 6.14d: ©Gary Ladd. Figure 6.14e: ©Tom Bean/DRK Photo. Figure 6.15a: S.C. Porter. Figure 6.15b: ©John S. Shelton. Figure 6.15c: Courtesy J.D. Griggs, U.S. Geological Survey. Figure 6.15d: ©Dr. Lee Tepley/Moonlight Productions. Figure 6.15e: ©Tui De Roy/Minden Pictures, Inc. Figure 6.15f: ©Reuters/Corbis-Bettmann. Figure 6.15g: ©S. Jonasson/F.L. /Bruce Coleman, Inc. Figure 6.15h: ©Roger Werth/Woodfin Camp & Associates. Figure 6.16a: ©William E. Ferguson. Figure 6.16b: Courtesy J.D. Griggs, U.S. Geological Survey. Figure 6.16c: ©Schofield/Gamma Liaison. Figure 6.17: ©Steve Vidler/Leo de Wys, Inc. Figure 6.18: ©Greg Vaughn/Tom

Stack & Associates. Figure 6.19: S.C. Porter. Figure 6.21: ©G. Brad Lewis/Gamma Liaison. Figure 6.22: ©David Hiser/The Image Bank. Figure 6.23: ©El Tiempo/Sipa Press. Figure 6.24: ©David Robert Austen. Figure 6.25: NRSC Ltd./Science Photo Library/Photo Researchers. Table 6.1: Brian J. Skinner. Box 6.3: ©Peter Turnley/Black Star.

## Chapter 7

Opener: ©Gordon Wiltsie/AlpenImages, Ltd. Figure 7.2: ©William E. Ferguson. Figure 7.3: ©Tony Waltham. Figure 7.4: ©Susan Rayfield/Photo Researchers. Figure 7.5: ©Stan Osolinski/Oxford Scientific Films Ltd. Figure 7.6: ©Kenneth W. Fink/Ardea London. Figure 7.7: ©Natural History Photographic Agency. Figure 7.8b: Brian J. Skinner. Figure 7.9: ©William E. Ferguson. Figure 7.10: Alan H. Strahler. Figure 7.11: ©Will & Deni McIntyre/Photo Researchers. Figure 7.15: ©Robin White/Fotolex Associates. Figure 7.16: S.C. Porter. Figure 7.17: Henry D. Foth. Figure 7.20: ©Hiroji Kubota/Magnum Photos, Inc. Figure 7.22a: Brian J. Skinner. Figure 7.22b,c: S.C. Porter. Figure 7.23: Courtesy George Plafker, U.S Geological Survey. Box 7.1: ©Frans Lanting/Minden Pictures, Inc. Box 7.2: ©Larry Lefever/Grant Heilman Photography.

## Chapter 8

Opener: ©Robert Azzi/Woodfin Camp & Associates. Figure 8.1: ©Nanci Kahn/Institute of Human Origins. Figure 8.2: ©Josef Muench. Figure 8.4a: ©Glenn M. Oliver/Visuals Unlimited. Figures 8.4b and 8.5: S.C. Porter. Figure 8.6c: ©Martin Miller. Figure 8.7: ©David Muench Photography. Figure 8.8: ©Lee Boltin/Boltin Picture Library. Figure 8.9: Richard J. Stewart. Figure 8.10: Brian J. Skinner. Figures 8.11a,d: S.C. Porter. Figure 8.11b: ©Stephen Trimble/DRK Photo. Figure 8.11c: ©Gordon Wiltsie/Bruce Coleman, Inc. Figure 8.12a: ©Carr Clifton. Figure 8.12b: ©Tom Bean. Figures 8.15 and 8.16: S.C. Porter. Figure 8.17: S.C. Porter. Box 8.1: Courtesy R. C. Murray.

## Chapter 9

Opener: ©Martin Bond/Science Photo Library/Photo Researchers. Figure 9.1: ©Doug Allan/Oxford Scientific Films. Figure 9.4a: ©Landform Slides. Figure 9.4b: ©Helmut Gritscher/Peter Arnold, Inc. Figure 9.5a: ©Stephen Dalton/Natural History Photographic Agency. Figure 9.5b: ©Mark E. Gibson/Visuals Unlimited. Figure 9.6: Courtesy Mervyn Paterson. Figure 9.15: ©Ole P. Rorvik/Aune Forlag. Figure 9.16a: ©Tom Bean. Figure 9.17a: ©Martin Miller/Visuals Unlimited. Figure 9.22: Courtesy Geological Survey of Canada, Ottawa. Figure 9.24a: Courtesy U.S. Geological Survey. Figure 9.25a: ©Westermann Schulbuchverlag GmbH. Figure 9.26a: ©Earth Satellite Corporation.

Figure 9.27: David Muench Photography. Box 9.1: Harrell and Brown, University of Toledo.

## Chapter 10

Opener: S. Pietro/Art Resource. Figure 10.1: ©Tony Waltham. Figure 10.3: ©Joe McDonald/Visuals Unlimited. Figure 10.4: ©Courtesy William Sacco. Figure 10.5: Brian J. Skinner. Figure 10.6: Brian J. Skinner. Figure 10.7c: ©Brenda Sirois. Figure 10.8: Courtesy William Sacco. Figure 10.9: Martin Miller. Figure 10.11: ©Barbara Cushman Rowell/DRK Photo. Figures 10.15, 10.16 and 10.17: Brian J. Skinner. Figure 10.18: ©Boltin Picture Library. Figure 10.19a (top left): ©Breck P. Kent. Figure 10.19a (top right): ©Runk/Schoenberger/Grant Heilman Photography. Figure 10.19b (center left): ©A.J. Copley/Visuals Unlimited. Figure 10.19b (center right): ©Craig Johnson. Figure 10.20: Courtesy William Sacco. Box 10.1c,d: Brian J. Skinner. Box 10.1e: ©Jake Rajs/Tony Stone Images/New York, Inc.

## Chapter 11

Opener: ©Gordon Wiltsie/AlpenImages, Ltd. Figure 11.1: Courtesy William Sacco. Figure 11.2: ©John S. Shelton. Figure 11.3: Courtesy Verkehrsamt der Stadt Nordlingen. Figure 11.4: ©John Sanford/Science Photo Library/Photo Researchers. Figure 11.6: Earth Satellite Corporation/Photo Researchers. Figure 11.10: ©William E. Ferguson. Figure 11.12: ©USGS/Tom Stack & Associates. Figure 11.14: Brian J. Skinner. Figure 11.16: ©David A. Jensen and Light Hawk. Figure 11.18: ©Dr. Bram Janse. Box 11.1: ©B. Murton/Southampton Oceanography Centre/Science Photo Library/Photo Researchers. Box 11.2: ©B. Murton, Southampton Oceanography Centre/Science Photo Library /Photo Researchers. Box 11.3: ©Emory/National Geographic Image Collection.

## Part 4

Opener: ©Joel W. Rogers.

## Chapter 12

Opener: ©Michele Burgess/The Stock Market. Figure 12.3b: ©Paul Steel/The Stock Market. Figure 12.3c: ©Milt Putnam/The Stock Market. Figure 12.3d: ©William E. Ferguson. Figure 12.3e: ©Jeff Gnass/The Stock Market. Figure 12.3f: ©Ken Wood/Photo Researchers. Figure 12.3g: ©Frans Lanting/Minden Pictures, Inc. Figure 12.4: ©G. Haling/Photo Researchers. Figure 12.5: ©Adam Jones/Dembinsky Photo Associates. Figure 12.7: S.C. Porter. Figure 12.8: Courtesy NASA. Figure 12.10: ©Martin Miller. Figure 12.11: ©Earth Satellite Corp./SPL/Photo Researchers. Figure 12.16: ©Earth Satellite Corporation. Figure 12.18: ©George Wuerthner. Figure 12.19: ©A. Holbrooke/Gamma Liaison. Figure 12.21: ©Scott Camazine/Photo Researchers. Figure 12.25a:

©John S. Shelton. Figure 12.25b: ©Michael Nichols/Magnum Photos, Inc. Figure 12.25c: ©Alex S. MacLean/Landslides. Figure 12.31: Courtesy U. S. Environmental Protection Agency. Box 12.1: ©David Turnley/Black Star.

## Chapter 13

Opener: European Space Agency/Science Photo Library /Photo Researchers. Figure 13.1: NASA/Science Source/Photo Researchers. Figure 13.14: ©Greg Scott/Masterfile. Figure 13.15: ©Arnulf Husmo/Tony Stone Images/New York, Inc. Figure 13.17: ©Jeff Divine/FPG International. Figure 13.18b: ©Nicholas DeVore III/Photographers Aspen. Figure 13.19: ©G. R. Roberts. Figure 13.21a: Gamma Liaison. Figure 13.21b: ©Spaceshots, Inc. Figure 13.21c: NASA/Grant Heilman Photography. Figure 13.22: ©Valeria Taylor/Ardea London.

## Chapter 14

Opener: Michael Pobst/AP Wideworld Photos. Figure 14.1: Courtesy NASA. Figure 14.3a: ©Yann Arthus-Bertrand/Photo Researchers. Figure 14.3b: ©Jim Brandenburg/Minden Pictures, Inc. Figure 14.3c: ©François Gohier/Photo Researchers. Figure 14.5c: S.C. Porter. Figure 14.6a: ©Jim Richardson/Woodfin Camp & Associates. Figure 14.6b: S.C. Porter. Figure 14.7: S.C. Porter. Figure 14.8b: ©Tom Bean/DRK Photo. Figure 14.9a: ©Georg Gerster/Comstock, Inc. Figure 14.9b: ©Ari Zhisheng. Figure 14.9c: Courtesy Aero Service Corporation, Litton Industries. Figure 14.9d: ©John S. Shelton. Figure 14.9e: ©ERIM, Ann Arbor, MI. Figure 14.10a: Library of Congress. Figure 14.10b: Alan H. Strahler. Figure 14.10c: ©Abril Imagens/Nani Gois/Editora Abril S/A. Figure 14.10d: ©Victor Englebert/Photo Researchers. Figure 14.11a: Courtesy John Price, USDA. Figure 14.11b: ©Breck P. Kent/Animals Animals. Figure 14.11c: Courtesy USGS. Figure 14.11d: ©Ric Ergenbright. Figure 14.11e: ©John Lythgoe/Planet Earth Pictures. Figure 14.12b: ©Thomas Ligon/Photo Researchers. Figure 14.16a,c: S.C. Porter. Figure 14.16b: Alan H. Strahler. Figure 14.17a: ©Michael Collier. Figure 14.17b: ©C. Wolinsky/Stock, Boston. Figure 14.17c: ©Fred Hirschmann Wilderness Photography. Figure 14.17d: S.C. Porter. Figure 14.18c: ©Stephen J. Krasemann/DRK Photo. Box 14.1: S.C. Porter.

## Part 5

Opener: ©Kevin Schafer/Kevin Schafer Photography.

## Chapter 15

Opener: ©Louis Psihoyos/Matrix. Figure 15.4: ©Reuters/Corbis-Bettmann. Figure 15.5: NASA/Science Photo Library/Photo Researchers. Figure 15.6: ©WHOI, D. Foster/Visuals Unlimited. Figure 15.8a: ©Dr. Jeremy Burgess/SPL/Photo Researchers. Figure 15.8b:

# •LINE ART, TABLE, AND TEXT CREDITS•

**Chapter 3 Figure 3.7:** Modified from P. R. Vail, J. Hardenbol, and R. G. Todd, "Jurassic Unconformities, Chronostratigraphy, and Sea-Level Changes from Seismic Stratigraphy and Biostratigraphy," in AAPG Memoir Series no. 36, John S. Schlee, ed., *Interregional Unconformities and Hydrocarbon Accumulation,* AAPG©1984. Reprinted by permission of the American Association of Petroleum Geologists.

**The Art of Geology Page 82:** From THE POETRY OF ROBERT FROST, edited by Edward Connery Lathem, Copyright 1936 by Robert Frost, ©1964 by Lesley Frost Ballantine, ©1969 by Henry Holt and Company, Inc., ©1997 by Edward Connery Lathem. Reprinted by permission of Henry Holt and Company, Inc. **Page 83:** Russell Baker, "The Cosmic Achoo," Copyright ©1974 by The New York Times Co. Reprinted by permission.

**Chapter 5** *Geology Around Us* box, pages 124-125: Ian Brown, "All Shook Up." *Saturday Night,* April 1994. Reprinted by permission of the author. **Figure 5.22:** Adapted from Patrick L. Abbott, *Natural Disasters,* ©1996, Wm. C. Brown, Publishers, by permission of The McGraw-Hill Companies.

**Chapter 8** *Geology Around Us* box, pages 228-229: From IRONS IN THE FIRE (Farrar, Straus & Giroux). ©1996 John McPhee. Originally in *The New Yorker.* All rights reserved.

**Chapter 9** *Geology Around Us* box, pages 260-261: From "Oldest Geologic Map Is Turin Papyrus," by J. A. Harrell and V. Max Brown. *Geotimes,* March 1989, pp. 10-11. Used with permission by the American Geological Institute.

**The Art of Geology Page 323:** Colin Fletcher, *The Man Who Walked Through Time,* pp. 142-143, 1967. Vintage Books Random House.

**Chapter 12 Figure 12.3, chart:** Adapted from "Safeguarding the world's water," UNEP Environmental Brief #6, Nairobi, 1988. **Pages 343-344:** Section on lakes adapted from Alan Strahler and Arthur Strahler, *Physical Geography: Science and Systems of the*

*Human Environment,* copyright ©1997 John Wiley & Sons, Inc. Reprinted by permission of John Wiley & Sons, Inc.

**Chapter 13 Figure 13.9:** Adapted from Stanley David Gedzelman, *The Science and Wonders of the Atmosphere,* copyright ©1980 John Wiley & Sons, Inc. Reprinted by permission of John Wiley & Sons, Inc. **Figure 13.18A:** H. J. de Blij and Peter O. Muller, *Physical Geography and the Global Environment,* copyright ©1996 John Wiley & Sons, Inc. Reprinted by permission of John Wiley & Sons, Inc.

**Chapter 14 Figure 14.8A:** Adapted from K. Pye and L. Tsoar, *Aeolian Sand and Sand Dunes,* Figure 7.1, Chapman & Hall, 1990. Reprinted by permission of International Thompson Publishing Services, Ltd. **Figure 14.9, drawing:** Adapted from John T. Hack, *The Geographical Review,* vol. 31, fig. 19, p. 260 by permission of the American Geographical Society. **Pages 419-420:** Section on periglacial landforms adapted from Alan Strahler and Arthur Strahler, *Physical Geography: Science and Systems of the Human Environment,* copyright ©1997 John Wiley & Sons, Inc. Reprinted by permission of John Wiley & Sons, Inc. **Figure 14.18A, B:** Adapted from A. H. Lachenbruch in Rhodes W. Fairbridge, ed., *The Encylopedia of Geomorphology,* Van Nostrand Reinhold, New York, Reprinted with permission. **Figure 14.19:** Understanding Atmospheric Change: A Survey of the Background Science and Implications of Climate Change and Ozone Depletion. Second edition 1995, Environment Canada SOE Report no. 95-2. **Figure 14.21:** After Nigel Calder, *The Weather Machine,* 1974. Used by permission of Nigel Calder. **Figure 14.22:** Illustration by Thomas Moore. First published in Richard Houghton and George M. Woodwell, "Global Climactic Change," *Scientific American,* vol. 260, no. 4, April 1989.

**The Art of Geology Page 430:** From *Lao Tzu: Tao Te Ching* by Ursula Le Guin, ©1997. Reprinted by arrangement with Shambhala Publications, Inc., Boston. **Page 431:** Chinua Achebe, *Things Fall Apart,* p. 35, 1959, Fawcett Books.

**Chapter 15 Figure 15.1:** OUR CHANGING PLANET, 2/E, by MacKenzie/Mackenzie, ©1998. Reprinted by permission of Prentice-Hall, Inc., Upper Saddle River, NJ. **Figures 15.2 and 15.7:** Adapted from G. Brum, L. McKane, and G. Karp, *Biology: Explaining Life,* copyright ©1994 John Wiley & Sons, Inc. Reprinted by permission of John Wiley & Sons, Inc. **14 Pages 419-420:** Section on periglacial landforms adapted from Alan Strahler and Arthur Strahler, *Physical Geography: Science and Systems of the Human Environment,* copyright ©1997 John Wiley & Sons, Inc. Reprinted by permission of John Wiley & Sons, Inc. **Figure 15.19:** Adapted from illustration by Bunji Tagawa in E. Banghoom and T. W. Schopf, "The oldest fossils," *Scientific American,* May 1971, p. 41. *Geology Around Us* box, page 452: Excerpted from Carl Zimmer/©1993, "Darwin's Atlantis," *Discover Magazine,* February 1993. Reprinted with permission of *Discover Magazine.*

**Chapter 16 Figures 16.13 and 16.14:** Adapted from John F. Bookout, "Two centuries of fossil fuel production," *Episodes,* vol. 12, no. 4, 1989, figs, 1, 14. Used with the permission of the International Union of Geological Sciences and the author.

**Chapter 17 Table 17.1 and material on pages 507 and 509:** Excerpted from American Geological Institute, <www.agiweb.org>, *Careers in the Geosciences,* 1997. Used with permission. **Figure 17.4:** OUR CHANGING PLANET, 2/E, by MacKenzie/Mackenzie, ©1998. Reprinted by permission of Prentice-Hall, Inc., Upper Saddle River, NJ. **Page 512:** Excerpted from Mary Beck Desmond, "Earth Scientist at the International Development Front," *Geotimes,* March 1992, pp. 21-23. Used with permission of the American Geological Institute. **Figures 17.6 and 17.7:** Adapted from Earl E. Brabb, "The World Landslide Problem," *Episodes,* vol. 14, no. 1, pp. 56, 57. Used with the permission of the International Union of Geological Sciences and the author. **Figure 17.11:** B. H. Luckman, "Global change and the record of the past," GOES 1989/3.

# • INDEX •